EMISSIONS FROM CONTINUOUS COMBUSTION SYSTEMS

PUBLISHED SYMPOSIA

Held at the
General Motors Research Laboratories
Warren, Michigan

Friction and Wear, 1959
Robert Davies, *Editor*

Internal Stresses and Fatigue in Metals, 1959
Gerald M. Rassweiler and William L. Grube, *Editors*

Theory of Traffic Flow, 1961
Robert Herman, *Editor*

Rolling Contact Phenomena, 1962
Joseph B. Bidwell, *Editor*

Adhesion and Cohesion, 1962
Philip Weiss, *Editor*

Cavitation in Real Liquids, 1964
Robert Davies, *Editor*

Liquids: Structure, Properties, Solid Interactions, 1965
Thomas J. Hughel, *Editor*

Approximation of Functions, 1965
Henry L. Garabedian, *Editor*

Fluid Mechanics of Internal Flow, 1967
Gino Sovran, *Editor*

Ferroelectricity, 1967
Edward F. Weller, *Editor*

Interface Conversion for Polymer Coatings, 1968
Philip Weiss and G. Dale Cheever, *Editors*

Associative Information Techniques, 1971
Edwin L. Jacks, *Editor*

Chemical Reactions in the Urban Atmosphere, 1971
Charles S. Tuesday, *Editor*

The Physics of Opto-Electronic Materials, 1971
Walter A. Albers, Jr., *Editor*

Emissions From Continuous Combustion Systems, 1972
Walter Cornelius and William G. Agnew, *Editors*

EMISSIONS FROM CONTINUOUS COMBUSTION SYSTEMS

Proceedings of the
Symposium on Emissions from
Continuous Combustion Systems
held at the
General Motors Research Laboratories
Warren, Michigan
September 27-28, 1971

Edited by
WALTER CORNELIUS and WILLIAM G. AGNEW
General Motors Research Laboratories

PLENUM PRESS • NEW YORK – LONDON • 1972

Library of Congress Catalog Card Number 72-80343

ISBN 0-306-30702-2

A Division of Plenum Publishing Corporation
227 West 17th Street, New York, N. Y. 10011

United Kingdom edition published by Plenum Press, London
A Division of Plenum Publishing Company, Ltd.
Davis House (4th Floor), 8 Scrubs Lane, Harlesden, London, NW10 6SE, England

Printed in the United States of America

PREFACE

This volume documents the proceedings of the *Symposium on Emissions from Continuous Combustion Systems* that was held at the General Motors Research Laboratories, Warren, Michigan on September 27 and 28, 1971. This symposium was the fifteenth in an annual series presented by the Research Laboratories. Each symposium has covered a different technical discipline. To be selected as the theme of a symposium, the subject must be timely and of vital interest to General Motors as well as to the technical community at large.

For each symposium, the practice is to solicit papers at the forefront of research from recognized authorities in the technical discipline of interest. Approximately sixty scientists and engineers from academic, government and industrial circles in this country and abroad are then invited to join about an equal number of General Motors technical personnel to discuss freely the commissioned papers. The technical portion of the meeting is supplemented by social functions at which ample time is afforded for informal exchanges of ideas amongst the participants. By such a direct interaction of a small and select group of informed participants, it is hoped to extend the boundaries of research in the selected technical field.

The subject of the 1971 symposium was considered to be an appropriate one in view of current wide-spread concern about the environmental effects of emissions from powerplants that burn fossil fuels. In the past, major effort has been expended by the technical community to maximize power with suitable efficiency. Now, a third and equally important factor must be considered, to minimize exhaust emissions without unduly penalizing performance and efficiency. The formation and control of emissions from cyclic combustion systems have been discussed in depth at many previous technical meetings; however, similiar discussions of the emissions from continuous combustion systems have been limited. The need for progress in this area is clear because of the current extensive use of continuous combustion powerplants in the aeronautical and stationary power generation fields and accelerated activity in developing such powerplants for vehicular transportation applications. It is hoped that this symposium will instigate future meetings in this technical discipline.

Justification for General Motors sponsoring this symposium is based on approximately fifteen years of research and development activity on continuous combustion systems at the Research Laboratories. During this time, combustion systems were developed for vehicular applications of gas turbine, Stirling and Rankine cycle engines. Basic studies of the formation and control of pollutant emissions were conducted on vehicular and aviation type gas turbine combustion systems. Specialized instrumentation and chemical analysis techniques were developed to measure accurately the extremely low exhaust concentrations of the pollutants that are typical of continuous combustion systems.

The main objective of the symposium was to stimulate the exchange of ideas and experiences among a small group of recognized authorities in the continuous combustion field; and by doing so, to generate new approaches for controlling the emissions of air pollutants from continuous combustion devices. To accomplish this goal, the symposium was organized into four closely-related technical sessions. The first session was concerned with the fundamentals of the combustion processes and the mathematical modeling of combustion systems. Chemical aspects of pollutant formation and destruction processes were the subjects of the second session. Effects of operating conditions and of fuel factors on these pollutant processes were discussed during the third session. In the final session, the discussions concerned the applications of practical techniques for reducing the emissions from actual aircraft, stationary and vehicular powerplants.

The small group of engineers and scientists invited to the symposium were chosen because of their expertise and current activity in the continuous combustion field. The number was limited both by the desire to keep the group intimate and by limitations of the physical facilities. Attendees represented academic, governmental and industrial communities in this country and in England and Canada. For each session, papers were presented by currently active authorities in that particular technical area. To assure competent direction during the technical sessions, four men of recognized international reputation in the continuous combustion field acted as the session chairmen. Ample time was alloted during the meeting for the presentation of the papers, and for prepared and informal discussions of the papers and allied investigations by the attendees. The proceedings of each session were summarized by the chairman. An overall summary of the symposium proceedings was presented by one of the symposium co-chairmen. The oral presentations were recorded on tape that was subsequently transcribed and edited lightly. These edited accounts were then submitted to the discussers for corrective action and approval. The submitted manuscripts of the papers, the formal and informal discussions presented at the meeting, and the symposium and session summaries have been reproduced in this volume.

We are especially indebted to Mr. E. N. Cole, President of the General Motors Corporation; who, because of his deep concern regarding the contribution of the automobile to air pollution, graciously consented to be the guest speaker at the symposium banquet. His address describing the difficult problems facing the automotive industry in achieving a suitably-low emission mass-produced automobile had a sobering influence on the audience.

We would be remiss if we did not acknowledge the valuable contributions of the following people: Dr. J. P. Longwell and Professors A. H. Lefebvre and G. C. Williams for assistance in organizing the technical program and acting as session chairmen; Dr. E. S. Starkmen for support in organizing the technical program; Professor P. S. Myers for acting as a session chairman; Dr. P. F. Chenea, Dr. R. F. Thomson and Mr. J. D. Caplan for authorizing and supporting the symposium; Mr. R. L. Scott for his efficient handling of the myriad details involving the physical arrangement of the symposium; Dr. W. H. Lipkea and Mr. D. N. Havelock for assistance in editing the symposium volume; Messers H. P. Fredriksen, W. R. Wade and R. M. Siewert for recording the symposium proceedings on tape; and Miss T. D. Nykanen for her dedicated and efficient secretarial support and for being a hostess at the technical and social functions.

Walter Cornelius and William G. Agnew

March 1972

CONTENTS

SESSION I

MODELING CONTINUOUS COMBUSTION

Session Chairman
J. P. LONGWELL

Esso Research and Engineering Company
Linden, New Jersey

MATHEMATICAL MODELS OF CONTINUOUS COMBUSTION

D. B. SPALDING

Imperial College, England

ABSTRACT

The Problem Considered – Economical design and operation of combustion systems can be greatly facilitated by prior predictions of performance by way of a mathematical model, incorporated into a digital-computer program. Morever, since the emission of pollutants is more sensitive to detailed design changes than is the overall heat transfer or power output of the system, the refined insight provided by predictions of concentration distributions is especially valuable nowadays. Suitable mathematical models exist for simple combustion systems and are being developed for others.

Purpose of the Paper – The difficulty of developing such models has two origins: mathematical and physical. The former is a result of the facts that: most combustion-chamber flows are three-dimensional; the differential equations describing them are numerous, simultaneous and non-linear; and neither digital computers of sufficient power nor knowledge of the optimal solution procedures are widely available. The physical source of difficulty is the complexity of the laws governing turbulent transport, the chemical kinetics of laminar and turbulent gases, formation and disappearance of condensed-phase particles, and thermal radiation through absorbing and scattering media.

The purpose of the paper is to illuminate these difficulties, to demonstrate that many useful predictions can already be made, and to indicate in what areas further research may be useful.

Topics Referred to – The paper will touch on the following detailed topics, and will illustrate them by examples of predictions:

References p. 16

1. Computational procedures for solving two-dimensional and three-dimensional steady-flow problems, both with and without recirculation.

2. Turbulence models for the prediction of the distribution through the combustion space of such statistical properties of turbulence as the kinetic energy of the fluctuating motion, the average length scale, the Reynolds stresses, and the root-mean-square fluctuations of concentration.

3. Hypotheses for the rates of the main combustion reaction, and of the NO_x-formation reaction, when turbulent fluctuations of temperature and concentration are present.

4. The calculation of the variation through a combustion space of the size distribution of a condensed-phase material.

5. The calculation of radiative transfer.

INTRODUCTION

There are three ways of finding out about combustion systems (and about any other pieces of engineering equipment, for that matter): (a) making measurements in real-life systems under realistic operating conditions; (b) making measurements on physical models, similar to the real combustion systems in important effects, but differing in linear scale, and perhaps in other ways; and (c) performing experiments with a computer program designed to simulate the combustion system.

The first method is very expensive, and often impractical. Usually this information is wanted before the full-scale equipment has been built.

The second method is less expensive; but still the cost of making, say, transparent models of power-station boilers is quite an expensive one. Also the information is seldom reliable; for the model often fails to simulate important components correctly, e.g. radiation.

The third method, the use of a mathematical model, is therefore becoming increasingly popular.

The subject is too large for comprehensive treatment in this short paper. Therefore only a few of its major features will be discussed. The main purpose is to provide a correct impression of how much has been already achieved, and of what still remains to be done. The following topics will be discussed: computational methods, turbulence models, particle-size calculations, radiative transfer, and turbulence-controlled combustion.

The main message about the computational techniques will be that these methods are well advanced even for three-dimensional flows. After this has been demonstrated, the physical processes which have to be correctly expressed in mathematical form will be considered, one by one, if the predictions are to be realistic.

Turbulence will be discussed in two places in the paper. First, it will be shown how turbulent flow and mixing can be calculated in the absence of chemical reaction; and secondly, how reaction and turbulence may interact. Many flames are two-phase in character. The fuel enters as a solid or a liquid, and solid carbon may be formed. Means exist for handling this process mathematically. Thermal radiation is another topic of great practical relevance. It will be indicated how it can be included in a mathematical model.

COMPUTATIONAL PROCEDURES

Computations of flame phenomena are still not performed frequently, and a few years ago they were never carried out at all. The reason is that there are substantial mathematical difficulties in the way. It is only recently that ways of surmounting them have been discovered.

The difficulties in part spring from the facts that the differential equations are numerous and non-linear, and that the flow situations are multi-dimensional. Digital computers are needed, and these computers must be very large to handle three-dimensional flows. But to have a digital computer is not enough. Economical, accurate and reliable computer programs are needed also and it has taken quite a long time to develop these programs. Even now, the practising engineer still lacks readily available procedures for the more complex of his tasks.

In addition to the development of the computer programs, there has had also to be an appropriate reformulation of physical knowledge. What is known about turbulence, for example, has had to be expressed in terms of differential equations. Verbal descriptions or quantitative findings about large-scale systems are not good enough.

Two-Dimensional Parabolic Flows – The computational problem is simplest when the flow is two-dimensional and without flow reversal in one direction. The last feature makes the differential equations parabolic. A standard program is available for this type of problem. (1)

Fig. 1 shows an example of a flame computation which has been made with this program. It concerns a turbulent propane-air diffusion flame. The diagram shows the computed shape of the reaction zone, with differing scales for the longitudinal and radial directions.

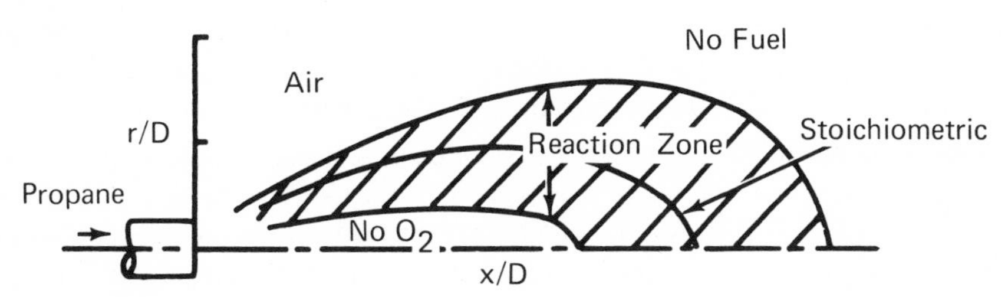

Fig. 1. Computation of a turbulent diffusion flame by the Patankar-Spalding (1) method.

References p. 16

To make such computations, six simultaneous differential equations must be solved for: velocity, turbulence energy, turbulence scale, stagnation enthalpy, fuel concentration, and concentration fluctuations. Nevertheless, the whole flame can be computed in less than 30 seconds on a CDC 6600. The underlying assumptions will be discussed in more detail later in the paper. The following questions raised by Fig. 1 will then be answered: how is it that the predicted reaction-zone thickness is finite, even though the chemical reaction rates in a diffusion flame may be taken as infinitely fast? Wherein lies the difference from the laminar diffusion flame, which always has a thin reaction zone?

Two-Dimensional Elliptic Flows – Most continuous-combustion systems exhibit recirculation. They are therefore not amenable to the kind of marching-integration process that is applicable to the 2D Parabolic flows mentioned. Instead iterative procedures are essential.

For *two*-dimensional flows with recirculation, the method of Gosman et al (2) is available in book form; and many people have used it for calculating combustion processes. Fig. 2 illustrates some early computations performed by a colleague, Dr. Pun. It shows the stream-line patterns and constant-temperature contours in an axi-symmetrical furnace supplied with gaseous fuel in a jet along the axis and with an annular swirling stream of oxidant gas.

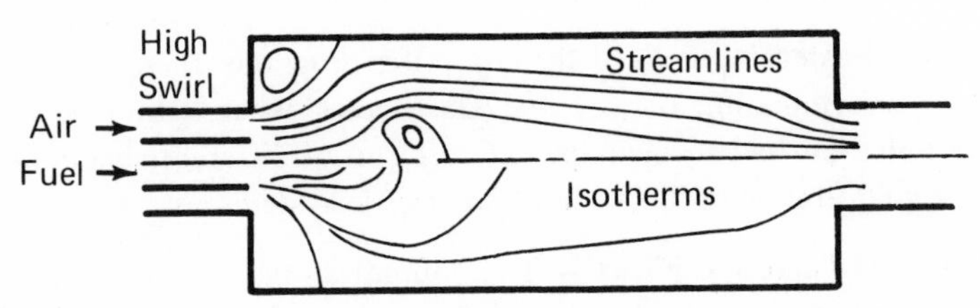

Fig. 2. Computation of a turbulent diffusion flame with recirculation by the method of Gosman, et al (2).

The predicted flow pattern agrees quite well, in its qualitative features, with what combustion engineers know from experience: that the swirl in the air stream causes a recirculation along the axis of the combustion chamber; and the resultant recirculation helps to stabilize and to shorten the flame.

Three-Dimensional Parabolic Flows – It is only comparatively recent that three-dimensional flows can be predicted economically. Therefore, there has been little time to apply the new skills to combustion processes. The example illustrated in Fig. 3 relates to a confined flow in which no chemical reaction is involved.

It will be appreciated that, even if there were no movement of the wall, the square cross section would not allow any axial symmetry: a two-dimensional problem must be solved at each section down the duct. A few of the results of such a computation, in terms of the axial-velocity distributions, are shown on the figure. Of course, the computer program can provide much more information.

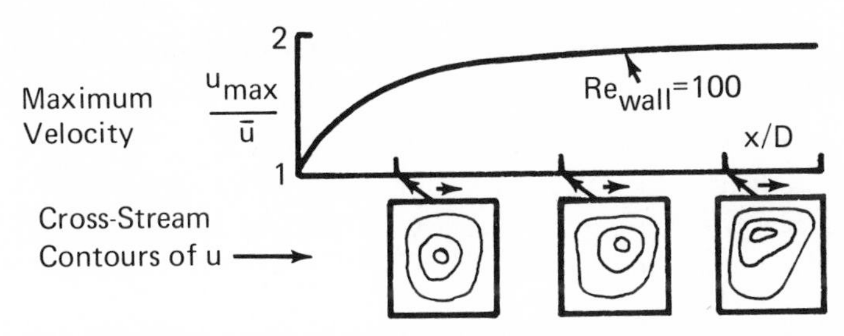

Fig. 3. Computation of flow in a square-sectioned duct with a moving wall (3).

The method of computation has not yet been published, but should appear shortly. It is now being adapted for problems with buoyancy, turbulence, and chemical reactions. A version is available for flows in the atmosphere and should be useful for predicting the flow of chimney effluents.

Three-Dimensional Elliptic Flows – The previous example of 3D parabolic flows was of a boundary-layer character. Therefore, although the flow was three-dimensional, marching integration could be used and only two-dimensional storage was needed in the computer. Fig. 4 by contrast illustrates a kind of flow which is fully three-dimensional. It is typical of flows in furnaces which are not axi-symmetrical.

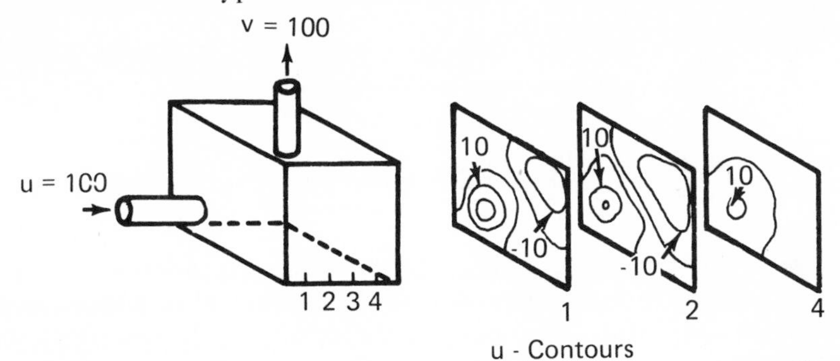

Fig. 4. Computation of a fully three-dimensional flow in a rectangular box (4).

Once again, the method is a new one. The calculation shown on the figure is an early demonstration of the method made by Dr. Caretto (4). It displays some features of the flow pattern which ensues when fluid is injected into a rectangular-sided box, large density variations being present. It can be seen that the predictions are qualitatively quite plausible.

There is much to be done before this method can be developed to the point where furnace designers can use it. For example, some means for dealing with turbulent flows has still to be incorporated into this program.

TURBULENCE MODELS

Special differential equations have to be solved when the flow is turbulent. The dependent variables of these equations are statistical properties of the turbulence, for

References p. 16

example, the kinetic energy of the fluctuating motion or the Reynolds stress across a particular plane.

The time-dependent Navier-Stokes equations cannot be used because important features of turbulent flows take place on a very small scale, for example the dissipation of energy. This implies that a very fine grid must be employed, so fine indeed as to be quite impracticable.

The only way forward is therefore to derive, rigorously in part, but with intuitively-guided additional hypotheses, differential equations for the important correlations. Such a set of equations constitutes a "turbulence model".

The important question to be settled is: how few equations are required to make adequate predictions? The current answer appears to be 2 or 3 equations.

The kWg Turbulence Model – The turbulence model which underlay the diffusion-flame computations of Fig. 1 involves the solution of the following three equations:

$$\rho\frac{Dk}{Dt}=\frac{1}{r}\frac{\partial}{\partial r}\left(\frac{r\mu_t}{\sigma_k}\frac{\partial k}{\partial r}\right)+\mu_t\left(\frac{\partial u}{\partial r}\right)^2-C_D\rho kW^{1/2} \tag{1}$$

$$\rho\frac{DW}{Dt}=\frac{1}{r}\frac{\partial}{\partial r}\left(\frac{r\mu_t}{\sigma_w}\frac{\partial W}{\partial r}\right)+C_1\mu_t\left(\frac{\partial^2 u}{\partial r^2}\right)^2-C_2\rho WW^{1/2} \tag{2}$$

$$\rho\frac{Dg}{Dt}=\frac{1}{r}\frac{\partial}{\partial r}\left(\frac{r\mu_t}{\sigma_g}\frac{\partial g}{\partial r}\right)+C_{g_1}\mu_t\left(\frac{\partial f}{\partial r}\right)^2-C_{g_2}\rho gW^{1/2} \tag{3}$$

The three correlations appearing as dependent variables are: turbulence energy (k), vorticity fluctuations (W) and concentration fluctuations (g). The equations are to be solved simultaneously with equations for: u (velocity), h (enthalpy) and f (concentration).

The differential equations are all similar in that each equations has a convection term on the left, and terms representing diffusion, generation, and dissipation appear on the right. It is to be noticed that each dissipation term is formed from the product of the main variable with $W^{1/2}$; this has the dimensions of frequency. The generation terms are formed from squares of gradients of time-mean quantities: velocity in the first case, vorticity in the second, and concentration in the third. The above equations are those relevant to boundary-layer circumstances, and their more general forms can be written out and solved.

Fig. 5 illustrates what happens when equations 1 through 3 are solved with initial and boundary conditions appropriate to the injection of a fluid through a nozzle of circular cross-section into a stagnant reservoir of different composition but equal density. Chemical reaction is supposed to be absent.

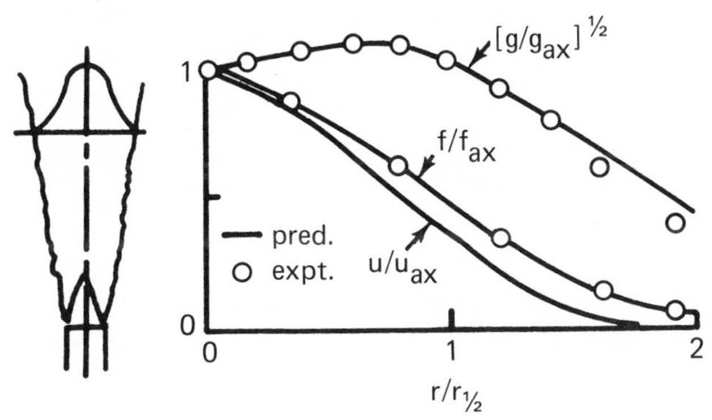

Fig. 5. Prediction of concentration fluctuations (5).

The left-hand sketch illustrates the jet which is formed. The right-hand diagram shows curves which result from solutions of the three equations, along with those for the velocity (u) and the time-mean concentration (f). Also shown, as circles, are the experimental data of Becker, Hottel and Williams (6). The agreement is rather satisfactory. It appears that the differential equation for $g^{1/2}$, the RMS concentration fluctuation, truly expresses the physical processes in the jet. This result is worth taking note of for two reasons: firstly, it may open the way to the use of these fluctuations as an indicator of turbulence behaviour; and secondly it has some revelance to the production of oxides of nitrogen.

In Fig. 6, it is supposed that the jet consists of hot oxygen, injected into a reservoir of nitrogen. The question arises as to how much nitric oxide is produced. It will be recognized that this is a simple example of the kind of question which has increasingly to be answered.

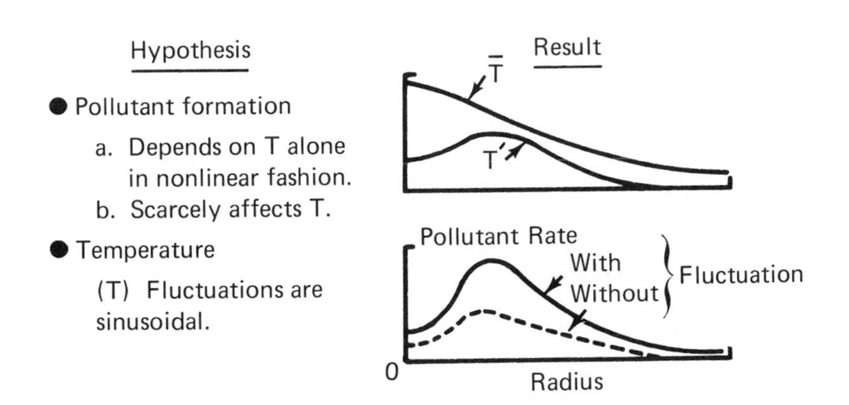

Fig. 6. Illustration of the influence of temperature fluctuations on the rate of pollutant formation.

References p. 16

To illustrate the role of fluctuations, a simple view of the chemical kinetics can be taken: it can be supposed that the reaction rate depends upon temperature alone. The dependence will be non-linear, of course.

If the flow were laminar, the proceeding would be simple, namely to solve the equations for momentum, enthalpy and nitric oxide, and hence deduce the overall production rate. However, because the flow is turbulent, the equation for the temperature fluctuations is needed also because the time-mean reaction rate depends upon the magnitude of these fluctuations as well as upon the time-mean temperature. The sketches on the right side of Fig. 6 are intended to illustrate this point.

PARTICLE-SIZE CALCULATIONS

The growth and disappearance of particles in flames present another problem which is intimately connected with air pollution from combustion systems. These particles may be of fuel or of products. Soot in flames can be considered either as the one or the other.

To determine the growth and disappearance of particles in the flame requires that the laws of nucleation, particle growth and particle disappearance be incorporated into the mathematical model. Then it should be possible to predict, at each point in the flame, the particle-size distribution. Obviously, this is more difficult than to predict simply the concentration of soot regardless of the size distribution. A curve rather than just a single value is needed for each point in the combustion chamber.

This task can also be performed. Of course, it is necessary to discretize the variations in particle-size space, just as is done for physical space. The graph in Fig. 7 illustrates how the particles are supposed to be distributed with respect to size at any point: in a series of plateaux, with a sloping distribution in the lowest-size interval. The measure of size may be the particle diameter, the particle mass, or any other convenient variable. The surface area of the particle has been found to be a

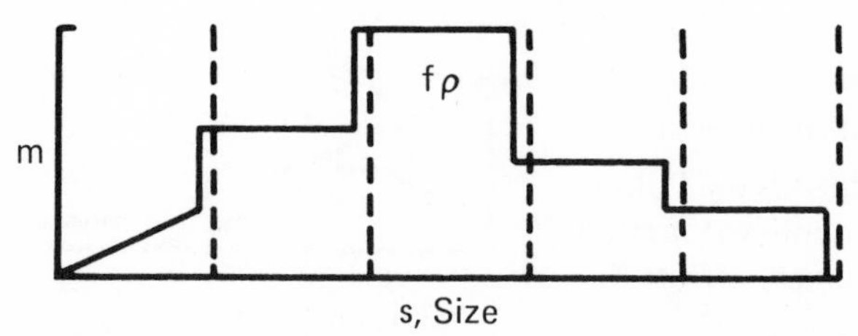

Differential equation for f_j in axi-symmetrical flows.

$$\frac{Df_j}{Dt} = \frac{1}{r}\frac{\partial}{\partial r}\frac{(r\,\partial f_j)}{\partial r} - \dot{s}(m_j - m_{j-1}) - \int_{S_{j-1}}^{S_j} \frac{3}{2}\frac{m}{s}\,ds$$

Fig. 7. Discretization of the particle-size-distribution curve. The areas beneath the segments (f's) became the variables of new differential equations.

useful dimension in some circumstances. The ordinate of the graph is the mass of solid material per unit volume of space per unit increment of size. Of course, in reality, the m vs. s variation should be smooth.

The areas beneath each segment of the curve, divided by the fluid density (ρ), are adopted as the variables which describe the distribution. The reason is that these quantities obey partial differential equations of the kind that can be solved. The terms of these equations express the processes of convective transport, diffusive transport, generation of particles, and decay of particles. Each can be expressed quantitatively in terms of local properties of the mixture. Once the equations for the particle-size variables have been set up, and the initial and boundary conditions have been provided, all that remains is to solve them, along with the equations for the various gas-mixture properties. This is now a routine matter.

Fig. 8 shows results of such computations for a particular combustion system in which a stream of gaseous and solid fuel, after partially mixing with an annular stream of air, is ignited by a stream of hot combustion products. The left-hand diagram illustrates the system, while the right-hand one shows the computed profiles of fuel and oxygen near the downstream end of the duct.

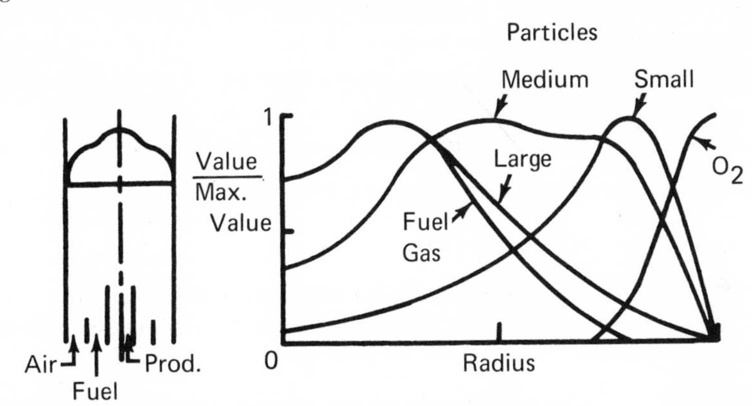

Fig. 8. Results of computation for injection of large fuel particles in the fuel-gas stream.

The concentrations of gaseous fuel and oxygen have profiles which are typical of diffusion flames. The curves for small, medium and large particles are of interest.

It will be seen that the small particles have their highest concentration in the oxygen-rich region; for it is only there that the large ones can be diminished in size. The large particles, by contrast, have a profile similar to that for gaseous fuel; but it does extend farther into the oxygen-rich region.

RADIATIVE TRANSFER

One reason for calculating the particle-size distributions is that the radiative properties of flames are greatly dependent on the amount and subdivision of the

References p. 16

solids suspended within them. It is therefore appropriate to turn next to the way in which radiative transfer is to be calculated.

It is not easy to calculate radiative transfer. Some of the reasons are listed below:

- Radiation is seven-dimensional (space-3, angle-3, wave length-1).
- Radiative properties of gases, solid surfaces and particles are complex.
- Governing equations are usually treated as integro-differential.

The seven-dimensionality of radiation and the complexity of the radiative properties of materials present severe difficulties. Further, at least in their most common formulation, the governing equations are integro-differential. This feature expresses the fact that radiation allows each point in the flame, at least in principle, to communicate with each other point.

The finite-difference expression of the latter feature is that each algebraic equation for temperature has a great many terms in it. By contrast, the finite-difference equations for the processes of momentum and mass transfer have few terms. Their matrices are "sparse." It would be desired to give the radiative equations the same quality of sparseness. A step-wise way of doing so is outlined as follows:

- Sub-divide into discrete wave-bands.
- Discretize angle (2, 4 or 6 fluxes).
- Formulate first-order differential equations for the fluxes.
- Derive second-order differential equations for the sums of flux pairs.
- Solve in a finite-difference manner.

Just as space was divided into a finite number of blocks, and particle size into a finite number of size intervals, so the wave-length and angle dimensions of radiation must be discretized.

The discretization of angle can be quite severe. Thus, for one-dimensional problems it often suffices to employ just two bundles of radiation, one outward and one inward. Two-dimensional problems can be handled by four fluxes, and so on.

Each of the fluxes is governed by a first-order differential equation. This expresses the diminution of a flux with distance in consequence of absorption and scattering, and its augmentation by emission and by scattering from other directions.

All of the other differential equations are second-order ones. It is therefore convenient to take the first-order radiation equations in pairs, and to make from each a single second-order equation. This latter equation can then be solved by the standard computer program because its finite-difference matrix is sparse.

This technique was not invented by the author. It has been known to astro-physicists for many years. Fig. 9 illustrates an example of computations that was carried out in this manner by a colleague, Dr. Lockwood (8). The problem is that of radiation from a turbulent diffusion flame confined in a duct. The walls are cooled. Only radially-inward and radially-outward fluxes are considered. The left-hand sketch illustrates the situation.

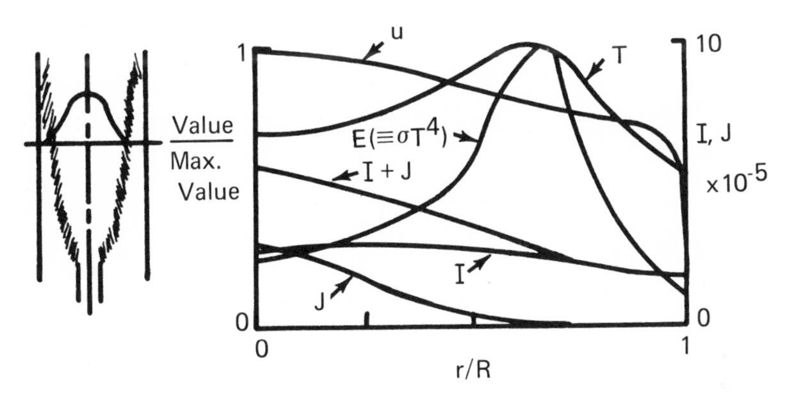

Fig. 9. Computation of radiative heat transfer from a turbulent diffusion flame confined to a duct.

The right-hand diagram represents computed profiles of various quantities along a radius some distance from the fuel nozzle. The familiar quantities are the gas velocity, the temperature, and the concentrations of fuel and oxygen. The quantity (E) is the emissive power of the gas, i.e. σT^4; and I + J is the sum of the inward and outward fluxes at any point. If the absorptivity of the gas were very great, I + J would be exactly twice E which is what the conductive approximation would give. The illustrated calculations show that I + J is actually much smaller than E. The technique can handle gases of large, small and intermediate emissivity.

Also shown are the inward and outward radiation fluxes. Their difference is of course the net heat flux by radiation.

TURBULENCE-CONTROLLED COMBUSTION

The concern of this section is not with the turbulent diffusion flame discussed above but with the spread of a turbulent flame through already pre-mixed gases. Turbulent flame spread, like flame propagation through a laminar pre-mixed gas, is the result of chemical reactions promoted by the mixing of burned with unburned gas. To be effective, this mixing must take place on the molecular scale, by way of laminar conduction and diffusion.

Often it appears, however, that there is another process which is rate-controlling. It might be thought of as the creation of contact surface by the breaking up of large eddies, or as the encirclement of lumps of unburned gas by a surrounding sea of burned gas. In these circumstances, the chemical-kinetic and laminar-transport

References p. 16

properties of the mixture appear to have little influence upon the rate of chemical reaction.

There is an analogy here with the dissipation of turbulence energy, and of concentration fluctuations. As was discussed earlier, their rates are affected primarily by the quantity ($W^{1/2}$).

In order to bring predictions into line with the experimental findings, it has been necessary to assume that the eddy breakdown is the controlling resistance to chemical reaction. A consequence has been the reaction-rate expression for premixed gases that appears below:

- Maximum possible rate of reaction = $C m_{fu} \rho W^{1/2}$ x intermittency factor.
- $W^{1/2}$ ($\equiv$turbulence frequency) and intermittency factor are calculated from differential equations.
- Chemical kinetics provides only an upper limit.

A major feature of the expression is that the rate of reaction is proportional to $W^{1/2}$ just like the dissipation rates, at least when the chemical-kinetic limit does not operate. It is however necessary to take note also of the intermittency of the turbulence, in the first instance, by multiplying by a factor lying between zero and one. The quantity ($W^{1/2}$) has been calculated earlier by employing a special differential equation for its square: the time-average fluctuating vorticity.

The intermittency factor, or something having very similar properties, may also be deduced from the solution of a differential equation, namely that for concentration fluctuations. The idea is that, when the RMS fluctuations greatly exceed the value of the time-mean concentration of fluid from the turbulent region, it is a sign that the turbulence is intermittent.

Fig. 10 illustrates some of the computations which have been made with this model. The top sketch illustrates the physical circumstances. A pre-mixed fuel-air

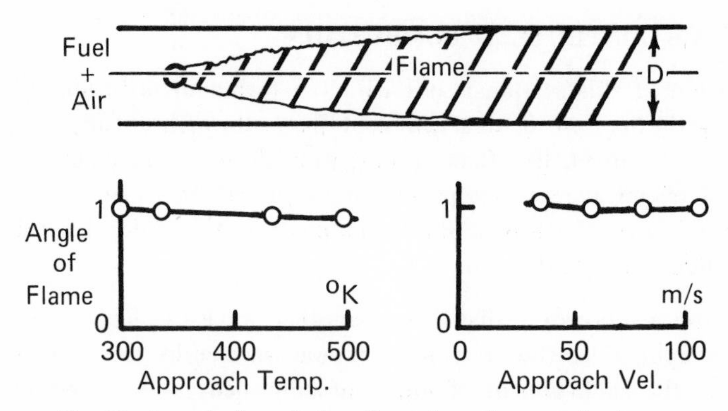

Fig. 10. Spread of a turbulent flame through a pre-mixed gas.

stream flows from left to right. A bluff-body stabilizer on the left holds the flame, which then spreads obliquely across the unburned stream, finally reaching the wall.

The experimental data show that the angle of spread of the flame depends very little on the mixture ratio, the initial temperature level and the inlet velocity. This would be very hard to explain on any hypothesis which allowed influence to the chemical kinetics.

The sketches in the bottom half of the figure show that the predictions based on the eddy-break-up hypothesis agree quite well with the experimental data. While the influence of turbulence on combustion rates is not as yet completely understood, it appears that some of its major features have been incorporated into the mathematical model.

CONCLUSIONS

Several objectives have been accomplished so far. The first point is that, for most practical circumstances, the computational problem can be taken as solved. There are still programs to produce for time-dependent three-dimensional flows, such as that in a gasoline-engine combustion chamber; but this is just a matter of program engineering. In a year or two, this work should be accomplished. Successes with turbulence models have also been very numerous. While not yet as reliably predictable as laminar ones, turbulent flows are no longer the insurmountable obstacles that they used to be. Particle-size distributions, radiative fluxes, and turbulence-reaction interactions are all now amenable to computation.

Of course, there is still much work to be done on the physical processes themselves. What are the laws according to which carbon particles nucleate, flocculate and break into fragments? How do the absorptivities and scattering coefficients depend upon wavelength, particle size and particle density? What is the true role of intermittency in turbulent reaction, and what are the correct formulae for reaction rate when both eddy-break-up and chemical kinetics exert comparable influences?

It will take a long time to answer these questions; for many hypotheses must be formulated, and tested against many experimental measurements, before the physical input to the mathematical models becomes wholly reliable. But the existence of the models will both stimulate and facilitate this research; for it allows the implications of the hypotheses to be swiftly expressed in practical terms; and these implications can then be verified or denied.

Yet there is no need for an equipment designer to decide that, before he begins to use the mathematical models, he should await the outcome of this research. The models are already fit to use; and their use will be a most powerful aid to their further development.

References p. 16

REFERENCES

1. *S. V. Patankar and D. B. Spalding, "Heat and Mass Transfer in Boundary Layers" – 2nd Edition. London, Intertext Books, 1970.*
2. *A. D. Gosman, W. M. Pun, A. K. Runchal, D. B. Spalding, and M. Wolfshtein, "Heat and Mass Transfer in Recirculating Flows", London, Academic Press, 1969.*
3. *L. S. Caretto, R. M. Curr, and D. B. Spalding, "Two numerical methods for three-dimensional boundary layers", to be published. 1971.*
4. *L. S. Caretto and S. V. Patankar, unpublished information. 1971.*
5. *D. B. Spalding, "Concentration fluctuations in a round turbulent free jet", Chemical Engineering Science, vol. 26, 1971, p. 95.*
6. *H. A. Becker, H. C. Hottel, and G. C. Williams, "The nozzle-fluid concentration field of the round turbulent free jet", Journal of Fluid Mechanics, vol. 30, 1967, p. 285.*
7. *D. B. Spalding, unpublished information. 1970.*
8. *F. C. Lockwood and D. B. Spalding, "Prediction of a turbulent reacting duct flow with significant radiation", Colloques d'Evian, Thermodynamics Session, 25-29 May 1971, Societe Francaise de Physique.*
9. *D. B. Spalding, "Mixing and chemical reaction in steady confined turbulent flames", Thirteenth Symposium (International) on Combustion, The Combustion Institute, 1971, p. 649.*

DISCUSSION

C. W. Shipman *(Worchester Polytechnic Institute)*

One conclusion to be drawn from Professor Spalding's paper is that he has done a remarkable job in developing methods of solving the very complex and vexing problem of computing flow, composition, and temperature patterns in a combustor. He implies that we no longer have an excuse for not anticipating these patterns in advance of prototype development. The ability to do this, regardless of whether or not one agrees with the particular reaction models used, will at the very least offer significant help in disclosing potential problems with emissions.

There are, however, some points discussed in the paper at some length which we must consider in some detail. I refer particularly to the models used to describe the reaction rates. It is important to recognize that for purposes of computing the flow, composition, and temperature patterns in the combustor the specific model chosen is not particularly critical. After all, the results are, in effect integrations of the reaction rates. When one is interested in emissions, however, much more attention to detail is required.

To illustrate the point, consider the premixed, turbulent flame burning in a duct. This is a situation for which the paper shows reasonable agreement between computed and measured profiles. As stated in the paper, the available data indicate that these flames are a dispersion of lumps of unburned gas in a sea of burned gas – or, as not stated in the paper, a dispersion of lumps of burned gas in a sea of unburned gas (1). One model of this burning zone has been presented (2, 3) which describes measured reaction rates by assuming that the dispersed lumps burn as laminar flames aided by comminution of the lumps at a rate proportional to the local shear. In considering emissions one must remember that the laminar combustion wave in a premixed gaseous flame produces carbon monoxide, and the carbon monoxide thus produced burns at a rate limited, apparently, by the chemical kinetics.

A second aspect of these flames which is of more than casual interest is the observed fact that turbulence can actually extinguish them. In complicated systems, and this includes nearly every system of practical interest, this phenomenon could take place in some parts of the flame and not in others. That such an occurrence is important in considering emissions from these flames is obvious, but the model described in the paper does not seem to admit the possibility.

Finally, a consideration which could lead, as I understand it, to an error in the compution. There is evidence that in regions of low shear in the flow, the lumps of burned (or unburned) gas do not break up (2, 4). The equations used to compute the break-up frequency, W, would have a decaying value in regions of low shear. The reaction model used by Professor Spalding for comparison with his computed values happens to be for a flame in which there are no regions of low shear of significant extent. There could be such regions in other geometrical situations and a consequent error in the computed results.

The paper does not mention the reaction model used in the calculation of the diffusion flame profiles, but some comments are, perhaps, in order. The complexities here are enormous, especially when one considers emissions. Evidence that fuel and air can be in contact without burning has been available for some time (5, 6). This results in some premixing and a consequent reduction in soot formation. Studies of ducted diffusion flames, in progress at WPI, have shown that such premixing is extremely sensitive to feed rates. Even more interesting, these same relatively soot-free diffusion flames indicate extensive soot formation when they are rapidly quenched by inspiration of cold air – a consideration of great importance in gas turbine combustors.

In view of the foregoing, the circumstances with respect to formation of oxides of nitrogen and soot will be complex indeed, requiring reaction models of significantly greater complexity than described in this paper.

In concluding this brief discussion, I should like to re-emphasize the principal point I want to make. This knit-picking could probably be done with the other models discussed in the paper – BUT. The development of the computational methods described by Professor Spalding is a significant achievement. The fact that it makes available to the design engineer good, quantitative estimates of the patterns of velocity, composition, and temperature means that regions of potential pollutant formation can be anticipated. The comments about the reaction models used in making the computations are intended to indicate that they are probably not sufficient by themselves for computation of the emissions.

References

1. *N. M. Howe, Jr., et al., Ninth Symposium (International) on Combustion, Academic Press, New York 1963, pg. 36.*
2. *N. M. Howe, Jr., and C. W. Shipman, Tenth Symposium (International) on Combustion, The Combustion Institute, Pittsburgh 1965, p. 1139.*
3. *B. S. Cushing et al., Eleventh Symposium (International) on Combustion, The Combustion Institute, Pittsburgh 1967, p. 817.*
4. *J. E. Faucher, PhD. Dissertation, Worchester Polytechnic Institute, Worchester, 1965.*
5. *W. R. Hawthorne et al., Third Symposium on Combustion and Flame and Explosion Phenomena, Williams and Wilkins, Baltimore, 1949, p. 266.*
6. *A. Vranos et al., Twelfth Symposium (International) on Combustion, The Combustion Institute, Pittsburgh, 1969, p. 1050.*

W. A. Sirignano *(Princeton University)*

I am interested in the applications of the numerical schemes to practical devices. I would expect that you would have to choose your mesh size to be small compared to say the air hole size or air hole diameter in a typical turbine. This would mean it would be extremely small compared to the diameter or the length of the burner. I wonder if you can treat such an involved program. While I only have rough estimates, I wonder about it. I would like to hear your opinion on it.

D. B. Spalding

I agree with Professor Shipman's remarks. What I have intended has been to take simple models to demonstrate that they can, when combined with the computational procedures, lead to informative comparisons with experiment. As soon as one makes these comparisons, one finds that the model differs from the facts; by examining these differences, one can usually find ways to improve the model. Professor Sirignano's question on the application of computational procedures to real devices is very important. How does one deal with phenomena which are smaller in scale than the size of the grid? Firstly, one cannot deal with them perfectly – not as well as one would wish. Suppose we are calculating a combustion-chamber flow with a coarse grid and that, at the wall, within the rectangle which represents the grid projection on the wall, there is one small hole. We can represent the effects of that hole only in the following way. We can make sure that the right mass flow rate enters the chamber through it; that the momentum fluxes are correct in each of the three directions; thirdly, we can make sure that what we put in is correct in concentration, and perhaps also in such turbulence properties as energy and eddy size. What I want to emphasize is that we have means of introducing into our computation only the gross effects of injection through a small hole; however we can not distinguish between a round hole and a square hole except insofar as the difference affects one of the things which I have enumerated. Of course, if that limitation becomes important, one must do a separate computation of the important region on a finer grid; or experimental data for discharge coefficients can be introduced.

W. A. Sirignano

Would you expect microscale shearing forces to be important in the area of the air jet for example? Do you really think the problem should be treated with a coarse mesh size?

D. B. Spalding

I think that is the best we can do. When some small-scale process is found to have especial importance, it is necessary to make a detailed study using a fine grid. It is the same with measurements: one can make overall measurements and get information with which one learns that there is an area of special sensitivity. One then modifies one's scheme of measurements in order to get more refined information.

F. V. Bracco *(Princeton University)*

You seem to be able to deal with a two-phase flow. Do you mean that drag and vaporization of fuel drops are understood so well, quantitatively, that you can include them in your model?

D. B. Spalding

I think that the drag and vaporization problems are sufficiently understood even under unsteady gas situations. More serious is the fact that the atomizer characteristics are seldom known well enough: one does not know what size distribution will result at a certain radius from a unit injector. Another process that deserves great attention is the joining up of particles, particularly when soot is being formed.

A. F. Sarofim *(Massachusetts Institute of Technology)*

The progress made by Professor Spalding towards the development of a generalized mathematical model of the complex processes occurring in combustion systems is most impressive. I wish to make one comment and raise one question on minor issues. The first concerns the statement by Professor Spalding that the gaps in our knowledge of radiative properties is one of the factors limiting the modeling of heat transfer in combustion systems. The extensive information presently available on the total and spectral emissivities of CO_2 and H_2O, although incomplete, is more than adequate for use in the usually simplified treatment of the radiative terms in the energy equation. Greater uncertainly is associated with the optical constants of soot at high temperatures but this cannot be considered to be a serious limitation at a time when there are no valid models for predicting soot concentrations. The question is directed at determining the method utilized by Professor Spalding for correlating fluctuatings in temperature with those in concentration.

D. B. Spalding

First, I agree with what you said concerning radiation. Concerning the correlation of the fluctuations of temperature and concentration, in my Utah paper they were assumed to be correlated. In reality correlations cannot be complete, but I do not see my way clearly here at all. When the reaction is very simple, or when there is no reaction at all, it is probably adequate to proceed with 100 percent correlation; one can make an accurate computation therefore of all of the processes. When there are competing reactions, I think it is these reactions which will uncorrelate the temperature and the various concentrations. I believe that all that one can do is to make the simplest hypotheses, assuming for example a degree of correlation which is less than 100 percent.

R. Shinnar *(New York City College)*

I am also very impressed with Prof. Spalding's work. I have also been interested in using modeling techniques where there is a lack of detailed knowledge about the flow field. In chemical reactors, we employ some very complex modeling procedures and I say that I am guilty of some of them. The result has been that some people seem to believe that we are able a priori to design a complex flow reactor for a complex reaction. This is not true. The only kinds of reactors that we are able to design, for complex flows, are those where either you do not care too much about the results or the results are insensitive to the flow. If the results are sensitive, I wouldn't advise it to anybody. In general, my procedure for design in complex reaction systems is to use a flow field which is as clearly defined as possible – unless I can find out in advance that the results are not very sensitive to my lack of knowledge about the flow field.

On the other hand, there are also cases where you want to get some feedback from the experiment itself, as to the kinetics; for instance in particulate processes and soot formation where the physics is not well known. In some of these emissions problems, I am not sure that the kinetic processes underlying it are as well known as our mathematical formulations seem to indicate. It is extremely important in such cases to have a good understanding of the flow field so that we do not have to use complex computational procedures to extract the kinetics. While such complex computational procedures are possible, they are not as unique and as clearly defined as some people would like to believe.

D. B. Spalding

I am not an expert in the models which are being used to describe chemical reactors by Dr. Shinnar, but my understanding is that it has not been possible until very recently to make computations of the flow field. I would like to stress the idea of *balance of complexity* in a mathematical model. It seems to me that it is wrong to have a highly complex kinetic model with 30 species and the corresponding number of reactions and yet have a primitive flow model such as the stirred reactor or the plug-flow one. It may be that some of the complex models employed in chemical reactor design have been unbalanced in this sense, the complexity of the kinetics being incommensurate with the excessive simplicity of the flow model.

CURRENT KINETIC MODELING TECHNIQUES FOR CONTINUOUS FLOW COMBUSTORS

A. M. MELLOR

Purdue University, Lafayette, Indiana

ABSTRACT

Analytical models existing in the open literature for gas turbine continuous flow combustors are reviewed and discussed from the point of view of predictions of pollutant emissions. Particular emphasis is placed on the kinetic aspects of the models involving liquid fuel droplet evaporation and/or combustion and homogeneous chemical kinetics for hydrocarbon/air combustion. A brief summary of the various flow models is also included. Comparisons with data obtained from experimental or practical combustors are made where appropriate, and suggestions for further research are listed.

INTRODUCTION

Emissions from continuous flow combustors have become of increased importance as attention has focused on standards for power generating equipment and aircraft, as well as possible replacement of spark ignition and diesel engines with devices operating on the Brayton, Rankine, or Stirling cycles. Because of comparatively long residence times and small surface to volume ratios for typical continuous flow combustors, the emissions of unburned hydrocarbons and carbon monoxide are usually not large at design conditions; however, large volumes of air are exposed to high temperatures for relatively long times in these burners, and thus nitric oxide can be formed in substantial amounts.

In this paper attention is centered on emissions of these pollutants from Brayton cycle combustors, that is, gas turbine combustors. In particular, the kinetics of liquid fuel droplet vaporization and combustion, and the homogeneous chemical kinetics of gaseous fuel-air reactions as appropriate to pollutant generation will be discussed.

References pp. 45-48

Because for transportation applications sulphur is not a significant constituent of most fuels(1), sulphur oxides will not be considered. In addition, since understanding of the formation of flame carbon is at an elementary stage (for contrasting theories of this process see Homann and Wagner(2), Chakraborty and Long(3), and Howard(4)), and since empirical design techniques exist for the alleviation of soot emissions (1,5-9), especially for aircraft combustors, soot will not be discussed. This latter area has recently enjoyed an excellent review by Linden and Heywood(10).

There are currently five models for estimating air pollutant emissions from gas turbine combustors: for nitric oxide, Fletcher and Heywood(11) and Roberts et al.(12); for nitric oxide and carbon monoxide, Hammond and Mellor(13,14); and for unburned hydrocarbons, nitric oxide, and carbon monoxide, Pratt et al.(15) and Edelman and Economos(16) have provided combustor models.

Various degrees of kinetic sophistication are in use in these models. Heywood and co-workers(11,17,18) assumed the carbon-hydrogen-oxygen chemistry to be in chemical equilibrium and used the resulting oxygen atom concentrations to kinetically determine nitric oxide emissions through a modified Zeldovich mechanism. No two-phase effects were considered.

Hammond and Mellor(13,14) also neglected the effects of droplet vaporization and burning, but included a more complete kinetic scheme to estimate carbon monoxide and nitric oxide emissions: it was assumed that the parent hydrocarbon fuel decomposes infinitely fast to H_2O and CO, which then react via the generally accepted moist CO mechanism. NO concentrations were predicted through combination of the kinetic O atom concentration with the Zeldovich mechanism.

Pratt et al.(15) concerned themselves with a combustor burning gaseous CH_4 or natural gas and used a complete kinetic mechanism for CH_4 oxidation to determine the emissions.

Both the models of Roberts et al.(12) and Edelman and Economos(16) included the possibly important phenomena of droplet vaporization and/or burning with finite rate hydrocarbon kinetics.

After a brief review of the flow portions of these models in the next section, the question of the influence of liquid fuel injection on emissions will be approached in a fundamental manner since disagreement regarding its importance exists in the literature. The appropriate homogeneous chemical kinetics for hydrocarbon fuel vapor/air mixtures will then be discussed, with particular regard to the approximate partial equilibrium method of Heywood and co-workers(11,17,18) for the calculation of NO emissions. Experimental data will be included where appropriate.

FLOW MODELS

In this section, the flow models for gas turbine combustors which have been presented in the open literature are briefly reviewed. This review is necessary for two reasons: firstly, some investigators have used different kinetic schemes for calculation of pollutant concentrations in different zones of their models, and secondly, it is desired to contrast these models with the more fundamental approach of Spalding(19).

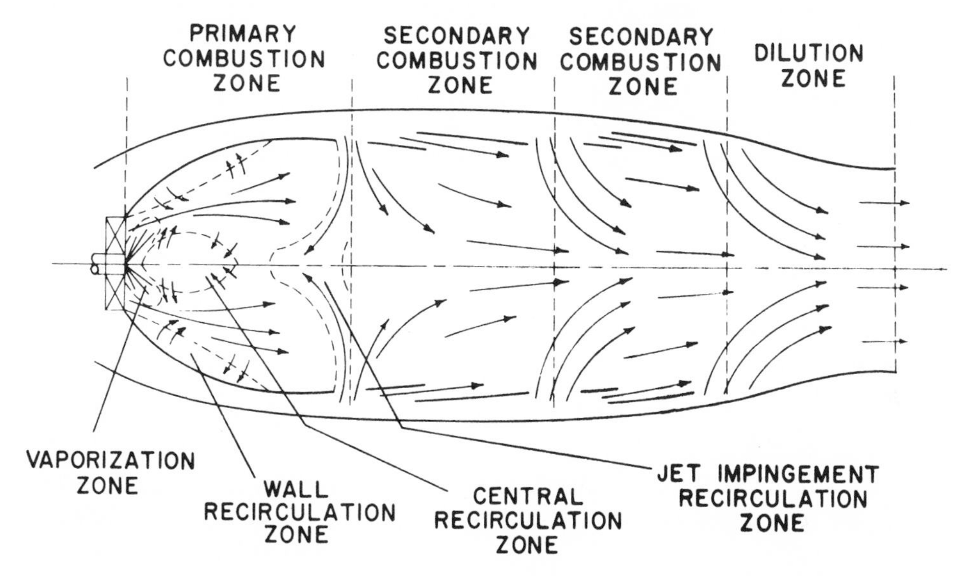

Fig. 1. Typical combustor cross-section showing the various flow regimes.

A typical combustor can be divided into three regions of differing flow and reaction characteristics, as shown in Fig. 1. The first or primary zone, located at the upstream end of the combustor before the first row of penetration holes, burns fuel at equivalence ratios not too far from unity, and possesses a strong recirculation pattern for the purpose of flameholding. In the secondary zone, air is added through the liner to the unburned gases and combustion products issuing from the primary; the function of this zone is to complete the combustion and begin to recover dissociation losses by lowering the bulk gas temperature. The remaining air, corresponding to the overall combustor equivalence ratio and a tolerable turbine inlet temperature, is added in the dilution zone.

Note that in these latter two zones the direction of flow is primarily downstream, with no large scale recirculation patterns. As a result most investigators have chosen to model the secondary and dilution zones as one-dimensional plug flow reactors (11,12,17,18) or as well stirred reactors in series (13-15; an infinite number of well stirred reactors in series is equivalent to a plug flow reactor). The air flow distribution is generally calculated with methods similar to those described in Ref. (20) and (21).

The main differences in the flow models occur in the primary zone, where essentially a modular approach has been used by every investigator, as opposed to the

References pp. 45-48

continuum approach discussed by Spalding(19). In Fig. 2 the primary zone of the combustor and the models which have been used to represent it are indicated schematically. Hammond and Mellor(14) have discussed the usual flow pattern which in addition to the droplet evaporation and/or combustion zone 1 possesses flow in the downstream direction (zone 2) and recirculation flow (zone 3), which may or may not entrain fresh air from the first row of penetration holes.

The simplest model of the primary zone is shown in Fig. 2b and is due to Pratt et al.(15), who modeled the entire zone as a single well stirred reactor. This assumption is equivalent to postulating uniform gas temperature and composition throughout the primary zone and is clearly a rather gross approximation (liquid fuel considerations are neglected in this model of combustors which burn only gaseous fuel).

A more sophisticated model was proposed by Fletcher and Heywood (11,18) and is shown schematically in Fig. 2c. Here also the presence of fuel droplets is neglected, but the primary zone consists of a macromixed partially stirred reactor.* In this way discrete fluid elements are allowed to react but retain their identity until the one-dimensional "lateral mixing" zone, in which the original equivalence ratio distribution $f(\phi)$ over these elements is relaxed to the mean value leaving the primary zone. An early primary zone model consisting of a single well stirred reactor used by Heywood et al.(17) is discussed by Heywood(18).

Hottel et al.(24) have investigated the stability characteristics of the recirculation pair of well stirred reactors shown in Fig. 2d and found that they qualitatively resembled those of a gas turbine combustor. For this reason, Hammond and Mellor(13,14) have chosen to represent the primary zone in this fashion with one reactor corresponding to zone 2 and the other to zone 3. Sizing of the volumes and an estimate of the recirculating flowrate R can be obtained from empirical information available in the literature(12). Zone 1 of droplet effects was also neglected in this model.

The scheme of Roberts and co-workers(12) replaces the first well stirred reactor of Hammond and Mellor with a one-dimensional or plug flow reactor. As will be discussed in the next section, fuel vaporization is allowed to proceed in this reactor as well as in the secondary zone (Fig. 2e), whereas only fresh air from the first row of penetration holes and fuel vapor are allowed to enter the recirculation zone (which is again modeled as a well stirred reactor and in which the C-H-O chemistry is assumed to be in chemical equilibrium).

Edelman and Economos(16) suggested a model consisting of a turbulent diffusion flame including finite-rate droplet evaporation (to correspond to zones 1 and 2), in which is embedded a well stirred reactor (zone 3); this model is shown schematically

Hammond and Mellor (22) and Pratt et al. (15) have described this nomenclature originally suggested by Levenspiel (23).

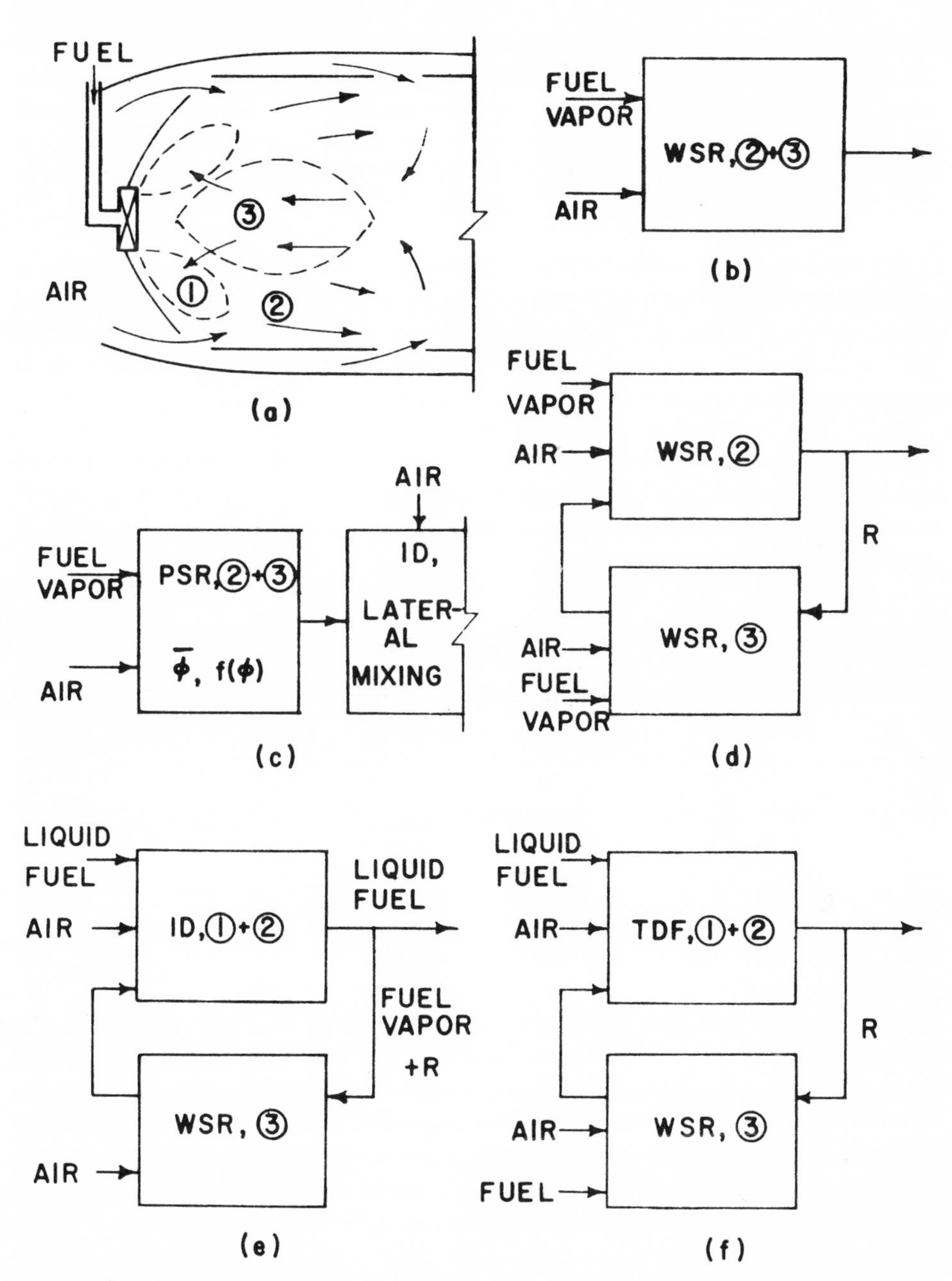

Fig. 2. Schematics of primary zone and flow models thereof. 1 = zone of droplet evaporation and/or combustion; 2 = zone with flow in downstream direction; 3 = recirculation zone. WSR = well stirred reactor; PSR = partially stirred reactor; 1D = one dimensional (plug) flow; TDF = turbulent diffusion flame; R = recirculating flow of combustion products. (a) Primary zone flow schematic; (b) after Pratt et al. (15); (c) after Fletcher and Heywood (11) and Heywood (18); (d) after Hammond and Mellor (13,14); (e) after Roberts et al. (12); (f) after Edelman and Economos (16).

in Fig. 2f. Fuel and air are also entrained in the recirculation zone. This last model is intuitively most pleasing and in a modular sense approaches the continuum technique of Spalding(19).

Because this paper is concerned with kinetic processes, including finite rate droplet evaporation and combustion, and because some of the flow models are in transient states, we will not discuss them further except to point out their rather arbitrary nature in terms of flow splits between the various zones, volumes and residence times in these zones, etc. Hopefully it will soon be possible to employ more fundamental approaches such as discussed by Spalding(19) to establish the flow characteristics, for the "macromodels" reviewed above will most likely at best only yield trend predictions for the emission of pollutants from actual combustors.

It has been seen in the above that some investigators have included zone 1 of droplet vaporization and/or combustion in their models; others have neglected it entirely. Thus in the next section an attempt is made to determine the importance of liquid fuel injection on emissions.

DROPLET EVAPORATION AND COMBUSTION

Most continuous flow combustors use liquid fuel which is injected at the head end of the combustor, and consequently the question of droplet evaporation and/or combustion must be considered as it pertains to the emission of air pollutants. (From a performance or efficiency point of view, droplet evaporation and burning times are usually negligible (25).) Parikh et al.(26) have demonstrated decreased emissions from a model combustor burning gaseous methane as opposed to liquid heptane, and natural gas as fuel is in use in stationary power generation applications because of similar low emission properties. Thus, heterogeneous processes are of importance in terms of air pollution from liquid-fueled combustors.

Unfortunately, very little information is available about the details of the injection process which concern us here, notably drop size distribution and drop trajectories in the primary zone. Semi-empirical correlations are available in the open literature for certain types of injectors (see for example Longwell(27) and Raushenbakh et al.(28)), and some success has been achieved in estimating size distributions and trajectories (28), but in the absence of detailed flow models for the primary zone, such calculations are of questionable utility. Therefore it is of interest to simply estimate evaporation and burning times for a single droplet from combustion theory for conditions which are typical of continuous flow combustors.

For a single spherical droplet either evaporating or burning in an oxidizing atmosphere, the "d^2 law" is well known to accurately correlate the droplet lifetime:

$$d^2 = d_o^2 - \lambda_i t. \tag{1}$$

Here λ_i, the evaporation coefficient, is taken as λ_e for evaporation or λ_c for combustion. (See the list of nomenclature in the Appendix for definition of other symbols.) The droplet lifetime is then given as:

$$\tau_i = d_o^2/\lambda_i. \tag{2}$$

The evaporation coefficient in a stagnant atmosphere λ_{oi} can be expressed as:

$$\lambda_{oi} = (8k/\rho_\ell C_p)\ \ell n\,(1 + B_i) \tag{3}$$

where B_i is the transfer number for evaporation

$$B_e = C_p(T - T_\ell)/L \tag{4}$$

or for combustion in the configuration of a droplet diffusion flame or envelope flame

$$B_c = \{C_p(T - T_\ell) + y_{ox}\ Q/i\}/L. \tag{5}$$

These expressions have been derived and/or experimentally verified by many investigators (see for example Spalding(29,30), Wise et al.(31), Wood et al.(32), Raushenbakh et al. (28), and Williams (33)).

Since it is expected that there will be some relative motion between droplet and gas at least initially in the primary zone of a combustor, the above expressions must be modified to account for the effect of forced convection. Three empirical correlations for this purpose have been proposed. The first is due to Frössling (see Ranz and Marshall(34)) and applies for $0 \leq Re \leq 1000$:

$$\lambda_i = \lambda_{oi}\,(1.0 + 0.276\ Re^{1/2}\ Sc^{1/3}); \tag{6}$$

Spalding(29,30) has suggested the following expression for $0.6 \leq B_i \leq 5$ and $800 \leq Re \leq 4000$:

$$\lambda_i = (2.12\mu/\rho_\ell)\ B_i^{3/5}\ Re^{1/2}; \tag{7}$$

for lower Reynolds number ranges, Eisenklam et al.(35) determined:

$$\lambda_e = \lambda_{oe}\,\frac{B_e}{1+B_e}\left(\frac{1.0 + 0.8\ Re^{1/2}}{\ell n\,(1 + B_e)}\right) \qquad 0.03 \leq Re \leq 25, \tag{8}$$

$$\lambda_c = \frac{19.6\ k\,(T - T_\ell)}{\rho_\ell L}\ \frac{Re^{0.4}}{1 + B_c} \qquad 0.1 \leq Re \leq 3. \tag{9}$$

In all cases the Reynolds number is based on relative velocity and drop diameter:

$$Re = \rho\ u\ d_o/\mu. \tag{10}$$

The Frössling correlation has been applied to gas turbine calculations by Longwell(27), Raushenbakh et al.(28), Lefebvre(36), and Roberts et al.(12).

References pp. 45-48

Heywood et al.(17) used the Eisenklam expression for evaporation, and Edelman and Economos(16) neglected convective effects and used an expression similar to that for λ_{oi} given above.

In Table 1 are listed values of the parameters used for calculations of droplet lifetimes in the primary zone. A gas temperature of 1000 °K and a pressure of 5 atm were assumed; JP-5 was taken as a representative liquid fuel. Gas phase properties are for pure air at 1000 °K.

TABLE 1

Parameters Used for Estimation of Droplet Lifetimes Listed in Table 2

Parameter	Value and Units	Source
d_o	80μ	Assumed*
p	5 atm	Assumed*
T	1000 °K	Assumed*
ρ	1.77×10^{-3} g/cm^3	Calculated
ρ_ℓ	.617 g/cm^3	(37)
Q	10, 150 cal/g	(37)
T_ℓ	573 °K	(37)
L	229 cal/g	(37)
i	3.48	(30)
c_p	.274 cal/g °K	(38)
B_e	.511	Calculated
B_c	3.44	Calculated
k	1.65×10^{-4} cal/cm sec °K	(39)
λ_{oe}	3.23×10^{-3} cm^2/sec	Calculated
λ_{oc}	11.6×10^{-3} cm^2/sec	Calculated
Pr = Sc	.71	(39)
μ	4.27×10^{-4} g/cm sec	Calculated**
u	50 m/sec	Assumed*
Re	165	Calculated

* *Assumed values were taken as typical on the basis of the work of Raushenbakh et al. (28).*
** *Obtained from* $\mu = k\,Pr/c_p$.

Estimated droplet lifetimes for an initial diameter of 80 μ are given in Table 2. Results of all the convective correlations are listed, as well as any results of other investigators. It is expected that those results using the correlation of Frössling ($\tau_e = 4.7$ msec and $\tau_c = 1.3$ msec) are most applicable, because the initial Reynolds number is estimated as 165 (Table 1). Spalding's correlation is also probably appropriate here, and as expected predicts somewhat longer times ($\tau_e = 5.1$ and $\tau_c = 1.6$ msec); it was found that for $800 \leq Re \leq 4000$ the lifetimes predicted by the Frössling correlation somewhat underestimated, while the Spalding correlation somewhat overestimated experimental values(29,30). The Eisenklam correlations have been used at a much higher Reynolds number than originally intended and thus are

TABLE 2

Estimates of Lifetimes in Primary Zone for Evaporating or Burning Droplets with Initial Diameter of 80 μ

Droplet	Atmosphere	Evaporation Coefficient, $\lambda_i \times 10^3$, cm^2/sec	Lifetime τ_i, msec	Source or Correlation
Evaporating	Stagnant	3.2	19.8	–
	Stagnant	11.6	5.5*	Edelman & Economos (16)
	Convective	13.5	4.7	Frössling
	Convective	12.6	5.1	Spalding
	Convective	29.8	2.2	Eisenklam
	Convective	65.9	1.0	Heywood et al. (17)
Burning	Stagnant	11.6	5.5	–
	Stagnant	16.0	4.0*	Edelman & Economos (16)
	Convective	48.5	1.3	Frössling
	Convective	39.6	1.6	Spalding
	Convective	17.0	3.8	Eisenklam

* *These values estimated from a graph in Edelman and Economos (16); λ_{oi} and τ_i thus obtained were multiplied by 2 and by 1/2 respectively to account for a discrepancy in the definition of λ_{oi} in this reference.*

probably inaccurate; in fact, the burning time is predicted to exceed the evaporation time. Heywood et al.(17), using this last correlation, estimated a lower τ_e than the present investigator primarily because they assumed too high a value of the stagnant-atmosphere evaporation coefficient (20×10^{-3} cm^2/sec). Because Edelman and Economos(16) ignored the mass transfer augmentation due to convection, longer droplet lifetimes were predicted by these investigators.

Thus we conclude that if the Reynolds number and ambient gas properties were constant during the droplet lifetime, an 80μ droplet would require 4 to 5 msec to evaporate or 1 to 2 msec to burn in a droplet diffusion flame in the primary zone. In actuality, the Reynolds number will decrease as the droplet penetrates into the primary zone, but this effect will be compensated for in part by an increase in the transfer number. Thus the total droplet lifetime is on the order of or greater than the primary zone residence time, and combustor models for emissions from liquid-fueled devices which neglect the finite rate evaporation or droplet combustion may be seriously in error.

It is much easier to come to this conclusion than to determine an appropriate approximate method which will account for the presence of evaporating or burning droplets. In a real engine configuration, there is a size distribution of droplets: for the lower end of the spectrum convection can probably be ignored, and these droplets may burn in vapor-phase diffusion flames; since the majority of reaction in such flames occurs near stoichiometric, large amounts of NO could be formed (such drops

References pp. 45-48

possess short burning times, however). On the other hand, the larger droplets may simply evaporate for a large portion of their lifetime because any relative velocity between the drop and the gas will be sustained longer and may prevent any flame from forming.* Here variable equivalence ratios would be experienced by the fuel vapor (depending on the turbulent mixing rate and trajectory of the droplet), and substantial amounts of unburned fuel and CO would persist into the secondary zone of the combustor. Thus the combustion process most likely consists of both droplet diffusion flame combustion at an equivalence ratio of unity and droplet vaporization and subsequent homogeneous reaction at variable equivalence ratios, with the relative importance of these heterogeneous and homogeneous reactions unknown. Similar qualitative discussions concerning the nature of the burning have been given in Ref. (1), (16), (26), (30), and (40).

To summarize, most likely both droplet burning and droplet vaporization (and subsequent turbulent mixing of the gaseous fuel) should be included in any complete model for emissions from continuous flow combustors which burn liquid fuels. Combustion in the former case will occur at near stoichiometric equivalence ratios and thus could be responsible for large NO emissions, although it has been surmised that only the smaller droplets with short lifetimes burn in this fashion. A large range of equivalence ratios could be involved in the homogeneous combustion, depending on the relative speed of the mixing of fuel vapor and air. Only complete knowledge of both the injector and flow characteristics will resolve the relative importance of these effects.

None of the combustor models in current use for gas turbines involve attempts to solve this formidable problem in the detail discussed above. Hammond and Mellor(13,14) were initially interested only in performance calculations and omitted droplet vaporization and burning from their model completely. Pratt et al.(15) attempted to model a combustor using methane or natural gas as fuel. Heywood and co-workers(17) have performed an analysis similar to that at the beginning of this section and as noted above estimated a droplet vaporization time of about 1 msec. Although we regard this conclusion to be erroneous, on this basis these investigators have neglected droplet vaporization and/or burning in their combustor models(11,17,18).

In the Pratt and Whitney combustor model, which has been described by Roberts et al.(12), it is assumed that all droplets evaporate according to Eqn. (1), (3), (5), and (6) above where B_c varies as y_{ox} and T change locally. A mean drop size and initial relative velocity are obtained from known injector characteristics. The droplet velocity is corrected by means of a droplet momentum equation as the drop moves downstream through the primary zone. Mixing of the fuel vapor with the ambient gas

** Using an empirical criterion developed by Spalding (29,30), Heywood et al. (17) have concluded that envelope flames cannot be sustained around 80 μ droplets in a typical combustor.*

is assumed to be infinitely fast, and reaction of the hydrocarbon then occurs via a homogeneous mechanism to be described in the next section of this paper. Note that the assumption of infinitely fast mixing precludes stoichiometric combustion (as in a droplet diffusion flame) unless the local equivalence ratio happens to be unity, and thus NO formation rates predicted by this method may be low.

In agreement with the conclusion reached above that typical droplet evaporation times are commensurate with primary zone residence times, Roberts et al.(12) found the mean diameter droplet to completely evaporate in about 20% of the overall combustor length; if the mean diameter was doubled or halved, then 80% or 10% respectively of the combustor length was required.

Edelman and Economos(16) proposed the use of a technique similar to that of Roberts et al.(12) except that the convective effect was ignored. Also, since the droplets presumably will supply fuel vapor for a gaseous turbulent diffusion flame, the mixing is not assumed instantaneous. Unfortunately no results have yet been presented for this model.

Homogeneous chemical kinetics as used by the various investigators in their models are described in the next section. These apply to premixed fuel vapor-air-burned gases at any desired equivalence ratio, with the source of the gaseous fuel as described above.

HOMOGENEOUS CHEMICAL KINETICS

Experimental studies of emissions from steady flow combustors have clearly shown that the concentrations of unburned hydrocarbons (HC), CO, and NO do not correspond in a simple way to chemical equilibrium values (5, 26, 40-43). In general, HC and CO freeze somewhere in the combustor, and their emissions exceed the equilibrium values at the exit. NO, however, can be more or less than its equilibrium value (see in particular the excellent study by Starkman and co-workers(43)). Thus any model for pollutant emissions must be based on chemical kinetics, because characteristic combustor residence times at a given local temperature are much less than the chemical reaction times required to approach equilibrium.

It should be stressed that the kinetics to be discussed apply in reality only to premixed, homogeneous systems, but have been used by various investigators to model emissions from gas turbine combustors. Spalding(19) has discussed the question of properly estimating reaction rates in turbulent mixing controlled situations.

Hydrocarbons – In general, kinetic mechanisms for the decomposition and partial oxidation of hydrocarbons are known only for the simpler species and for realistic fuels are intractable. A discussion of some of the available mechanisms was given by Hammond and Mellor(22). Thus approximations must be made in order to estimate the rate of production of partially oxidized species.

References pp. 45-48

TABLE 3
Reactions for Hydrocarbon – Air Combustion

Number		Reaction
I.	Overall Partial Oxidation (Initiation)	
	1.	$C_nH_m + \frac{n}{2} O_2 \rightarrow nCO + \frac{m}{2} H_2$
	2a.	$C_8H_{16} + O_2 \rightarrow 2C_4H_8O$
	2b.	$2C_4H_8O + 3O_2 \rightarrow 8CO + 8H_2$
	3.	$C_nH_m + (\frac{n}{2} + \frac{m}{4}) O_2 \rightarrow nCO + \frac{m}{2} H_2O$
II.	CH_4 Decomposition and Partial Oxidation	
	4a.	$CH_4 + M \rightleftarrows CH_3 + H + M$
	4b.	$CH_4 + O \rightleftarrows CH_3 + OH$
	4c.	$CH_4 + H \rightleftarrows CH_3 + H_2$
	4d.	$CH_4 + OH \rightleftarrows CH_3 + H_2O$
	4e.	$HCO + OH \rightleftarrows CO + H_2O$
	4f.	$CH_3 + O_2 \rightleftarrows HCO + H_2O$
	4g.	$HCO + M \rightleftarrows H + CO + M$
III.	CO Oxidation	
	5.	$CO + OH \rightleftarrows CO_2 + H$
IV.	O-H Bimolecular Propagation	
	6.	$OH + H_2 \rightleftarrows H_2O + H$
	7.	$O_2 + H \rightleftarrows OH + O$
	8.	$O + H_2 \rightleftarrows OH + H$
	9.	$O + H_2O \rightleftarrows OH + OH$
	10.	$H_2 + O_2 \rightleftarrows OH + OH$
V.	O-H Termolecular Termination	
	11.	$H + H + M \rightleftarrows H_2 + M$
	12.	$O + O + M \rightleftarrows O_2 + M$
	13.	$O + H + M \rightleftarrows OH + M$
	14.	$H + OH + M \rightleftarrows H_2O + M$
VI.	NO Formation *	
	15.	$O + N_2 \rightleftarrows NO + N$
	16.	$N + O_2 \rightleftarrows NO + O$
	17.	$N + O + M \rightleftarrows NO + M$
	18.	$N + OH \rightleftarrows NO + H$
	19.	$N_2 + O_2 \rightleftarrows NO + NO$
	20.	$N_2 + OH \rightleftarrows NO + NH$
VII.	Involving NO_2	
	21.	$N_2 + O_2 \rightleftarrows N + NO_2$
	22.	$NO + NO \rightleftarrows N + NO_2$
	23.	$NO_2 + M \rightleftarrows O + NO + M$
	24.	$NO_2 + M \rightleftarrows O_2 + N + M$
	25.	$NO + O_2 \rightleftarrows NO_2 + O$
VIII.	Involving N_2O	
	26.	$H + N_2O \rightleftarrows N_2 + OH$
	27.	$O + N_2O \rightleftarrows N_2 + O_2$
	28.	$O + N_2O \rightleftarrows NO + NO$

** Zeldovich mechanism = reactions 15 + 16.*

Two schemes have been proposed in the literature for gas turbine combustors burning arbitrary fuels: these are indicated in Table 3 as reactions 1 and 3. Edelman and Economos(16) used a partial oxidation of the parent hydrocarbon (C_nH_m) to H_2 and CO (reaction 1) which proceeds at an empirically determined finite rate(44). Reactions 2a and 2b (specifically for JP-5) were used by Roberts et al.(12) for compatibility with their computer program, but the rate coefficients were adjusted to correspond to the same correlation; thus the models of Roberts et al.(12) and Edelman and Economos(16) for hydrocarbon partial oxidation are equivalent. Hammond and Mellor(14) used reaction 3 and assumed that its rate is infinite. Good agreement with experimental data for C_3H_8 burning in a well stirred reactor was obtained with this scheme(22). Obviously this latter model is unable to predict hydrocarbon emissions, however.

For a combustor which burns CH_4 or natural gas, Pratt et al.(15) used the mechanism shown in Table 3 under II, which was proposed by Seery and Bowman(45) on the basis of shock tube studies of CH_4/O_2 kinetics. Marteney(46) also used this mechanism as representative of a typical hydrocarbon for the purpose of calculating NO emissions in his fifteen reaction scheme, as shown in Table 4. In addition, this last investigator used a twelve mechanism scheme with equivalent H_2 and CO as input for CH_4, that is, used reaction 1 with an infinitely fast forward rate. No absolute comparisons were made, but Marteney(46) stated that "the NO concentrations computed with this (fifteen reaction) program exhibited the same trends and levels as in the previous (twelve reaction) calculations," which lends some support to the use of reaction 3 with infinite forward rate by Hammond and Mellor(14). The various approximations are summarized in Tables 3 and 4.

In lieu of a satisfactory mechanism and the appropriate rate coefficients for a realistic liquid fuel, it is recommended that reaction 1 be used for the purpose of estimating HC emissions, as in the combustor models of Edelman and Economos(16) and Roberts et al.(12). Such estimates are expressed in terms of unburned fuel(16) and can give no information concerning other species such as aldehydes.

Carbon Monoxide – Due to subsequent reactions, not all of the CO produced in reaction 1 is emitted by a combustor. Most investigators have modeled these reactions by use of the moist CO combustion mechanism shown as reactions 5 through 14 in Table 3; the specific reactions used in each study are listed in Table 4. Note that all investigators limit conversion of CO to CO_2 to reaction 5, which is the generally accepted step in moist CO combustion, and that only minor differences exist in the remainder of the mechanisms. Reaction 10 was found to be insignificant by Hammond and Mellor(14) under the conditions listed in Table 5; many other investigators have excluded it entirely. Roberts et al.(12) found that reaction 13 was also negligible in the temperature range from about 1100 to 2500 °K (unfortunately the other conditions for their calculations were not given), but Hammond and Mellor(47) have reached the opposite conclusion.

TABLE 4
Kinetic Mechanisms for HC, CO and NO Emissions

Reference / Reaction Number	Edelman and Economos (16)	Roberts et al. (12)	Hammond and Mellor (14)	Pratt et al. (15)	Seery and Bowman (45)	Marteney (46) (12 Reactions)	Marteney (46) (15 Reactions)	Bowman (49)
I. Overall Partial Oxidation (Initiation)								
1.	•					• ∞		
2.		•						
3.			• ∞					
II. CH_4 Decomposition and Partial Oxidation								
4.				•	•		•	
III. CO Oxidation								
5.	•	•	•	•	•	•	•	
IV. O-H Bimolecular Propagation								
6.	•	•	•	•	•	•	•	•
7.	•	•	•	•	•	•	•	•
8.	•	•	•	•	•	•	•	•
9.	•	•	•	•	•	•	•	•
10.			*			•		•
V. O-H Termolecular Termination								
11.	•	•	•			•		•
12.	•	•	•			•		•
13.	•	*	•					
14.	•	•	•	•	•	•		•
VI. NO Formation								
15.	•	•	•	•		•	•	•
16.	•	•	•	•		•	•	•
17.	•		*	•		•	•	*
18.		•						*
19.	•							
20.								?
VII. Involving NO_2								
21.	•					*	*	
22.	•					*	*	
23.	•					*	*	
24.	•					*	*	
25.	•					*	*	
VIII. Involving N_2O								
26.		•				*	*	*
27.		•				*	*	
28.		•				*	*	*

** Reactions found to be negligible under conditions listed in Table 5.*

TABLE 5

Conditions for which Mechanisms
Listed in Table 4 Were Obtained or Applied

Reference	Fuel/Oxidizer	ϕ	p atm	T °K	t msec
Edelman and Economos (16)	C_9H_{20}/Air	1.0-1.5	2-20	T_o = 460*	0.6-5.7**
Roberts et al. (12)	JP-5/Air	NA	NA	800-2500	NA
Hammond and Mellor (14)	C_3H_8/Air	0.15-1.34	1-10	1200-2750	0-13.7
Pratt et al. (15)	CH_4/Air	0.8,1.0	1,15	T_o = 700*	0.2-2000**
Seery and Bowman (45)	CH_4/O_2	0.2-5.0	1.5-4.0	1350-1900	0-1.0
Marteney (46)	CH_4/Air	0.8-1.25	1.0-10	1000-3079	0.0001-10
Bowman (49)	H_2/Air	0.18-3	2.2	2150-2800	0-1.0

* T_o = *initial unburned mixture temperature.*
** *Estimated assuming* $\rho = 1.99 \times 10^{-3}$ *g/cm*^{3} *at 15 atm.*

Thus for the purpose of estimating CO emissions, most likely reactions 5 through 9 and 11 through 14 are satisfactory.

Nitric Oxide – The various reactions involving nitrogen oxides which have been suggested for use in combustor models are shown in Table 3 under VI for NO formation; those reactions involving NO_2 and N_2O are listed under VII and VIII, respectively. The Zeldovich mechanism (48) consists of reactions 15 and 16. In Table 4 the kinetic mechanisms which were used by the various investigators are given, and in Table 6 partial equilibrium schemes (in which the C-H-O chemistry is assumed to be in equilibrium) with nitrogen oxide kinetics, as originally applied by Heywood et al.(17) to gas turbines, are summarized. Roberts et al.(12), Bowman(49), Pratt et al.(15), and Hammond and Mellor(14,47) all have compared results of the partial equilibrium technique with the kinetic technique.

Before proceeding to these results, let us discuss if nitrogen oxides other than NO should be included in turbine combustor models. Reactions 21 through 25 involving NO_2 (Table 3) were included by Edelman and Economos(16), but many authors have concluded that this oxide does not form in appreciable amounts in combustors (1,26,41,42,46,50). Reactions involving N_2O have also been shown unimportant by Heywood(18) and Marteney(46), and reactions 26 and 28 were found to be negligible by Bowman(49). Conditions under which these conclusions were reached are listed in

TABLE 6

Partial Equilibrium Schemes for NO Emissions

Reference / Reaction Number	Heywood et al. (17)	Fletcher and Heywood (11)	Heywood (18)	Roberts et al. (12)	Bowman (49)	Pratt et al. (15)	Hammond and Mellor (14,47)
VI. NO Formation							
15.	●	●	●	●	●	●	●
16.	●	●	●	●	●	●	●
17.			*			●	
18.	●	●	*	●			
19.	*						
20.							
VII. Involving NO_2							
21.							
22.							
23.							
24.							
25.							
VIII. Involving N_2O							
26.	●	●	*	●			
27.	●	●	*	●			
28.	●	●	*	●			

** Reactions found to be negligible under conditions listed in Table 7.*

Tables 5 and 7 and are seen to correspond roughly to gas turbine combustors. Thus we exclude reactions 21 through 28 from further discussion.

Some reactions which form NO can also be eliminated: in kinetic calculations Bowman(49) and Hammond and Mellor(14) have shown the termolecular reaction 17 to be unimportant; Heywood(18) reached the same conclusion on the basis of his partial equilibrium technique. In addition, reaction 18 was shown to be unimportant by both Bowman(49) and Heywood(18), but these conclusions were obtained from H_2/air and lean HC/air investigations, respectively. Reaction 19 has been omitted by most investigators because of its slow rate. Finally, Bowman(49) was unable to determine if reaction 20 should be included; a parametric study was performed for the rate coefficient which is unknown. The effect of this reaction seemed most important in the early stages ($t < 0.2$ msec) of the reaction for fuel rich H_2/air mixtures. In internal combustion engines the discrepancy between measured NO and that calculated with reactions 15, 16, and 18 was greatest for rich mixtures(51,52); perhaps reaction 20 is responsible.

TABLE 7

Conditions for which Mechanisms
Listed in Table 6 Were Applied

Reference	Fuel/Oxidizer	ϕ	p atm	T °K	t msec
Heywood et al. (17)	C_nH_{2n}/Air	0.4-1.4	15	1250-2500	0-10
Fletcher and Heywood (11)	C_nH_{2n}/Air	0.3-1.2	15	1500-2500	0-11
Heywood (18)	C_nH_{2n}/Air	0.8-1.0	1-20	2000-3000	0.1-100
Roberts et al. (12)	JP-5/Air	NA	NA	800-2500	NA
Bowman (49)	H_2/Air	0.18	2.07	2560	0-0.6
Pratt et al. (15)	CH_4/Air	1.0	15.0*	∿2500	0-4.2
Hammond and Mellor (14,47)	C_3H_8/Air	0.15-1.34	5.8	1200-2750	0-13.7

** This reference lists 1 atm for this case.*

In summary, it appears that for lean and stoichiometric mixtures NO concentrations can be estimated by use of the Zeldovich mechanism, reactions 15 and 16. Similar conclusions have been reached by Bowman(49) and Newhall and Shahed(53) for H_2/air mixtures. Bowman and Seery(54) are extending their study to CH_4/air mixtures. For modern gas turbines which operate with somewhat lean primary zone equivalence ratios the Zeldovich mechanism is probably appropriate.

As was noted above, Heywood and co-workers(11,17,18) used a partial (C-H-O) equilibrium computation to obtain the O atom concentration, which was then used to estimate NO emissions via the mechanism shown in Table 6. Bowman(49) has cautioned against this computational technique for the early stages of the H_2/air reaction at low pressures (Table 7). It can lead to an underestimation of NO concentration by a factor of about two as a result of superequilibrium O atom concentration and was probably responsible for Fenimore's (55) observation of "prompt NO" in hydrocarbon/air flames (49).

Apparently, however, for typical primary zone residence times (which exceed Bowman's calculated time of 0.6 msec by almost one order of magnitude), the non-equilibrium O atom concentration and resulting excess NO over that predicted by the partial equilibrium method are negligible. In results published by Roberts et al.(12) kinetic O atom concentration exceeds the equilibrium prediction by almost four orders of magnitude at one point in the primary zone. Yet the residence time here appears to be too short and the kinetically calculated temperature too low for substantial NO formation, because the NO calculated with the partial equilibrium scheme exceeds the kinetic prediction everywhere. At the combustor exit the kinetic value is lower by a factor of two, which the authors consider negligible. Note that both at the exit and elsewhere in the combustor the full kinetic value was lower than the partial equilibrium value for NO, whereas Bowman was concerned with the opposite possibility.

Hammond and Mellor(14) also compared the results of kinetic and partial equilibrium estimations for the conditions listed in Table 7; the two values were within two percent of each other throughout the combustor. Significant deviations of O atom concentration from equilibrium occurred only when the temperature was less than about 2000°K, but below this temperature the NO kinetics were also frozen: thus both computational techniques gave the same result.

Data presented by Pratt et al.(15) do not show the results of the two calculational techniques to be the same, as is shown in Table 8. We(47) have repeated the calculations using the same rate coefficients where possible, and these values are also listed in the table. It is seen that our kinetic and partial equilibrium results are equivalent and agree with the trend shown by the data of Fletcher and Heywood as their mixing parameter s_o approaches zero. We have no explanation for the

TABLE 8

Comparison of NO Concentrations Calculated by Various Methods (After Pratt et al. (15))

Conditions are representative of a Pratt and Whitney JT8D combustor primary zone with $\phi = 1$, $T_o = 700$ °K, p = 15 atm*, $\dot{m}_{air} = 1565$ g/sec, $V_{primary} = 3.28 \times 10^3$ cm^3, $\tau \simeq 4.2$ msec.

Flow Model**	Chemistry	T, °K	NO Mass Fraction, ppm	Reference
Micro	Partial Equilibrium	2502	2019	Pratt et al. (15)
Micro	Partial Equilibrium	2513	1650	Hammond and Mellor (47)
Macro, $s_o = 0.05$†	Partial Equilibrium	2506	1540	Fletcher and Heywood (11)
Macro, $s_o = 0.1$	Partial Equilibrium	2477	1270	Fletcher and Heywood (11)
Macro, $s_o = 0.2$	Partial Equilibrium	2394	850	Fletcher and Heywood (11)
Micro	Kinetic with Reaction 4	2474	1250	Pratt et al. (15)
Micro	Kinetic with Reaction 1††	2507	1590	Hammond and Mellor (47)
Micro	Kinetic with Reaction 3, ∞Rate	2514	1643	Hammond and Mellor (47)
Micro	Full Equilibrium	2502	3840	Hammond and Mellor (47)

** Erroneously listed by Pratt et al. (15) as 1 atm.*

*** All are single micro or macro well stirred reactors.*

† s_o is a mixing parameter with $s_o = 0$ for a micro well stirred reactor; see Fletcher and Heywood (11).

†† Empirical overall finite rate law after Edelman (56) as follows: $d(CH_4)/dt = -6 \times 10^9 p^{-0.815} T (CH_4)^{1/2} (O_2) e^{-12,200/T}$ where units are moles/cm^3, sec, atm, and °K.

disagreement between the two micro-reactor partial equilibrium results, but we give the following interpretation for our kinetic result exceeding that of Pratt et al.(15). It has been shown elsewhere that reaction 3 (with infinitely fast forward rate) when applied to CH_4 combustion overpredicts the overall CO_2 formation rate and thus the heat release rate(22). Therefore for a given residence or reaction time, a higher NO concentration results; and the agreement of the kinetic and partial equilibrium schemes is fortuitous.

The data of Roberts et al.(12) do not discount this possibility as in this study a slower and finite rate hydrocarbon oxidation step was used (Table 3, reaction 2). As might be expected and as discussed above, the kinetic results predict lower emissions than the partial equilibrium results.

To summarize, it appears that the partial equilibrium technique used by Heywood and co-workers(11,17,18) for calculating NO concentrations will give the correct trends. However, if applied in short residence time applications (<1 msec), its predictions may be low due to superequilibrium O atom concentrations(49). For longer times, as for the JT8D primary zone conditions listed in Table 8, the calculated NO concentration depends on the method of calculation in the fashion shown in Table 9: as the hydrocarbon partial oxidation rate is increased (from reaction 4 to

TABLE 9

Dependence of Calculated NO Emissions on Method of Calculation (Primarily for JT8D Primary Zone Conditions Listed in Table 8)

Calculational Method	Rate of Heat Release	NO Mass Fraction, ppm	Reference
Partial Equilibrium	Infinite	2019 1650	Pratt et al. (15) Hammond and Mellor (47)
Reaction 3 with Infinite Rate	Infinite	1643	Hammond and Mellor (47)
Reaction 2 with Finite Rate	Finite*	Less than Partial Equilibrium, See Text	Roberts et al. (12)
Reaction 1 with Finite Rate	Finite*	1590	Hammond and Mellor (47)
Reaction Mechanism 4	Slow*	1250	Pratt et al. (15)

***** *Empirical overall finite rate coefficient was not obtained from CH_4, but rather with higher hydrocarbons (56); thus when applied to CH_4 it overpredicts the burning rate.*

References pp. 45-48

reaction 3 with infinite forward rate or the partial equilibrium method), the rate of heat release is increased. The resulting more rapid temperature rise allows more NO formation for a given residence time. Evidence for this conclusion is listed in Tables 8 and 9. Bowman and Seery(54) have performed similar calculations as a function of reaction time and also have concluded that the assumption regarding the rate of CH_4 decomposition determines how much NO forms. The relative concentrations given by the various computational techniques depend on both temperature and reaction time. However, this effect is probably exaggerated for slow-burning CH_4 and is of less importance for common aviation and automotive fuels.

The real test of all of these models and hypotheses must come from comparison of their predicted results with experimental data. In the next section some available data are described.

EXPERIMENTAL COMPARISONS WITH THE OVERALL MODELS

Unfortunately it is difficult to make meaningful comparisons of experimental results obtained with gas turbine combustors (or engines) with the overall models for two reasons: firstly, some of the models have not been completely assembled or programmed, and very few results of the working models have been presented at this time; and secondly, most experimental data were obtained by gas sampling at the combustor (or turbine) exit plane and thus yield little insight into the processes occurring within the combustor (for typical results of this genre see Ref. 1, 5, 40, 41, 50, and 57-60). In fact, by comparison of NO emissions calculated via various primary zone models with the experimental data of Cornelius and Wade(50) for standard and modified (early quench) GT-309 combustors, Heywood(18) has shown that the several different primary zone flow models all give the measured trends. These data from Heywood(18) are listed in Table 10; the flow models used were a well stirred reactor with integration of the NO kinetic equations to the mean primary zone residence time $\bar{\tau}$ ($t = \bar{\tau}$), a well stirred reactor with residence time distribution:

TABLE 10

Comparison of NO Emission from GT-309
Standard Burner and Two Early Quench Designs
(from Heywood, (18))

Modification**	$\bar{\tau}/\bar{\tau}_{ST}$*	Measured	$[NO]/[NO]_{ST}$* at full load $t = \bar{\tau}$	$s_o = 0$	$s_o = 0.5$
#1	0.66	0.78	0.65	0.71	0.83
#2	0.44	0.58	0.47	0.50	0.63

** ST is standard combustor.*

*** Design modifications as listed by Cornelius and Wade (50).*

$$\psi(t) = (1/\bar{\tau})\, e^{-t/\bar{\tau}} \tag{11}$$

(s_o = o), and a partially stirred reactor with residence time distribution given by Eq. 11 and s_o = 0.5 (an equivalence ratio distribution with standard deviation half the mean).

Heywood also showed that a value of s_o = 0.5 in his model reproduced the rather constant exit NO concentration for 70 to 100% gasifier speed range as reported by Cornelius and Wade(50) (whereas smaller values of s_o could not) in an attempt to support the partially stirred reactor flow model. However more detailed information than exit measurements is required to establish the validity of the overall models.

Only recently have detailed internal measurements become available in the open literature(26,42,43). These investigators have probed both axially and radially a model combustor of 7.6 cm diameter and 35.6 cm length operating at pressures slightly in excess of atmospheric. Room temperature air and either liquid n-C_7H_{16}(42,43) or gaseous CH_4(26) were burned at overall equivalence ratios of 0.119, 0.205, 0.281 and 0.20 and 0.27, respectively. Local concentrations of HC (as C_6H_{14}), CO, CO_2, NO, and for the latter fuel O_2 and local temperature were reported. These data were then used to calculate local equivalence ratios, which in turn were used to calculate the equilibrium gas composition and adiabatic flame temperature. Emissions measured at the combustor exit demonstrated trends entirely consistent with other results published in the literature.

For both fuels the equivalence ratio was varied experimentally by varying the air flow rate at constant fuel flow rate. Thus the residence time (as well as mean temperature) in the combustor was increased as the equivalence ratio was increased. As a result, exhaust emissions of CO and HC decreased while NO increased with increasing ϕ. Temperature profiles inferred from internal measurements for the combustor burning liquid C_7H_{16}(43) revealed reaction to occur both in a recirculation zone in the corner of the combustor dome and on the centerline, with the latter mode becoming more important with increasing air flow rate. Peak HC and CO concentrations were observed just ahead of these maximum temperature zones. Most NO was formed near these zones where the local equivalence ratio was slightly less than unity; since the corner recirculation zone was fuel rich, it did not contribute significantly to NO formation at high air flow rates.

Because of the lack of axial symmetry, data reduction was more difficult for the combustor burning gaseous CH_4(26). However, exhaust emissions of HC, CO, and NO were found to be lower than those if liquid C_7H_{16} were used at a comparable equivalence ratio. This result was attributed to a lower total combustion time and a shorter flame length in the case of CH_4, due primarily to the lack of heterogeneous droplet effects.

Since none of the overall combustor models have been applied to this laboratory combustor, no absolute comparisons are possible. However, Hammond and Mellor

(14) have observed that local NO concentrations can be below equilibrium values in the primary zone, but above equilibrium at the combustor exit, as was found at high air flow rates by Starkman et al.(43).

In studies with their model Roberts et al.(12) have investigated the influence of primary zone equivalence ratio, combustor residence time, and mean fuel droplet diameter on NO emissions. Although there was little effect of overall residence time, the baseline configuration was shown to have higher NO emissions than those resulting from increased or decreased primary zone equivalence ratio and initial drop diameter. The authors noted that other aspects of performance dictated the particular baseline configuration (for which no specifications were given), and that the model predicted "absolute values of nitric oxide concentration which are in general agreement with available experimental data."

In general all of the overall combustor models most likely can yield accurate trend predictions, but insufficient detailed data exist at this time to distinguish between the various flow models and kinetic calculational schemes.

SUMMARY AND CONCLUSIONS

Overall models for pollutant emissions of hydrocarbons, CO and NO from gas turbine combustors have been reviewed. In terms of the flow portions of these models, the significant differences exist in the primary zone where modular rather than continuum approaches have been applied. Heywood(18) has shown that exhaust plane NO concentration measurements are too gross to distinguish between various primary zone models, and thus more detailed data must be obtained inside combustors in order to better define the appropriate approach in lieu of more fundamental flow solutions as discussed by Spalding(19).

Without such solutions there is little hope that liquid fuel drop trajectories and lifetimes can be predicted accurately. Order of magnitude analysis for evaporating or burning 80 μ diameter droplets indicates that these lifetimes are commensurate with primary zone residence times, and there are clear indications that heterogeneous droplet effects are responsible in part for the generation of pollutants. Large droplets may slowly evaporate rather than burn and thus produce CO as well as residual hydrocarbons, whereas smaller drops may burn in stoichiometric envelope flames and generate NO. Therefore, depending on the injector and flow characteristics for a particular combustor, both homogeneous and heterogeneous combustion may occur.

Nevertheless, some investigators have ignored droplet effects for combustors which burn liquid fuels(11,13,14,17,18), whereas others include only droplet evaporation, with the droplet diameter characterized by a mean value(12,16). It would be of interest to incorporate a droplet burning model with finite rate kinetics, such as discussed by Bracco(61), into a combustor model to determine its effect on emissions. A mean droplet diameter could be assumed and spray penetration could be treated similarly to the technique of Roberts et al.(12).

Experimentally, the influence of heterogeneous effects could be determined by injecting the fuel either as liquid or as a vapor, with care taken to maintain the injector characteristics and primary zone flow pattern as constant as possible. Parikh et al.(26) have attempted this type of study but changed the fuel as well. An alternative technique is to increase the initial Weber number(62) in the primary zone by for example decreasing the surface tension of the liquid fuel, which could augment aerodynamic droplet shattering after injection and decrease droplet lifetimes.

For premixed homogeneous systems the chemical kinetics can be modeled reasonably well for the purpose of estimating unburned fuel, CO and NO emissions. Use of an overall finite rate hydrocarbon decomposition (reaction 1, Table 3) is recommended for fuels other than natural gas, with reactions 5 through 9 and 11 through 14 accounting for the remaining C-H-O chemistry. Since overall gas turbine equivalence ratios are less than unity, the Zeldovich mechanism (reactions 15 and 16) may be sufficient for NO predictions, but further work is required to clarify NO formation in fuel rich mixtures. Recommended rate coefficients for most of these elementary reactions have been provided by the Leeds group(63). In addition, experiments to test this hydrocarbon/air mechanism and values of the rate coefficients in well-understood aerodynamic flows with residence times from one to fifteen msec are extremely desirable: the stirred reactor may prove particularly useful here.

Although the partial (C-H-O) equilibrium scheme for NO emissions used by Heywood and co-workers(11,17,18) yields proper trend predictions, it must be applied with caution. In general, the NO concentrations calculated from this and other approximate methods depend on the approximation as well as temperature and reaction time. Further experimental studies are required here, but practical combustors will not yield this type of information because here flow effects can easily mask kinetic effects.

The overall models which have been discussed can only be verified or modified through detailed internal measurements in actual combustors. Important initial data of this type have been published recently in the open literature(26,42,43), and attempts should be made to obtain additional data under more realistic gas turbine operating conditions.

ACKNOWLEDGEMENTS

The author wishes to express his gratitude to D.C. Hammond, Jr., C.W. Owens, and R.D. Anderson for discussions of some of the material presented in the paper.

REFERENCES

1. Anon., "Nature and Control of Aircraft Engine Exhaust Emissions," Northern Research Eng. Corp. Report No. 1134-1 (PB 187-711), 1968.

2. K. H. Homann and H. G. Wagner, "Chemistry of Carbon Formation in Flames," Proc. Roy. Soc., Vol. 307A, 1968, pp. 141-152.
3. B. B. Chakraborty and R. Long, "The Formation of Soot and Polycyclic Aromatic Hydrocarbons in Diffusion Flames. III. Effect of Additions of Oxygen to Ethylene and Ethane Respectively as Fuels", Comb. Flame, Vol. 12, 1968, pp. 469-476.
4. J. B. Howard, "On the Mechanism of Carbon Formation in Flames," Twelfth Symposium (International) on Combustion, The Combustion Institute, Pittsburgh, 1969, pp. 877-887.
5. T. Durrant, "The Control of Atmospheric Pollution from Gas Turbine Engines," SAE Paper 680347, 1968.
6. J. J. Faitani, "Smoke Reduction in Jet Engines through Burner Design," SAE Paper 680348, 1968.
7. K. Gradon and S. C. Miller, "Combustion Development on the Rolls-Royce Spey Engine," Combustion in Advanced Gas Turbine Systems, Pergamon, Oxford, 1968, pp. 45-76.
8. A. H. Lefebvre, "Design Considerations in Advanced Gas Turbine Combustion Chambers," Combustion in Advanced Gas Turbine Systems, Pergamon, Oxford, 1968, pp. 3-19.
9. B. Toone, "A Review of Aero Engine Smoke Emission," Combustion in Advanced Gas Turbine Systems, Pergamon, Oxford, 1968, pp. 271-296.
10. L. H. Linden and J. B. Heywood, "Smoke Emission from Jet Engines," Comb. Sci. Tech., Vol. 2, 1971, pp. 401-411.
11. R. S. Fletcher and J. B. Heywood, "A Model for Nitric Oxide Emissions from Aircraft Gas Turbine Engines," AIAA Paper No. 71-123, 1971.
12. R. Roberts, L. D. Aceto, R. Kollrack, J. M. Bonnell, and D. P. Teixeira, "An Analytical Model for Nitric Oxide Formation in a Gas Turbine Combustion Chamber," AIAA Paper No. 71-715, 1971.
13. D. C. Hammond, Jr. and A. M. Mellor, "A Preliminary Investigation of Gas Turbine Combustor Modelling," Comb. Sci. Tech., Vol. 2, 1970, pp. 67-80.
14. D. C. Hammond, Jr. and A. M. Mellor, "Analytical Calculations for the Performance and Pollutant Emissions of Gas Turbine Combustors," Revised Version of AIAA Paper No. 71-711, 1971, Vol. 4, 1971, pp. 101-112.
15. D. T. Pratt, B. R. Bowman, C. T. Crowe, and T. C. Sonnichsen, "Prediction of Nitric Oxide Formation in Turbojet Engines by PSR Analysis," AIAA Paper No. 71-713, 1971.
16. R. Edelman and C. Economos, "A Mathematical Model for Jet Engine Combustor Pollutant Emissions," AIAA Paper No. 71-714, 1971.
17. J. B. Heywood, J. A. Fay, and L. H. Linden, "Jet Aircraft Air Pollutant Production and Disperson," AIAA Paper No. 70-115, 1970.
18. J. B. Heywood, "Gas Turbine Combustor Modeling for Calculating Nitric Oxide Emissions," AIAA Paper No. 71-712, 1971.
19. D. B. Spalding, "Mathematical Models of Continuous Combustion," Emissions from Continuous Combustion Systems, Plenum, New York, 1972.
20. Anon., "Computer Program for the Analysis of Annular Combustors. Vol. I: Calculational Procedures," Northern Research Eng. Corp. Report No. 1111-1 (NASA Cr 72374), 1968.
21. R. R. Tacina and J. Grobman, "An Analysis of Total Pressure Loss and Airflow Distribution for Annular Gas Turbine Combustors," NASA TN D-5385, 1969.
22. D. C. Hammond, Jr. and A. M. Mellor, "An Investigation of Gas Turbine Combustors with High Inlet Air Temperatures. Part I: Combustor Modelling," U.S. Army Tank-Automotive Command Tech. Rep. 11321, 1971.
23. O. Levenspiel, "Chemical Reaction Engineering," Wiley, New York, 1962.
24. H. C. Hottel, G. C. Williams, and A. H. Bonnell, "Application of Stirred Reactor Theory to the Prediction of Combustor Performance," Comb. Flame, Vol. 2, 1958, pp. 13-34.

25. A. H. Lefebvre, "Theoretical Aspects of Gas Turbine Combustion Performance," Note Aero. No. 163, College of Aeronautics, Cranfield, 1966.
26. P. G. Parikh, R. F. Sawyer, and A. L. London, "Pollutants from Methane Fueled Gas Turbine Combustion," College of Eng. Rep. No. TS-70-15, U. Cal. Berkeley, 1971,
27. J. P. Longwell, "Combustion of Liquid Fuels," Combustion Processes, Princeton Univ. Press, Princeton, 1956, pp. 407-443.
28. B. V. Raushenbakh, S. A. Belyy, I. V. Bespalov, V. Ya. Borodachev, M. S. Volynskiy, and A. G. Prudnikov, "Physical Principles of the Working Process in Combustion Chambers of Jet Engines," English Translation, Wright-Patterson Air Force Base FTD-MT-65-78, 1964.
29. D. B. Spalding, "The Combustion of Liquid Fuels," Fourth Symposium (International) on Combustion, Williams and Wilkins, Baltimore, 1953, pp. 847-864.
30. D. B. Spalding, "Some Fundamentals of Combustion," Butterworths, London, 1955.
31. H. Wise, J. Lorell, and B. J. Wood, "The Effects of Chemical and Physical Parameters on the Burning Rate of a Liquid Droplet," Fifth Symposium (International) on Combustion, Reinhold, New York, 1955, pp. 132-141.
32. B. J. Wood, W. A. Rosser, Jr., and H. Wise, "Combustion of Fuel Droplets," AIAA J., Vol. 1, 1963, pp. 1076-1081.
33. F. A. Williams, "Combustion Theory," Addison-Wesley, Reading, 1965.
34. W. E. Ranz and W. R. Marshall, Jr., "Evaporation from Drops," Chem. Eng. Prog., Vol. 48, 1952, pp. 141-146 (Part I) and 173-180 (Part II).
35. P. Eisenklam, S. A. Arunachalam, and J. A. Weston, "Evaporation Rates and Drag Resistance of Burning Drops," Eleventh Symposium (International) on Combustion, The Combustion Institute, Pittsburgh, 1967, pp. 715-728.
36. A. H. Lefebvre, "Factors Controlling Gas Turbine Combustor Performance at High Pressure," Combustion in Advanced Gas Turbine Systems, Pergamon, Oxford, 1968, pp. 211-226.
37. H. C. Barnett and R. R. Hibbard, "Properties of Aircraft Fuels," NACA TN 3276, 1956.
38. D. R. Stull, Editor, "JANAF Thermochemical Tables," PB 168 370, 1965.
39. W. M. Kays, "Convective Heat and Mass Transfer," McGraw-Hill, New York, 1966.
40. D. S. Smith, R. F. Sawyer, and E. S. Starkman, "Oxides of Nitrogen from Gas Turbines," Air Poll. Control Assn. J., Vol. 18, 1968, pp. 30-35.
41. R. F. Sawyer and E. S. Starkman, "Gas Turbine Exhaust Emissions," SAE Paper 680462, 1968.
42. R. F. Sawyer, D. P. Teixeira, and E. S. Starkman, "Air Pollution Characteristics of Gas Turbine Engines," ASME Trans., J. Eng. Power, Vol. 91, 1969, pp. 290-296.
43. E. S. Starkman, Y. Mizutani, R. F. Sawyer, and D. P. Teixeira, "The Role of Chemistry in Gas Turbine Emissions," ASME Paper 70–GT–81, 1970.
44. R. B. Edelman and O. F. Fortune, "A Quasi-Global Chemical Kinetic Model for the Finite Rate Combustion of Hydrocarbon Fuels with Application to Turbulent Burning and Mixing in Hypersonic Engines and Nozzle," AIAA Paper No. 69-86, 1969.
45. D. J. Seery and C. T. Bowman, "An Experimental and Analytical Study of Methane Oxidation behind Shock Waves," Comb. Flame, Vol. 14, 1970, pp. 37-48.
46. P. J. Marteney, "Analytical Study of the Kinetics of Nitrogen Oxide in Hydrocarbon-Air Combustion," Comb. Sci. Tech., Vol. 1, 1970, pp. 461-469.
47. D. C. Hammond, Jr. and A. M. Mellor, Unpublished Data.
48. Ya. B. Zeldovich, P. Ya. Sadovnikov, and D. A. Frank-Kamenetskii, "Oxides of Nitrogen in Combustion," Acad. Sci. USSR, Inst. Chem. Phys., Moscow-Leningrad (M. Shelef, Translator), 1947.
49. C. T. Bowman, "Investigation of Nitric Oxide Formation Kinetics in Combustion Processes: the Hydrogen-Oxygen-Nitrogen Reaction," Comb. Sci. Tech., Vol. 3, 1971, pp. 37-45.
50. W. Cornelius and W. R. Wade, "The Formation and Control of Nitric Oxide in a Regenerative Gas Turbine Burner," SAE Paper 700708, 1970.

51. G. A. Lavoie, "Spectroscopic Measurements of Nitric Oxide in Spark Ignition Engines," Comb. Flame, Vol. 15, 1970, pp. 97-108.
52. J. B. Heywood, S. M. Mathews, and B. Owen, "Predictions of Nitric Oxide Concentrations in a Spark-Ignition Engine Compared with Exhaust Measurements," SAE paper 710011, 1971.
53. H. K. Newhall and S. M. Shahed, "Kinetics of Nitric Oxide Formation in High Pressure Flames," Thirteenth Symposium (International) on Combustion, The Combustion Institute, Pittsburgh, 1971, pp. 381-389.
54. C. T. Bowman and D. J. Seery, "Investigation of NO Formation Kinetics in Combustion Processes; the Methane-Oxygen-Nitrogen Reaction," Emissions from Continuous Combustion Systems, Plenum, New York, 1972.
55. C. P. Fenimore, "Formation of Nitric Oxide in Premixed Hydrocarbon Flames," Thirteenth Symposium (International) on Combustion, The Combustion Institute, Pittsburgh, 1971, pp. 373-380.
56. R. B. Edelman, General Applied Science Laboratories, Personal Communication, July 15, 1971.
57. R. E. George, J. A. Verssen, and R. L. Chass, "Jet Aircraft: a Growing Pollution Source," Air Poll. Control Assn. J., Vol. 19, 1969, pp. 847-855.
58. K. W. Porter and L. H. Williams, "Gas Turbines for Emergency Vehicles," SAE Paper 650460, 1965.
59. W. Cornelius, D. L. Stivender, and R. E. Sullivan, "A Combustion System for a Vehicular Regenerative Gas Turbine Featuring Low Air Pollutant Emissions," SAE Paper 670936, 1967.
60. M. W. Korth and A. H. Rose, Jr. "Emissions from a Gas Turbine Automobile," SAE Paper 680402, 1968.
61. F. V. Bracco, "A Model for the Diesel Engine Combustion and NO Formation," Paper Presented at the 1971 Meeting, Central States Section/The Combustion Institute, 1971.
62. J. A. Nicholls, "Aerodynamic Shattering and Combustion of Fuel Drops and Films in I. C. Engines," Paper Presented at the 1971 Meeting, Central States Section/The Combustion Institute, 1971.
63. D. L. Baulch, D. D. Drysdale, D. G. Horne, and A. C. Lloyd, "Critical Evaluation of Rate Data for Homogeneous, Gas Phase Reactions of Interest in High-Temperature Systems. Parts 1 through 4," Dept. of Phys. Chem., The University, Leeds, May 1968 – Dec. 1969.

NOMENCLATURE

B_i = transfer number for droplet evaporation (i=e) or combustion (i=c)
C_p = gaseous specific heat at constant pressure, cal/g °K
d = instantaneous droplet diameter, cm
d_o = initial droplet diameter, cm
$f(\phi)$ = equivalence ratio distribution function
i = stoichiometric gravimetric oxidizer/fuel ratio
k = gaseous thermal conductivity, cal/cm sec °K
L = sensible enthalpy of liquid fuel from 15° C to temperature T_ℓ and latent heat of evaporation at T_ℓ, cal/g
$\dot{m}$ = mass flow rate, g/sec
p = pressure, atm
Pr = Prandtl number
Q = heat of combustion, cal/g
R = universal gas constant, cal/mole °K; recirculating flowrate ratio

Re = Reynolds number
Sc = Schmidt number
s_o = mixing parameter of Fletcher and Heywood (11)
t = time, sec
T = temperature of ambient gas, °K
T_ℓ = boiling point temperature of liquid fuel at pressure p, °K
T_o = initial unburned mixture temperature, °K
u = droplet velocity relative to gas, m/sec
V = volume, cm^3
y_{ox} = ambient oxidizer mass fraction
λ_i = evaporation coefficient in forced convection for droplet evaporation (i=e) or combustion (i=c), cm^2/sec
λ_{oi} = evaporation coefficient in stagnant ambient for droplet evaporation (i=e) or combustion (i=c), cm^2/sec
μ = gaseous viscosity, g/cm sec
ρ = gaseous density, g/cm^3
ρ_ℓ = density of liquid fuel at T_ℓ, g/cm^3
τ_i = lifetime of droplet in evaporation (i=e) or with combustion (i=c), sec
ϕ = equivalence ratio

DISCUSSION

J. B. Heywood *(Massachusetts Institute of Technology)*

I think that Prof. Mellor has given us a very good summary of where we are in trying to model the combustion process inside a turbine burner or in an atmospheric pressure furnace burner which has similar characteristics. I would like to comment at this time on flow modeling as distinct from kinetics.

From the models that Prof. Mellor has described, it is clear that the statistical approach to modeling the flow has been taken, as distinct from the fluid mechanic approach which Prof. Spalding described earlier. I think that one of the troubles why we have such a long way to go is because nobody has a good physical description of the details of how such combustors perform. Now Prof. Mellor picked out fuel vaporization as an important process that has escaped scrutiny or perhaps has been underrated today. I want to make the point that fuel vaporization is just one of many processes which in sequence determines the mixing of the fuel and the air, position of the flame, and the state of the burned gases. Examples of other processes are the fuel injector type, the breakup of the spray, the droplet trajectories, evaporation, the fuel vapor and air mixing processes, the flame structure, whether any pre-mixed burning takes place, whether the flame is of the diffusion type, whether there is volume combustion as implied in some of the stirred reactor models. I do not think that we have a good enough detailed physical description of this sequence of events, at this stage, to determine adequate models.

I would like to stress one feature that I think is very important, that is, the sequence of steps resulting in burned gases that are not uniform in temperature or composition. I think that real combustors do not approach well-stirred reactors in practice. So I feel that simple approaches using perfectly stirred reactors to model what is happening in combustors will not prove valuable. These comments apply to liquid and gaseous fueled burners. The only major difference between gaseous and liquid fuel injection is the droplet vaporization process. Fuel vapor still has to mix with the air and burn, and again that is going to result in non-uniform burned gases.

Now coming back to the statistical as against the fluid mechanic models of the flow, I think that both of these directions are worth exploring for these reasons. I have seen several high-speed color movies of the burning process inside a gas turbine combustor and an industrial furnace. Now sometimes there is a well-defined mean flow pattern with fluctuations about the mean. Sometimes there seems to me to be very substantial fluctuations with a poorly defined mean flow path. Now if the latter is the case, it is not clear to me that fluid mechanic-type modeling will give us a sufficiently accurate representation of the position and structure of the flame and the composition and temperature of the burned gas. If the fluctuations about the mean are relatively small, obviously the fluid mechanic approach is going to be very valuable. So I believe it is worth exploring both these two approaches, because for practical situations it is not yet clear which one of them is likely to give us the simplest models with which to predict the combustion characteristics.

W. A. Sirignano *(Princeton University)*

Prof. Mellor, you talked about the secondary zone being modeled by virtually everyone as a one-dimensional phenomena. I thought that you were implying that it was all right, because you didn't talk any further about it. I really wonder if it is all right. There are data existing to show that you get temperature profiles at the end of the burner. Therefore you would expect that it is not one dimensional, and you would need a multi-dimensional theory to predict this.

A. M. Mellor

You are correct, I believe, that the one-dimensional flow is an obvious oversimplification. Currently some work that was done this past summer by EPA is showing that significant amounts of NO_2 are present in the exhaust of combustors. One place where such species could be formed could be in the little recirculation zones behind the penetration jets where you have long residence times and relatively cold air. One-dimensional modeling would ignore such zones and may predict zero NO_2 emissions. I myself don't feel that the secondary zone can be handled too much better without getting too complicated. Now Dr. Edelman, for example, does not propose to model in this fashion.

E. K. Bastress *(Northern Research and Engineering Corp.)*

In view of this being a survey-type paper, I am a bit surprised that a few other problems are not included which certainly should be included as goals for the future. The first one that comes to mind is the problem of carbon formation. To my knowledge, it is not being treated in any gas turbine model. It is very important now not only from the standpoint of carbon emission but also from the standpoint of heat transfer which is a very difficult problem for the designer. Now I would suggest that this at least be included as a goal for the future for the sake of completeness. Another problem which has been appearing very recently is the conversion of NO to NO_2. In many cases, this is not important because we assume that NO goes to NO_2 eventually in the atmosphere if not in the engine. However, there are some circumstances where the distribution between NO and NO_2 is important and it is becoming clear that this distribution varies considerably with operating condition.

A. M. Mellor

In regard to your second comment on NO_2; unfortunately when I wrote the paper, this apparently carefully documented information about NO_2 being found was not available to me. Consequently, I ignored NO_2 because no one as yet had decided it to be that important. From a fundamental point of view, I ignored soot primarily because the formation process is not well understood. However, I do believe that one can model the oxidation process once the soot is formed. The other reason that I ignored soot was simply because there seems to be reasonable well-defined empirical techniques for alleviating the problem that exists.

W. Bartok *(Esso Research and Engineering Co.)*

We have attempted to formulate a somewhat similar type of macroscopic mixing model in which we have two series of parallel stirred reactors with mass and transport exchange between parallel reactor pairs. In other words, instead of having a single pair of stirred reactors at the inlet, we have two series of parallel reactors. Eventually, in a system like this one, which is equivalent to two plug flow reactors in parallel, the concentrations and conditions asymptotically become the same. We have not yet been able to test the model that would be meaningful from the point of view of predictions. The question that I would like to raise is whether you have compared a number of different models and their predictions and how would they compare with actual experimental data?

A. M. Mellor

I am very glad that you asked that question. The experimental information, which is available in the nature of exhaust plane measurements, cannot be used very well for comparison with the models. While some of the models are being programmed and assembled, there are no results available as yet. In the written version of the paper, I made an attempt to make comparisons where appropriate. About all that I can say is that in the paper describing the Pratt-Whitney model, the investigators conducted some parametric studies on a baseline configuration and then simply said that the absolute values of nitric oxide which are predicted were in reasonable agreement with experimental values. There isn't much else available at this time on the subject.

R. Shinnar *(New York City College)*

I would like to refer to Prof. Heywood's comments on the distinction between statistical and fluid dynamic models which are not really that contradictory. There are two ways that one can use stirred tank models or more refined statistical methods such as residence time distributions obtained by tracer experiments. One of them is a bounding procedure that one uses to evaluate what type and magnitude of deviations from ideal mixing would have physical importance and how close one has to come to an ideal stirred tank before this idealization becomes meaningful. Or one can estimate the magnitude of possible effects due to imperfect mixing.

There are, however, other more direct modeling techniques based on networks of stirred tanks which are used for reactor design and these techniques are only valid when one can get the parameters of one's statistical models from well defined experiments. There are methods by which one can get those parameters in a quite sensible way either from fluid dynamic calculations, or from some type of special tracer experiments. Just assuming various flow splits are normally of little value unless one deals with bounding methods.

D. T. Pratt *(Washington State University)*

I would like to make two observations. First, if you look at the progression in complication of these chemical reactor-type models – starting with one stirred reactor and then going to two in parallel, and eventually as Dr. Bartok has mentioned to presumably six or a dozen – you eventually arrive at the generalized tank and tube formulation that Dr. Spalding has discussed. The second comment, in regard to these simple models, is a possible third version of your zones 2 and 3 where zone 2 is a one-dimensional upstream flow with entrainment or distributed mass addition, and zone 3 is the same thing downstream. In this case, the governing equations are hyperbolic in character, which is typical of large scale turbulence. In other words, if gross patches of material are being transported across zone boundaries, the governing mass conservation equations are hyperbolic in character. So a third type of differential equation has to be kept in mind when dealing with large-scale as opposed to small-scale mixing.

J. P. Longwell

I have one comment to make in regard to the problem of modeling carbon formation. While it is true that you can solve the problem by leaning the fuel-air mixture to preclude carbon formation, one of the important solutions to the nitric oxide problem is to burn very rich. This is particularly useful where you have organic nitrogen in the fuel. If you burn lean, this organic nitrogen is converted rather efficiently to nitric oxide. On the rich side, this is not true. The organic nitrogen disappears but does not reappear as nitric oxide. Therefore when burning fuels like coal or some petroleum fractions that contain quite a bit of organic nitrogen, I think there is a strong incentive to burn extremely rich. When you burn extremely rich, you are up against the carbon formation problem again and so there may be a delicate balance between the two. In that case, modeling becomes more interesting.

SOME OBSERVATIONS ON FLOWS DESCRIBED BY COUPLED MIXING AND KINETICS

R. B. EDELMAN, O. FORTUNE and G. WEILERSTEIN

General Applied Science Laboratories Inc., Westbury, New York

ABSTRACT

This paper presents some aspects of analytical formulations and numerical solutions pertaining to modeling of practical combustion systems. Emphasis is on the gas turbine combustor problem but related flows involving turbulent mixing and reaction kinetics are discussed. The concepts are oriented toward the description of the origin and disposition of species relevant to combustion efficiency and pollutant emissions.

INTRODUCTION

Until recently the single most important goal in the design of practical combustion chambers was the achievement of maximum combustion efficiency. This involved, to a large extent, providing conservative residence times limited only by size constraints dictated by the particular system and its application. Achieving relatively large residence times would tend to insure sufficient time for mixing of the fuel and oxidizer while also providing ample time for the completion of combustion.

More recently the increased demands for higher performance in combustion systems has required a more precise definition of the structure of flames and the parameters that control the heat release distribution. Furthermore, increasing public concern regarding pollutant emissions has made the problem of modeling flames an absolute requirement for the definition of the relationships between the combustion and pollutant formation mechanisms. Such modeling is now needed to ultimately define design guidelines for combustion efficiency with minimum pollutant output.

To accomplish this the variables of the flame must be examined within a framework which accounts for the coupling of the relevant processes governing the

References pp. 86-87

degree of completion of combustion and the formation of pollutants. Such a mathematical framework must include representations for the fluid mechanics which involves convective and diffusive transport of mass, momentum, and energy; and chemical production of the species relevant to the particular chemical system. To these basic mechanisms one must in general add radiative heat transfer and multiphase flow processes.

To completely model such coupled physical-chemical processes as they occur in most practical combustion chambers is beyond the capability of existing analytical tools. Computer storage limitations combined with prohibitive costs preclude numerical solutions of the full Navier-Stokes system. Thus modeling is required if any feasible computations are to be made.

Perhaps the most widely used models for continuous flow systems are those based upon one-dimensional, or plug flow, concepts. Such models can provide insight into the gross trends of only the most idealized configurations and consequently are limited in their lack of ability to describe flame structure. However, several models can be postulated which characterize the mechanisms above and can provide insight into the relationships among the parameters controlling the fluid mechanical and chemical kinetic processes as they occur in real systems.

The combustion process will generally initiate with the discrete injection of fuel and oxidizer in the vicinity of an ignition source. The subsequent development of this flame will depend upon the combined effects of the mixing rates, reaction rates, and residence times. In some applications these factors can be augmented by the use of swirl and various mixing enhancers to increase the heat release rate and minimize zones of off-design stoichiometry. For example, jet engine combustors and certain burners used in stationary heating and power plants are designed to provide intense back mixing zones to improve flame stability and combustion performance. Within such zones the structure of the flame tends to be dominated primarily by chemical reaction rates and less so by the rate of mixing. In the ideal limit, infinitely fast mixing occurs and the recirculating zone behaves as a perfectly stirred reactor. Thus, the stirred reactor is not only a valuable research tool but it also represents a very important element of practical combustion systems. Of course this model represents only one of the possible flow regimes which can be encountered in a given combustion system. Flow field non-uniformities with the attendant turbulent exchange of mass, energy, and momentum are encountered in real flames. Thus, a model which is designed to provide quantitative predictions of flame behavior in real burner configurations must include at least the mechanisms of convection, diffusion and generation of mass, energy, and momentum. A model which contains appropriate representation for these processes will inherently contain the controlling parameters which are basic to the description of most flames. Such a model will provide the link between flame characteristics and the initial and boundary conditions which are in the control of the designer.

MATHEMATICAL MODEL

Even with simplifying assumptions, a model which will predict burner flows with engineering accuracy will still be complex. First of all, non-linear partial differential equations are involved and secondly the bulk of practical flows are turbulent. The former presents a challenge in numerical analysis and the latter presents a challenge of the imagination. Of course, progress has been made over the years and where turbulence is concerned empirically determined representations have evolved which permit computations to be made with a useful degree of confidence. It should be emphasized here that even the simplest of flows involving kinetics alone (plug flows and perfectly stirred reactors) require empirical information for definition of the specific rate constants.

Because of these inherent weaknesses, the model we shall describe is constructed in modular form. This approach lends itself to the study of each mechanism either separately or coupled and provides a means of updating aspects of the model which depend upon empirical data for definition.

Modular Form – The modular concept proposed here includes a rather general set of discrete mechanisms which control combustion efficiency and pollutant emissions. These mechanisms can be conveniently represented in terms of chemical kinetic and fluid mechanical processes as shown in Fig. 1. However, it should be noted that these ingredients are coupled in the actual combustor. Thus, Fig. 1 shows that the reaction kinetics include droplet combustion and involves the exothermic reactions associated with the production of CO_2 and H_2O, the non-exothermic reactions involving combustion intermediates, as well as NO_x and SO_x (appropriate for sulfur bearing fuels), and finally CO, C_xH_y, and soot (C_s) formation which can be associated with combustion efficiency. In terms of pollutants, the net output of the chemical reaction processes are the concentrations of CO, C_xH_y, C_s, NO_x and SO_x. Of course, the reaction processes are occurring within a complex flow field involving fuel droplets (and soot) suspended in the gas. Thus, the multiphase fluid mechanics include both gas and droplet mixing in the presence of swirl and recirculation.

Multi-Phase Mixing and Combustion Model – In general, non-uniformities exist within the combustor giving rise to diffusive transport between the regions of differing velocity, temperature and concentrations. Thus, transport of momentum, energy and mass occurs between rich and lean regions, as well as between hot interior and cool liner wall regions. Depending upon the relative rates of mixing and kinetics, high concentrations of CO, C_xH_y, NO_x and C_s can be transported by turbulent mixing into cool regions and quench at levels much higher than would be predicted by local equilibrium. Then high levels can persist throughout the combustor if the mixing rate is not properly controlled.

An analysis appropriate for the description of this coupled mixing and combustion process has been detailed in Ref. 1. This analysis includes consideration of thermal

References pp. 86-87

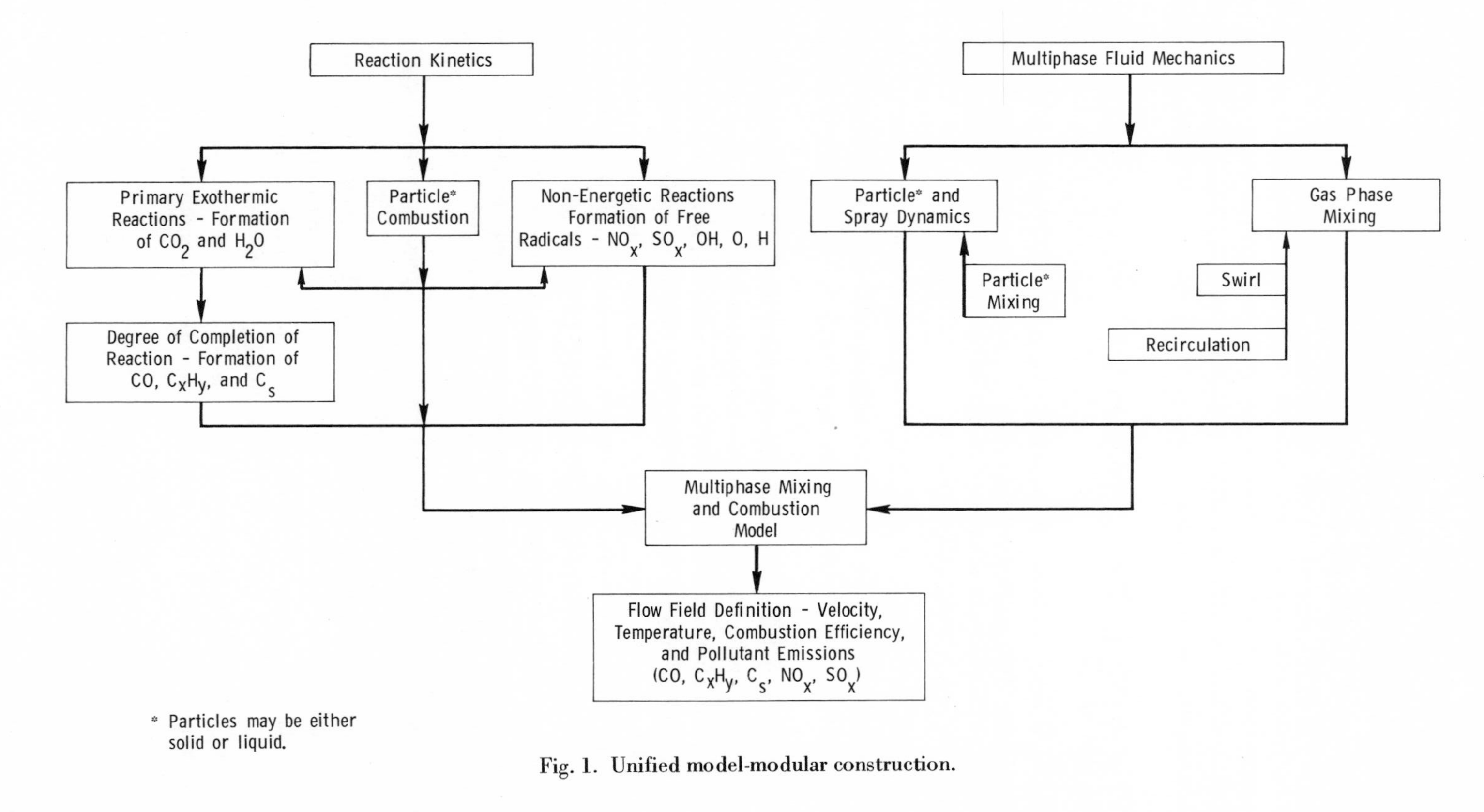

Fig. 1. Unified model-modular construction.

and dynamic non-equilibrium for a dispersion of droplets suspended in a turbulent multi-component gaseous carrier undergoing homogeneous and heterogeneous chemical and phase transition kinetics. A set of working equations has evolved from this study which is applicable to the flow regimes bounding the recirculation zones as shown in Fig. 2. The primary considerations are:

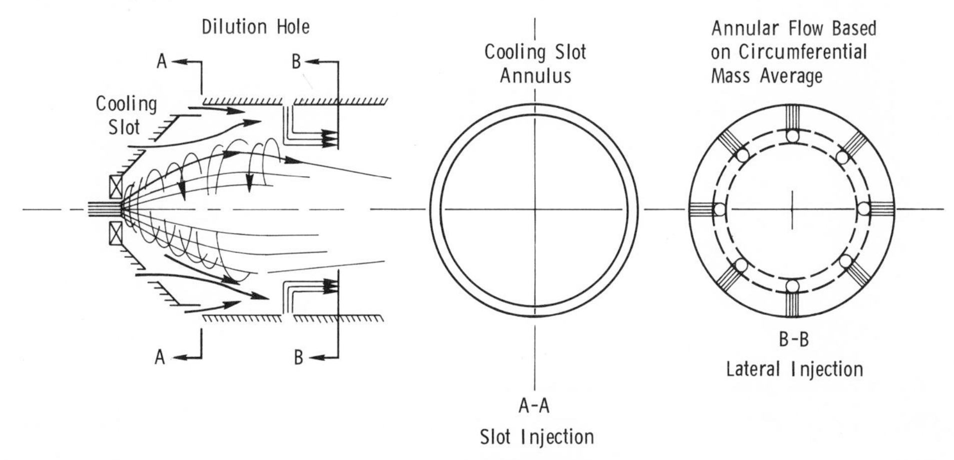

Fig. 2. Mixing zones.

1. The fuel droplets and other particulates form a dilute, continuum-like suspension in the multi-component gas phase carrier.
2. The volume occupied by the droplets and solids is negligible.
3. Turbulent transport is important only in the direction normal to the burner flow.
4. The droplets and particulates are classified in terms of size and composition, and each class mixes (or spreads) at a rate determined by its ability to respond to the gas phase motion.

The describing conservation equations are parabolic and for steady axisymmetric turbulent flow are given as follows:

Global Continuity:

$$\frac{\partial y \rho u}{\partial x} + \frac{\partial y \rho v}{\partial y} = 0 \tag{1}$$

Species Diffusion:

i^{th} gas phase species

$$\rho u \frac{\partial \beta_{gi}}{\partial x} + \rho v \frac{\partial \beta_{gi}}{\partial y} = \frac{1}{y} \frac{\partial}{\partial y}$$
$$\left\{ y \rho \epsilon_{Dg} \left[\frac{\partial \beta_{gi}}{\partial y} - \beta_{gi} \left(\frac{1 - \sum_j \theta_j \beta_{pj}}{\beta_g} \right. \frac{\partial \beta_g}{\partial y} + \right. \right. \tag{2}$$

References pp. 86-87

$$\sum_j \theta_j \frac{\partial \beta_{pj}}{\partial y}\Big)\Big]\Big\} + \dot{w}_{gi} + \sum_j \dot{w}^F_{pij}$$

j^{th} particle class

$$\rho u \frac{\partial \beta_{pj}}{\partial x} + \rho v \frac{\partial \beta_{pj}}{\partial y} = \frac{1}{y}\frac{\partial}{\partial y}$$

$$\Big\{ y \rho \epsilon_{Dg} \theta_j \Big[\frac{\partial \beta_{pj}}{\partial y} - \beta_{pj} \Big(\frac{1 - \sum_\mu \frac{\theta_\mu \beta_{p\mu}}{\theta_j}}{\beta_g} \frac{\partial \beta_g}{\partial y} + \tag{3}$$

$$\sum_\mu \frac{\theta_\mu}{\theta_j} \frac{\partial \theta_{p\mu}}{\partial y}\Big)\Big]\Big\} + \dot{w}^F_{pj}$$

where

$\dot{w}_{gi}$ = production of ith gas specie due to homogeneous gas phase reactions;

$\dot{w}^F_{pij}$ = production of ith gas phase specie from j^{th} particle class due to evaporation of hetergeneous reactions;

$\dot{w}^F_{pj}$ = production of the jth particle class.

Momentum:

$$\rho u \frac{\partial u}{\alpha x} + \rho v \frac{\partial u}{\partial y} = -\frac{dp}{dx} + \frac{1}{y}\frac{\partial}{\partial y}\Big[(y \rho \epsilon_{vg} (\beta_g + \sum \sigma_j \beta_{pj}) \frac{\partial u}{\partial y} \Big] \tag{4}$$

Energy:

$$\rho u \frac{\partial H}{\partial x} + \rho v \frac{\partial H}{\partial y} = \frac{1}{y}\frac{\partial}{\partial y} \Big\{ y \rho \epsilon_{vg} \Big[\frac{1}{Pr}\frac{\partial H}{\partial y} + \Big(h_g \frac{1}{Sc} \sum_j \theta_j \beta_{pj} - \frac{H_g}{Pr} -$$

$$\frac{1}{Pr}\Big(\frac{Pr}{Sc} - 1\Big) h_g - \frac{1}{Sc} \sum_j \theta_j h_{pj} \beta_{pj} \frac{1 - \sum_\mu \frac{\theta_\mu}{\theta_j} \beta_{p\mu}}{\beta_g} \Big) \frac{\partial \beta_g}{\partial y} +$$

$$\frac{1}{Pr} \sum_j \beta_{pj} (\delta_{pj} - 1) \frac{\partial H_{pj}}{\partial y} + \frac{1}{Pr}\Big(\frac{Pr}{Sc} - 1\Big) \sum_j h_{ig} \frac{\partial \beta_{gi}}{\partial y} + \tag{5}$$

$$\sum_j \Big[\theta_j \frac{(h_{pj} - h_g \beta_g)}{Sc} - \frac{H_{pj}}{Pr} \Big] \frac{\partial \beta_{pj}}{\partial y} + \Big[\beta_g \Big(1 - \frac{1}{Pr}\Big) -$$

$$\sum_j \beta_{pj} \frac{\delta_j}{Pr} \Big] \frac{\partial u^2/2}{\partial y} - \frac{1}{Sc} \Big(\sum_\mu \theta_\mu \frac{\partial \beta_{p\mu}}{\partial y} \Big) \Big] \Big\}$$

where

$\theta_j = \frac{\epsilon_{Dj}}{\epsilon_{Dg}}$, $\delta_j = \frac{\epsilon_{Tj}}{\epsilon_{Tg}}$, $\sigma_j = \frac{\epsilon_{vj}}{\epsilon_{vg}}$ are the ratios of particle to gas mass, thermal and momentum diffusivities; and Pr and Sc are the gas phase Prandtl and Schmidt numbers, respectively.

The above energy equation may not, in general, be sufficient to define the thermal state of the system.

The temperature of the particles can be sensibly different from the gas phase temperature depending upon the process occurring on the particle scale. For evaporating/burning droplets, the local saturation (wet bulb) temperature is appropriate. In cases where the thermal equilibrium assumption does not apply, this equation must be supplemented by either the individual gas-phase or particulate phase energy equations. The jth particulate phase energy equation is given by:

$$\beta_{pj}\,\rho u \frac{\partial h_{pj}}{\partial x} + \beta_{pj}\,\rho v \frac{\partial h_{pj}}{\partial y} = q_{pj} +$$

$$\frac{1}{y}\frac{\partial}{\partial y}\left(y \frac{\sigma_{pj}}{Pr}\rho\,\epsilon_{vg}\,\beta_{pj}\frac{\partial h_{pj}}{\partial y}\right) +$$

$$\rho\,\epsilon_{Dg}\,\theta_j \left\{ \frac{\partial \beta_{pj}}{\partial y} - \left[\frac{1 - \sum_\mu \frac{\theta_\mu \beta_{p\mu}}{\theta_j}}{\beta_{pj}} \right] \frac{\partial \beta_g}{\partial y} + \right.$$

$$\left. \sum_\mu \frac{\theta_\mu}{\theta_j}\frac{\partial \beta_{p\mu}}{\partial y} \right\} \frac{\partial h_{pj}}{\partial y}$$

where q_{pj} is the heating rate of the interior of the jth particulate phase, and h_{pj} is the static enthalpy of the particles.

Swirl Effects – In practical combustors, it is desirable to complete the combustion process efficiently in a short length while minimizing pollutant emissions. In addition, stable flames are required. The optimum situation is to maximize the mixing rate providing sufficient residence time for stable and complete combustion, while hopefully attaining a residence time that does not permit an undesirable amount of pollutants.

Swirl is employed in turbine combustors and offers a means of controlling mixing rates, residence times, and flame stabilization. Swirl affects the flow field in two distinct ways; first it increases turbulent diffusivities in burner type flames of interest here, and second it is coupled to the momentum flux and affects the mean properties of the flow. The former relates to the turbulent transport rate, and the latter relates specifically to the decelerating effect that large swirl can have upon the axial component of velocity.

The extension of the above analysis to include swirl is straight forward. It can be shown that the inclusion of swirl requires the momentum equations for the radial and swirl components of velocity to be added to the system. The above single momentum equation is replaced by three momentum equations:

$$x:\ \rho u \frac{\partial u}{\partial x} + \rho v \frac{\partial u}{\partial y} = \frac{1}{y}\frac{\partial}{\partial y}\left[y\,\rho\,\epsilon_{vg}\,(\beta_g + \sum_j \sigma_j \beta_p)\frac{\partial u}{\partial y}\right] - \frac{\partial p}{\partial x} \tag{7}$$

$$y:\ \rho \frac{w^2}{y} = \frac{\partial p}{\partial y} \tag{8}$$

$$\varphi:\ \rho u \frac{\partial wy}{\partial x} + \rho v \frac{\partial wy}{\partial y} = \frac{1}{y}\frac{\partial}{\partial y}\left\{ y\,\epsilon_{vg}\,(\beta_g + \Sigma\, \sigma_j \beta_p)\left[\frac{\partial wy}{\partial y} - 2w\right]\right\} \tag{9}$$

Detailed examination of this analysis shows that a non-uniform pressure field is developed within the swirling flame. In fact, the pressure within the flame will be *below* the pressure level in the immediate surroundings of the swirling zone. Thus, as the flame develops and mixes out, the pressure will rise toward this level. If the pressure rise is large enough, a stagnation point will appear and a recirculation zone will develop. Although swirl is not necessarily the principal generator of recirculation in a turbine combustor, it can be used to control the primary zone behavior.

The particular effect of swirl is dependent upon the swirl parameter. This is defined in terms of the ratio of the axial flux of swirl momentum to the axial flux of linear momentum as follows:

$$S = \frac{\int \rho u w r^2 dr}{R[\int \rho u^2 r dr + \int p r dr]} \tag{10}$$

This ratio is evaluated by integrating the above momentum equations across the flow.

The characteristic value of the swirl parameter, S, is its initial value at the burner exit plane,

$$S = \frac{\int_0^R {}_o\rho_o u_o w_o r^2 dr}{R[\int_0^R \rho_o u_o^2 r dr + \int_0^R p_o r dr]} \tag{11}$$

$$\frac{w_o}{v_o} < 1 \rightarrow S_o < \qquad \frac{\partial p}{\partial y} \sim 0 \text{ (Small to Moderate Swirl)} \tag{12}$$

$$\frac{w_o}{v_o} > 1 \rightarrow S_o \geq 1 \qquad \frac{\partial p}{\partial y} \neq 0 \text{ (Large Swirl)} \tag{13}$$

The effect of increased w/u is seen in the denominator of the equation for S. Thus, S increases rapidly with w/u as w/u → 1.

In terms of large swirl, the resulting deceleration is caused by an induced pressure rise in the axial direction. In this case, the combined effects of increased turbulence intensity and the mean axial flow deceleration can provide a large increase in relative mixing rate and residence time. It should be noted that even for small to moderate swirl the effect of swirl on turbulent transport is retained in all the equations even though induced pressure effects become negligible. The present analysis differs from that of Ref. 2 in two important ways:

1. Ref. 2 uses an integral technique requiring an assumed velocity profile, and
2. the flow is "incompressible" with density gradient effects being accounted for by an assumed density variation. Thus, chemistry is not included.

The present model is not restricted in either of these ways and a summary of this analysis is given in Table I. In general, the model will predict the details of the flow including the local velocity components, concentrations of gas and droplets, temperature and pressure (for a specified liner contour), and mean properties within a recirculation zone are predicted. As shown in Table I, these variables are displayed in Column I while Column II shows their initial conditions (which are arbitrarily specified), and Columns III and IV detail the boundary conditions. Note that the area contour or wall pressure distributions may be specified. This option is of particular interest for advanced high speed combustors where sufficient dynamic head is available for significant pressure conversion. Thus, it may be desirable to determine a liner contour for some specified pressure distribution that provides control of the velocity field as well as of the kinetics. In particular, control of mixing rates, residence times, and combustion rates can be studied with the options as outlined in Column III. Column IV outlines the treatment of slots and holes. The "lateral" injection process through the holes is treated as a unit problem for the determination of the penetration into the main flow at the particular combustor cross section. The depth of penetration defines the radial location at which the injected flow becomes essentially parallel to the main combustor flow. Here, the flow is circumferentially mass averaged after taking into account the acceleration of the main combustor flow due to the area occupied by the injected air. The "new" profiles of the flow variables provide the information required to continue the solution downstream.

Recirculation Model – The existence of recirculating regions in the primary zone of turbine combustors provides a mechanism for rapid mixing. In the limit, some point of the primary zone will become homogeneously mixed and will behave as an ideal stirred reactor as shown in Fig. 3. As the primary zone approaches this stirred reactor limit, the principal mechanism defining the state of flow is the chemical kinetics of the fuel oxidation process. Thus, the residence time becomes the controlling parameter for a given fuel and air state entering the primary zone.

A mathematical model for this stirred reactor must relate the conservation of mass and energy to the generation (or depletion) of the various chemical species within this

References pp. 86-87

TABLE 1

Computations in Mixing Zones

I VARIABLE MIXING ZONES	II SPECIFIED INITIAL CONDITIONS	III BOUNDARY CONDITIONS ALONG IMPERMEABLE PART OF WALL
Linear Velocity, $u(x,r)$	$V = u(O,r)$ for $O \leq r \leq y_w$	Specify wall skin friction *or* shear, $C_f(x)$ or $T_w(x)$
Swirl Velocity, $w(x,r)$	$w = w(O,r)$ for $O \leq r \leq y_w$	
Temperature, $T(x,r)$	$T = T(O,r)$ for $O \leq r \leq y_w$	Specify heat transfer or wall temp. $(\frac{\partial T}{\partial r})_w$ or $T_w(x)$
Concentration, $a_i(x,r)$ including droplet, where size, $a_j(x,r)$ and number density, $N_j(x,r)$ are auxiliary variables	$a_i = a_i\,(O,r)$ for $O \leq r \leq y_w$ $a_j\ (O,r)$ for $O \leq r \leq y_w$ $N_j = N_j\,(O,r)$ for $O \leq r \leq y_w$ for $i = j$	Specify impermeable wall, $(\frac{\partial a_i}{\partial r})_w = O$
Pressure, $P(x,r)$	$P = P\,(O,r)$ for $O \leq r \leq y_w$	Specify $P_w(x,r_w)$ *or* if $r_w(x)$ is specified then $P_w(x,r_w)$ is given as part of the solution
Wall contour, $y_w(x)$	$y_w = y_w(O)$ for $O \leq r \leq y_w$	Specify wall contour, $r(x)$ *or* if $P_w(x,r)$ is specified then $y_w(x)$ is given as part of the solution

IV
AIR INJECTION ALONG THE WALL

Variable	Slots	Holes
V_I T_I P_I A	When the combustor flow reaches a slot, the problem is reinitialized. The "new" initial conditions involve the extension of the profiles at $x = x_s$, as given in Column II, across the slot where the slot width is known by the specification of $y_w = y_w(x)$.	When the combustor flow reaches a hole, the penetrations are determined and the problem is reinitialized and the flow properties at $x = x_H$ are specified as shown in Col. II. For both slots and holes, the state of the air flow $(P_I,\ T_I,\ v_I)$ must be specified on the *upstream* side of the openings. Then, given an estimate for discharge coefficients, the state of flow entering the combustor is determined. The distribution of slots and holes are input information.

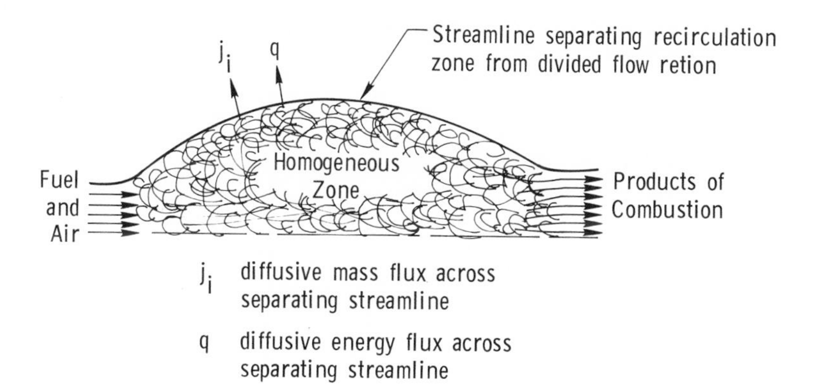

Fig. 3. Recirculation zone.

part of the combustion chamber volume. In particular, the describing equations are given by:

Conservation of Mass:

$$m = \text{constant} = \sum_i \dot{m}_i^I = \sum_i \dot{m}_i^o \tag{14}$$

where i is the ith species and I and o refer to inflow and outflow, respectively.

Conservation of Energy:

$$\sum_i \alpha_i^I h_i(T_i^I) + \dot{Q}/\dot{m} = \sum_i \alpha_i h_i(T) = h(T) \tag{15}$$

where $\dot{Q}$ is the net rate of external heat addition to the recirculation zone and the inflow of enthalpy is written to allow each component to enter the zone with an arbitrary temperature.

Species Generation:

$$\alpha_i = \alpha_i^I + \frac{V}{\dot{m}} \dot{W}_i \qquad i = 1, \ldots N \tag{16}$$

where w_i represents the production rate of each of the N species occurring in the chemical system:

$$\dot{w}_i = W_i \sum_{p=1}^{R} (\nu''_{ip} - \nu'_{ip}) k_{f_p} \rho^{m_p} \prod_{i=1}^{N} \left(\frac{\alpha_i}{w_i}\right)^{\nu'_{ip}} \left[1 - \frac{\rho^{n_p}}{k_{c,p}} \prod_{i=1}^{N} \left(\frac{\alpha_i}{W_i}\right)^{\nu''_{ip} - \nu'_{ip}}\right]$$

$$m_p = \sum_{j=1}^{N} \nu'_{ip} \qquad n_p = \sum_j^N (\nu''_{jp} - \nu'_{jp}) \tag{17}$$

References pp. 86-87

for a chemical system containing N species entering into R elementary reversible reactions given by:

$$\sum_{i=1}^{N} \nu'_{ip} M_i \underset{k_{b_p}}{\overset{k_{f_p}}{\rightleftharpoons}} \sum_{i=1}^{N} \nu''_{ip} M_i \quad , \quad p = 1, \ldots, R \tag{18}$$

Equations of State:

$$\rho = \frac{p}{RT \sum_i \alpha_i / W_i} \tag{19}$$

$$h_i = h_i(T) \tag{20}$$

The above formulation shows that for a given injection state and heat transfer per unit mass flowing through the reactor, the reactor state is a function only of $\dot{m}/V$. This ratio is directly related to the residence time, τ, within the reactor:

$$\tau = \frac{\rho V}{\dot{m}} \tag{21}$$

Because τ is dependent upon density, it is determined as part of the solution for specified inlet conditions, $\dot{Q}/\dot{m}$, and $V/\dot{m}$. It is of interest to note, however, that between equilibrium and blowout ρ generally does not vary nearly as much as $\dot{m}/V$. Thus, if one desires, an estimate of τ can be made a priori of making a calculation by using the equilibrium density and the particular value of $\dot{m}/V$.

Under steady operation, the reactor state is fully defined in Eqs. 15, 16, 19 and 20. In particular, for specified α_i^I, p, $\dot{m}/V$, $\dot{Q}$, $\dot{m}$ and T_i^I, the 3+i equations define the basic variables ρ, T and α_i, where $h = \sum_i \alpha_i h_i(T)$ is employed as a convenient working variable. The residence time, $\tau = \frac{\rho V}{\dot{m}}$, is implied by the definition of the reactor state and specified operating conditions.

In their basic form, Eqs. 15, 16, 19 and 20, represent 3+i coupled non-linear algebraic equations. Although there exist more or less standard techniques for the solution of such equations (3,4), the iterations required and the tracking and exclusion of the non-physical solutions renders such an approach extremely cumbersome. An alternative approach is that of seeking the asymptotic (steady state) solution to a transient problem wherein the "boundary conditions" are held fixed at the desired steady state values. This leads to a "marching on" problem which eliminates the difficulties encountered in the solution of the algebraic equations. In essence, our approach involves the use of the non-steady form of the species conservation equations:

$$\frac{d\alpha_i}{dt} = \frac{\dot{m}}{\rho V}(\alpha_i^I - \alpha_i) + \frac{\dot{W}_i}{\rho} \qquad (22)$$

where t is the time variable of significance only during the transient period. Thus, Eq. 22 is identical to Eq. 16 when:

$$\frac{d\alpha_i}{dt} \rightarrow 0 \qquad (23)$$

The remaining working equations 15, 19, and 20 retain their steady state form. To initiate a calculation requires the specification of the α_i's at $t = o$. We start with the equilibrium state although this "initial" composition may be chosen arbitrarily.

Input to the computer program includes specification of the reactor volume, pressure level and entrance conditions. Initial conditions for the transient calculation are conveniently taken to be the species concentrations at the equilibrium state corresponding to the particular choice of influx enthalpy, element composition and pressure. The computer output includes the temporal variations of the outflow composition and temperature from which the desired steady state values as well as the average residence time can be determined. The blowout condition which provides a direct measure of the ignition delay time can be predicted by parametric variation of the volume-to-mass flow rate ratio which is equivalent to varying the residence time. Thus, decreasing $V/\dot{m}$ is analogous to increasing the flow-thru velocity. When the associated residence time is sufficiently small, stable combustion will no longer be possible.

The primary virtue of the proposed model lies in its inherent ability to respond to changes in many of the operating conditions and geometric factors which are of practical interest.

The major uses of the overall analysis will be that of establishing the sensitivity of the combustor outflow properties to (a) changes in operating conditions and (b) the accuracy required in specifying transport properties, kinetic rate constants, and recirculation zone volumes. The latter item can serve as an aid in defining the most important aspects of combustor flows which require experimental data.

Some of the features of the various elements of the model are discussed later in terms of examples of isolated mixing and combustion problems.

Chemical Kinetics – It is generally agreed that in high performance combustion systems the definition of the structure of the flame and effluent gases cannot be adequately defined by the equilibrium assumptions. In general, the residence times available in practical systems are insufficient for all species to attain their equilibrium levels. This is particularly true for species of interest in pollution such as NO. For instance, equilibrium concentrations of CO and NO are shown in Fig. 4 for a range of temperatures and equivalence ratios. Such predictions are useful provided they are

References pp. 86-87

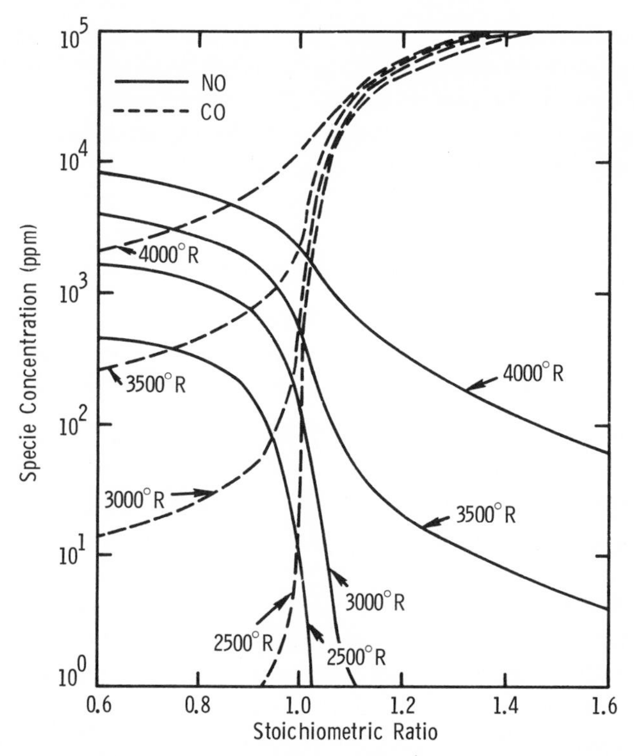

Fig. 4. Equilibrium composition of a C_nH_{2n}/air system, P = 1 atm.

regarded as limiting values and to some extent as indicating potential trends. Unfortunately, these levels and even the relative trends are rarely, if ever, observed in practical systems. In addition to the fluid mechanics of the flow, the kinetics of the system are a major cause for this observation.

Perhaps the single most important consideration in the formulation of appropriate kinetic mechanisms is the recognition of the spectrum of local states that are encountered in various combustor flow fields. This includes fuel/air ratios, temperatures, residence times and pressure levels. This means that a rather general consideration must be given to species and reactions each of which may become important under different conditions of local equivalence ratio, temperature and pressure. The importance of this has been recognized in the kinetic models for hydrogen (5), methane (6), and in Ref. 7, wherein the oxidation of a number of straight-chain hydrocarbons, aldehydes, alcohols and other partially oxidized hydrocarbon species was described by a chemical kinetic mechanism consisting of 69 reactions involving 31 species. That work marked the first time that such detailed kinetics had been successfully investigated analytically.

The experience gained in developing and comparing this complex system with experimental results, made possible the development of the GASL quasi-global finite rate hydrocarbon combustion model (8, 9, 10). Its advantages are adaptability to different fuels while retaining the ability to predict ignition delay time as well as reaction time (unlike the older one step global models), and rapidity of numerical solution. The quasi-global model has been used successfully for high energy fuels, such as Shelldyne-H and H-Methylcyclopentadiene, as well as for standard fuels such as propane and JP-7. The C-H-O reactions in the model are presented in Table 2. The rate data as given here are kept under review and are updated as new kinetic data become available. (See, for example, Ref. 11.)

TABLE 2

C-H-O Chemical Kinetic Reaction Mechanism

$$k_f = AT^b \exp(-E/RT)$$

REACTION	FORWARD			Ref.
	A	b	E/R	
1. $C_nH_m + \frac{n}{2}O_2 \rightarrow \frac{m}{2}H_2 + nCO$	$\frac{5.52 \times 10^8}{p^{.825}} \cdot C_{C_nH_m}^{1/2} C_{O_2}$	1	12.2×10^3	
2. $CO+OH = H+CO_2$	5.6×10^{11}	0	$.543 \times 10^3$	11
3. $OH+H_2 = H_2O+H$	2.19×10^{13}	0	2.59×10^3	11
4. $OH+OH = O+H_2O$	5.75×10^{12}	0	$.393 \times 10^3$	11
5. $O+H_2 = H+OH$	1.74×10^{13}	0	4.75×10^3	11
6. $H+O_2 = O+OH$	2.24×10^{14}	0	8.45×10^3	11
7. $M+O+H = OH+M$	1×10^{16}	0	0	12
8. $M+O+O = O_2+M$	9.38×10^{14}	0	0	13
9. $M+H+H = H_2+M$	5×10^{15}	0	0	12
10. $M+H+OH = H_2O+M$	1×10^{17}	0	0	11

Reverse reaction rate, k_r, is obtained from k_f and the equilibrium constant, K_c.

units: $[A] \equiv \frac{cc}{mole\text{-}sec}$ for bimolecular reactions

$\equiv \left(\frac{cc^2}{mole}\right) / sec$ for termolecular reactions

An adequate description of the combustion kinetics represents only a part of what is needed to predict pollutant emissions from flames. For example, the N-O reaction process must be considered and one mechanism proposed here is given in Table 3. This mechanism is included as part of a total reaction scheme comprised of the 19 reactions given in Tables 2 and 3. The simultaneous coupled solution of the N-O and C-H-O reactions is believed to be more realistic than the usual assumptions of

References pp. 86-87

hydrocarbon equilibrium with N-O finite rate reactions, and provides a capability to evaluate wide variations of local conditions of potential interest inside various combustors.

TABLE 3

N-O Chemical Kinetic Reaction Mechanism

$k_f = AT^b \exp(-E/RT)$

REACTION	FORWARD			Ref.
	A	b	E/R	
11. $O+N_2 = N+NO$	1.36×10^{14}	0	3.775×10^4	11
12. $N_2+O_2 = N+NO_2$	2.7×10^{14}	-1.0	6.06×10^4	14
13. $N_2+O_2 = NO+NO$	9.1×10^{24}	-2.5	6.43×10^4	13
14. $NO+NO = N+NO_2$	1.0×10^{10}	0	4.93×10^4	11
15. $NO+O = O_2+N$	1.55×10^9	1.0	1.945×10^4	11
16. $M+NO = O+N+M$	2.27×10^{17}	-0.5	7.49×10^4	11
17. $M+NO_2 = O+NO+M$	1.1×10^{16}	0	3.25×10^4	14
18. $M+NO_2 = O_2+N+M$	6.0×10^{14}	-1.5	5.26×10^4	11
19. $NO+O_2 = NO_2+O$	1×10^{12}	0	2.27×10^4	14

Reverse reaction rate, k_r, is obtained from k_f and the equilibrium constant, K_c.

At the present time the quasi-global model is being extended to include additional carbon oxidation reactions to improve the modeling for certain temperature ranges, and additional coupled C-H-O-N reactions, as shown in Table 4. Such extensions may be relevant under certain conditions and are of particular interest in studies involving chemical enhancement of ignition and combustion. For example, it has been experimentally observed that large initial concentrations of NO (e.g., 5,000 ppm for the H_2/Air system of Ref. 19) can significantly shorten the ignition delay period.

Other reaction systems with pollution applications now being integrated with the quasi-global model include those for ozone and sulfur oxides and hydrides.

Another combustion phenomena of interest from both a combustion efficiency and pollution standpoint is the formation and subsequent oxidation of soot. Although this problem has been attacked and apparently resolved to a certain extent in jet engines, the generality of the solution is not obvious. Thus, to guarantee that the problem will not be reintroduced upon making other modifications of the combustor, an understanding of soot formation and oxidation is still required.

TABLE 4

Additional C-H-O-N Chemical Kinetics Reactions

$k_f = AT^b \exp(-E/RT)$

REACTION	FORWARD			Refs.
	A	b	E/R	
20. $CO+O_2 = CO_2+O$	3×10^{12}	0	25.2×10^3	15
21. $CO+O+M = CO_2+M$	2.2×10^{15}	0	1.86×10^3	16
22. $N+OH = H+NO$	1.4×10^{12}	.5	0	17,18
23. $H+NO_2 = NO+OH$	3×10^{13}	0	0	18
24. $CO+NO_2 = CO_2+NO$	2×10^{11}	-.5	2.52×10^3	16
25. $CO_2+N = CO+NO$	2×10^{11}	-.5	4.03×10^3	16

Soot forms in high temperature fuel-rich regions, but the process may be arrested if oxygen is mixed into the region in time. Once soot is formed, high temperatures are required to oxidize it in a reasonable time.

Experiments show that most hydrocarbons yield soot as well as hydrocarbon fragments for equivalence ratios on the order of $\varphi = 1.2$. Data indicate that the rate of decomposition history is extremely sensitive to temperature (20). A correlation of these data results in the first order reaction:

$$\frac{C_f}{C_{f_o}} = \exp\left[-6.30\text{x}10^{13} \exp\left(-\frac{64{,}120}{RT}\right)\right] t \qquad (24)$$

where C_{f_o} is the initial fuel concentration. The sensitivity to temperature at a given time is indicated by the double exponential dependence upon T. For example, propane, which is typical of the higher hydrocarbons, behaves as shown in Fig. 5. The residence time required to produce significant quantities of H/C fragments and soot is actually small in comparison to the scale shown.

Fig. 6 from Ref. 20, shows the variety of measured species within the time required to decompose the first 10 percent of the fuel. It is worth noting that soot is not formed immediately but accompanies the formation of acetylene (C_2H_2) after about 7 percent of the fuel has decomposed. This suggests that if oxygen could be brought into this region, without quenching the reactions, then the H/C fragments could be partially oxidized here prior to the formation of soot. Of course, this requires optimum mixing, but it is one explanation for the reduction of soot by use of mixing enhancement. Thus the formation of soot depends upon the mixing and combustion rates in terms of the resulting residence time, composition and temperature existing in the fuel-rich regions.

References 86-87

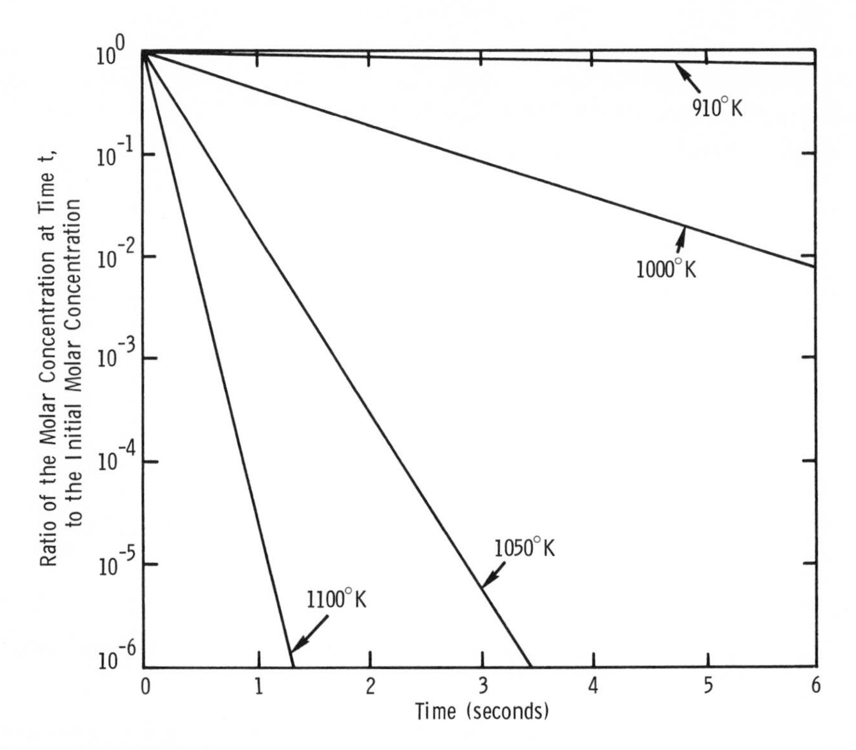

Fig. 5. Effect of pyrolysis temperature on the disappearance of propane.

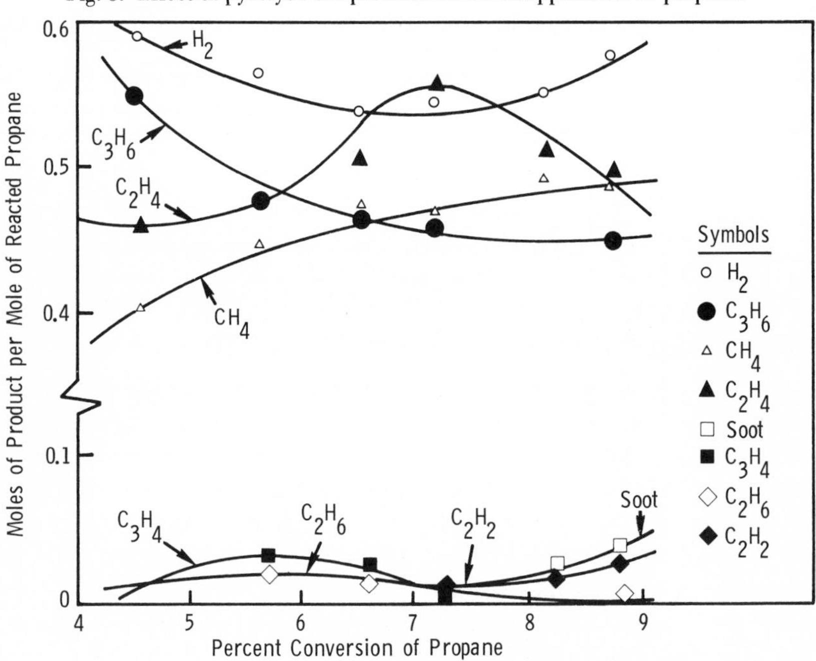

Fig. 6. Propane product distribution (T_{max} = 1514°F).

In general, some soot will be formed and is responsible for the flame luminosity that results. Thus, it is of interest to estimate its subsequent history upon coming in contact with oxygen. Figs. 7 and 8 show the history of soot particles as deduced from a correlation based on the work of Ref. 21:

$$\frac{C_c}{C_{c_o}} = \left\{ 1-.2713\text{x}10^{10}\left(\frac{400}{d_c}\right) t \frac{P_{O_2}}{\sqrt{T}} e^{-39,399/RT} \right\} \tag{25}$$

where C_{c_o} is the initial concentration of soot, d_c is the soot particle diameter, T is the temperature, and t is time.

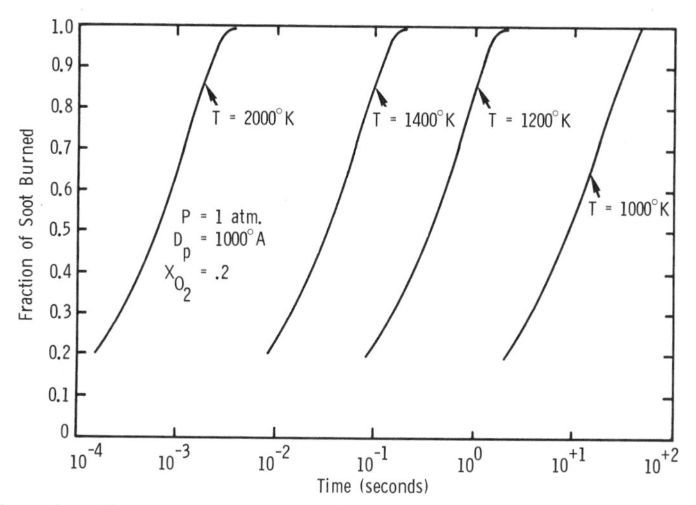

Fig. 7. Effect of ambient temperature upon the soot combustion process in a fuel-rich mixture using the model of Lee, Thring and Beer.

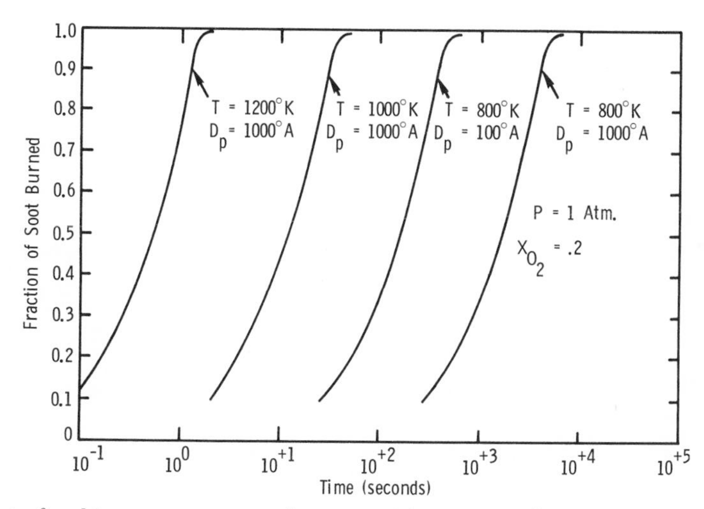

Fig. 8. Effect of ambient temperature and mean particle size upon the soot combustion process.

These results assume that the concentration of oxygen is that available in pure air. The size range (100 Å ≤ 1000 Å) is based on a variety of experimental observations (21). In terms of the time scale of interest it appears that once the soot is formed, it tends to persist except at very high temperatures.

TYPICAL RESULTS

Mixing with Finite Rate Kinetics — To show quantitatively the structure of turbulent diffusion flames, a flow field calculation has been carried out for a basic non-swirling pilot ignited concentric burner configuration of 1 atmosphere as shown in Fig. 9. For purposes of illustrating the trends that can be expected, conditions were

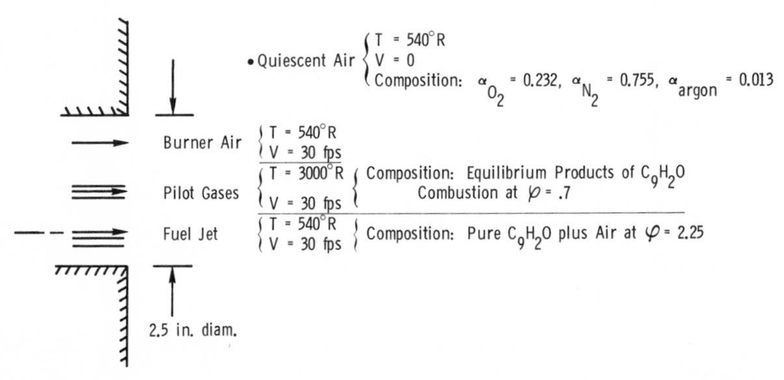

Fig. 9. Burner exit and boundary conditions for finite rate/mixing calculation.

chosen where the fuel is all vapor and some initial penetration and mixing of fuel and air has already occurred. The fuel, nonane (C_9H_{20}) was chosen to simulate the combustion of kerosene and the effect of the igniter is represented here by its effect in producing a local combustion zone. The predictions are based upon the finite difference solution of the coupled mixing and kinetics equations presented in the section discussing the Mathematical Model — Multi Phase Mixing and Combustion Model.

Fig. 10 shows the profiles of the equivalence ratio φ, at two axial stations. Until sufficient mixing has occurred, there will be a rich zone within the field bounded by the stoichiometric surface which extends to a distance L_{fc} from the burner exit plane. Beyond this point the entire flow field will be fuel lean. To illustrate the characteristics of the flame in both regions, profiles at $X/L_{fc} < 1$ and $X/L_{fc} > 1$ are shown. The profiles show the extent of mixing that has occurred over the respective parts of the flow field but of more direct interest are the temperature and composition profiles in these regions.

The temperature profiles corresponding to the same axial stations are shown in Fig. 11. At $X/L_{fc} = 0.68$, the temperature reaches a peak at a radial point corresponding to where the local equivalence ratio is essentially unity. Within the fuel-rich zone, the temperature is lower but nevertheless intense burning is indicated by the temperature levels. Outside the stoichiometric point, the temperature drops rapidly toward the ambient value. At $X/L_{fc} = 1.33$, the temperature is monotonic having a maximum on the axis where $\varphi = 0.9$. The extent of mixing is indicated by the spread of the hot gases out into the environment.

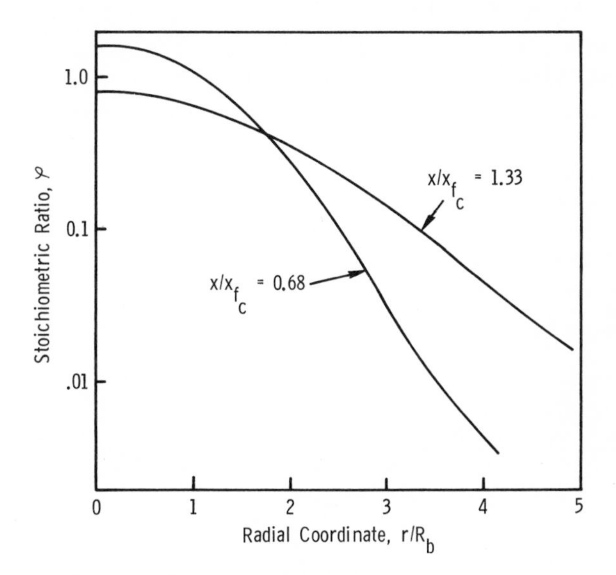

Fig. 10. Finite rate chemistry – equivalence ratio.

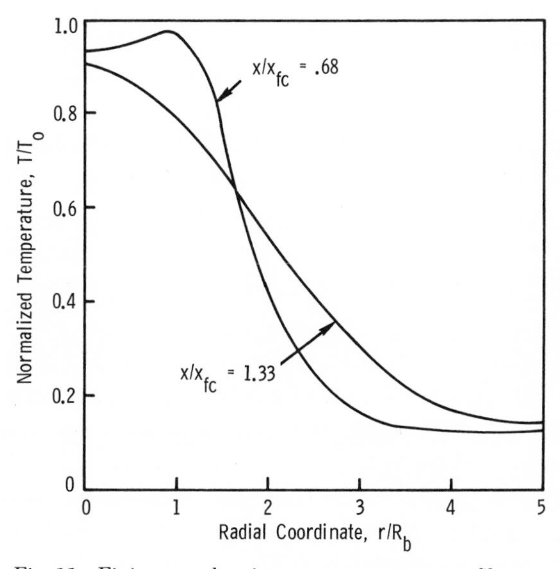

Fig. 11. Finite rate chemistry – temperature profiles.

References pp. 86-87

Fig. 12 shows the radial profiles of the concentration of CO in ppm. The maximum concentrations occur in the hot fuel-rich regions. In general, these predicted levels of CO are higher than the corresponding local equilibrium values. This is particularly true out in the lean regions where the temperature is dropping rapidly. This demonstrates the importance of relative mixing and kinetic rates and shows that too rapid mixing with a cool environment can quench the CO at high levels associated with both fuel-rich and high-temperature zones within the flow.

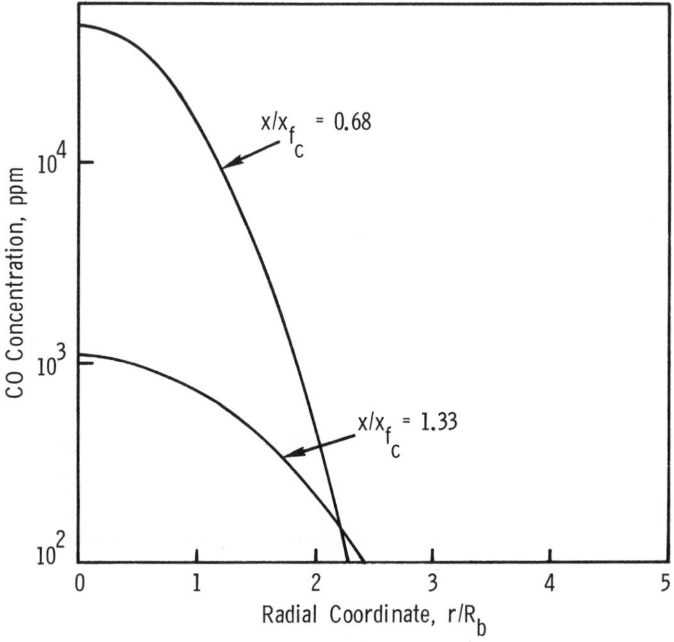

Fig. 12. Finite rate chemistry – CO profiles.

The CO_2 profiles in ppm are shown in Fig. 13. The behavior is similar to the temperature profiles exhibiting a peak at the stoichiometric point. Here, the CO_2 is essentially in local equilibrium. It should be pointed out that since CO is generally at least one order of magnitude smaller than CO_2 (see Fig. 12) its being out of equilibrium in the lean outer region does not affect the thermal field.

The hydrocarbon is represented by unburned fuel which in the present example is C_9H_{20}. A profile in ppm is shown in Fig. 14. Due to the coupled effects of mixing and combustion, the concentration decays rapidly outward from the fuel-rich levels in the inner region. It is interesting to note that the concentration levels of the hydrocarbon at $X/L_{fc} = 1.33$ (flow is fuel lean everywhere) are less than 10 ppm. This is due to the assumed conditions for this calculation wherein the flame is developing in a pure air environment. When combustion gases are assumed for the environment, higher levels of C_nH_m would be present.

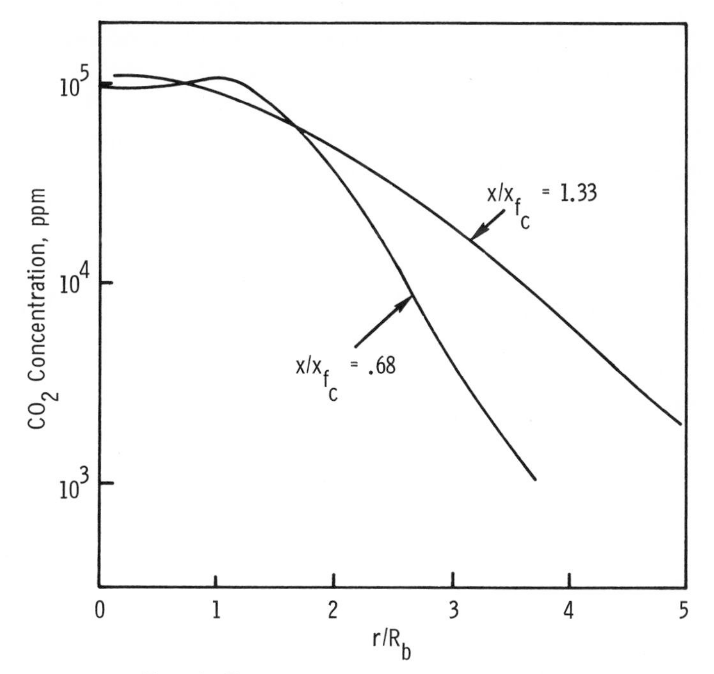

Fig. 13. Finite rate chemistry – CO_2 profiles.

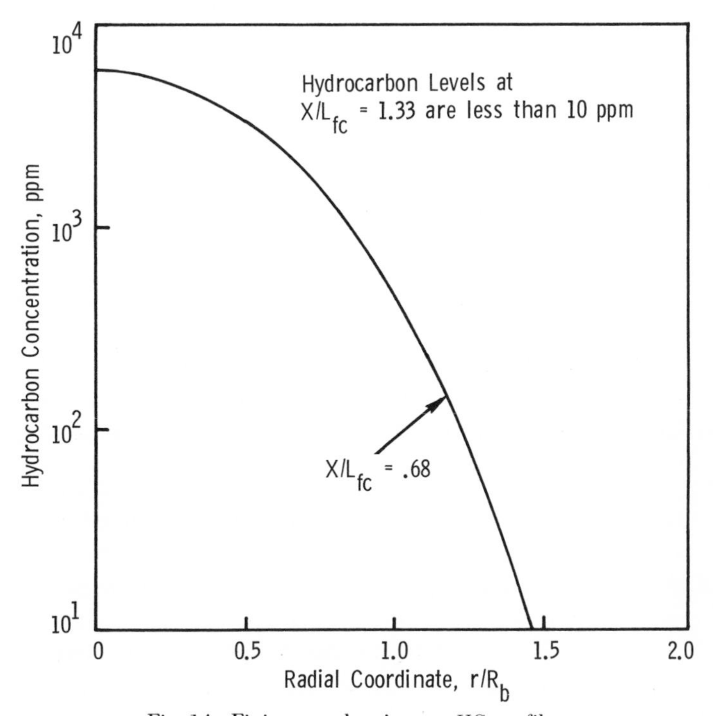

Fig. 14. Finite rate chemistry – HC profile.

References pp. 86-87

The profiles of NO in ppm are shown in Fig. 15. At $X/L_{fc} = .68$, a peak concentration of about 200 ppm occurs in the lean side of the flame zone. At this particular point the NO level is about an order of magnitude lower than the corresponding equilibrium value. It is also important to note that the temperature at this point is not the peak temperature which occurs at $\varphi = 1$. In the lean outer regions at $X/L_{fc} = .68$ and across the entire flow at $X/L_{fc} = 1.33$, there is a tendency for the NO concentrations to be higher than the local equilibrium values. This result illustrates the extreme importance of the coupled effects of mixing and kinetics. Only by including the combined effects of mixing and kinetics can a consistent set of local values of residence times, equivalence ratios and temperatures be obtained upon which the local concentrations of NO crucially depend. Neither mixing nor kinetics alone can provide this all important consistency requirement.

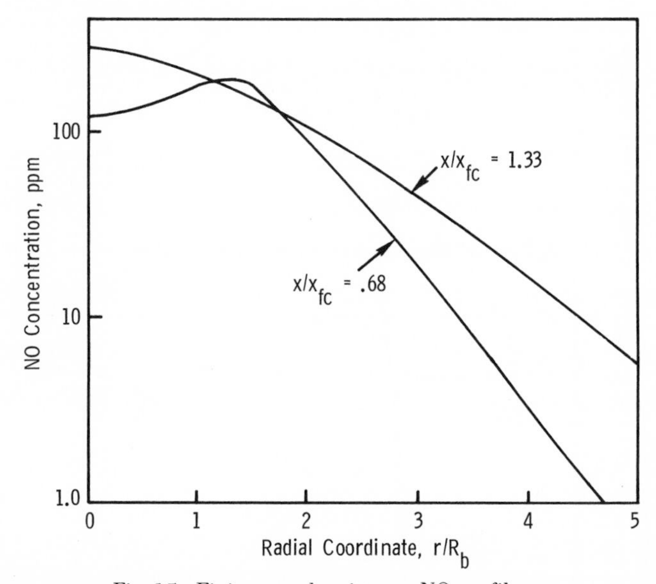

Fig. 15. Finite rate chemistry – NO profiles.

The profiles described above were at axial stations downstream of the point where the flame becomes self-propagating. To illustrate the streamwise development of the flame, axial distributions of some of the relevant flow properties are shown in Figs. 16 and 17.

Fig. 16 shows the equivalence ratio and temperature distributions. At $X/L_{fc} = .3$, the hot gases first reach the axis and between $.3 \leq X/L_{fc} \leq .5$, the fuel-rich mixture is ignited and attains near equilibrium temperatures corresponding to the local $\varphi = 2.1$. Since φ represents the relative amount of total oxygen present (independent of the combined form), the decay of φ along the axis indicates the mixing rate. At the point where $\varphi = 1$, the flame front closes on the axis and the hot combustion gases continue to mix and react as the flow moves downstream.

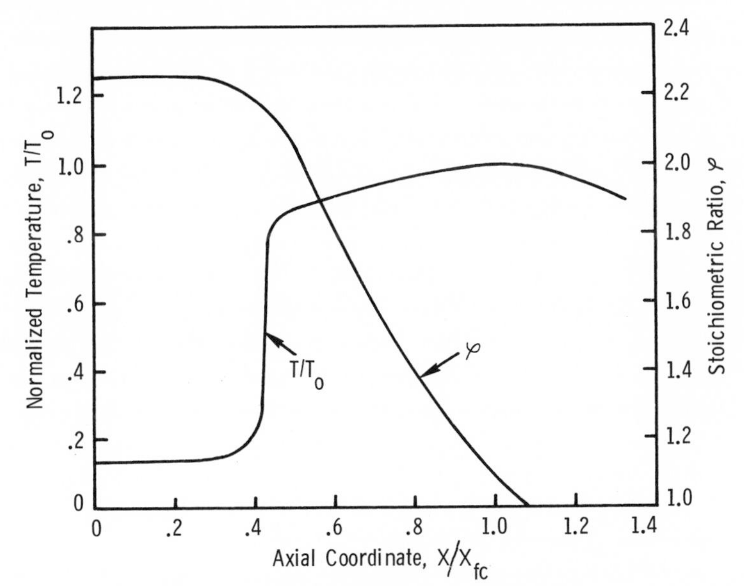

Fig. 16. Finite rate chemistry — axial histories.

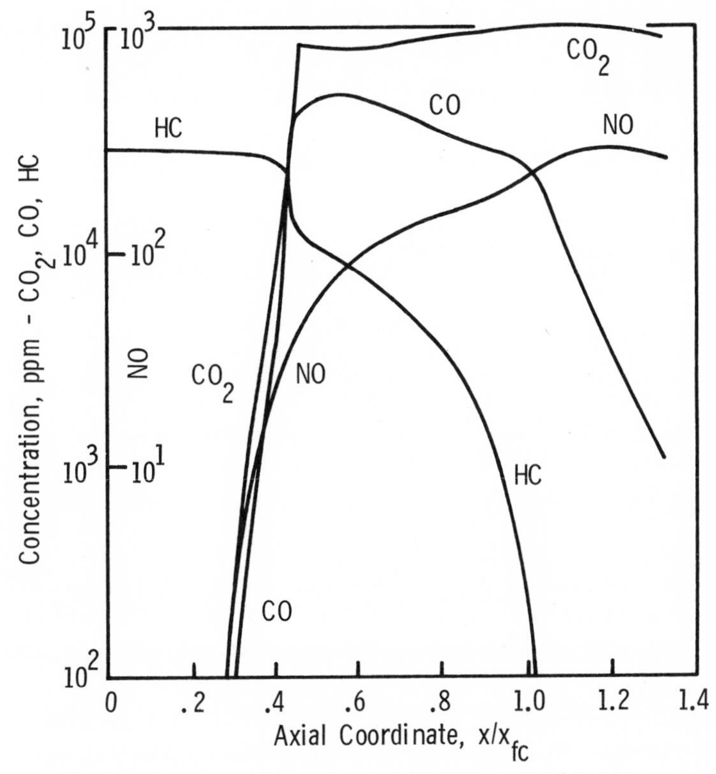

Fig. 17. Finite rate chemistry — axial concentration histories.

References pp. 86-87

Fig. 17 shows the corresponding concentration histories of the relevant species. The rapid rise of CO_2 and CO between X/Lf of .3 and .5 is associated with the ignition of the fuel-rich mixture. The NO rises but at a slower rate and remains at levels well below equilibrium values until the flow on the axis begins to cool due to continued mixing. However, from $X/L_{fc} = 1.0$ and beyond, the rate of decrease of NO is slower than that corresponding to shifting, or local, equilibrium. This "freezing" effect will result in higher than equilibrium concentrations of NO corresponding to the completely mixed temperature. The downstream behaviors of CO and the hydrocarbon are similar to the behavior exhibited by the profile in the lean outer regions. The hydrocarbon decreases rapidly while the CO decreases but tends to remain at levels above the corresponding equilibrium values.

Some "Pure" Kinetics Results – The previous example represents perhaps the most meaningful type of calculation for non-uniform reacting flows. Predicting the effects of the extent of fuel-rich regions, peak temperatures, and the interaction of "hot" and "cool" zones is essential for the definition of combustion efficiency and pollutant emissions. However, it is also true that no matter how detailed a fluid mechanical model may be its utility for predicting the species field depends crucially upon the adequacy of the kinetics mechanism. Furthermore, if the kinetics mechanism is sufficiently detailed it will be applicable to any flow configuration.

The "basic" kinetic mechanism that we have been working with to date shows excellent promise in terms of representing the combustion characteristics of a variety of hydrocarbon fuels. For example, Fig. 18 shows a comparison of our predictions of

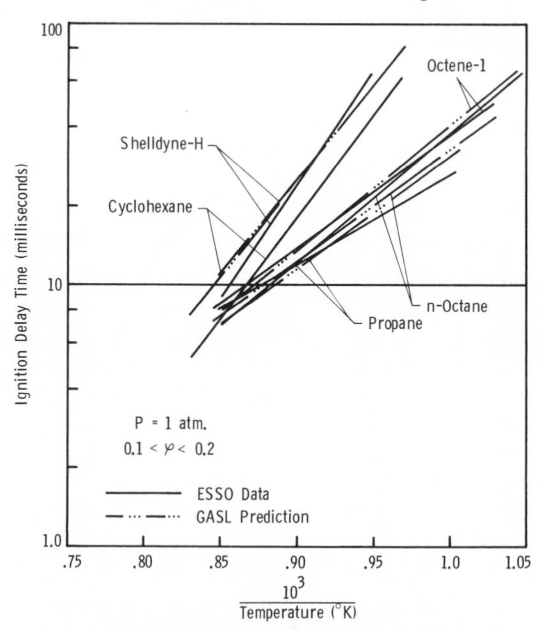

Fig. 18. Comparison of GASL theoretical quasi-global combustion model with Esso experimental data for ignition delay times in hydrocarbon-air systems.

ignition delay time with data obtained in a plug flow reactor. The variety of fuels includes both cyclic and long chain type hydrocarbons. Although there is a lack of data of this kind over wider ranges of conditions, this comparison is indeed encouraging.

One of the features of the present kinetics mechanism is that it does not assume that the combustion reactions are uncoupled from the NO_x reactions. Most workers in the field have made this assumption based upon the observation that the time required to reach combustion equilibrium is short compared with the NO formation time. However, this appears to be in contradiction with the observation that some NO forms early in the combustion process (22). This "prompt" NO appears to be related to the characteristic free radical overshoot which occurs at the tail end of the ignition delay period and persists during the bulk of the reaction zone. For hydrocarbons, such superequilibrium values of atomic oxygen can account for an "early" formation of NO in the milli-second range that would not be predicted assuming the equilibrium level of atomic oxygen existed throughout. Figs. 19 and 20 give an example of this

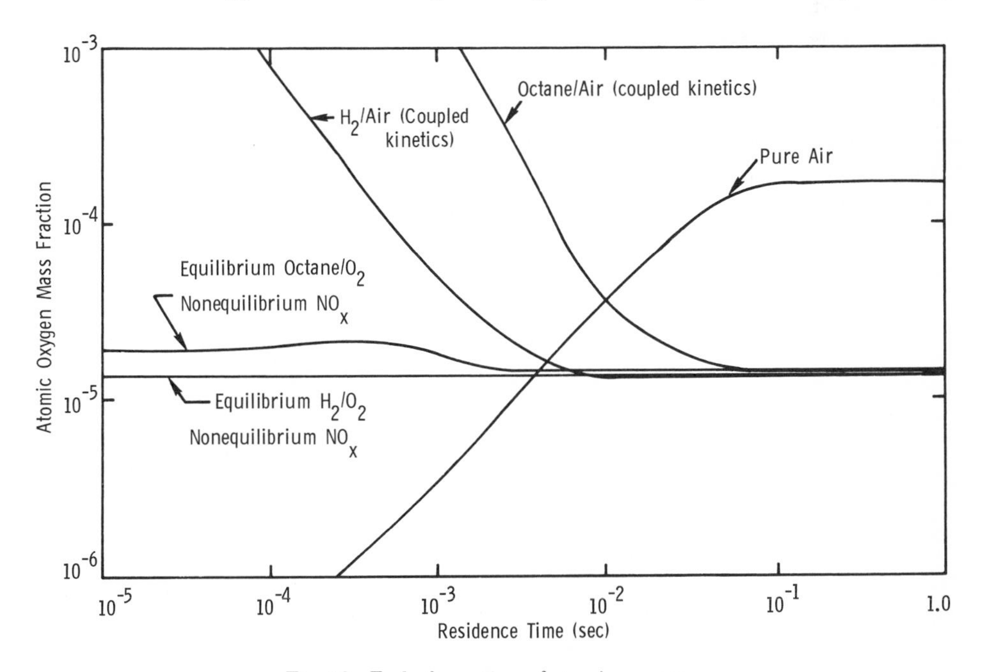

Fig. 19. Early formation of atomic oxygen.

for isothermal plug flow of octane and hydrogen assuming two models for the combustion process. In both models, the full system given in Tables II and III was employed. However, the "coupled kinetics" results were obtained by initiating the calculation with pure fuel and air whereas the "Equilibrium Fuel/O_2 – non-equilibrium NO_x" results were obtained by initiating the calculation assuming equilibrium Fuel/O_2 wherein the N_2 was assumed inert. Fig. 19 shows that the atomic oxygen overshoot for the hydrogen/air occurs earlier than in the corresponding octane/air case. Of course, for pure air the growth of O is monotonic. The fact that hydrogen is more reactive than the hydrocarbon is the reason for the "shift" in the overshoot and Fig. 20 shows the consequences of this behavior. Thus, the early formation of NO is significant for the hydrocarbon but not so for hydrogen (or pure air). The higher "final" pure air level of NO is due to the available O_2 that would otherwise be consumed by the fuel. Although this result does not cover a wide range of possible states that would be encountered in a practical system, it demonstrates, for a typical condition, that a coupled combustion/NO_x mechanism is necessary to explain and predict certain observations.

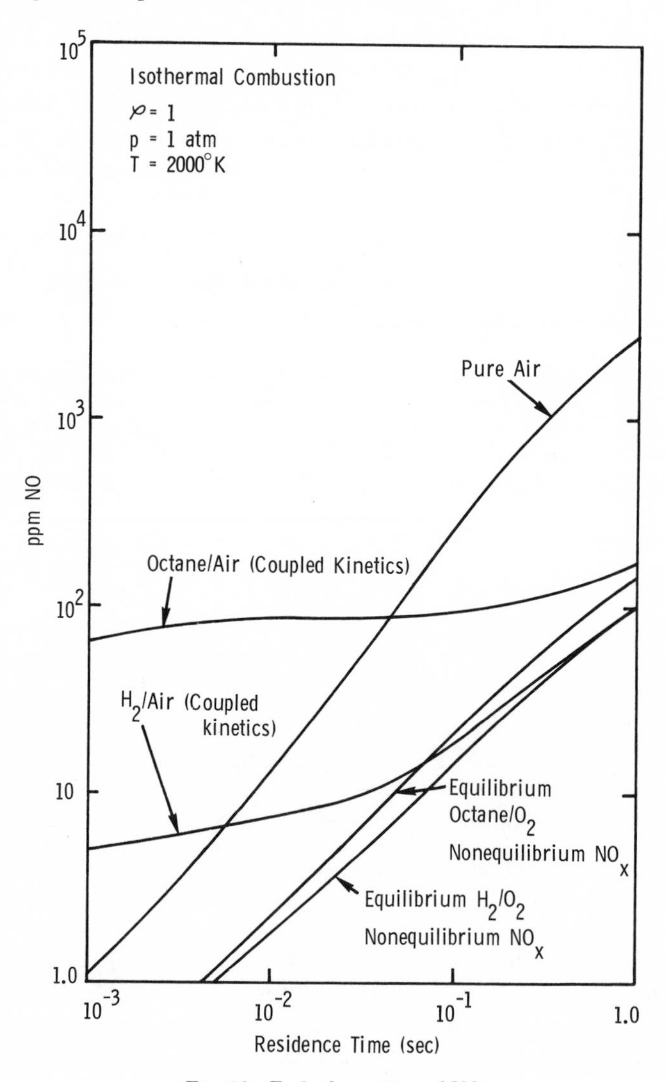

Fig. 20. Early formation of NO.

It was previously cited that the stirred reactor is not only a research tool but also represents the limiting behavior of well mixed zones of interest in practical combustion systems. An example of the application of the basic kinetics mechanism to a stirred reactor configuration is shown in Figs. 21 and 22.

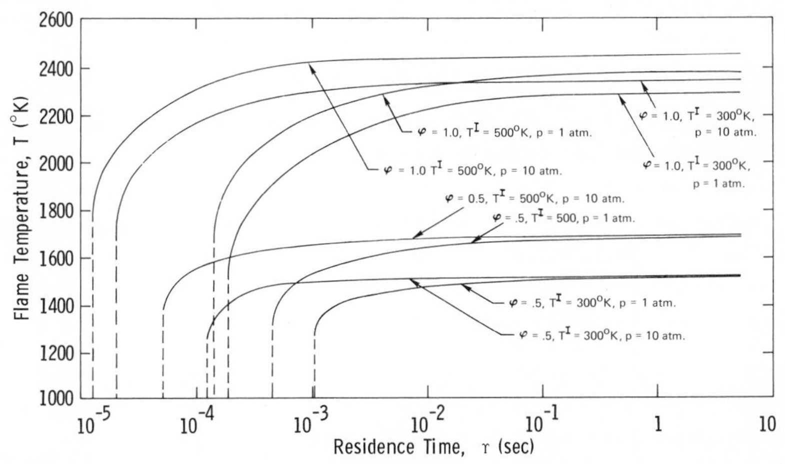

Fig. 21. Flame temperature as a function of residence time for various φ's, p's and T^I's.

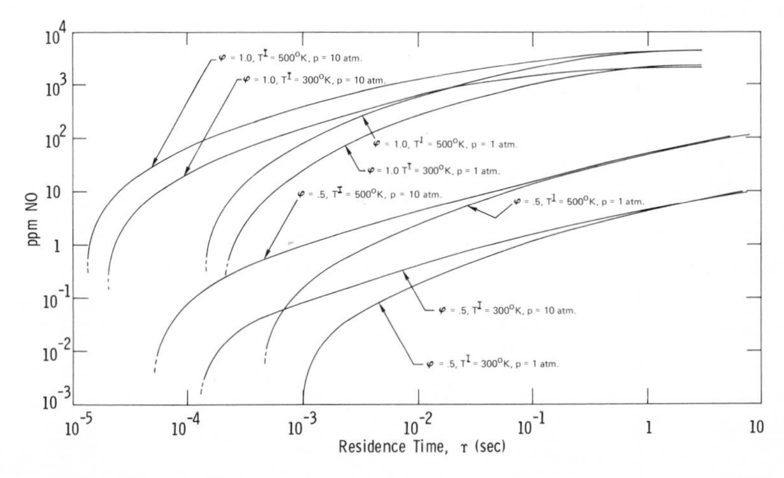

Fig. 22. Concentration of NO as a function of residence time for various φ's, T^I's, and p's – adiabatic reactor.

Fig. 21 shows the reactor temperature as a function of residence time. In all cases, the discrete values of residence time were chosen to cover a range extending from the blowout condition (ignition delay time) to well beyond the time required for combustion equilibrium. In fact, equilibrium for temperature is attained in times ranging from about 1 order of magnitude for $\varphi = .5(T^I = 300°K)$ to about 2 orders of

References 86-87

magnitude for φ = 1.0(T^I = 500°K) longer than the respective ignition delay times. Thus, combustion equilibrium is attained within a maximum of a little over 10 msecs for the cases considered here. Some additional observations are:

1. Blowout, or ignition delay time, varies roughly inversely with the pressure to the unity power.
2. Increased pressure tends to increase the temperature at blowout.
3. Increased initial temperature (T^I) also increases the temperature at blowout.

These results illustrate the trends which accompany variations in primary parameters controlling the combustion process. It would be of interest, however, to attempt in the future to establish a correlation for ignition delay times as a function of T^I as well as of p and φ. This would require a wider ranger of parameter variations than were presently considered and could provide a composite representation useful in aid of combustion system design where ignition and pollutant emission characteristics must be jointly considered.

In terms of the question of pollutant emissions, Fig. 22 shows the NO concentrations as a function of residence time for the adiabatic reactor. There are several observations to be drawn from these results:

1. For φ = .5, NO remains highly out of equilibrium even after a 10 second residence time. In fact, at this residence time the NO level at T^I = 300°K is about two orders of magnitude below the equilibrium value.
2. For φ = 1.0, the NO levels approach their equilibrium values between 0.1 and 1.0 seconds, depending mostly upon the initial temperature.
3. For both φ = 0.5 and 1.0, the respective NO levels are weakly dependent upon pressure. This is particularly evident as residence time increases and the system moves away from blowout. As blowout is approached, or more specifically, as the system gets further out of equilibrium, the effect of pressure becomes more pronounced. This behavior suggests that the apparent pressure effect upon the NO levels at shorter residence times is actually a result of the direct effect of pressure upon the combustion mechanism rather than upon the NO_x reactions. It should be noted that the crossover in NO concentration (as a function of pressure) with increased residence time is to be expected on equilibrium grounds.

The sum total of these observations reduces to a verification of the extreme sensitivity of the NO formation rate to the temperature. Finally, it should be noted that the NO_2 levels are at least two orders of magnitude less than the corresponding NO levels for all conditions considered here.

Additional results of this type are presented in Ref. 23. These include a comparison between predicted and measured NO concentration obtained in a stirred reactor as well as the effects of cooling.

Additional discussions outlining the unified model formulation may be found in Ref. 24.

SUMMARY

A unified model for combustion system analysis is presented which includes the effects of:

a) gas phase convection and mixing

b) gas phase ignition

c) gas phase combustion

d) droplet convection and mixing

e) droplet evaporation

f) droplet combustion

Examples of the application of various elements of the model are given which serve to demonstrate the importance of mixing while also isolating and showing the crucial importance of an appropriate coupled kinetics mechanism.

LIST OF SYMBOLS

A slot or hole area
a particle radius
h static enthalpy
H total enthalpy
$\dot{m}$ mass flow rate
p pressure
S swirl parameter
T temperature
u streamwise component of velocity
v radial component of velocity
V volume
w swirl component of velocity
$\dot{w}_i$ production rate of ith specie
x axial coordinate
y radial coordinate
α_i mass fraction of ith specie
β_i mass fraction of ith component in multiphase system
ϵ diffusivity
ρ density

References 86-87

Subscripts

i i^{th} specie
g gas
H hole
p droplet
I injection value
j j^{th} particulate class
S slot
μ dummy variable

REFERENCES

1. *R.B. Edelman, C. Economos, and J. Boccio, "An Analtyical and Experimental Study of Some Problems in Two-Phase Flows Involving Mixing and Combustion with Application to the B-O-H-N System," AIAA Paper 70-737.*
2 *H.L. Morton, "Effect of Swirl on Turbulent Flow in the Front End of Gas Turbine Main Burners," AIAA Paper No. 71-2.*
3. *D.R. Jenkins, V.S. Yumlu, and D.B. Spalding, "Combustion of Hydrogen in a Steady-Flow Adiabatic Stirred Reaction" 11th Symposium (International) on Combustion, The Combustion Institute, 1967, p. 779.*
4. *D.C. Hammond, Jr., and A.M. Mellor, "An Investigation of Gas Turbine Combustors with High Inlet Air Temperature," Jet Prop. Center, Purdue University, TM-70-2.*
5. *P.A. Libby, H.S. Pergament, and M.H. Bloom, "A Theoretical Investigation of Hydrogen-Air Reactions," Part I, GASL TR-250, August 1961.*
6. *W. Chinitz, Pyrodynamics, 3, 1966, p. 196.*
7. *W. Chinitz and T. Baurer, "An Analysis of Non-Equilibrium Hydrocarbon-Air Combustion," Paper 65-19, presented at 1965 Fall Meeting, Western States Section, The Combustion Institute.*
8. *R. Edelman and O. Fortune, "Mixing and Combustion in the Exhaust Plumes of Rocket Engines Burning RP-1 and Liquid Oxygen," General Applied Science Laboratories Report TR-631, Nov. 1966.*
9. *O. Fortune and R. Edelman, "The Effect of Mixing, Radiation and Finite-Rate Combustion Upon the Flow Field and Surroundings of the Exhaust Plumes of Rocket Engines Burning RP-1 (Kerosene) and Liquid Oxygen," General Applied Science Laboratories Report TR-681, Dec. 1967.*
10. *R. Edelman and O. Fortune, "A Quasi-Global Chemical Kinetic Model for the Finite Rate Combustion of Hydrocarbon Fuels," AIAA Paper 69-86.*
11. *D.L. Baulch et al., "Critical Evaluation of Rate Data for Homogeneous, Gas Phase Reactions," Vols. 1 through 5, The University, Leeds, U. K., 1969, 1970.*
12. *G. Moretti, "A New Technique for the Numerical Analysis of Nonequilibrium Flows," AIAA J. 3, 1965.*
13. *P.R. Ammann and R.S. Timmins, "Chemical Reactions During Rapid Quenching of Oxygen Nitrogen Mixtures from Very High Temperatures," A. I. Ch. E. Journal, 12, 1966, p. 956.*
14. *Pyrodynamics, 5, No. 1 and 2, pp. 54-143, January 1967.*
15. *R.S. Brokaw, "Ignition Kinetics of the Carbon Monoxide-Oxygen Reaction," 11th Symposium (International) on Combustion, The Combustion Institute, 1967, p. 1063.*

16. *M.L. Carnicom, "Reaction Rates for High Temperature Air with Carbon and Sodium Impurities," Sandia Laboratories, SC-R-68-1797, May 1968.*
17. *I.M. Cambell and B.A. Thrush, "Reactivity of Hydrogen to Atomic Nitrogen and Atomic Oxygen," Trans. Faraday Soc. Vol. 64, 1968, p. 1265.*
18. *Pyrodynamics, 5, Nos. 1 and 2, pp. 54-143, Jan. 1967.*
19. *A.D. Snyder et al., "Shock Tube Studies of Fuel-Air Ignition Characteristics," Air Force Aero Propulsion Laboratory, TR-65-93, WPAFB, 1965.*
20. *K.J. Laidler, Chemical Kinetics, 2nd Edition, McGraw Hill, New York, 1965.*
21. *K.B. Lee, M.W. Thring, and J.M. Beer, "On the Rate of Combustion of Soot in a Laminar Soot Flame," Combustion and Flame, Vol. 6, 1962, p. 137.*
22. *C.P. Fenimore, "Formation of NO in Premixed Hydrocarbon Flame," 13th Symposium (International) on Combustion, The Combustion Institute, 1971, p. 373.*
23. *R. Edelman and G. Weilerstein, "A Theoretical Study of Combustion and NO Formation in Well Mixed Regions," General Applied Science Laboratories Report TR-758, May 1971.*
24. *R. Edelman and C. Economos, "A Mathematical Model for Jet Engine Combustor Pollutant Emissions," AIAA Paper No. 71-714.*

DISCUSSION

D. T. Pratt *(Washington State University)*

On your figure showing the quasi-global hydrocarbon pyrolysis model, isn't the pre-exponential factor "A". that was listed as the rate constant, really the entire reaction rate expression?

R. B. Edelman

Yes.

C. T. Bowman *(United Aircraft Research Laboratories)*

Dr. Edelman, you have compared nitric oxide formation rates, calculated assuming isothermal plug flow and no diffusion, with Fenimore's experimental observations in a flame. Such comparisons can be, at best, only qualitative since thermal conduction and diffusion effects, present in flames, may play a significant role in the nitric oxide formation process.

R. B. Edelman

Yes, this is quite true. By performing an isothermal calculation, the importance of the atomic oxygen is clearly demonstrated. While it tends to explain the observation, it is not a model of the flat flame burner that I believe Dr. Fenimore used in making his observation. It turns out, however, that similar results are obtained for adiabatic plug flow predictions. This result is not as dramatic as the isothermal one but both models illustrate the importance of the free radicals.

C. T. Crowe *(Washington State University)*

I was wondering, Ray, on your third slide, in which you show a tremendous block diagram, everything that goes into your computer is . . . (**R. B. Edelman:** That is in the paper, so you don't have to dwell on it) I just wanted to ask you a specific question about it, and that is, why didn't you include the effect of the recirculation and swirl on the droplet and spray dynamics? Why did you leave them decoupled?

R. B. Edelman

A major feature of the model is that it does couple various fluid dynamics, chemical kinetics and multiphase processes. The block diagram (Fig. 1) to which you refer attempts to indicate the distinct mechanisms that are contained in the model. Fig. 1 should be viewed together with the paper and not taken out of context.

W. Bartok *(Esso Research and Engineering Co.)*

In connection with the calculations that show the importance of O-atom overshoot for the hydrocarbon/air system, you referred to Dr. Fenimore's work on hydrogen/air flames. Have you attempted to calculate what would happen with carbon monoxide as the fuel? If I recall correctly, Dr. Fenimore reported that CO behaved similarly to hydrogen, that is, no "prompt" NO was produced in flames of either one of these fuels.

R. B. Edelman

No, I have not, as yet, made a study of pure CO oxidation. The reactions involving CO which are apparently of potential importance are:

$$CO + O_2 \rightleftarrows CO_2 + O$$

$$CO + O + M \rightleftarrows CO_2 + M$$

where M is a general third body. Now, in the case of hydrogen oxidation, the "overshoot" leading to superequilibrium levels of the free radicals can be traced to the two body shuffling reactions. The comparison to be made is between either hydrogen or CO systems and a hydrocarbon system. Hydrogen reacts significantly faster than hydrocarbons and therefore the atomic oxygen overshoot occurs earlier. Apparently, the NO formation mechanism cannot respond in this time scale whereas for hydrocarbons the NO response time is compatible with the increase in induction time. The predictions given in our paper seem to validate this explanation. Now, a pure CO/Air system would require significantly longer induction times than hydrogen and perhaps even longer than some hydrocarbons. However, the CO oxidation mechanism does not appear to provide for a significant atomic oxygen overshoot. Thus, for these two different reasons I would expect the NO to behave in a similar way for both the H_2 and CO systems. We hope to examine and verify this behavior in the near future.

C. P. Fenimore *(G.E. Research and Development Center)*

Our CO flames were not dry; they contained moisture and therefore were probably not fundamentally different from hydrogen flames.

Dr. Edelman's calculations seem to blunt the force of one argument for my suggestion that "prompt" NO does not involve an attack of O atoms on N_2. I found "prompt" NO in hydrocarbon but not in hydrogen or moist CO flames of equal mixture strength and temperature, and I supposed that O atoms were no less prevalent in the latter flames. However, his calculations show that O atoms are generated at higher temperature in hydrocarbon flames and this may be an important difference.

The other reason for suggesting that "prompt" NO does not involve an attack of O atoms on N_2 is that more "prompt" NO was found in fuel-rich than in stoichiometric or fuel-lean hydrocarbon flames of equal temperature. I supposed this was contrary to the trend of [O] with mixture strength. It will be interesting if his calculations can support the supposition, or show that other considerations are involved in this argument too.

R. B. Edelman

Our intention is to study CO oxidation with added amounts of both hydrogen and water. Hopefully, this will help to explain your observations as well as establish some general trends in chemical systems whose basic reaction mechanisms are believed to be fairly well established. Regarding fuel rich operation, we believe that both the reduced oxygen levels combined with the crucial reductions in temperature make "staged" combustion techniques, initiating with some fuel rich burning, appear very attractive from an NO emissions standpoint.

A. H. Lefebvre *(Cranfield Institute of Technology)*

Prof. Mellor apologized for the lack of sophistication in his model, but I would like to argue against introducing more complexity. I believe this is the wrong way to go, because as far as the production of NO is concerned, and I am talking now about gas turbine combustion, the overriding factor is flame temperature, or reaction temperature, and the distribution of temperature. These are the key considerations that govern the rate of formation of NO. The only way in which other factors such as pressure, inlet air temperature and fuel-air ratio affect NO formation is indirectly through their effect on reaction temperature.

One need not say much about exhaust smoke because for one thing engineering solutions are available and, in any case, we know that a small change in fuel atomizer characteristics can have a profound effect on smoke output, so that it isn't a profitable emission to try and model anyway. CO and hydrocarbons are again very dependent on the temperature in the combustion zone. Other complications arise from the effects of film-cooling air, which are very difficult to deal with and which none of the proposed models take into account. Fuel vaporization is not important. It is, of course, conceivable that someone could design a fuel injector where the atomization quality is so bad that it could become important. However, vaporization should not be a significant factor in practical combustors in the range of operating conditions where one is worried about NO. Neither are chemical reaction rates come to that. At high pressures they are not limiting to the temperature in the combustion zone and therefore they are irrelevant to the problem of emissions.

Many of the factors that we have discussed this morning do not add to our understanding of the emissions problem and they certainly do not contribute towards engineering solutions. Now it doesn't necessarily follow that modeling is a waste of time. I think if modeling techniques can be developed which allow us to estimate in advance the temperature in the reaction zone, the distribution of temperature in the reaction zone and, as a secondary consideration, residence time, they will then make a useful contribution to the emissions problem.

R. B. Edelman

The turbine combustor presents a highly complex coupling of fluid dynamics and kinetics processes. Our model attempts to characterize this system with sufficient detail so that the relevant behavior in terms of combustion performance and pollutant emissions can be related to the parameters in control of the designer. This is why we have included the effects of mixing, kinetics and droplet processes. Perhaps one most important application of the model will be to evaluate modifications from the standpoint of ensuring that a new problem is not introduced or an old one reintroduced such as soot formation. Of course, data interpretation and reduced hardware testing programs are always a major value of mathematical models.

GENERAL DISCUSSION

J. P. Longwell

We have recently carried out some experimental and kinetic modeling work on well stirred reactors of interest in defining some of the requirements for modeling in intensely mixed combustion zones.

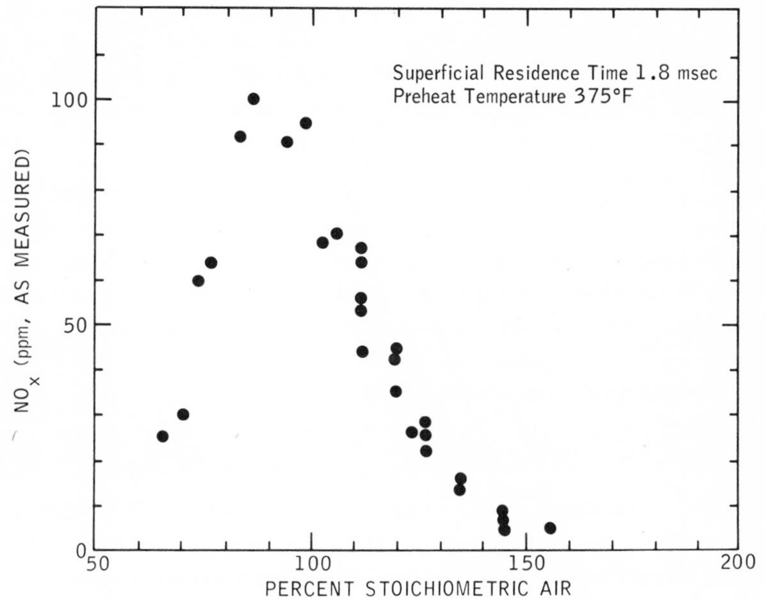

Fig. 1 NO_x emissions in jet-stirred combustor, methane-air.

Fig. 1 shows experimental results of burning a hydrocarbon-air mixture in a small well stirred reactor. In this case, there was substantial heat abstraction by the reactor walls and nitric oxide levels are quite low. This work was done by Dr. Bartok and Dr. Engleman in our Government Research Labs under contract with EPA.

It is clear that a wide range of results can be obtained as air/fuel ratio is varied and that very low NO levels are possible with proper control. As a matter of perspective, NO concentrations of 100 ppm or less can be taken as a reasonable goal for stoichiometric combustion in stationary combustion systems and in automotive gas turbine systems.

Mathematical modeling of this system is being done by Dr. Edelman of the General Applied Science Laboratory, who mentioned this work in the third paper in this session. A quasi-global mechanism was assumed where combustion proceeded rapidly to hydrogen and CO. A rather complete kinetic scheme was then used for the combustion of these materials and for the formation of NO. These calculations gave results low by about a factor of 2; however, they are useful for discussion of the general characteristics of NO formation under these conditions.

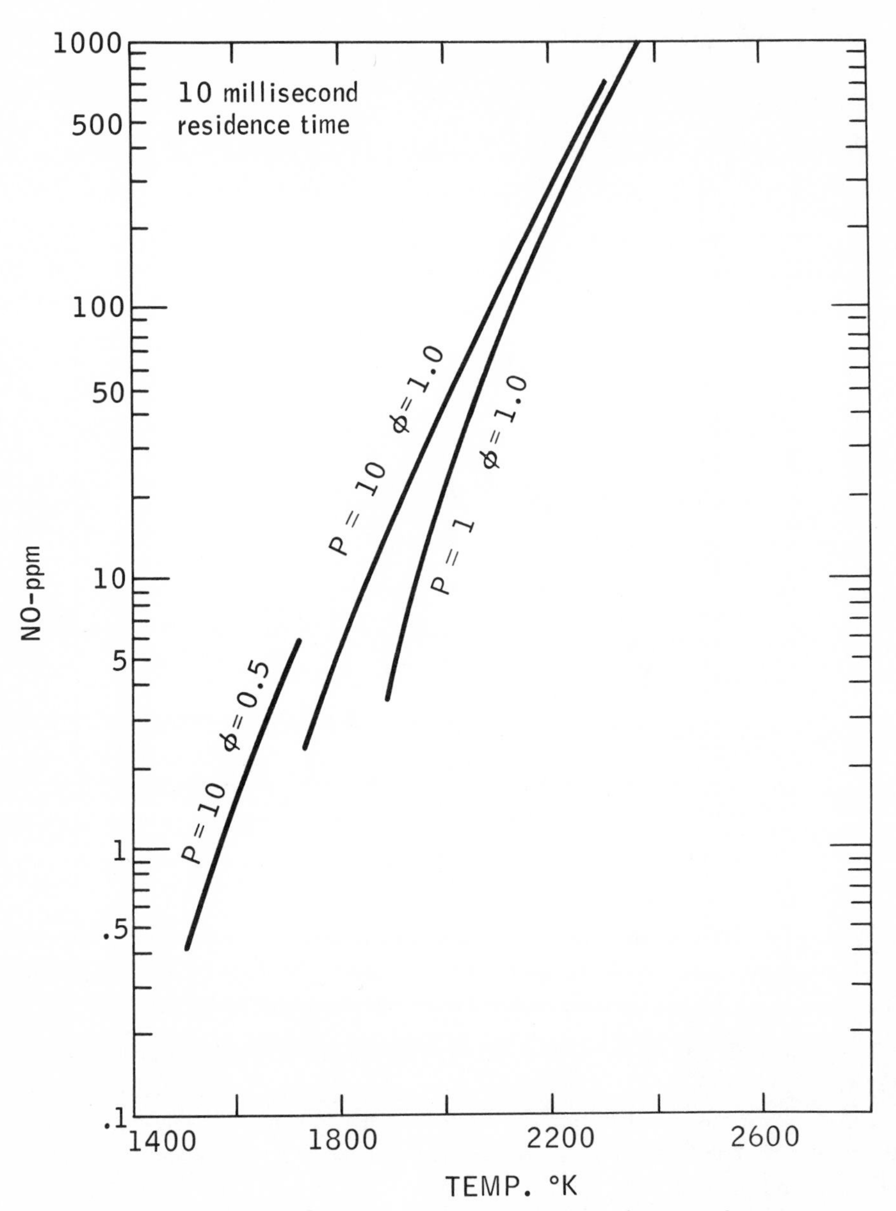

Fig. 2 Calculated NO production in a well stirred reactor as a function of temperature.

Fig. 2 shows the calculated NO formation as a function of temperature. In this case temperature was varied by varying heat abstraction – the upper end of the curve corresponding to adiabatic combustion. Extreme sensitivity to temperature is illustrated while the effect of pressure or equivalence ratio is relatively unimportant. NO levels are very high, even at 10 milliseconds, when temperature is not controlled. It is therefore clear that accurate modeling of temperature is essential.

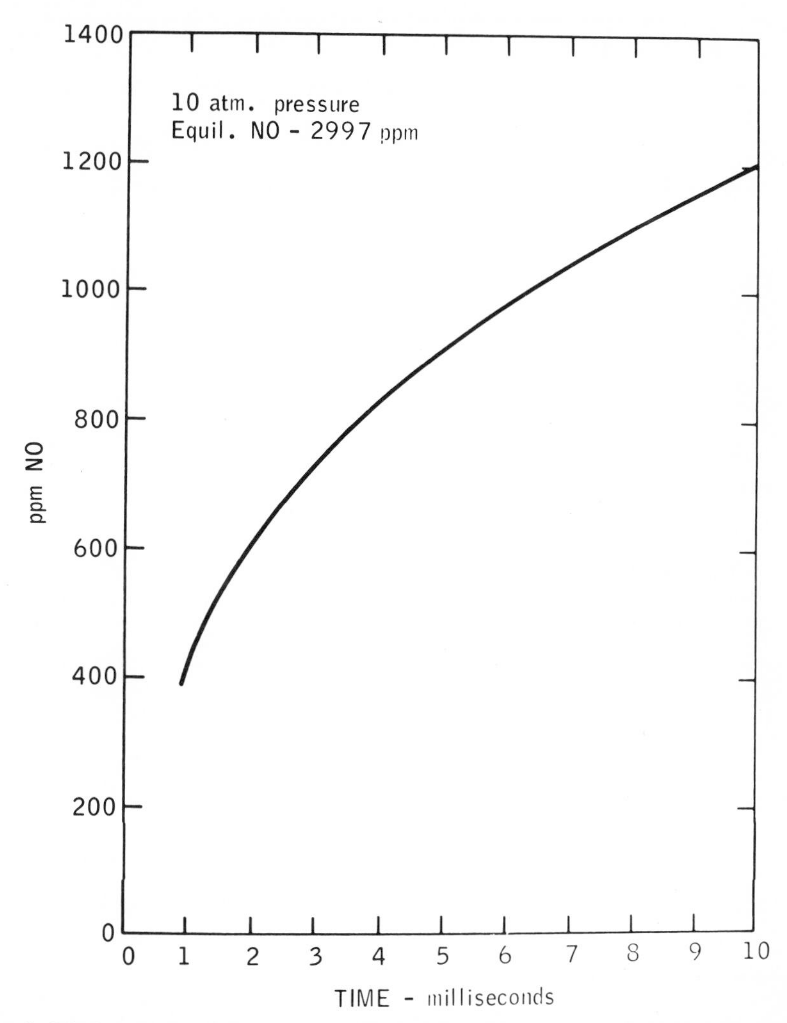

Fig. 3 Calculated NO production in a well stirred reactor as a function of residence time.

Fig. 3 shows the effect of residence time. In this case there was no heat abstraction and NO levels were again high, even at short times. The decrease of NO formation as time is decreased was less than we expected.

The reason for this is apparent in Fig. 4 where the calculated ratio of O atoms to their equilibrium concentration of 200 ppm is shown. The large concentration of O atoms at short times accounts for the high rates when short residence times are used. Since O atom concentration depends strongly on the combustion of carbon and hydrogen containing species, it is clear that rather complete kinetic schemes are required in zones where combustion is taking place and that substantial amounts of NO can be formed in these zones.

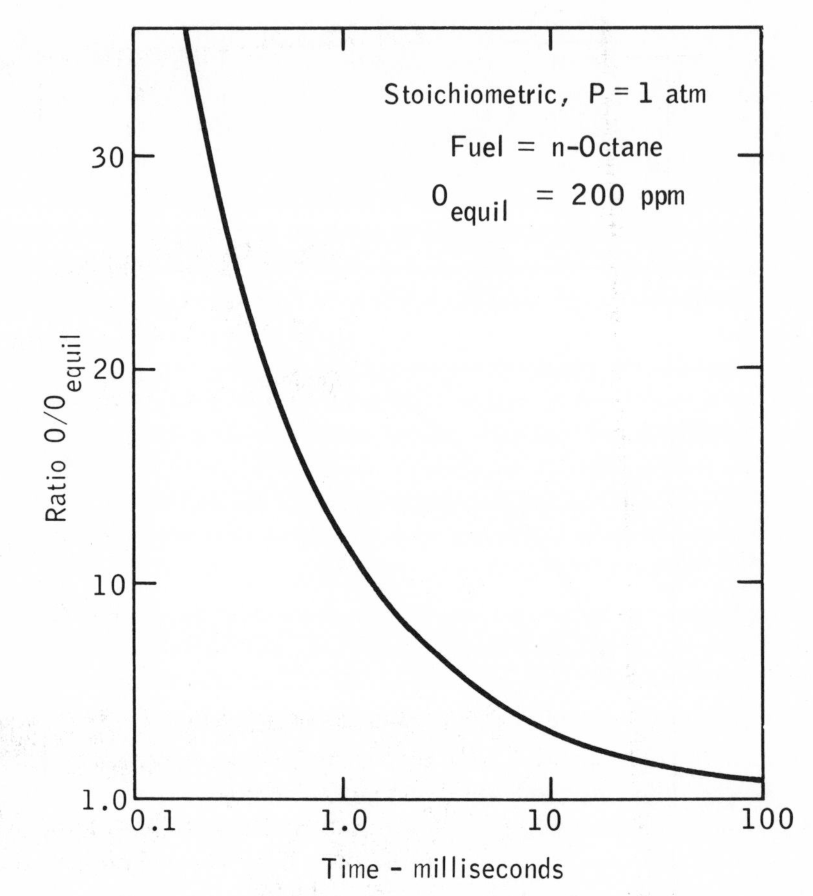

Fig. 4 Calculated O atom concentration in a well stirred reactor.

J. M. Beér *(University of Sheffield)*

I think that models in general serve two purposes. Firstly they can be regarded as maps for research work and secondly in fortunate cases they can be used for engineering design calculations. The models which we have heard discussed today are satisfying the former objective already at present and have varying potentials towards application in engineering practice. Because of the wide ranging requirements of the engineer for prediction methods it seems to me that there is amply scope for the development of both the detailed and the statistical models. The detailed mathematical model that Prof. Spalding and his group developed at Imperial College has given great stimulation to research workers active in the field of fluid flow and combustion. It has reoriented a large amount of research effort towards postulating and testing new turbulence hypotheses. It has also shown some of the limitations of this procedure for complex three dimensional engineering systems, difficulties which no doubt will be steadily overcome as the method is further developed.

In the mean time, statistical models serve a useful purpose where the objective is limited to the determination of simpler approximate flow and heat release patterns. In many cases these will suffice for the calculation of combustion performance, the emission of pollution or radiative heat transfer and for the optimisation of a process when a choice between varying degrees of performance and efficiencies has to be made. Statistical models can also conveniently be combined with physical, experimental modeling. Physical modeling has been mentioned by Prof. Spalding in passing, but I think it deserves more of our attention. Physical modeling is based on the same partial differential equations which form the basis of the detailed prediction procedure but they do not need to be solved for physical modeling, instead ratios of significant terms in the equations are formed and these are then maintained constant in the physical model and prototype. Because of the rate determining nature of flow in most combustion processes, the use of small scale isothermal models has been highly successful in combustion chamber design. The justification of the use of isothermal modeling lies in the fact that the effect of the flow and mixing pattern on the heat release pattern is controlling while the effect of combustion upon the flow pattern is much less important and often negligible.

Physical models can be used for visualization of the flow, for determining residence time distributions for statistical model calculations, and for determining the detailed distributions of velocities and of local mixing ratios. This then permits also the detailed heat release pattern to be computed with good accuracy.

In the short term, physical modeling combined with statistical mathematical models have perhaps the best chance to assist the design engineer in calculations aimed at reducing the emission of pollutants or determining combustor performance, but they are valuable also in connection with long term development of the detailed mathematical prediction procedures.

W. G. Burwell *(United Aircraft Research Laboratories)*

I would like to make one point that backs off a little bit from the rejection of modeling that I thought I heard from Prof. Lefebvre. That is to say, it is clear from experimental observations that neither carbon formation nor NO formation in practical systems is an equilibrium phenomenon. The measured levels of these emissions are not coincidental with those that would be predicted on the basis of the equilibrium behavior of the gases. Therefore, there is at least one other parameter that is involved and that is time. As suggested by Prof. Lefebvre, temperature is not the only factor and, in fact, I would guess you would have to include compositions in a direct way, not just as they influence temperature. So I think there has to be some liberalization of the implications of Prof. Lefebvre's remarks. That is, there are a number of related factors that have to be taken into account. Maybe they are most suitably taken into account by experimentation and not by modeling. But I don't think that one should go to the extreme of saying temperature is the only important parameter in the systems.

R. B. Edelman

If Prof. Lefebvre said that temperature is the most important parameter, then I do agree with his statement. Residence time is not quite as important.

C. W. Shipman *(Worchester Polytechnic Institute)*

I want to make a remark about what I think is an assumption. It seems that the nitrogen oxides do not correspond to what one would expect from thermodynamic equilibrium. Therefore, the assumption is made that we are dealing with a kinetic phenomenon, and we then become interested in the oxygen atoms and start looking at details of the kinetics. I do not object to that specifically, but what I wonder is if perhaps the non-equilibrium nitrogen oxides cannot also be a consequence of local inhomogenieties. This gets us back to the mixing problem. Now with this possibility I don't want to let the Chair off the hook – he spoke about burning rich in order to avoid the problem. If you burn rich locally in a system that is lean overall, as a gas turbine combustor is, then you may be postponing the problem to later on in the combustor and form the NO anyway.

J. P. Longwell

I think I need to amplify my statement on richness. In a furnace, one can use two-stage combustion with some heat abstraction between stages. This offers a promising solution for the coal burning problem, for example. In the gas turbine, this heat abstraction is generally not practical; however, by proper control of air dilution the second stage combustion can in principle be carried out at a sufficiently low temperature to give low nitric oxide emissions.

E. S. Starkman *(Environmental Activities Staff, GMC)*

Prof. Mellor was kind enough to make reference to the work that's going to be reported on a little more thoroughly by Bob Sawyer in Session III. I think we have to take into context both what Prof. Lefebvre had to say and what Drs. Burwell and Bill Shipman had to say. You can't explain on the basis of kinetics alone what is happening inside a gas turbine combustion chamber as we have probed them. You cannot explain, or come close to explaining, even by adjusting the kinetics appropriately. As far as I am concerned, I think that the work that Brian Spalding is doing, is going to lead to a better understanding of nonhomogeneties within the combustion chamber and help us to the answer. As for simple models, I disagree that a simple model is going to provide the answer.

F. V. Bracco *(Princeton University)*

I would like to add to the objections raised against the statements of Prof. Lefebvre. By saying that the NO formation is a function mostly of the temperature,

Prof. Lefebvre did not say anything new. The trouble is that the temperature itself is not uniform in space and time and depends on mixing, vaporization, geometry, initial conditions, etc. There is no simple way of calculating this temperature distribution. It has been proven in diesel and gas turbine engine studies that an average temperature calculation does not lead to reasonable NO estimates. To calculate the detailed temperature field in space and time, one must then consider mixing and vaporization. As far as vaporization is concerned, Prof. Lefebvre tends to discount its importance on account of the fact that the vaporization time decreases as the pressure increases in the gas turbine and the trend is toward higher pressures. While vaporization time decreases faster than the residence time as pressure increases, it still does not mean that the vaporization time is negligible. Indeed, the estimates of Prof. Mellor and similar estimates by myself indicate that the life time of the droplets, for certain operating conditions, is of the same order as the residence time in the primary zone. Therefore, vaporization and mixing are equally important. The only way then to calculate the temperature field and NO formation is to take into account all of these aspects through combustion modeling.

E. B. Zwick *(The Zwick Company)*

I want to add a little fuel to both sides of the fire or maybe to calm it down. First, I am a firm believer in modeling. I think it is ridiculous to think that the world should depend only on experimental efforts. We cannot afford it. It just costs too much to try everything. Only by some reasonable modeling can you find which way to go. On the other hand, I recently completed some experimental measurements on a burner in which mixing of the fuel and air had nothing to do with the result because they were well mixed before they got into the burner. The measured oxides of nitrogen looked quite similar to the results that Dr. Longwell showed earlier, trending to very low levels with excess air. But the thing that disturbed me about our measurements was that temperature was the only parameter I could correlate with the oxides of nitrogen. Residence time and oxygen concentration, at least the overall values, did not appear to relate directly to the experimental measurements whereas temperature did correlate with the data. I do not think that is a reason to quit modeling. Somebody should now improve the model so that it shows why that was so. It might not really have been temperature, it might have actually been fuel-air concentration and had something to do with the prompt NO_x, for example.

J. P. Longwell

As you pointed out, Mr. Zwick, determining the temperature is quite a problem in itself; however, knowledge of the temperature – time history is a good start since calculations using the quasi-global model indicate that for fuel lean conditions, nitric oxide production rate depends primarily on temperature. For fuel rich conditions, the rate decreases rapidly with equivalence ratio at constant temperature.

C. T. Bowman *(United Aircraft Research Laboratories)*

Earlier this morning, Prof. Mellor reviewed the kinetic aspects of current analytical models for continuous flow combustors, emphasizing the limitations of approaches to modeling gas-phase hydrocarbon combustion and droplet burning. I would like to offer some additional comments on the limitations of these models in predicting nitric oxide emissions.

In the way of review, current combustor models employ one of the gas-phase hydrocarbon combustion mechanisms shown schematically on Fig. 1. For the purposes of the present discussion, methane was chosen as a representative hydrocarbon fuel principally because detailed information on the combustion mechanism is available.

The first mechanism is a detailed kinetic mechanism, in which methane reacts to form methyl radicals, which, in turn, react through intermediate species to form products. Two quasi-global models were discussed in Session I and both involve the reaction of fuel with oxygen to form CO and H_2 or H_2O at either a finite or infinite rate. The CO and H_2 then react via the known CO/H_2 oxidation mechanism to form products. The final model shown is called the C-H-O equilibrium model, where it is assumed that the combustion chemistry is in equilibrium. These combustion mechanisms, coupled with nitric oxide reactions, have been used to calculate NO formation rates at constant volume for a range of initial temperatures, pressures and fuel/air ratios, with the objective of determining how well the various combustion mechanisms model the NO formation process.

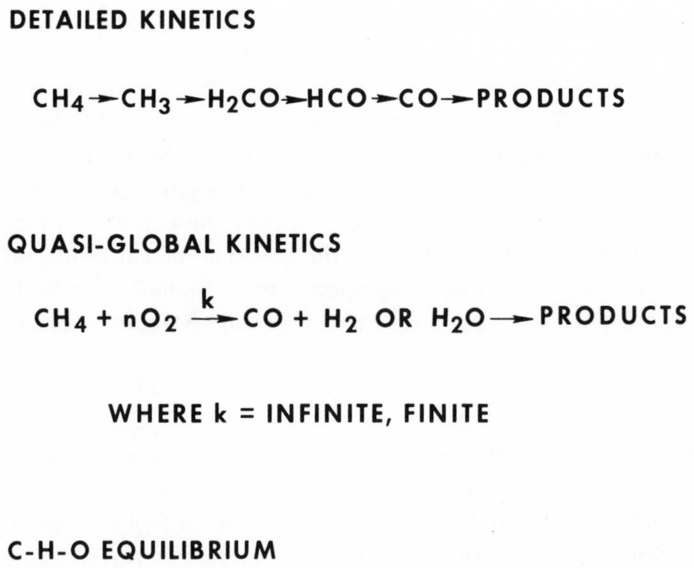

DETAILED KINETICS

$$CH_4 \rightarrow CH_3 \rightarrow H_2CO \rightarrow HCO \rightarrow CO \rightarrow \text{PRODUCTS}$$

QUASI-GLOBAL KINETICS

$$CH_4 + nO_2 \xrightarrow{k} CO + H_2 \text{ OR } H_2O \rightarrow \text{PRODUCTS}$$

WHERE k = INFINITE, FINITE

C-H-O EQUILIBRIUM

Fig. 1 Schematic representation of various combustion mechanisms for methane.

Two factors known to strongly influence NO formation rates are temperature and oxygen atom concentration. Fig. 2 is a typical comparison of temperature profiles calculated for fuel-rich methane/air using the different combustion mechanisms. The temperature, calculated using the detailed combustion mechanism, remains nearly constant for ∿ 1 msec and then rapidly increases to its equilibrium value. In the C-H-O equilibrium mechanism, the temperature is, by assumption, constant and equal to the equilibrium temperature. In the quasi-global combustion mechanisms, partial oxidation of methane is assumed to occur via $CH_4 + 1/2\, O_2 \rightarrow CO + 2H_2$. If the rate of the partial oxidation reaction is assumed to be infinite, the calculated temperature instantaneously increases from its intial value to a value determined by the exothermicity of the partial oxidation reaction. Following this instantaneous increase, the calculated temperature rapidly approaches its equilibrium value. If the partial oxidation reaction is assumed to have a finite rate, then it is possible, by a judicious choice of the rate constant, to obtain a calculated temperature profile which closely approximates that calculated using the detailed combustion mechanism.

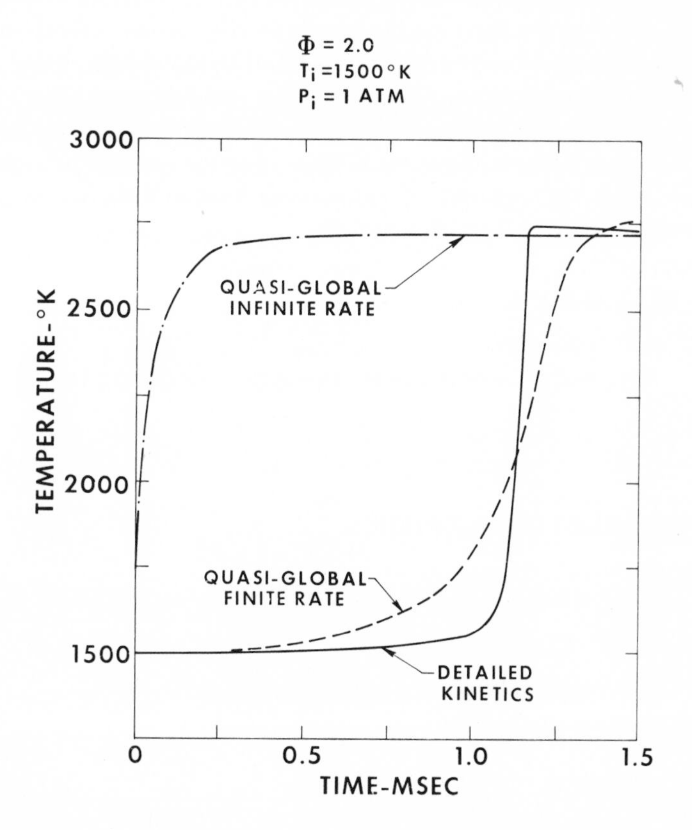

Fig. 2 Comparison of calculated temperature profiles obtained using various gas-phase mechanisms to model NO formation in constant-volume methane-air combustion.

Fig. 3 shows the ratios of the oxygen atom and nitric oxide concentrations to their equilibrium values as functions of time. The detailed combustion mechanism predicts a rapid increase in oxygen atom concentration coincident with the rapid increase in temperature. The predicted oxygen atom concentration overshoots its equilibrium value by a factor of 2.5 and then rapidly approaches equilibrium. The predicted nitric oxide concentration does not increase significantly until late in the reaction, when the temperature has attained its equilibrium value. This delayed appearance of nitric oxide has been cited as evidence for assuming C-H-O equilibrium in kinetic models for nitric oxide formation in combustion processes. However, as illustrated in the figure, nitric oxide concentration profiles, calculated assuming C-H-O equilibrium, can be significantly different from those calculated using a detailed combustion mechanism. The rapid nitric oxide formation rate, predicted by the equilibrium mechanism, is a direct result of the assumption that temperature instantaneously attains its equilibrium value. The nitric oxide formation rate, calculated using an infinite-rate quasi-global combustion mechanism, also exceeds that predicted by the detailed mechanism. In this case, the rapid nitric oxide formation rate is a result of the mechanism over-estimating the oxygen atom concentration and the rate of increase of temperature. The nitric oxide concentration profile, obtained from the finite-rate quasi-global mechanism, is qualitatively similar to that predicted by the detailed mechanism. In both cases the nitric oxide concentration does not increase significantly until late in the reaction, when the temperature begins to increase. However, the nitric oxide formation rate, calculated using the finite-rate quasi-global mechanism, is faster than the rate obtained from the detailed mechanism. This faster formation rate results from a significant overestimation of the oxygen atom concentration by the finite-rate quasi-global mechanism.

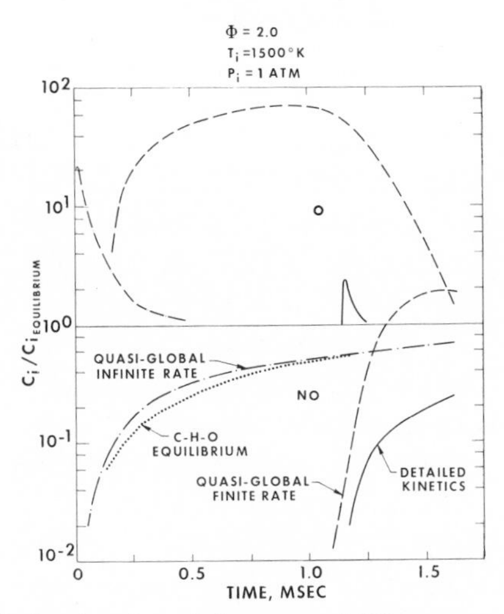

Fig. 3 Comparison of calculated ratios of O-atom and NO concentration to equilibrium values obtained using various gas-phase mechanisms to model NO formation in constant-volume methane-air combustion.

These calculations demonstrate the importance of temperature and oxygen atom concentration in nitric oxide formation kinetics. Observed differences in nitric oxide concentration profiles, calculated using various combustion mechanisms, can be related directly to differences in predicted temperature and oxygen atom concentration profiles. For the present calculations, the relative differences between the results obtained from the various combustion mechanisms decrease as temperature and pressure increase or as leaner mixtures are used. Of the various mechanisms, the detailed combustion mechanism is most accurate from the standpoint of modeling nitric oxide formation kinetics. Under certain conditions (high temperature and pressure, lean mixtures), the finite-rate quasi-global mechanism predicts nitric oxide formation rates in close agreement with those obtained from the detailed mechanism. However, for low temperature and pressures and especially for rich mixtures, nitric oxide formation rates, calculated using the finite-rate quasi-global mechanism, differ significantly from those predicted by the detailed mechanism. As noted earlier, these differences are the direct result of differences in predicted oxygen atom concentration. Hence, to accurately model nitric oxide formation kinetics, a combustion mechanism must accurately predict not only the energy release rate but also the oxygen atom concentration history. For a given set of conditions, the nitric oxide formation rates predicted by the C-H-O equilibrium mechanism and by the infinite-rate quasi-global mechanism are approximately the same and are always greater than the rates predicted by the detailed mechanism or by the finite-rate quasi-global mechanism. From the standpoint of modeling nitric oxide formation kinetics, the relatively complex infinite-rate quasi-global mechanism appears to have few, if any, advantages over the C-H-O equilibrium mechanism.

In several existing combustor models, heterogeneous combustion is modeled by considering droplet vaporization, followed by instantaneous mixing and homogeneous gas phase reaction. While such variable equivalence ratio homogeneous burning, no doubt, occurs in combustors burning liquid fuels a substantial portion of the combustion may take place as diffusion flames around droplets and as gaseous diffusion flames. Kesten (1) and Bracco (2) have modeled NO formation in single droplet combustion and Williams, et al. (3) have examined NO formation in gaseous diffusion flames. All of these investigators have concluded that diffusion effects play a significant role in the NO formation process. Some results from Kesten's droplet study are shown on Fig. 4. This figure shows a comparison of the NO emission indices for ethanol droplet burning in an infinite air environment and ethanol-air homogeneous gas-phase combustion. The emission indices for gas-phase combustion for residence times equal to the lifetimes for three droplet sizes are plotted as a function of equivalence ratio. The three corresponding droplet emission indices are indicated on the ordinate axis. It is only in the narrow region around an equivalence ratio of one that nitric oxide formation in homogeneous combustion is more important than nitric oxide formation in droplet combustion. From results such as these, one concludes that diffusion effects can be important in NO formation in heterogeneous combustion and should be considered in analytical models for combustors burning liquid fuels.

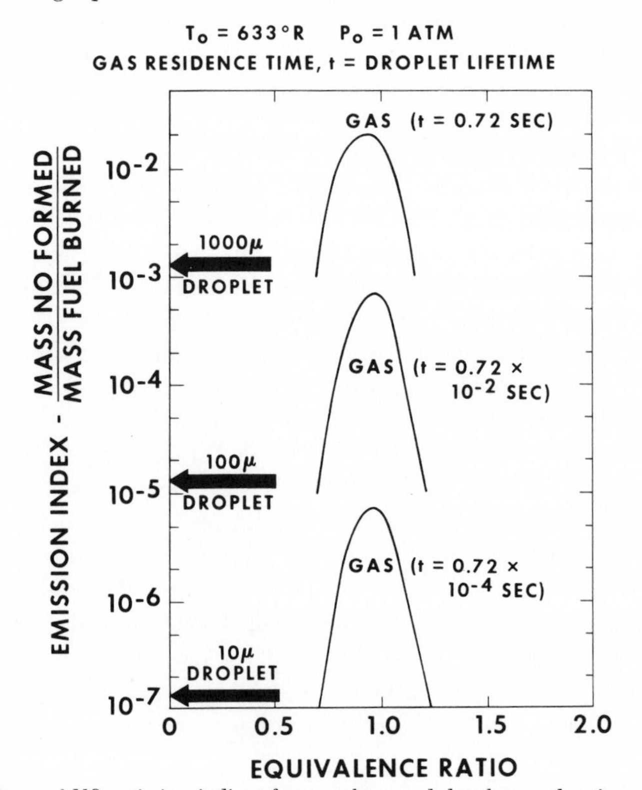

Fig. 4 Comparison of NO emission indices for gas-phase and droplet combustion of ethanol-air.

References

1. *A S. Kesten, "Analysis of NO Formation in Single Droplet Combustion", to be published.*
2. *F. V. Bracco, "NO Formation in Droplet Diffusion Flames", presented at the 1971 Fall Meeting of the Western Section of the Combustion Institute Irvine, California, October, 1971.*
3. *G. C. Williams, A. F. Sarofim and D. H. Fine, "Nitric Oxide Formation in Premixed and Diffusion Flames", Paper 37c presented at the 70th National A.I.Ch.E. Meeting, Atlantic City, New Jersey, August, 1971.*

J. P. Longwell

Dr. Bowman, what was the range of equivalence ratio over which you calculated this effect?

C. T. Bowman

In examining the various gas-phase combustion mechanisms, I considered equivalence ratios in the range 0.5 (lean) to 2.0 (rich).

J. P. Longwell

Did you find as large an effect on the excess air side?

C. T. Bowman

No, for leaner mixtures and higher temperatures and pressures the differences between the various combustion mechanisms decreased.

B. P. Breen *(KVB Engineering, Inc.)*

I would like to emphasize that same point from an engineering point-of-view: that the stoichiometry effects are large and everyone admits to the magnitude of these effects (i.e., not much NO is formed in fuel-rich mixtures but large amounts are formed in fuel-lean mixtures). However, when in the field working on real equipment (such as atmospheric-pressure utility boilers), we are interested in reducing the high levels which do commonly occur. When levels in the range of one thousand parts per million occur it is always when we have intense, fuel-lean combustion. Then we find that the kinetics of practical importance can be calculated using the assumptions of Starkman and Sawyer, thus the kinetics can be calculated by hand as well or better than the engineering or modificaiton work can be accomplished. The kinetics of the fuel-rich region ($\phi \gtrsim 2.0$) are much more complicated, as Dr. Bowman has indicated, but since the NO formed in these regions may be three or even four orders of magnitude lower, these complications may have only academic importance.

J. B. Heywood *(Massachusetts Institute of Technology)*

One of the points raised was whether the assumption that the carbon-hydrogen-oxygen system is in equilibrium during the period when most of the NO is forming is valid. Since I have been making this assumption for some time, let me make some comments that I think will help to put it in context and show under what conditions it is an appropriate assumption to make. Now obviously it is not correct because in the gases coming out of the flame, the carbon-hydrogen-oxygen system does not immediately attain equilibrium. That's the wrong question. The real question is whether at the conditions that are important in practical devices is it a useful approximation. Now I do not know the full answer to that question as we come down in pressure level. Let me try and explain why.

When we started working on this problem, we were looking at spark-ignition engines where one has a pre-mixed charge and very high gas pressures and temperatures. There, it appears that onc can neglect the nonequilibrium oxygen atom effects because the recombination processes coming out of the flame are very rapid at the 30 or 40 atmosphere pressures of interest. We have been able to match the measured NO emissions going all the way from fuel-lean through fuel-rich mixtures over a range of exhaust gas recycle fractions with calculations based on this simple model. Now the question is, if one drops the pressure and slows down the recombination processes that bring the oxygen atoms into equilibrium as the gases come out of the flame, at what pressure do these nonequilibrium effects become important in practical systems? I think that one must be very careful when one attempts to make general conclusions regarding the temperature regime, the pressure regime, residence time at these high temperature and high pressure conditions, and the fuel-air ratios under which this equilibrium oxygen atom assumption would be a good or poor approximation.

Now we have moved down in pressure, and we have tried to model the NO formation process in gas turbine combustors where the fuel and air, of course, are not pre-mixed, but burning is, I think, mainly diffusion controlled. The pressure there is about 10 atmospheres. It appears from my own work and some of the Northern Research studies that are just being completed that calculations of NO concentrations using this simple model again give reasonably good correlation with the data over a range of air temperatures coming into the gas turbine burner and a range of fuel-air ratios as the fuel flow into the burner is changed. This is so provided one makes allowance for the fact that the primary zone is not uniform; though on average it may be lean, parts of the flow are going to be close to stoichiometric where NO forms much faster.

Now to come down in pressure level to an atmospheric pressure burner, again the recombination processes have slowed down still further, and one would expect them to be more important. I think that in pre-mixed systems there is enough evidence now to show that for typical residence times in practical flames, nonequilibrium effects can be important. It is not quite so clear to me that in heterogeneous combustion systems they are as important. I will discuss a slide during Session II which shows that we can predict NO formation in an atmospheric pressure burner with a rather simple model that takes account of the nonuniformities in the flow, but assumes that when most of the NO is being formed, oxygen atoms are in equilibrium.

One can attribute one's inability to predict NO emissions to incorrect oxygen atom concentrations. Another reason may be flow nonuniformities; one may not have correctly weighted that fraction of the flow which is close to stoichiometric where most of the NO is produced. I think the real question is: Are these simpler models useful for practical devices? They are obviously not correct in the strict sense. The local conditions reached in the particular burner one is examining have a very important bearing on the conclusion.

P. D. Agarwal *(Research Laboratories, GMC)*

Combustion is a very complex process, both microscopic and macroscopic, with actual flow of matter and with all the transients that are taking place. We have heard about modeling this complex process and solving the differential equations by digital computer. How does one choose the grid size that would produce convergence or a solution within reasonable computer time? Our experience in solving much simpler electro-magnetic field problems by this technique has proven them to be quite complex and time consuming. Even when the program is made to converge by under or over-relaxation techniques, one does not know for sure that the right answer has been obtained. The complexity of modeling and solving the combustion process would be at least an order of magnitude greater than the three-dimensional field problems I mentioned.

SESSION SUMMARY

J. P. LONGWELL
Esso Research and Engineering Co., Linden, New Jersey

We have just heard three excellent summaries and many excellent floor discussions of the progress being made in mathematical modeling of continuous combustion systems. It is clear that a powerful set of tools is being applied to this problem and that the current and past work in fluid mechanics, kinetics and computional techniques offer opportunities for the study of combustion equipment that were not available for use during the early design period of the gas turbine.

The presentations and discussions have shown that modeling for the gas turbine is still in the formulation stage where available knowledge of chemical kinetics and fluid mechanics is being combined with simplified models of various processes such as droplet combustion and turbulent transport to represent combustion in practical combustors. Such models are of necessity complex – sufficiently so that, as Prof. Mellor points out, simple measurement of the output of combustion products provides little information concerning the validity of the simplifications incorporated in the model.

Of particular interest in the discussion was the debate concerning the utility and promise of mathematical modeling in the development of improved gas turbine combustors when emission restrictions are imposed. Extremes of optimism and pessimism were presented and it must be concluded that the ultimate usefulness of modeling to the combustor developer is at present a moot point. If success requires complete characterization of combustor performance making use only of basic data then success is indeed unlikely. However, modeling can exist without attaining this ultimate success by establishing limiting performance to help in understanding the degree of improvement possible by continued research, by providing a sound framework for correlations of experimental results, by indicating routes to improved performance and by defining the research needed to usefully characterize the component processes. It would appear that increased orientation toward these more limited and specific goals might be productive.

SESSION II

POLLUTANT FORMATION AND DESTRUCTION PROCESSES

Session Chairman
G. C. WILLIAMS

Massachusetts Institute of Technology
Cambridge, Massachusetts

NITRIC OXIDE FORMATION IN LAMINAR DIFFUSION FLAMES

A. D. TUTEJA and H. K. NEWHALL

University of Wisconsin, Madison, Wisconsin

ABSTRACT

A vertical coaxial burner has been used to study nitric oxide formation in diffusion flames. Localized combustion gas samples extracted from the diffusion flame by means of a quartz microprobe have been analyzed for NO as well as for the species CO, CO_2 N_2, O_2 and CH_4, the fuel employed. In this manner it has been possible to establish detailed radial profiles for each of the above species. Concomitant temperature profiles have been established through use of a 0.001 inch diameter Pt/Pt-10%Rh thermocouple.

Experimental results indicate that in diffusion flames of this type NO formation occurs in a narrow region corresponding to that of maximum temperature. It is of particular interest that so-called "prompt" NO formation within the highly reactive flame zone is not observed.

INTRODUCTION

Combustion generated oxides of nitrogen are presently recognized as major pollutants which react in the atmosphere to form photochemical smog. The predominant oxide of nitrogen associated with combustion products (1,2) is NO, which is later oxidized to NO_2 in the atmosphere.

Previous investigations of NO emissions have for the most part been related to spark-ignition engines in which premixed flames occur. Studies relating to the behavior of NO after its formation in the spark-ignition engine have demonstrated that once formed in the combustion process, NO remains fixed throughout the expansion and exhaust processes.(3,4)

References p. 118

Recent studies investigating the formation of NO in a combustion vessel (5,6) and in the cylinder of an engine (7) have shown that NO is formed primarily in post flame combustion gases and that the time required for formation is significant. The combustion vessel studies have shown that NO formation rates can be predicted with reasonable accuracy through consideration of the two elementary reactions:

$$N_2 + O \rightleftarrows NO + N$$

$$O_2 + N \rightleftarrows NO + O$$

Based on steady flow burner measurements the formation of "prompt NO" immediately within the flame front in premixed flames has been postulated recently (8). However, this finding has not been observed by other investigators. (5,6,7)

Combustion systems based on the burning of fuel sprays, such as diesel engines and gas turbines, essentially involve diffusion flames in which mixing of the fuel with oxidizer occurs simultaneously with combustion. It is generally recognized that these systems can be designed to emit smaller quantities of hydrocarbons and CO as compared with conventional spark-ignition engines. However, they may emit significant quantities of NO. Studies of NO from gas turbines and from diesel engines have for the most part been related to determination of measured exhaust concentrations as influenced by engine operating parameters. (2,9,10,11,12) More detailed studies of combustion in gas turbines have demonstrated that the concentrations of NO in exhaust gases are always below the chemical equilibrium level. (13,14)

In one case, experimental results obtained with a laboratory combustor have been interpreted as indicative of significant NO decomposition late in the combustion process. (13) However, this cannot be accounted for by known kinetic mechanisms.

There have been only limited studies concerning details of the formation of NO in diffusion flames. Most previous work with such flames has been devoted to study of flame heights and flame structure as well as overall combustion kinetics and heat release rates.

The purpose of the work presented here has been the experimental determination of NO concentration profiles within a laminar diffusion flame. Local combustion gas samples were extracted from a flame using microprobes and analyzed for NO by a modified Saltzman technique. In addition to NO profiles, concomitant temperature and concentration profiles for CO, CO_2, O_2, N_2 and the fuel, CH_4, have been obtained. Further work involving theoretical analysis of results obtained for varied flame conditions is in progress and will be reported later.

EXPERIMENTAL APPARATUS

A vertical coaxial diffusion flame burner, shown schematically in Fig. 1, has been designed to produce an enclosed steady laminar diffusion flame at pressures ranging

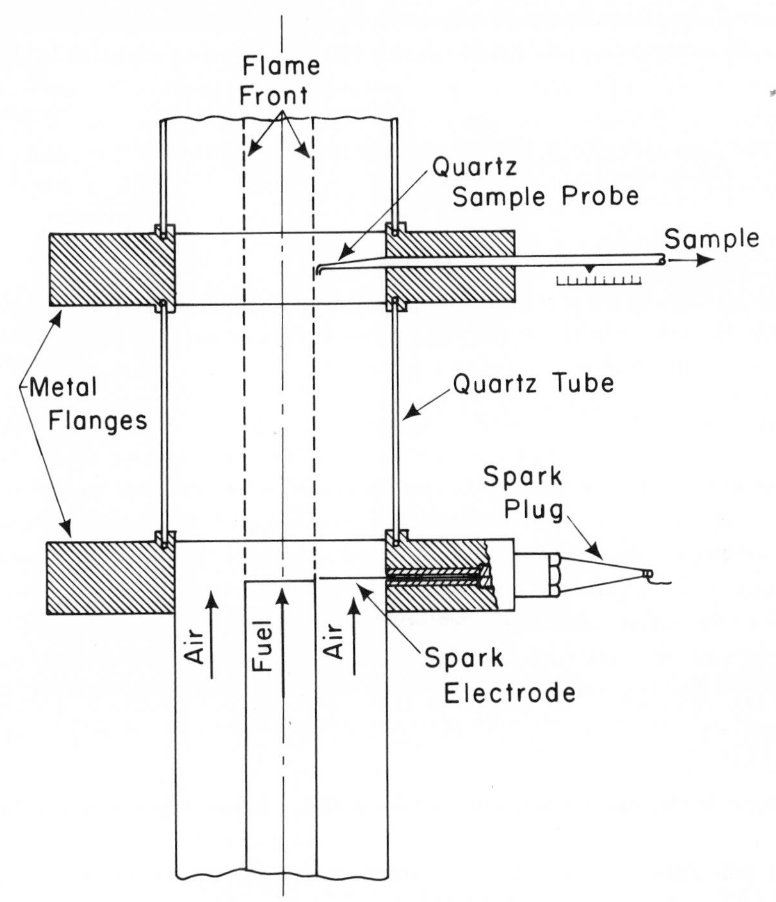

Fig. 1. Laboratory diffusion burner.

from one to ten atmospheres. The fuel, CH_4 in the present case, is metered through a sonic nozzle and fed to the burner through the central fuel tube which is 16 mm. in diameter. Air is metered in a similar manner and is fed to the burner through the outer tube of 43 mm. diameter, surrounding and concentric with the fuel tube. Fuel and air flow rates can be regulated or varied by adjusting pressures upstream of the sonic nozzles. An automotive-type ignition system has been modified to produce a continuous spark discharge as an ignition source. The flame is enclosed in vertical quartz tubing elements interposed with steel flanges providing for sample probe insertion at a number of prescribed axial locations.

Combustion gas samples are extracted from localized positions within the flame by means of a quartz microprobe introduced through the flange. The microprobe has an inside diameter of the order of 0.002" and outside diameter of 0.020" at the probe tip. Radial position of the sampling probe is adjusted and measured by means of a micromanipulator having a least count of 0.1 mm.

References p. 118

The sampling system is shown schematically in Fig. 2. Gas samples are collected in a previously purged and evacuated variable volume sampling vessel. Operation of a

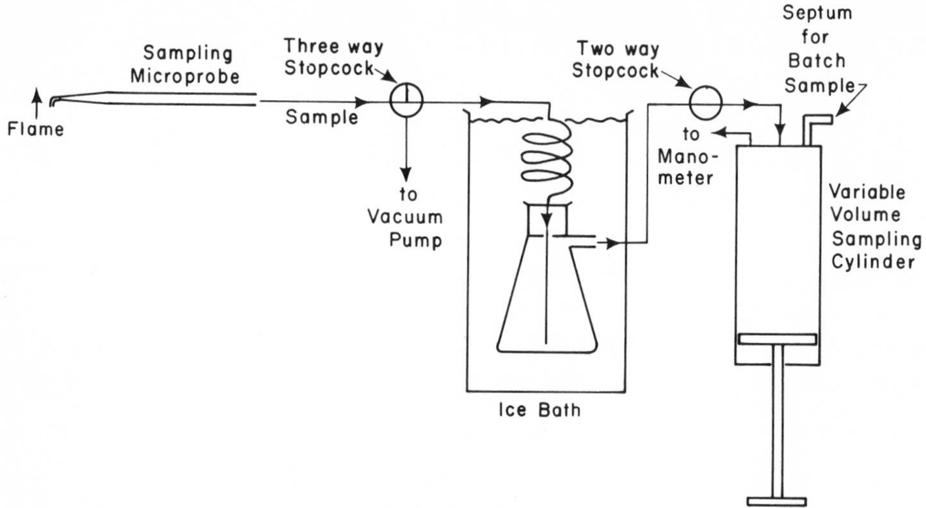

Fig. 2. Combustion gas sampling system.

valve diverts the continuous sample flow from the vacuum system to the sampling vessel. Water is removed from the sample by means of an ice bath located in the sampling line upstream of the sampling vessel. Sonic flow conditions are maintained at the probe tip by maintaining a prescribed vacuum in the sampling line. Thus, rapid freezing of reactions involving chemical species of interest is ensured. From observed sampling rates, the residence time of sample gases in the high temperature portion of the probe is found to be of the order of 0.1 msec.

Combustion gas samples are analyzed for NO using a modified Saltzman technique to be described later, and for CO, CO_2, N_2, O_2, and CH_4, through use of a thermal conductivity gas chromatograph. The chromatograph was calibrated using standard analyzed mixtures of appropriate gases supplied by the Matheson Company.

Radial flame temperature profiles at fixed axial locations are measured with a 0.001" diameter silica coated Pt/Pt-10%Rh thermocouple. As in the case of the sample probe, the thermocouple is introduced radially through the flange, and its radial position is adjusted and measured by means of the micro-manipulator.

Temperatures reported in this work are direct thermocouple readings without correction for losses due to radiation or thermal conduction. Thus absolute accuracy of reported temperature measurements is not claimed. However, the structure of temperature profiles and the location of temperature maxima as indicated by these results should be of acceptable precision.

MEASUREMENT OF NITRIC OXIDE CONCENTRATION

During the initial phase of this study, NO concentrations were measured by dispersive ultraviolet absorption after first oxidizing NO to NO_2. It was found that the sensitivity of the ultraviolet technique was not sufficient to accomodate the entire concentration range of interest, and consequently these determinations could not be considered reliable. However, the initial work did establish the order of magnitude of concentration of NO in the flame as ranging from 10 to 100 ppm.

The Saltzman analytical technique, often used for measurement of atmospheric concentrations of NO_2 up to concentrations of not more than 5 ppm and similar concentrations of NO by first oxidizing to NO_2, appeared amenable to modification for measurement of higher concentrations in the range of interest in this work. While developing a modified technique, it was kept in mind that in the interest of minimizing sampling time the total amount of sample required for analysis should be limited.

The following NO analysis procedure was finally adopted:

1. A measured quantity (usually 10 cc.) of Saltzman reagent is pipetted into a 250 cc. Erlenmeyer flask.

2. The flask is evacuated.

3. A measured quantity of sample (usually 50 cc.) is introduced into the flask through a septum using a syringe. The flask is then filled with pure O_2 bringing it to atmospheric pressure.

4. The resulting mixture of sample, O_2 and reagent is stirred with a teflon coated motor-driven magnetic stirrer for 15 minutes.

5. The optical density of the reagent is measured against that of unreacted reagent using a Beckman Model B spectrophotometer set at 500 mμ wave-length. The NO concentration is determined from calibration curves prepared earlier using standard analyzed NO-N_2 mixtures.

Standard analyzed NO-N_2 mixtures supplied by the Matheson Company containing 20, 53, 105, and 310 ppm of NO were used initially for establishing the method and later for calibration.

The above procedure yielded generally satisfactory results. Repeatability of measurements was better than $\pm$ 5% in optical density for samples in the range of 10-105 ppm. Fig. 3 is a typical calibration curve for the 10-105 ppm range using 10 cc. of Saltzman reagent and 50 cc. of sample. The quantity of Saltzman reagent has to be varied depending upon the NO concentration in order to maintain the optical density within reasonable limits.

References p. 118

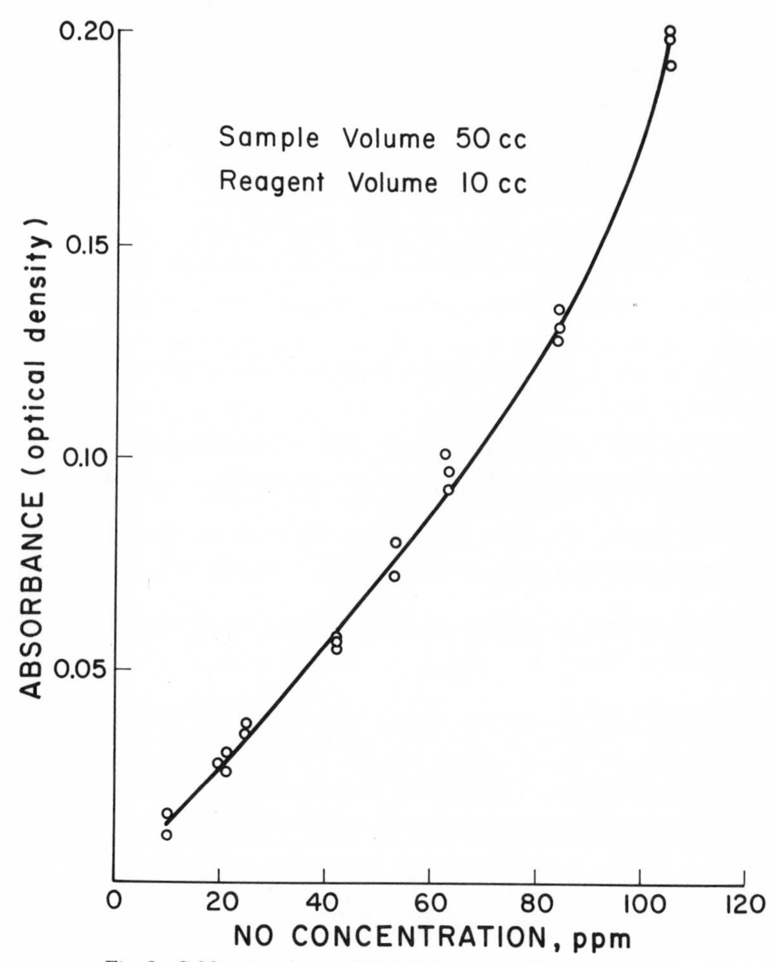

Fig. 3. Calibration for modified Saltzman analysis technique.

PROCEDURE

Temperature and concentration profiles for NO, CO_2, CO, O_2, N_2 and CH_4 were determined for an overall fuel-air equivalence ratio of 0.5755 corresponding to an air-fuel velocity ratio of 1.841. These conditions correspond to average velocities of 1.33 ft./sec. and 0.721 ft./sec. and mass flow rates of 7.95×10^{-2} lb_m/min. and 2.67×10^{-3} lb_m/min. for air and fuel respectively. Measurements were made at two axial locations, one 1-7/8" (47.5 mm.) and the other 4-5/8" (117.5 mm.) above, (downstream of), the burner entrance. At each axial position, samples were drawn from about 40 radial positions extending from the outer wall of the burner to the centerline or axis of the system. Samples were taken in 0.2 mm. radial increments in the region of steep gradients, while nearer the wall, where gradients were not severe, samples were taken at 1 mm. increments. In a narrow annular zone in the fuel-rich region, samples could not be withdrawn, a result of excessive sooting of the sample probe.

Temperature profiles were generated by recording temperatures at each of the points from which samples were withdrawn.

Reproducibility of results was checked by repeated sampling and analysis on different days. The reproducibility for CO, CO_2, O_2, N_2 and CH_4, the species determined by chromatographic analysis, was high, generally better than 5%. NO concentrations were reproducible generally within 5% for concentrations above 10 ppm and within 20% for concentrations below 10 ppm. Measured temperatures were reproducible to within ± 10°F. In general, reproducibility in the fuel-rich region was poorer than that in the O_2 rich region.

RESULTS AND INTERPRETATION

Concentration profiles for NO, CO, CO_2, O_2, N_2 and CH_4 along with concomitant temperature profiles are presented in Figs. 4 and 5 for the two axial positions noted

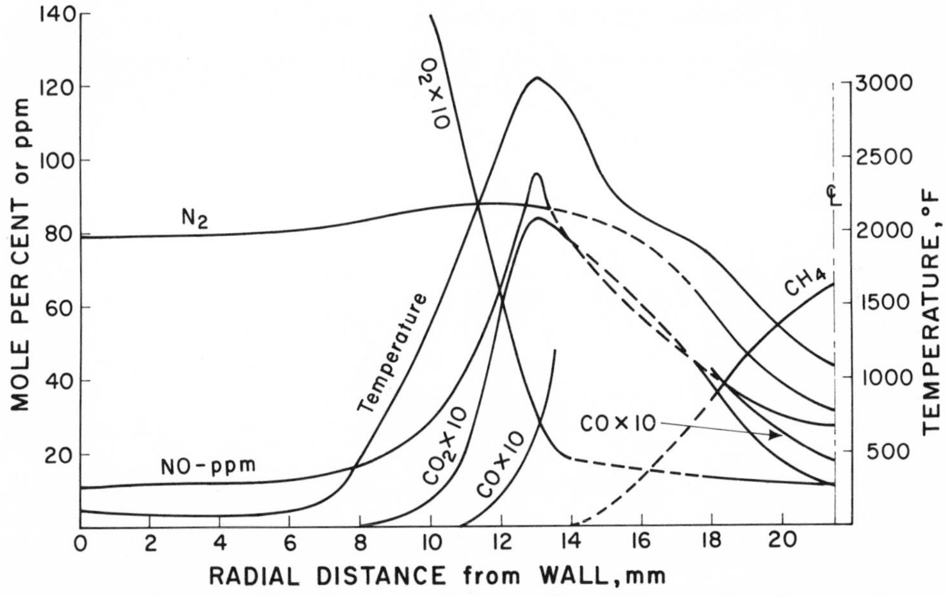

Fig. 4. Temperature and species concentration profiles 47.5 mm. above burner entrance; pressure, one atmosphere; fuel, CH_4.

previously. The concentration profiles are presented on a dry basis. The burner was operated at atmospheric pressure for these tests. These data are replotted in Figs. 6 and 7 to show the variation of NO and temperature profiles with variation in axial position. In the soot formation region, where samples could not be taken, the data have been graphically interpolated. This region is represented by dashed lines.

References p. 118

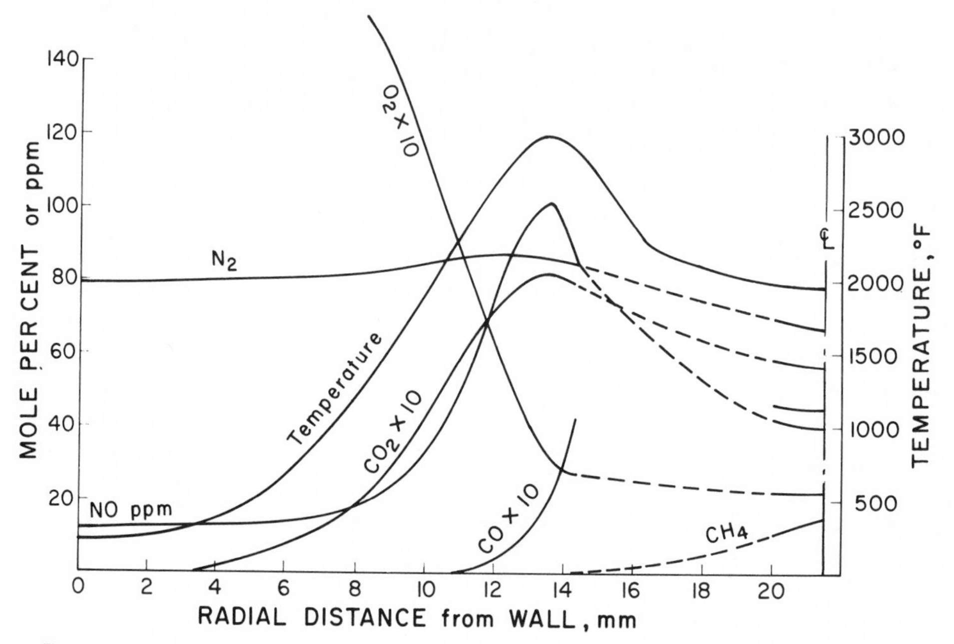

Fig. 5. Temperature and species concentration profiles 117.5 mm. above burner entrance; pressure, one atmosphere; fuel, CH_4.

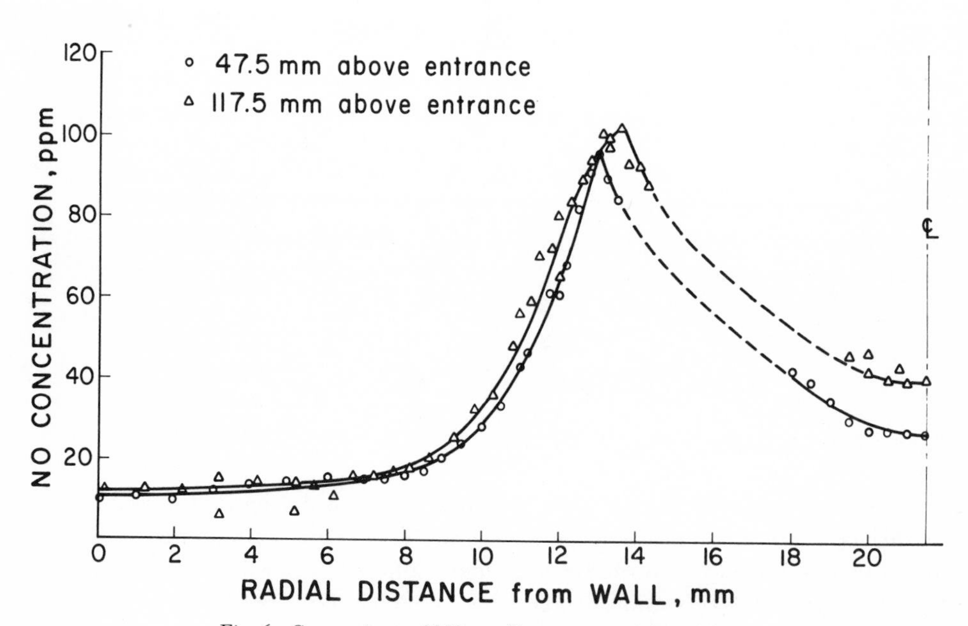

Fig. 6. Comparison of NO profiles at two axial positions.

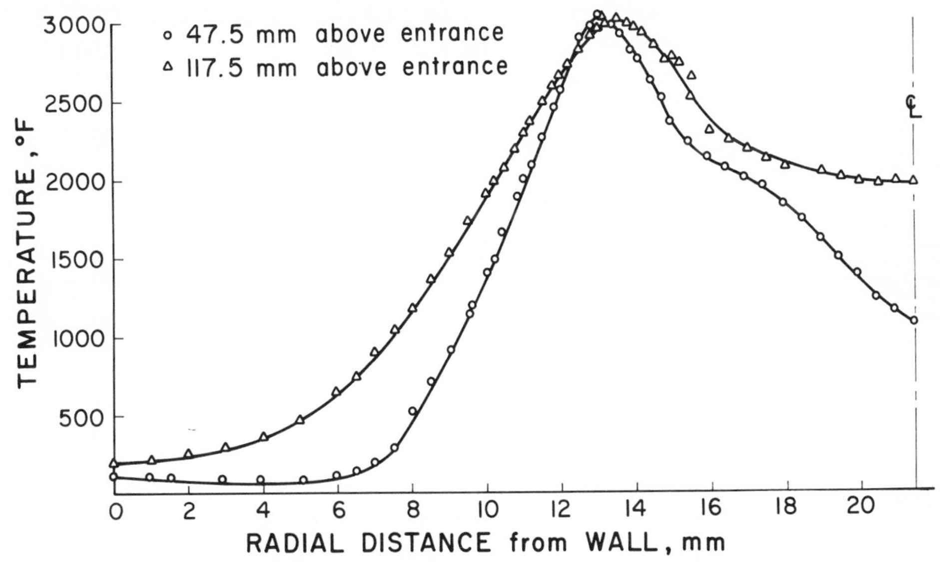

Fig. 7. Comparison of temperature profiles at two axial positions.

Examination of the above results leads to a number of definite observations.

Nitric oxide exhibits a fairly sharp peak in concentration in the intense temperature zone external to the fuel-rich region. This peak coincides with that of temperature and approximately with that of CO_2.

It is well known that the rate of formation of NO becomes insignificant for temperatures below about 2800°F. Examination of the recorded temperature profiles suggests that NO formation should be confined to a very narrow region of maximum temperature and that the NO formed therein diffuses to both the fuel-rich and fuel-lean zones.

The sharpness of the nitric oxide peak further suggests highly localized nitric oxide formation. It is worthy of note that no evidence of so-called "prompt NO formation" in the reactive fuel-rich flame zone is apparent.

Fig. 6 shows that with increasing downstream distances the peak concentration of NO increases. It is also observed that the radial position of this peak shifts inward as one moves downstream. Fig. 7 demonstrates that this observed shift in the NO_x peak coincides with corresponding shift in the temperature peak that results simply from depletion of the fuel core.

The fact that NO profiles on the air side of the combustion region change little with increasing downstream distance further indicates that NO formation is confined to a very narrow region in the vicinity of the peak temperature point.

References p. 118

It is interesting to note that for both axial positions represented by Fig. 6, the points of peak NO formation coincide with nearly identical O_2 concentrations of about 3 mole percent.

CONCLUSIONS

Experimental results indicate that in laminar diffusion flames NO formation is confined to an extremely narrow region exhibiting an appropriate combination of temperature and O_2 availability. The slopes of the concentration profiles suggest that the formation of NO in the fuel-rich region involving fuel derived species is probably not significant.

ACKNOWLEDGEMENTS

This work has been supported by the Office of Research Grants, Air Programs, Environmental Protection Agency, through Grant AP-00582.

REFERENCES

1. *R. M. Campau and J. C. Neerman, "Continuous Mass Spectrometric Determination of Nitric Oxide in Automobile Exhaust", SAE Trans., Vol. 75, 1967, Paper 660116.*
2. *E. R. Lozano, W. W. Melvin Jr. and S. Hochleiser, "Air Pollution Emissions from Jet Engines", J. of APCA, Volume 18, 1968.*
3. *D. B. Wimmer and L. A. McReynolds, "Nitric Oxides and Engine Combustion", SAE Trans., Vol. 70, 1962.*
4. *H. K. Newhall and E. S. Starkman, "Direct Spectroscopic Determination of Nitric Oxide in Reciprocating Engine Cylinders", SAE Trans., Vol. 76, 1968.*
5. *H. K. Newhall and S. M. Shahed, "Kinetics of Nitric Oxide Formation in High Pressure Flames". Thirteenth Symposium (International) on Combustion, The Combustion Institute, Pittsburgh, 1971, p. 381.*
6. *S. M. Shahed and H. K. Newhall, "Kinetics of Nitric Oxide Formation in Propane-Air and Hydrogen-Air-Diluent Flames", Combustion and Flame, in press.*
7. *G. A. Lavoie, J. B. Heywood and J. C. Keck, "Experimental and Theoretical Study of Nitric Oxide Formation in Internal Combustion Engines", Comb. Sci. and Tech., Vol. 1, 1970.*
8. *C. P. Fennimore, M. B. Hilt and R. H. Johnson, "Formation and Measurement of Nitrogen Oxides in Gas Turbines", ASME Paper 70-WA/GT-3, 1970.*
9. *D. S. Smith, R. F. Sawyer and E. S. Starkman, "Oxides of Nitrogen from Gas Turbines", J. of APCA, Vol. 18, 1968.*
10. *R. F. Sawyer and E. S. Starkman, "Gas Turbine Exhaust Emissions", SAE Paper 680462.*
11. *V. S. Yumlu and A. W. Carey Jr., "Exhaust Emission Characteristics of Four-Stroke, Direct Injection Compression Ignition Engines", SAE Paper 580420, 1966.*
12. *W. F. Marshall and R. W. Hurn, "Factors Influencing Diesel Emissions", SAE Paper 680528.*
13. *L. S. Carreto, L. J. Muzio, R. F. Sawyer and E. S. Starkman, "The Role of Kinetics in Engine Emission of Nitric Oxide", Comb. Sci. and Tech., Vol. 3, 1970.*
14. *W. Cornelius and W. R. Wade, "The Formation and Control of Nitric Oxide in a Regenerative Gas Turbine Burner", SAE Paper 700708, 1970.*

DISCUSSION

C. P. Fenimore *(G.E. Research and Development Center)*

In a discussion of this interesting paper with John Moore, three main points occurred to us.

1. The authors find that the maximum concentration of NO coincides with the maximum temperature and conclude that NO forms in the hottest regions of the gas. The conclusion is certainly reasonable. One should note that it is an inference from the distribution of NO after its formation has almost ceased. This is evident because the weighted area under the composition traverse (weighted for the cylindrical geometry) is proportional to the total NO. The curves indicate that at the first level of observation, 48 mm above the fuel port, NO was already around 85 percent of that found at the downstream level 118 mm above the port.

Nitric oxide must have formed at a much faster rate somewhere in the first 48 mm than it did in the next 70 mm. Possibly the peak temperatures were higher in the unobserved first 48 mm since – as appears from the CH_4 traverse at 48 mm – most of the fuel was also consumed in the unobserved region. One hopes the authors will get traverses farther upstream, and will give actual corrected temperatures.

2. It is not clear how the authors know whether or not a significant amount of so-called "prompt" NO is formed. I used the term at the 13th Combustion Symposium to indicate excess NO over that expected under the assumptions that:

a. O atoms are equilibrated with O_2 molecules.

b. The concentration of N atoms is steady.

c. NO is formed at the known rate of the reactions $O + N_2 \rightleftharpoons N + NO$, $N + O_2 \rightleftharpoons O + NO$.

To say if "prompt" NO is significant, one must compare quantitatively the NO actually formed with that expected on assumptions a - c. But the authors do not yet know how rapidly, and under what conditions of temperature and concentrations of O_2 and N_2, most of their NO was formed.

3. It may turn out that "prompt" NO is less significant in steady hydrocarbon diffusion flames than in premixed ones. In premixed flames of constant temperature, I found rather more "prompt" NO in somewhat fuel-rich mixtures than in stoichiometric or somewhat fuel-lean mixtures. Now a premixed flame can be constrained to react at an imposed, fuel-rich mixture strength. But a diffusion flame chooses its own mixture strength, the reactants diffusing to form whatever mixture is the fastest burning. It is often supposed that the reaction zones of overventilated diffusion flames contain essentially stoichiometric mixtures. If so, diffusion flames may not possess reactive regions of the best temperature and mixture strength to form significant "prompt" NO.

A. D. Tuteja

We agree that it will be valuable to have composition and temperature profiles farther upstream than those already available. We will report these in some future communications.

The absence of "prompt NO" is suggested by the concentration profiles of nitric oxide but is not conclusive. The reason for this apparent absence of "prompt NO" may well be the same as mentioned by Dr. Fenimore.

F. V. Bracco *(Princeton University)*

I would like to provide some information in support of that which has just been presented. I made some theoretical calculations of NO formation within the diffusion flame around fuel droplets burning in air. I came up with the same ppm of NO concentration, both in magnitude and in distribution, that was shown by Tuteja and Newhall. For droplets burning in air under ambient conditions of temperature and pressure, I found the maximum NO concentration to be of the order of 100 ppm. But for the same droplets burning in air at higher temperature and pressure, I computed NO concentrations of the order of 1,000 ppm. Thus, diffusion flames around droplets in diesel engines and gas turbines could be important sources of NO. My work on a spherically symmetrical system differs somewhat from this work of Tuteja and Newhall. In spite of these differences, the main element is the diffusion flame itself, and therefore the concentration profiles I calculated could have been expected to resemble those measured by Tuteja and Newhall.

E. B. Zwick *(The Zwick Company)*

I want to know if you measured or analyzed the water that dropped out in your ice bath. In making some measurements of the emissions of oxides of nitrogen from a burner, we used a long line between the sampling point and the point where we collected exhaust samples for Saltzman analyses. We discovered that when we were measuring on the order of say 10 parts per million of total NO_x, that if we collected samples close to the sampling point we measured 10 parts per million, while samples collected at the end of the line measured one or two parts per million. It turned out that the water in the sample had dropped out in the sample line and then the NO_2 had dissolved in the water. Putting this water into Saltzman solution produced a strong NO_2 reading. So with your ice bath technique, I wonder whether you looked to see if you weren't losing some NO_2 before you got around to making the measurement.

A. D. Tuteja

We did not specifically measure or analyze condensed water in question. However in early stages during preliminary runs we did some nitric oxide measurements of the order of 100 ppm at a particular flame position without using an ice bath. Later on during the actual runs, using an ice bath, the values of nitric oxide at the same location were found to be not significantly different from those of preliminary runs. We would like to mention here that our sampling line was not too long.

W. R. Aiman *(Research Laboratories, GMC)*

Would Mr. Tuteja comment on the following:

1. Was the flame diffuse or could a definite flame sheet be observed?

2. Would the authors discuss the effect of the presence of the probe tip on the gradients they attempted to observe? The scale of the flow disturbance caused by the probe tip (OD = 0.5 mm) seems large compared to the scale of the measurements (0.1 mm increments).

3. Where did the authors expect to observe "prompt NO"? Fenimore observed this "prompt NO" in fuel-rich flames (ϕ = 1.3 - 1.4) whereas in this experiment soot obscured the fuel-rich region of the flame.

A. D. Tuteja

In reply to Dr. Aiman's questions:

1. It is not clear as to what is meant by diffuse flame. However we would like to indicate that there was a definite reaction zone and not just a flame surface. This is also indicated by concentration profiles of CO and CO_2.

2. The effects of presence of a probe tip in gradients in flames have been discussed in detail by Fristrom and Westenberg(1). Their studies indicate for small probes with proper taper, disturbances are very small and can be accounted for.

We would like to add that the minimum difference in sampling was not 0.1 mm but 0.2 mm.

3. Dr. Fenimore observed "prompt NO" in premixed flames and implied that this concept could be used for explaining NO formation in gas turbine combustion, hence the reason to look for "prompt NO" in diffusion flames.

Though we do not have measurements in the fuel rich region, the graphically interpolated concentration profile suggests only diffusion and no generation of nitric oxide in any manner.

---

1. R. A. Fristrom and A. A. Westenberg, Flame Structure, McGraw Hill, 1965.

INVESTIGATION OF NO FORMATION KINETICS IN COMBUSTION PROCESSES: THE METHANE-OXYGEN-NITROGEN REACTION

C. T. BOWMAN and D. J. SEERY

United Aircraft Research Laboratories, East Hartford, Connecticut

ABSTRACT

An experimental and analytical investigation of the kinetics of formation of NO in shock-induced combustion of methane-oxygen-nitrogen mixtures diluted by argon has been carried out. Concentration histories of NO, OH and CO_2 were measured during reaction behind reflected shock waves using spectroscopic techniques. Experimental concentration profiles were obtained for an oxidizer-rich and a fuel-rich mixture for initial post-shock temperatures in the range 2600-3200°K and for an initial post-shock pressure of 3.5 ± 0.5 atm. Time rates of change of species concentrations and thermodynamic properties during reaction were calculated by numerically integrating the coupled reaction kinetic, state and energy equations. Calculated concentration profiles were compared with experimental profiles to obtain information on the reaction mechanism for formation of NO. Observed NO formation rates in both the oxidizer-rich and fuel-rich mixture were consistent with a three-reaction mechanism for nitrogen chemistry,

$$O + N_2 \rightleftharpoons NO + N \qquad (1)$$
$$N + O_2 \rightleftharpoons NO + O \qquad (2)$$
$$N + OH \rightleftharpoons NO + H \qquad (3)$$

Reactions 1 and 2 were found to be the principal NO formation reactions, with reaction 3 being of minor importance for the fuel-rich mixture.

INTRODUCTION

An essential feature of analytical models for predicting nitric oxide emissions from combustion devices is a reaction mechanism for NO formation. Early investigations of NO formation kinetics by Zeldovich and his co-workers (1) indicated that NO formation, during combustion of fuel-oxygen-nitrogen mixtures, is governed by two reactions,

$$O + N_2 \rightleftharpoons NO + N \quad (1)$$
$$N + O_2 \rightleftharpoons NO + O \quad (2)$$

and that the combustion reactions are equilibrated prior to the onset of NO formation. Recent experimental studies of NO formation in the post-reaction zone of flames (2-5) support this mechanism. However, Fenimore (2) has noted that, in the reaction zone of hydrocarbon-air flames, NO formation rates exceed those predicted by the Zeldovich mechanism, and he attributed the faster rates to reactions other than 1 and 2. In a recent shock tube study of NO formation during combustion of hydrogen-oxygen-nitrogen mixtures (6), observed NO formation rates were found to exceed those predicted by the Zeldovich mechanism. The rapid NO formation rates in the reaction zone were shown to be consistent with reactions 1 and 2 coupled with finite-rate combustion chemistry.

The objective of the present investigation was to determine the mechanism for NO formation during combustion of a hydrocarbon fuel. In the experimental phase of the investigation, combustion of methane-oxygen-nitrogen mixtures diluted by argon was initiated by reflected shock waves in a shock tube. During reaction, concentration time-histories of NO, OH and CO_2 were measured using spectroscopic techniques. Experimental concentration profiles were compared with those obtained from calculations using an assumed kinetic model to obtain information on the NO formation mechanism.

EXPERIMENTAL AND ANALYTICAL DETAILS

The shock tube used in the experimental study has been described in detail (7). In the present investigation, spectroscopic measurements were made behind reflected shock waves at a position 1.6 cm from the end wall. The shock tube was evacuated to less than 3×10^{-4} torr prior to filling with the combustible gas mixture. The leak and outgassing rate was less than 1×10^{-3} torr/min.

Conditions behind the reflected shock wave were calculated using measured incident and reflected shock velocities and the ideal, one-dimensional shock equations. In these calculations, it was assumed that the gas behind both incident and reflected shock waves was vibrationally equilibrated and unreacted. Uncertainties in the measured shock velocities resulted in ±50°K uncertainties in reflected shock temperature.

During reaction, the concentrations of NO, OH and CO_2 were measured using spectroscopic techniques. The concentration of NO in the v=1 vibrational state was determined by monitoring absorption of radiation by the γ (0,1) band of the molecule at 2358 Å. The OH concentration was obtained by measuring absorption of radiation by the Σ - Π (0,0) band of the molecule at 3080 Å. The CO_2 concentration was measured by monitoring infrared emission from the ν_3-fundamental band of the molecule at 4.3 μ. The optical system used to make these concentration measurements and the calibration procedures have been described (6,8).

The methane used in the experimental study was Matheson Research Grade. The oxygen and argon were Matheson Ultrapure Grade, and the nitrogen was Matheson Prepurified Grade. All gases were used without further purification. Gas mixtures were prepared manometrically and stored in glass vessels for at least 48 hrs. prior to use. Mixture composition was verified by mass spectrometric analysis.

To obtain the desired information on the NO formation mechanism, species concentration profiles measured during combustion of $CH_4/O_2/N_2$ mixtures diluted by Ar were compared with those obtained from calculations using an assumed kinetic model. In this analytical study, the time rates of change of species concentrations and thermodynamic properties during reaction were calculated by numerically integrating the coupled reaction kinetic, state and energy equations. These calculations were carried out using a computer code which models reacting systems behind reflected shock waves (constant volume, adiabatic).

RESULTS AND DISCUSSION

As a preliminary step in the investigation, the methane-oxygen reaction mechanism, which was to be used in the analytical study of NO formation, was checked by observing the reaction of $CH_4/O_2/Ar$ mixtures behind reflected shock waves. The OH and CO_2 concentration profiles measured during reaction were compared with those calculated using the mechanism shown in Table 1. This mechanism has been proposed in several shock tube studies of the methane-oxygen reaction (8-10). The rate constants, k_f, in Table 1 are given for the forward reaction. Rate constants for the reverse reaction were calculated using k_f and the equilibrium constant obtained from the JANAF Thermochemical Tables (11).

Experiments were conducted using an oxidizer-rich (0.5% CH_4 - 2.0% O_2 - 97.5% Ar) and a fuel-rich mixture (1.2% CH_4 - 2.0% O_2 - 96.8% Ar) for initial post-shock conditions in the range 2100-2500°K and 2-3 atm. These mixtures have the same mole percent of fuel and oxygen as the mixtures used in the NO formation experiments. A typical comparison between measured and calculated OH and CO_2 concentration profiles is shown in Fig. 1. The vertical bars indicate the uncertainty in measured concentration due to uncertainties in the emission or absorption measurement and the optical system calibration. The solid lines are calculated concentration profiles. Although there are slight differences between the observed and calculated concentration profiles, the proposed mechanism appears to adequately model the methane-oxygen chemistry for the conditions of the present investigation.

References pp. 134-135

TABLE 1

Methane-Oxygen Reaction Mechanism

Reaction	Rate Constant, k_f*	Reference
$CH_4 + M \rightleftarrows CH_3 + H + M$	$2.0 \times 10^{17} \exp(-44500/T)$	8
$CH_4 + OH \rightleftarrows CH_3 + H_2O$	$2.8 \times 10^{13} \exp(-2500/T)$	12
$CH_4 + O \rightleftarrows CH_3 + OH$	$2.0 \times 10^{13} \exp(-4640/T)$	13
$CH_4 + H \rightleftarrows CH_3 + H_2$	$6.9 \times 10^{13} \exp(-5950/T)$	14
$CH_3 + O_2 \rightleftarrows HCO + H_2O$	2×10^{10}	8
$CH_3 + O \rightleftarrows HCO + H_2$	1×10^{14}	8
$HCO + OH \rightleftarrows CO + H_2O$	1×10^{14}	8
$HCO + M \rightleftarrows H + CO + M$	$2.0 \times 10^{12}\, T^{1/2} \exp(-14400/T)$	8
$CO + OH \rightleftarrows CO_2 + H$	$5.6 \times 10^{11} \exp(-545/T)$	15
$H + O_2 \rightleftarrows O + OH$	$2.2 \times 10^{14} \exp(-8340/T)$	15
$O + H_2 \rightleftarrows H + OH$	$1.7 \times 10^{13} \exp(-4750/T)$	15
$O + H_2O \rightleftarrows 2OH$	$5.8 \times 10^{13} \exp(-9070/T)$	15
$H + H_2O \rightleftarrows H_2 + OH$	$8.4 \times 10^{13} \exp(-10100/T)$	15
$H + OH + M \rightleftarrows H_2O + M$	$1.0 \times 10^{19}\, T^{-1.0}$	16

** Units – cm, cal, °K, mole, sec; M is any collision partner.*

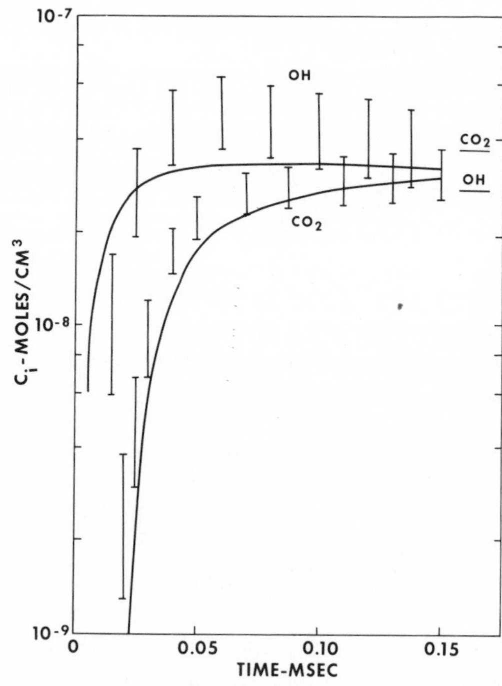

Fig. 1. Typical experimental and calculated OH and CO_2 concentration profiles during combustion of methane-oxygen-argon mixtures: 0.5% CH_4-2.0% O_2-97.5% Ar. Initial temperature = 2480°K; initial pressure = 2.31 atm. Calculated equilibrium OH and CO_2 concentrations are indicated on the right side of the figure.

The NO formation experiments were carried out behind reflected shock waves using an oxidizer-rich (0.5% CH_4 - 2.0% O_2 - 17.5% N_2 - 80% Ar) and a fuel-rich (1.2% CH_4 - 2.0% O_2 - 16.8% N_2 - 80% Ar) mixture for initial post-shock temperatures in the range 2600-3200°K and an initial post-shock pressure of 3.5 ± 0.5 atm. NO concentration profiles could not be obtained for temperatures lower than 2600°K due to the limited sensitivity of the NO concentration measuring technique ($\approx$ 1 x 10^{-9} mole/cm^3). The large dilution by argon and nitrogen reduces temperature and pressure changes during reaction and simplifies interpretation of the experimental data. Since only one optical path was available for absorption measurements, it was necessary to run several experiments to obtain the desired concentration data. The variation in initial post-shock conditions in these experiments was held to approximately one percent.

Typical experimental traces for combustion of the oxidizer-rich mixture are shown in Fig. 2. The two absorption traces were obtained by duplicating runs as discussed

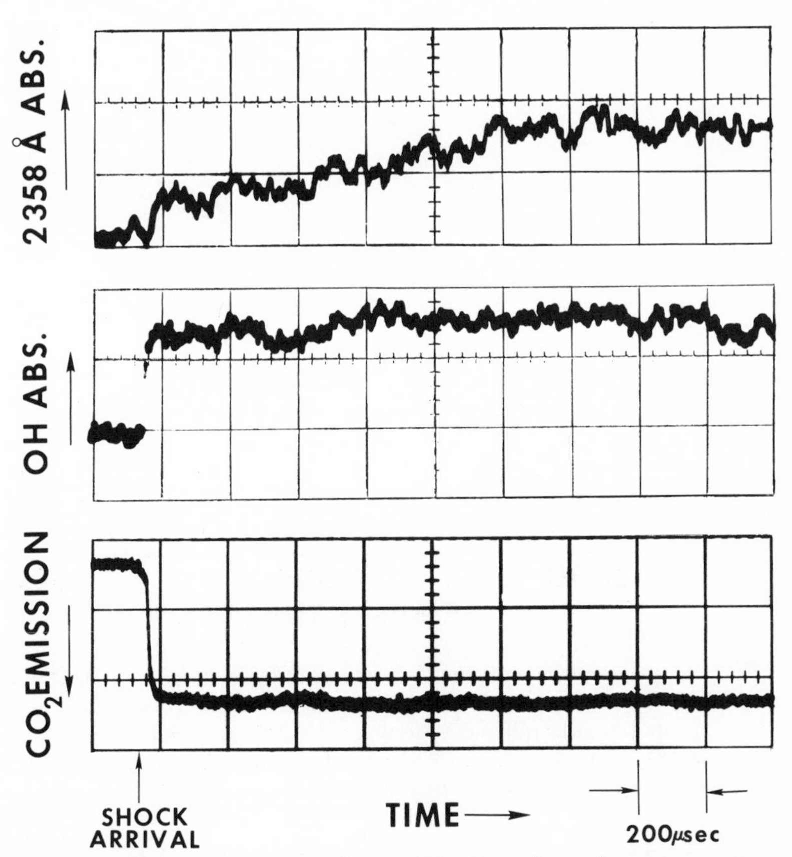

Fig. 2. Typical experimental traces for combustion of methane-oxygen-nitrogen-argon mixtures: 0.5%CH_4-2.0%O_2-17.5%N_2-80%Ar. Initial temperature = 2950±35°K; initial pressure = 3.24±0.04 atm.

previously. For reasons of simplicity the zero absorption traces are not shown. The 2358 Å absorption trace shows absorption immediately behind the shock front. This initial absorption is due to the Schumann-Runge band of oxygen and the ultraviolet continuum of CO_2. Fig. 3 shows the effective absorption cross-sections (σ_{eff}) of NO,

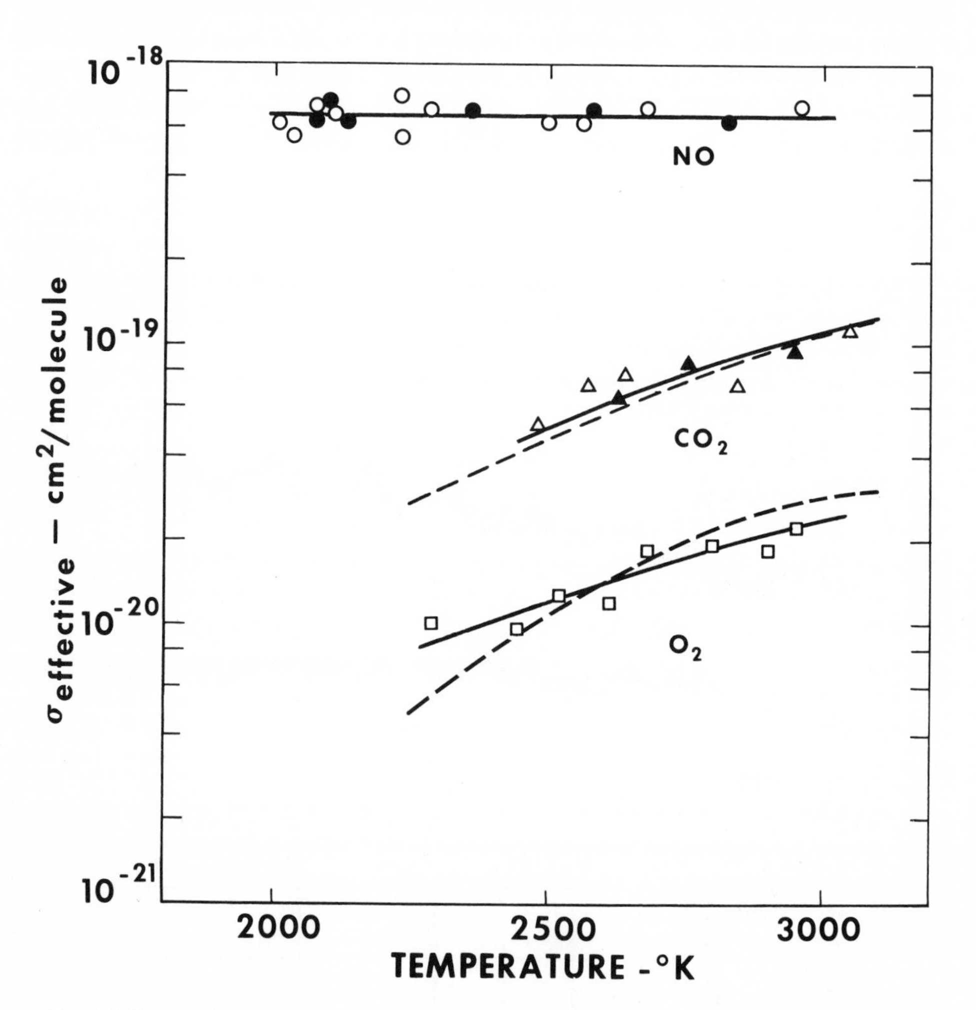

Fig. 3. Effective absorption cross-section at λ = 2358 Å ($\Delta\lambda$ = 12 Å) for NO, CO_2 and O_2 as functions of temperature. Open symbols - P = 2.5 atm; solid symbols – P = 4.0 atm. The dashed line through the CO_2 data is the absorption cross-section at λ = 2380 Å taken from Ref. 17. The dashed line through the O_2 data is a Sulzer-Wieland calculation (18) for λ = 2358 Å using the room-temperature absorption cross-section from Ref. 19.

CO_2 and O_2 at 2358 Å as functions of temperature. These data were obtained by shock-heating known mixtures of NO, CO_2 or O_2 diluted by Ar and measuring absorption of incident radiation by the vibrationally-equilibrated gas. For the conditions of the present study, the σ_{eff} of CO_2 is approximately four times larger than the σ_{eff} of O_2 and is seven to ten times smaller than the σ_{eff} of NO. These data indicate that observed 2358 Å absorption traces must be corrected for absorption by CO_2 and O_2 if an accurate measure of absorption by NO is to be obtained. This correction was made by shock-heating gas mixtures identical to those used in the NO formation experiments, with the N_2 replaced by Ar*, and observing absorption at 2358 Å. Initial post-shock conditions in the correction experiments were matched to those used in the NO formation experiments to within one percent. Absorption traces at 2358 Å, obtained from the correction experiments, were subtracted from the 2358 Å absorption traces obtained from the NO formation experiments. The 2358 Å absorption traces, corrected in this manner, were used to determine the NO concentration in the v=1 vibrational state. In the NO formation experiments, the combustion reactions equilibrated prior to the end of the observation time (see Fig. 2). Following equilibration of the combustion reactions, observed 2358 Å absorption traces also could be corrected for absorption by CO_2 and O_2 using the absorption cross-section data from Fig. 3 and calculated equilibrium properties. NO(v=1) concentration profiles obtained using both techniques were the same within the experimental uncertainty.

Typical experimental results for the oxidizer-rich $CH_4/O_2/N_2/Ar$ mixture are shown in Fig. 4, where the concentrations of CO_2, OH and NO (v=1) are plotted as functions of time for the conditions of Fig. 2. The OH and NO (v=1) concentration data are an average of the concentrations measured in three runs at each wavelength. Concentration data for CO_2 were obtained in all six runs, and the plotted CO_2 data are an average of the concentrations measured in the six runs. As before, the vertical bars indicate the uncertainty in the concentration measurements. The lines are concentration profiles obtained from kinetics calculations, and are discussed in detail later in the paper.

Typical experimental results for the fuel-rich $CH_4/O_2/N_2/Ar$ mixture are shown in Fig. 5, where the concentrations of CO_2, OH and NO (v=1) are plotted as functions of time. As before, the OH and NO (v=1) concentration data are average values obtained in three runs at each wavelength, and the CO_2 data are average values obtained in six runs.

Several conclusions can be drawn from the experimental data, as typified by Figs. 4 and 5: (a) For the conditions of the present study, the initial rate of formation of NO exceeds the rate later in the reaction. (b) At the higher temperatures investigated, the combustion reactions rapidly equilibrate, with radical concentrations departing only slightly from equilibrium values. However, at lower temperatures, radical concentrations exceed the equilibrium values during the combustion reaction, with the

---

** Gas mixtures used in these correction experiments were the same as were used to check the CH_4/O_2 reaction mechanism.*

References pp. 134-135

departure from equilibrium increasing as the temperature decreases. (c) For a given temperature and pressure, radical concentrations in the fuel-rich mixture exceed equilibrium values by a greater amount than in the oxidizer-rich mixture.

To obtain information on the NO formation mechanism, the experimental concentration profiles were compared with profiles obtained from kinetics calculations. The reaction mechanism used in these calculations consisted of methane-oxygen mechanism (Table 1) and various reactions for the oxidation of nitrogen.

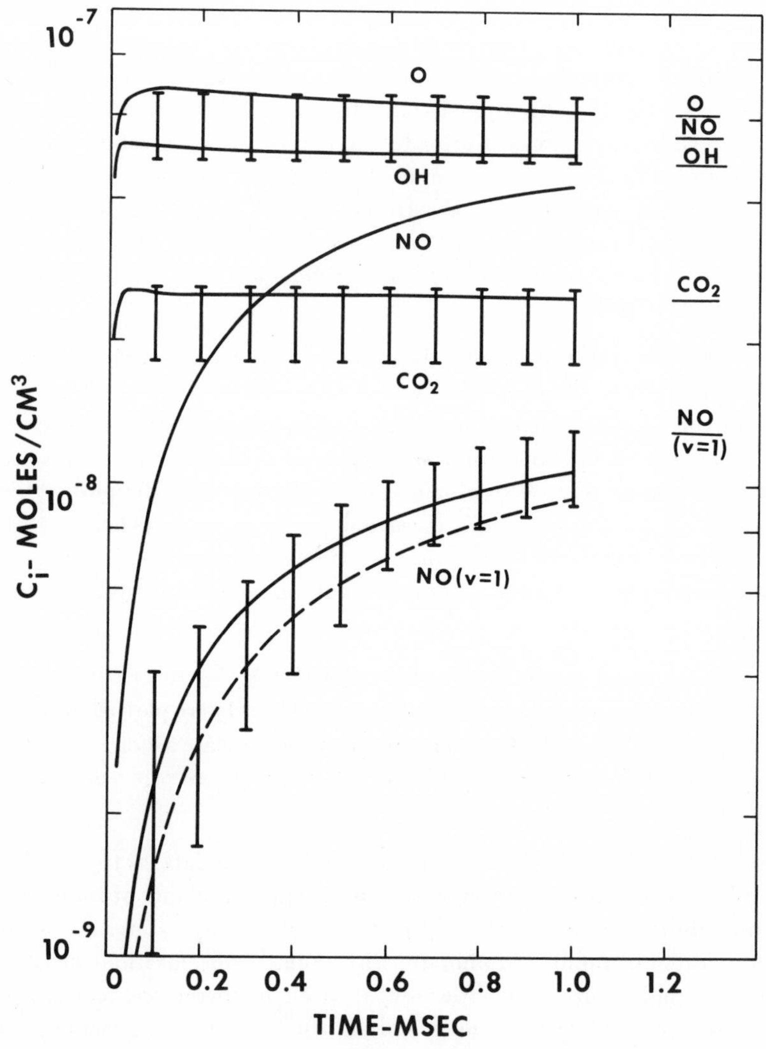

Fig. 4. Typical experimental and calculated concentration profiles during combustion of 0.5%CH_4-2.0%O_2-17.5%N_2-80%Ar mixture. Initial temperature = 2950±35°K; initial pressure = 3.24±0.04 atm. Calculated equilibrium concentrations are indicated on the right side of the figure.

Three nitrogen oxidation reactions were considered in the present study (Table 2). Previous studies (1-6) have shown that reactions 1 and 2 are the principal NO formation reactions in combustion of hydrogen, carbon monoxide and hydrocarbons. Heywood (22) has suggested that reaction 3 also may be of importance in NO formation in the post-reaction zone. Previous analytical studies of NO formation in combustion reactions (6,23,24) have shown that for temperatures in the range

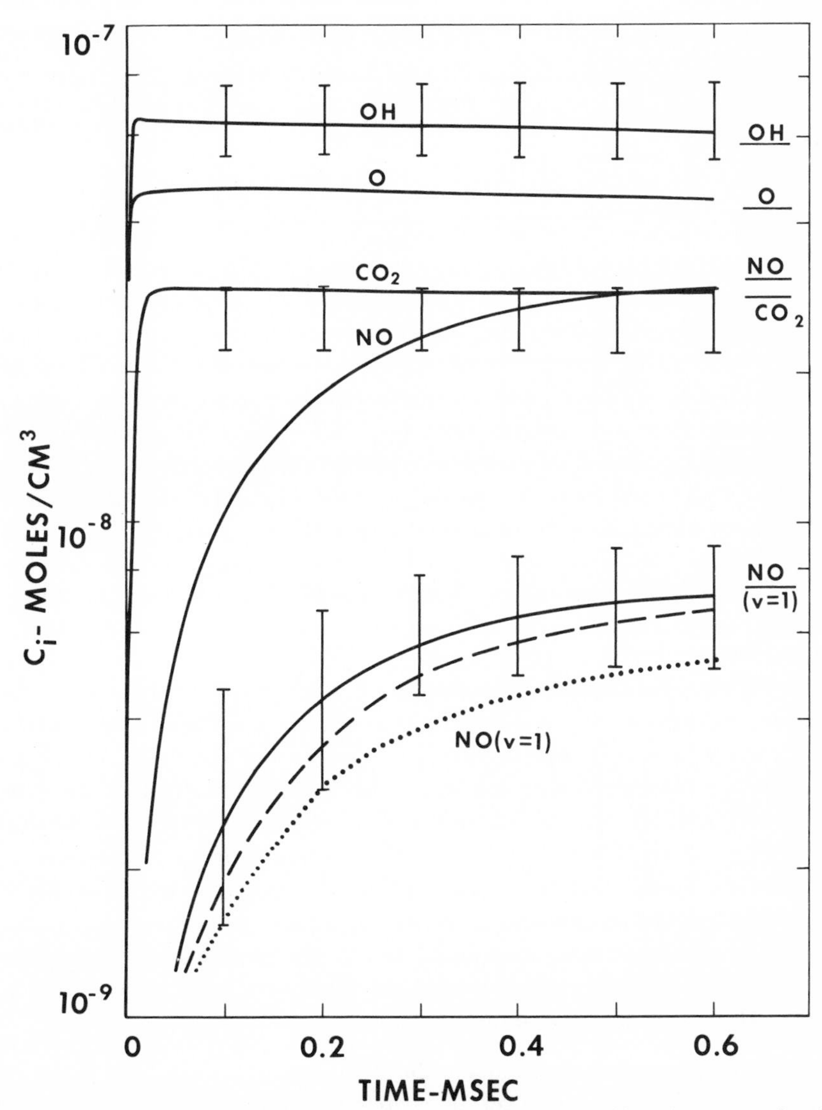

Fig. 5. Typical experimental and calculated concentration profiles during combustion of 1.2%CH_4-2.0%O_2-16.8%N_2-80%Ar mixture. Initial temperature = 2970±35°K; initial pressure = 3.11±0.05 atm. Calculated equilibrium concentrations are indicated on the right side of the figure.

References pp. 134-135

2000-3000°K additional reactions involving NO, N_2O or NO_2 are unimportant in the NO formation mechanism.

TABLE 2

Nitrogen-Oxygen Reaction Mechanism

Reaction	Rate Constant, k_f*	Reference
1. $O + N_2 \rightleftarrows NO + N$	$1.4 \times 10^{14} \exp(-37900/T)$	20
2. $N + O_2 \rightleftarrows NO + O$	$6.4 \times 10^{9}\ T \exp(-3140/T)$	20
3. $N + OH \rightleftarrows NO + H$	4.0×10^{13}	21

** Units – cm, cal, °K, mole, sec.*

Initial calculations were carried out using a reaction mechanism consisting of the methane-oxygen reactions (Table 1) and reactions 1 – 3 from Table 2. Temperature, pressure and concentration profiles were calculated for the mean initial post-shock conditions of each experiment. The calculated concentration profiles were compared with the experimental profiles to determine how well the proposed mechanism modeled the reaction. Typical results from this comparison are shown in Figs. 4 and 5 for the oxidizer-rich and fuel-rich mixtures, respectively. The solid lines in these figures are the calculated concentration profiles. The concentration profiles for NO (v=1) were obtained from the calculated profiles for total NO concentration and temperature, assuming local thermal equilibrium. This assumption was verified in a previous study of NO formation during combustion of $H_2/O_2/N_2$ mixtures (6). The calculated and measured profiles for both oxidizer-rich and fuel-rich mixtures agree within the precision of the experiment and the uncertainty of the rate constants.

To determine the relative importance of reaction 3 in the NO formation mechanism, calculations were carried out using the methane-oxygen mechanism (Table 1) and reactions 1 and 2 from Table 2. For the conditions of the present study, including reaction 3 in the mechanism had a negligible effect on calculated NO concentration profiles for the oxidizer-rich mixture. However, for the fuel-rich mixture, NO concentrations calculated with reaction 3 in the mechanism exceed those calculated using reactions 1 and 2 alone. The extent of the difference in NO concentration profiles is illustrated in Fig. 5, where the dotted line is the NO concentration calculated using reactions 1 and 2. The greater relative importance of reaction 3 in the NO formation mechanism for fuel-rich mixtures is due to the increased OH/O_2 concentration ratio.

Several simplified analytical models for NO formation have been based on the assumption that the combustion reactions are equilibrated prior to the onset of NO formation. NO concentration profiles calculated using one of these "equilibrium" models for NO formation (25) have been compared with results from the present study. In Ref. 25, the rate equation for NO was formulated using the three reactions in Table 2, assuming equilibrium temperature and O and OH concentrations and using a steady-state approximation for N concentration. NO concentration profiles calculated using the "equilibrium" rate equation are plotted as dashed lines in Figs. 4 and 5. The "equilibrium" NO concentration profiles are lower than the measured NO profiles. At elevated temperatures, the discrepancy between "equilibrium" and measured NO profiles is small. As the temperature decreases, the discrepancy between "equilibrium" and measured NO profiles increases. These observations may be explained by the fact that at elevated temperatures the combustion reactions rapidly equilibrate, and radical concentration overshoots are relatively small. Hence, "equilibrium" approximations should be reasonably accurate. As the temperature decreases, radical concentration overshoots increase, and "equilibrium" approximations become less accurate. For a given temperature and pressure, radical concentration overshoots in the fuel-rich mixture exceed those in the oxidizer-rich mixture. Hence, "equilibrium" approximations should be less accurate for the fuel-rich mixtures. In the present study, temperature variations were relatively small ($<$ 100°K). Hence, the effects of temperature non-equilibrium on NO formation could not be assessed.

CONCLUSIONS

1. For the conditions of the present investigation, observed NO formation rates are consistent with a three-reaction mechanism for NO chemistry

$$O + N_2 \rightleftharpoons NO + N \tag{1}$$

$$N + O_2 \rightleftharpoons NO + O \tag{2}$$

$$N + OH \rightleftharpoons NO + H \tag{3}$$

Reactions 1 and 2 are the principal NO formation reactions, with reaction 3 being of minor importance for fuel-rich mixtures.

2. Discrepancies between observed NO formation rates and those calculated using an "equilibrium" model for NO formation are explainable in terms of radical concentration overshoots in the reaction zone. There is no evidence to suggest that reactions other than 1 – 3 are of importance in the NO formation mechanism.

ACKNOWLEDGEMENTS

The authors thank Mr. R. L. Poitras for his assistance in the experimental investigation, Mr. L. D. Aceto for his efforts in developing the computer program used for the kinetics calculations and Mr. R. L. Leighton for designing the absorption optical system.

REFERENCES

1. *Ya. B. Zeldovich, P. Ya. Sadovnikov and D. A. Frank-Kamenetskii, "Oxidation of Nitrogen in Combustion." Academy of Sciences of USSR, Institute of Chemical Physics, Moscow-Leningrad (trans. by M. Shelef), 1947.*
2. *C. P. Fenimore, "Formation of Nitric Oxide in Premixed Hydrocarbon Flames." Thirteenth Symposium (International) on Combustion: The Combustion Institute, Pittsburgh, 1971, pp. 373-379.*
3. *H. K. Newhall and S. M. Shahed, "Kinetics of Nitric Oxide Formation in High-Pressure Flames." Thirteenth Symposium (International) on Combustion: The Combustion Institute, Pittsburgh, 1971, pp. 381-388.*
4. *S. M. Shahed and H. K. Newhall, "Kinetics of Nitric Oxide Formation in Propane-Air and Hydrogen-Air-Diluent Flames." Combustion and Flame, Vol 17 (in press).*
5. *M. E. Harris, V. R. Rowe, E. B. Cook and J. Grumer, "Reduction of Air Pollutants from Gas Burner Flames." Bureau of Mines Bulletin 653, 1970.*
6. *C. T. Bowman, "Investigation of Nitric Oxide Formation Kinetics in Combustion Processes: The Hydrogen-Oxygen-Nitrogen Reaction." Combustion Science and Technology, Vol. 3, 1971, pp. 37-45.*
7. *D. J. Seery and C. T. Bowman, "Dissociation of HCl Behind Shock Waves." J. Chem. Phys., Vol. 48, 1968, pp. 4314-4317.*
8. *C. T. Bowman, "An Experimental and Analytical Investigation of the High-Temperature Oxidation Mechanisms of Hydrocarbon Fuels." Combustion Science and Technology, Vol. 2, 1970, pp. 161-172.*
9. *D. J. Seery and C. T. Bowman, "An Experimental and Analytical Study of Methane Oxidation Behind Shock Waves." Combustion and Flame, Vol. 14, 1970, pp. 37-48.*
10. *R. M. R. Higgin and A. Williams, "A Shock-Tube Investigation of the Combustion of Lean Methane and n-Butane Mixtures with Oxygen." Twelfth Symposium (International) on Combustion: The Combustion Institute, Pittsburgh, 1969, pp. 579-587.*
11. *D. K. Stull (ed.), "JANAF Thermochemical Tables." Midland, Michigan: Dow Chemical Company, 1965.*
12. *W. E. Wilson, "A Critical Review of the Combustion Reactions of the Hydroxyl Radical." Paper presented at the Spring Meeting, Western Section, The Combustion Institute, LaJolla, California, April 1967.*
13. *A. A. Westenberg and N. deHaas, "Reinvestigation of the Rate Coefficients for $O+H_2$ and $O+CH_4$." J. Chem. Phys. Vol. 50, 1969, pp. 2512-2516.*
14. *M. J. Kurylo and R. B. Timmons, "ESR Study of the Kinetics of the Reaction of H Atoms with Methane." J. Chem. Phys. Vol. 50, 1969, pp. 5076-5087.*
15. *D. L. Baulch, D. D. Drysdale and A. C. Lloyd, "Critical Evaluation of Rate Data for Homogeneous Gas-Phase Reactions of Interest In High-Temperature Systems." Dept. of Phys. Chem. Report Nos. 1-3, Leeds, England, 1968-69.*
16. *G. L. Schott and P. F. Bird, "Kinetics Studies of Hydroxyl Radicals in Shock Waves. IV. Recombination Rates in Rich Hydrogen-Oxygen Mixtures." J. Chem. Phys., Vol. 41, 1964, pp. 2869-2876.*
17. *N. A. Generalov, S. A. Losev and V. A. Maksimenko, "Absorption of Ultraviolet Radiation by Strongly Heated Carbon Dioxide Gas." Optics and Spectroscopy, Vol. 15, 1963, pp. 12-14.*
18. *P. Sulzer and K. Wieland, "Intensity Distribution of a Continuous Absorption Spectrum as a Function of Temperature and Wavenumber." Helv. Phys. Acta, Vol. 25, 1952, pp. 653-676.*
19. *A. J. Blake, J. H. Carver and G. N. Haddad, "Photo-absorption Cross-sections of Molecular Oxygen between 1250 Å and 2350 Å." J.Q.S.R.T., Vol. 6, 1966, pp. 451-459.*

20. *D. L. Baulch, D. D. Drysdale, D. G. Horne and A. C. Lloyd, "Critical Evaluation of Rate Data for Honogeneous Gas-Phase Reactions of Interest in High-Temperature Systems." Dept. of Phys. Chem. Report No. 4, Leeds, England, 1969.*
21. *I. M. Campbell and B. A. Thrush, "Reactivity of Hydrogen to Atomic Nitrogen and Atomic Oxygen." Trans. Faraday Soc., Vol. 64, 1968, pp. 1265-1274.*
22. *J. B. Heywood, "Gas Turbine Combustor Modeling for Calculating Nitric Oxide Emissions." Paper 71-712 presented at AIAA/SAE 7th Propulsion Joint Specialist Conference, Salt Lake City, Utah, June 1971.*
23. *R. S. Fletcher and J. B. Heywood, "A Model for Nitric Oxide Emissions from Aircraft Gas Turbine Engines." Paper No. 71-123 presented at AIAA 9th Aerospace Sciences Meeting, New York, January 1971.*
24. *C. T. Bowman, unpublished results.*
25. *A. A. Westenberg, "Kinetics of NO and CO in Lean, Premixed Hydrocarbon-Air Flames". Combustion Science and Technology, Vol. 4, 1971, pp. 59-64.*

DISCUSSION

F. Kaufman *(University of Pittsburgh)*

Before I briefly discuss the results of this excellent shock tube paper, I should like to make a few general remarks on the present state of chemical kinetics and on its relevance to this fine Symposium. I can not help but feel just a bit isolated as one of very few physical chemists/kineticists in attendance here, and probably as the only attendee whose major professional concern is the study of elementary reaction kinetics. Why, then this isolation,when it must be clear to everyone that the topic of this Symposium is directly concerned with our understanding of which reactions produce and which remove undesirable emissions and how to make the source steps slower and/or the sink steps faster? There are various answers to this question: (1) For many of the important reactions involving H, O, N, and their simplest molecules, we have been so successful that the kinetics are now well known (e.g. for the 'Zeldovich' reactions $O + N_2 \rightleftarrows NO + N$ and $N + O_2 \rightleftarrows NO + O$, for $CO + OH \rightleftarrows CO_2 + H$, and for many others) (2) For much more complex systems such as those involving larger hydrocarbons, the detailed reaction scheme is so complicated as to be considered kinetically out of reach; (3) there is, unfortunately, a growing reluctance in industry and government to support long-range fundamental research.

Regarding the recent successes in elementary reaction kinetics it is most important that users of such information be made aware of literature summaries and critical evaluations, and that they be discouraged from doing their own reviewing unless nothing else is available. A partial listing of recent reviews includes: High Temperature Reaction Rate Data by D. L. Baulch, D. D. Drysdale and D. G. Horne, Department of Physical Chemistry, The University, Leeds 2, England (Nos. 1 through 5); various monographs of the NSRDS-NBS Series such as No. 9, Tables of Bimolecular Gas Reactions by A. F. Trotman-Dickenson and G. S. Milne and its supplement by E.

Ratajczak and A. F. Trotman-Dickenson, University of Wales Institute of Science and Technology, Cardiff, Wales; Gas Phase Reaction Kinetics of Neutral Oxygen Species, NSRDS-NBS 20 by H. S. Johnston; Rate Constants of Gas Reactions by V. N. Kondratiev, Academy of Sciences USSR, Moscow, 1970 (to be translated by the Applied Physics Lab, Johns Hopkins University); specific review papers in such series as Progress in Reaction Kinetics, Annual Review of Physical Chemistry, Advances in Chemical Physics, and others; and increasing use of computer searching of the recent literature with and without subsequent critical analysis. The latter is done extensively by the Chemical Kinetics Information Center of the National Bureau of Standards and results in highly useful compilations such as NBS Report 10,241, A Bibliography of Kinetic Data on Gas Phase Reactions of Nitrogen, Oxygen and the Nitrogen Oxides by F. Westley.

Regarding (2) above, it should not be necessary to establish the complex sequence of elementary steps in all its detail, because its rate will be controlled by a few initiation or branching reactions, several of which are already well characterized, such as O + RH and OH + RH. Important secondary reactions to be studied include $R \cdot + O_2$, O, or OH, some methylene and CH reactions such as Fenimore's 'prompt NO' reaction $CH + N_2 \rightarrow HCN + N$, and others. There have been such spectacular advances in the field of atom and radical reaction kinetics, especially in low pressure flow systems using mass spectrometry, electron spin resonance, optical absorption or fluorescence, catalytic probes, chemiluminescence, and other detection methods, that virtually any elementary reaction is open to experimental study. It is particularly ironic that the above-mentioned reluctance on the part of industry and government to support such research has come at a time when our arsenals of research tools are bulging, so that we could do the job far more quickly and efficiently than five or ten years ago when support was more plentiful.

I should also mention that historically it was combustion research which provided the main thrust for the study of elementary reaction kinetics in its early days. The Combustion Symposium volumes are gold mines of kinetic data obtained from studies of flames, explosion limits, and shock waves. Only in the last few years has the emphasis on and support of such work on elementary processes shifted from the field of combustion to that of planetary atmospheres and space. I certainly hope that important engineering, industrial, and societal problems such as the topic of this Symposium will be attacked in a way which recognizes their chemical content and which results in a real understanding of their principal processes.

Now, to the discussion of the present paper. This is a first-rate piece of experimental work and provides a fine example of a well-planned and executed shock tube study using the reflected shock technique. Three species, NO, OH, and CO_2 are measured spectroscopically, the first two in the ultraviolet and the last in the infrared, and their time histories are obtained. The work whose main aim is the study of NO emission in methane-air combustion is first checked for internal consistence by measuring OH and CO_2 in CH_4 - O_2 - Ar mixtures and finding good agreement with a numerically integrated 14 step mechanism in which most rate constants are reasonably well known. Then, NO is measured in lean and rich CH_4 - O_2 - N_2 - Ar mixtures and three reactions are added to the mechanism, viz. the 'Zeldovich' reactions and the very fast step N + OH $\rightarrow$ NO + H. Good agreement is again obtained with the postulated reaction scheme and it is shown (1) that the N + OH reaction is a necessary addition especially under fuel-rich conditions, where it increases NO formation appreciably; and (2) that the assumption of equilibrated NO reactions throughout the course of oxidation leads to an underestimate of NO, i.e. that there are small but non-negligible overshoots in atom and radical concentrations which are properly accounted for by the numerical integration.

It is only in the interpretation of these results, as implying that there are no other NO-forming reactions, that I find myself in some disagreement with the authors. Two facts should be borne in mind: (1) The temperatures in the reflected shocks are very high, e.g. 2950 and 2970°K in Figs. 4 and 5. This greatly increases the formation of NO, i.e. an equilibrium concentration of more than 2000 p.p.m. is calculated even under somewhat fuel-rich conditions; (2) There is considerable dilution of the CH_4 – air reactants by excess N_2 and Ar, so much so that on the basis of only methane plus air, the calculated final NO amounts to more than 10,000 p.p.m. for the fuel-rich case. It is clear that dilution by excess inert components must favor attainment of thermal equilibrium.

Fenimore (13th Combustion Symposium, p. 373) claims to have observed "prompt NO" at 2100 to 2300°K (600 to 800°K cooler than in this work), undiluted by inert gas, and his intercept (prompt) ppm NO was in the 50 to 100 range. Such an amount is so small compared to the presently measured quantities as to be hopelessly in the noise of measurement. My statement here does not, of course, suggest that "prompt NO" exists, but only that the present experiments can not decide the question. I am very doubtful that any shock tube experiment can make this decision, because (a) observation time is too short when realistic combustion conditions prevail; (b) dilution by inert carrier is usually required which reduces non-equilibrium processes; and (c) transport processes such as diffusion and thermal conduction, which play an important role in combustion, are virtually eliminated.

Considering the attractive "fix" of reducing NO emission by staged combustion in which a fuel-rich first stage might be followed by a very lean final composition, it is important that this experimental problem be resolved promptly.

C. T. Bowman

Prof. Kaufman has suggested that the differences in conditions between our experiments and those of Fenimore preclude our drawing valid conclusions regarding the origins of "prompt NO". In our experiments, we have observed a rapid rate of formation of NO in the combustion zone, qualitatively similar to that reported by Fenimore, at temperatures as low as 2600°K. Under certain conditions, most notably rich mixtures, Fenimore has reported levels of "prompt NO" approximately equal to the equilibrium concentration in the post-flame zone. His observations suggest that the processes resulting in formation of "prompt NO" are not merely perturbations on the overall NO formation process, but are, in some instances, the dominant processes. If this is so, then these processes should have some effect on our experimental results. In our experiments, the rapidly formed NO, in both lean and rich mixtures, was shown to be consistent with reactions [1]-[3] if non-equilibrium radical concentrations during combustion were correctly evaluated. Hence, it seems reasonable to suggest that the "prompt NO", observed in flames, also is due to such non-equilibrium effects. It is possible that reactions other than the three discussed in our paper are involved in the NO formation process and, indeed, may be significant for conditions substantially different from those of the experimental study.

A. A. Quader *(Research Laboratories, GMC)*

I would like Dr. Bowman to answer the following questions:

1. In Figure 2 of your paper, what percentage of the incident intensity do the maximum absorptions of NO and OH represent?

2. Why was the $\gamma(0,1)$ band of NO at 2360Å selected in favor of the $\gamma(0,2)$ band near 2465Å where no Schuman-Runge absorption was observed by one of the authors in his previous work (Ref. 6 of the paper)?

3. At room temperature, the $\gamma(0,0)$ band of NO at 2260Å can be readily identified, but practically no absorption occurs at 2360Å. At what temperature is the NO absorption measurable at this wave-length?

4. Since the calibration information in Figure 3 of the paper can be valuable for similar work on other apparatus, will the authors outline the equations used to transform the absorption intensity into the effective absorption cross section?

C. T. Bowman

In answer to the first question, the percentage of incident intensity absorbed varied from experiment to experiment, and depended on the concentration. In the case of nitric oxide, maximum absorption was somewhere around 8 to 9 percent; the minimum absorption observable, before problems with the signal to noise ratio arose, was approximately 1 percent. In the case of the hydroxyl radical, maximum absorption was approximately 15 percent. With regard to the reason for using the $\gamma(0,1)$ band rather than the $\gamma(0,2)$ band – the effective absorption cross-sections of these two absorption bands are approximately equal. Hence, the absorptivity of either band depends only on the concentration in the particular vibrational state corresponding to the band. For the conditions of our experiment, approximately 25 percent of the NO molecules were in the v=1 vibrational state (corresponding to the $\gamma(0,1)$ band) and approximately 10 percent were in the v=2 state (corresponding to the $\gamma(0,2)$ band). Hence, the $\gamma(0,1)$ band could be used to measure lower NO concentrations than could be obtained using the $\gamma(0,2)$ band. The temperature at which one can begin to see absorption at 2360 Å depends, of course, on the NO concentration. In our experiments, I would guess that temperatures in excess of 2000°K would be required before enough NO would be formed to be detectable. A conventional procedure was used to obtain the absorption cross-sections (see, for example, Ref. 19).

H. K. Newhall *(University of Wisconsin)*

About a year ago, we presented a paper dealing with some combustion vessel experiments on nitric oxide formation in hydrogen-air combustion. The combustion vessel was operated at high pressures on the order of 10 to 20 atmospheres. The results of the studies indicated that from an engineering standpoint, the rate of nitric oxide formation could be predicted quite well with an equilibrium model, (for species other than nitric oxide), under the conditions investigated. There was some indication that an atom overshoot or a radical overshoot occurred early during the process. This of course led to speculation that if when we move on from hydrogen to hydrocarbon fuels we would see a number of things happening, mainly large divergences of experimental results from equilibrium based calculations. The results that we have gotten and that are presently in press with relation to propane-air mixtures, at pressures on the order of 10 to 20 atmospheres, indicate that in fact the overshoot is less of a problem with hydrocarbons or at least with propane than it was with hydrogen. We're running propane-air mixtures at various equivalence ratios, with pressures of 10 to 20 atmospheres, and the equilibrium model at these pressures does an excellent job of predicting the measured nitric oxide formation rates. We speculate that if we drop down in pressure to levels on the order of atmospheric or slightly above that we would begin to see some of the divergences that have been discussed here today. The upshot of this is that this question of atom overshoot, and divergence from equilibrium calculations is really a matter of the particular conditions under which the combustion process is being carried out. Particularly as regards pressure, we know that at 10 atmospheres we do fairly well with the equilibrium assumption. I think there is an indication that at pressures as low as 3 atmospheres, from an engineering standpoint the equilibrium assumption is not too bad. Apparently it is at the 1 atmosphere level that you begin to run into significant problems with this assumption.

NITRIC OXIDE FORMATION AND CARBON MONOXIDE BURNOUT IN A COMPACT STEAM GENERATOR

G. C. WILLIAMS, A. F. SAROFIM, and N. LAMBERT

Massachusetts Institute of Technology, Cambridge, Massachusetts

ABSTRACT

In a proposed design of a compact-boiler for a steam-engine, a premixed, vaporized-fuel/air mixture is fed through a perforated feed-plate into the combustion chamber and the products of combustion are withdrawn through a compact, low-pressure-drop heat exchanger. This paper describes measurements of carbon monoxide (CO) and nitric oxide (NO) emissions from a prototype combustion chamber in which the vaporized fuel is simulated by propane. The fuel/air mixture is fed at rates up to 4625 cu.ft./hr. through 6500, 0.033 in. I.D. holes corresponding to 20% open area in a 6 in. O.D. plate. The combustion products are withdrawn through a flat, coiled heat exchanger positioned 3 in. downstream from the burner plate. The maximum average volumetric heat release rate is about 6×10^6 Btu/ft^{3}hr. Samples withdrawn through a water-cooled probe from different positions above the plate are analyzed for hydrocarbons (HC), CO and NO. The measured emission rates are compared with values calculated assuming that the overall reaction rates are limited by the reaction of CO with OH and N_2 with O, and that the free radical concentrations are determined by the combination of the reactions

$$OH + H_2 = H_2O + H, \; H + O_2 = OH + O, \text{ and } O + H_2 = OH + H.$$

The results are used to define combustion chamber volume for different desired fuel rates and emission rates.

INTRODUCTION

Compact direct-fired heaters have potential for a number of applications ranging from domestic heating systems to boilers in steam engines. The volume of a combustion chamber for such heaters can be reduced to a minimum by using

References p. 154

premixed fuel-air mixtures with either a gaseous fuel or a prevaporized liquid fuel. This study is concerned with the pollutant emission levels from such a unit where the premixed fuel-air system is fed through a perforated burner plate and the combustion products are exhausted through a compact heat exchanger opposed and parallel to the burner plate. A measure of control of the rate of emission from a heater of this type is provided by the freedom to adjust the gap width between the burner plate and the heat exchanger. Increases in gap width will result in a decrease in the emissions of HC and CO at the expense of an increase in the emission of NO. Emphasis in this study is placed on the quantitative determination of the effect of the gap width on pollutant emissions and on the determination of the rate controlling steps in the formation and burnout of the pollutants.

EXPERIMENTAL APPARATUS AND PROCEDURE

The active area of the burner plate, used in the tests, consisted of a 6 in. diameter stainless-steel plate, 0.030 in. thick, perforated with 0.033 in. diameter holes with a density of 225 holes/sq. in., corresponding to an open area of 20%. The burner plate was held between two flanges, the upper of which was welded to a stainless-steel cylinder 8.0 in. I.D., 18 in. high. Most of the runs were performed with a heat exchanger positioned three inches above the burner plate. The inner walls of the stainless-steel cylinder were covered with a one-inch thickness of insulation thus defining a combustion chamber 6 in. I.D. and 3 in. high. Additional runs were performed with the heat exchanger positioned at 2.0, 4.0, and 8.0 inches above the burner plate. The heat exchanger was formed from 0.5 in. O.D. stainless steel tubing, finned with stainless-steel stubs extending 0.25 in. from the outer surface. The finned tubing with an effective external diameter of 1.0 in. was coiled in two layers providing a two-inch deep heat exchanger blocking the entire six-inch I.D. cross section of the combustion chamber with the exception of a small opening at the center through which a sampling probe or thermocouple could be inserted. The heat exchanger was water cooled, and a water flow rate of 4 gallons per minute was selected to maintain the water temperature rise below 100°F. The lower flange holding the burner plate was welded to a one-foot long duct, with an internal diameter varying from 8 inches at the burner plate to 6 inches at the outlet of a variable-speed air blower. The propane fuel selected for these studies was diffused into the air stream through 224, 0.03 in. holes, evenly distributed across the outlet of the air blower in an attempt to obtain a fuel/air mixture of uniform concentration at the burner plate.

The fuel flow rate to the burner was metered by a Brooks rotameter, and the air rate was determined from flue-gas analysis. Samples of the combustion products along the center-line were obtained with a water-cooled, stainless-steel probe, 0.18 in. O.D. and 0.016 in. I.D. Temperatures along the center-line were measured with a thermocouple constructed from 0.008 in. 80% Pt/20% Rh, 60% Pt/40% Rh wires with a junction diameter of 0.012 in. The thermocouple junction was coated with a fired silicone.

The flue gases were either pumped continuously through a chemiluminescent NO detector and a paramagnetic O_2 analyzer or were collected in an evacuated glass flask for subsequent analysis for HC on a F. and M. Scientific 700 laboratory chromatograph equipped with a flame ionization detector and for CO_2, O_2, N_2, CO, and H_2 on a Fisher Hamilton gas partitioner Model 29 equipped with a thermal conductivity detector. Helium was used as a carrier gas with the exception of the H_2 analyses for which argon was used.

RESULTS

The concentrations profiles, dry basis, along the axis above the burner plate are presented in Figs. 1, 2 and 3 for energy release rates of 200,000, 300,000, and 400,000 Btu/hr. and for 25 percent excess air. The heat exchanger was positioned at 3.0 inches* for most of the data taken at energy release rates of 200,000 Btu/hr. (Fig. 1) and 400,000 Btu/hr. (Fig. 3) and at 8.0 inches for 300,000 Btu/hr. (Fig. 2). The data points at 8.0 inches in Figs. 1 and 3 refer to gas samples that had been quenched to 1058°K (1900°R) and 1523°K (2750°R) by the heat exchanger. The H_2 concentrations between 3.0 and 8.0 inches, however, were taken with the exchanger positioned at 8.0 inches. The quenching of the reactions across the heat exchanger was studied at the 300,000 Btu/hr. energy release rate by obtaining, in addition to the profiles shown in Fig. 2 with the exchanger positioned at 8.0 inches, additional measurements with the probe at 8.0 inches and the exchanger positioned at 2.0, 3.0, and 4.0 inches. The concentrations of NO, CO, and H_2 at these positions are

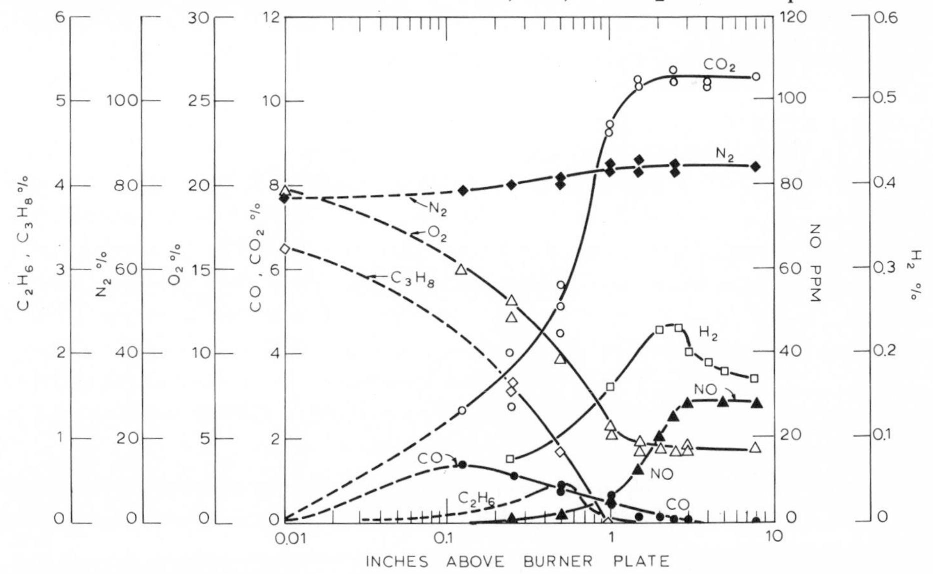

Fig. 1. Concentration profiles, dry basis, for the products of combustion of a premixed propane/air mixture; energy release rate = 200,000 Btu/hr; excess air = 25%; burner diameter = 6.0 in.

* *The leading edge of the fins on the first row of the tubes is used to define the exchanger position.*

References p. 154

compared with those that would have been obtained if the cooling rate in the heat exchanger was sufficient to freeze† the concentration at the value occurring in the undisturbed gases at the position of the leading edge of the exchanger.

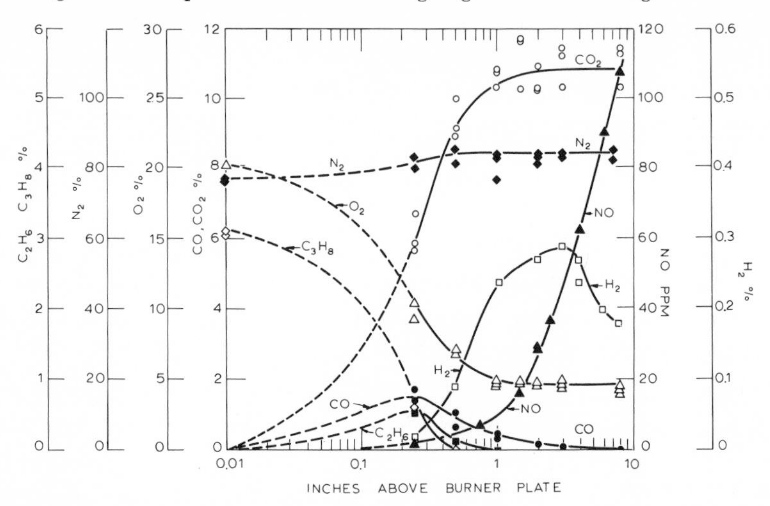

Fig. 2. Concentration profiles, dry basis, for the products of combustion of a premixed propane/air mixture; energy release rate = 300,000 Btu/hr; excess air = 25%; burner diameter = 6.0 in.

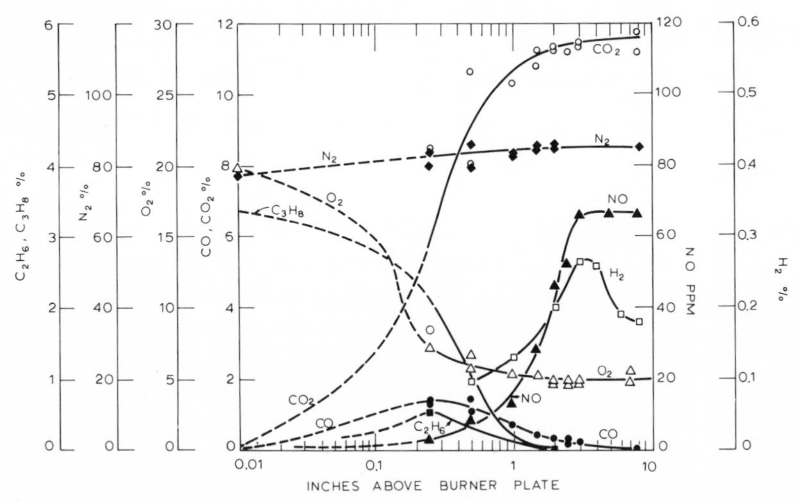

Fig. 3. Concentration profiles, dry basis, for the products of combustion of a premixed propane/air mixture; energy release rate = 400,000 Btu/hr; excess air = 25%; burner diameter = 6.0 in.

† *The concentrations at any postion correspond to the frozen concentrations emerging from the sampling probe and include the effect of reactions within the probe. The cooling rate in the entrance region of the probe, averaged across the probe cross-section, is estimated to be 3 x 10^7°K/sec.*

TABLE I

Comparison of Post-Exchanger Concentrations (dry basis) with the Corresponding Frozen Concentrations

Heat Exchanger Position	Post Exchanger Concentration			Corresponding Frozen Concentration		
inches from burner	NO ppm	CO %	H_2 %	NO ppm	CO %	H_2 %
2.0	27	0.051	n.a.*	29	0.18	0.27
3.0	40	0.005	0.29	40	0.13	0.29
4.0	70	n.d.†	0.27	63	n.a.	0.27

* *not available.*

† *not detectable; the lower limit resolution for CO of the chromatograph used in the study was 50 ppm.*

The temperature distribution above the plate was calculated from an enthalpy balance, using the measured concentration distribution to determine the enthalpy release and making allowance for energy loss through the side walls. (The calculated energy loss was small, varying from 1,220 Btu/hr./in. height at 200,000 Btu/hr. to 1,410 Btu/hr./in. height at 400,000 Btu/hr. The calculated temperatures for the three firing densities ranged from 750 to 1435°K at 0.25 in., 1390 to 1920°K at 0.5 in., 1930 to 1993°K at 1.0 in., 1990 to 2000°K at 2.0 in., and 1991 to 1998°K at 3.0 in., with incomplete combustion providing the major cause for departure from the calculated adiabatic temperature of 2020°K. The corrected measured temperatures agreed well with the calculated temperatures when a value of 0.6 was assigned the thermocouple emissivity.

DISCUSSION

Examination of Table 1 shows that the NO concentration is frozen at the value entering the heat exchanger but that the CO continues to react within the exchanger. This demonstrates the potential for control of CO and NO emissions of a cooling rate fast enough to quench the NO reactions but slow enough to permit the CO reactions to proceed. The concentration profiles in Figs. 1, 2 and 3 therefore provide a valid measure of the effect of combustion chamber volume on NO emission but grossly overestimate the CO emission. In order to place the concentration figures in better perspective, Figs. 2 and 3 have been used to calculate the emissions from a steam engine driven vehicle with a test weight of 4600 lbs., a product of drag coefficient and frontal area of 13 ft^2, with an anticipated energy consumption of 7100 Btu/mile at a cruising speed of 50 mph (1). The emissions are reported in Fig. 4 as ratios of the 1975 Federal Standards of 3.4 gm/mile for CO and 0.41 gm/mile for HC and the 1976 standard of 0.40 gm/mile for NO_x reported as NO_2. It must be emphasized that the vehicle emissions have been evaluated for predicted performance for just one road speed, and not a driving cycle. From Figs. 1, 2 and 3 it is, however, expected that the emissions per unit fuel consumption will increase with the rate of firing of the boiler.

References p. 154

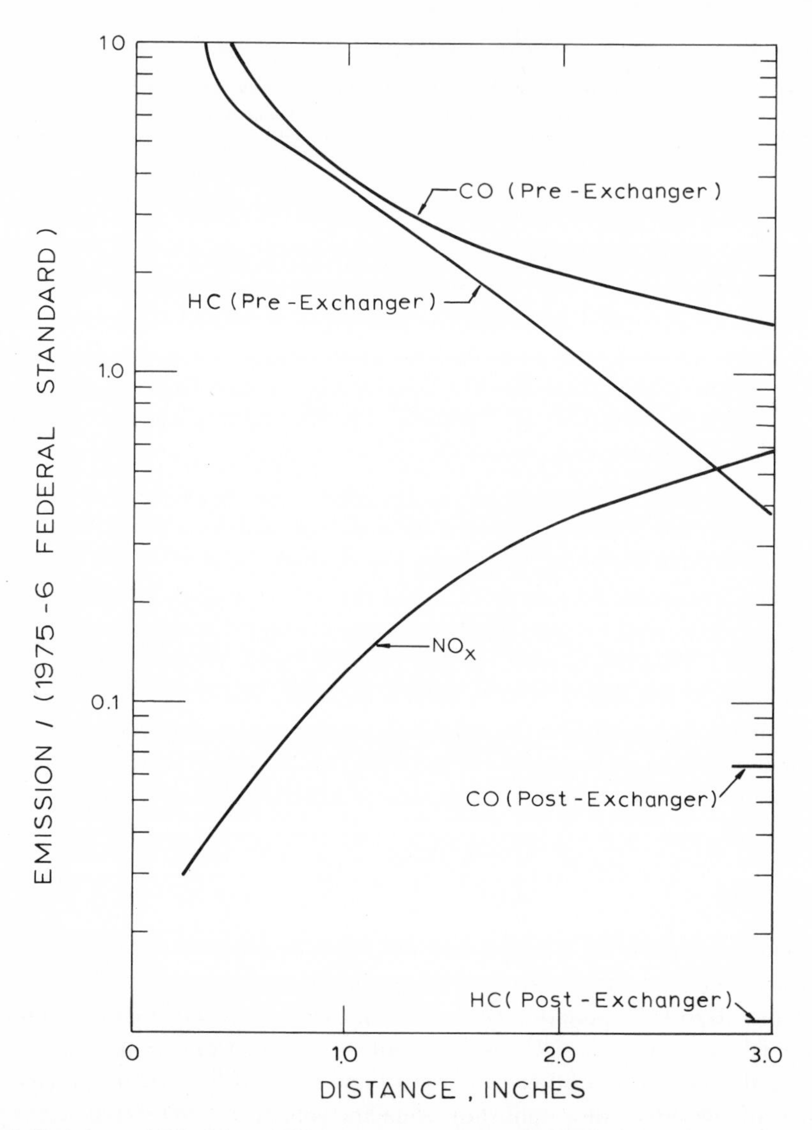

Fig. 4. Emissions of HC, CO, and NO_x expressed as ratios of 1975-6 Federal Standards as a function of distance from burner plate; values calculated for road speed of 50 mph.

For the conditions selected, it can be seen that the NO_x levels at 3.0 inches from the burner plate have risen to 0.6 times the 1976 standards, that the CO at 3.0 inches within the combustion chamber has fallen to a value 1.4 times the 1975 standard but that the CO at the exit of the exchanger has fallen far below the standard. A reduction of the distance of separation of the exchanger and burner plate to 2.0 inches would reduce the NO_x emission to 0.36 times the 1976 standard but the CO and HC emissions would rise. The results for CO in Table 1 suggest that with the exchanger at 2.0 inches, the CO concentration leaving the exchanger would be less than 0.4 times the 1975 standard. Selection of the optimum spacing would require a more detailed study of the reactions of CO and HC in the heat exchanger. These preliminary results are encouraging but data on a prototype operating under realistic conditions are needed to establish valid emission values.

Different models for explaining the data are examined next.

Kinetics of Nitric Oxide Formation – The NO profiles show no sign of the prompt NO reported by Fenimore (2) for premixed hydrocarbon-air laminar diffusion flames. Most of the NO is formed in the post-flame zone, here defined as the region in which the temperatures have risen to within 5% of their peak values (at approx. 0.8 in. above the burner plate). That the NO is formed in the post-flame zone is consistent with the early conclusions of Zeldovich (3) who also proposed the following mechanism for the formation process:

$$M + O_2 \rightleftharpoons O + O + M \qquad (1)$$

$$N_2 + O \rightleftharpoons NO + N \qquad (2)$$

$$N + O_2 \rightleftharpoons NO + O \qquad (3)$$

This sequence of reactions yields an expression for the rate of NO formation given by

$$\frac{d\,(NO)}{dt} = 2k_2\left(\frac{k_1}{k_{-1}}\right)^{1/2} (N_2)\,(O_2)^{1/2} = 1.36x10^{15} \exp\left[\frac{-134{,}700}{RT}\right](N_2)\,(O_2)^{1/2} \frac{mole}{(cm3)\,(sec)} \qquad (4)$$

at NO concentrations low enough to permit neglecting the reverse reaction, and with the rate constant k_2 selected from the compilation of Baulch et al. (4). The rate of NO formation calculated from Eq. 4 at the peak temperatures in the system is 800 ppm/sec, compared with measured peak values of 5,000, 9,000, and 18,000 ppm/sec at energy release rates of 200,000, 300,000, and 400,000 Btu/hr., respectively. Such discrepancies between measured and calculated rates of NO formation have been attributed by several investigators to the occurrence of superequilibrium free radical concentrations in the flame and post-flame zones.

Thompson et al. (5) have proposed using the H_2 concentration in lean flames to predict the non-equilibrium radical concentration by postulating that the following three reactions are sufficiently fast at flame temperatures to be equilibrated:

$$H + O_2 \rightleftharpoons OH + O \qquad (5)$$

$$O + H_2 \rightleftharpoons OH + H \qquad (6)$$

References p. 154

$$H_2 + OH \rightleftharpoons H_2O + H \tag{7}$$

Eqs. 5 and 7 yield

$$H_2 + O_2 \rightleftharpoons H_2O + O \tag{8}$$

and

$$(O) = K_8 \frac{(H_2)(O_2)}{(H_2O)} \tag{9}$$

which combined with Eqs. 2 and 3 can be used to derive the following expression for the rate of NO formation:

$$\frac{d\,(NO)}{dt} = 2k_2 K_8 \frac{(N_2)(H_2)(O_2)}{(H_2O)} \tag{10}$$

or

$$\frac{dP_{NO}}{dt} = 4.44 \times 10^{12} \frac{\exp\,[-73,700/RT]}{T} \frac{P_{N_2} P_{H_2} P_{O_2}}{P_{H_2O}} \text{ atm/sec} \tag{11}$$

The measured concentrations of N_2, O_2, and H_2, the temperature derived from an enthalpy balance, and the water vapor concentration derived from an oxygen balance were inserted in Eq. 11 to evaluate the rates of NO formation for the conditions corresponding to Figs. 1, 2 and 3.

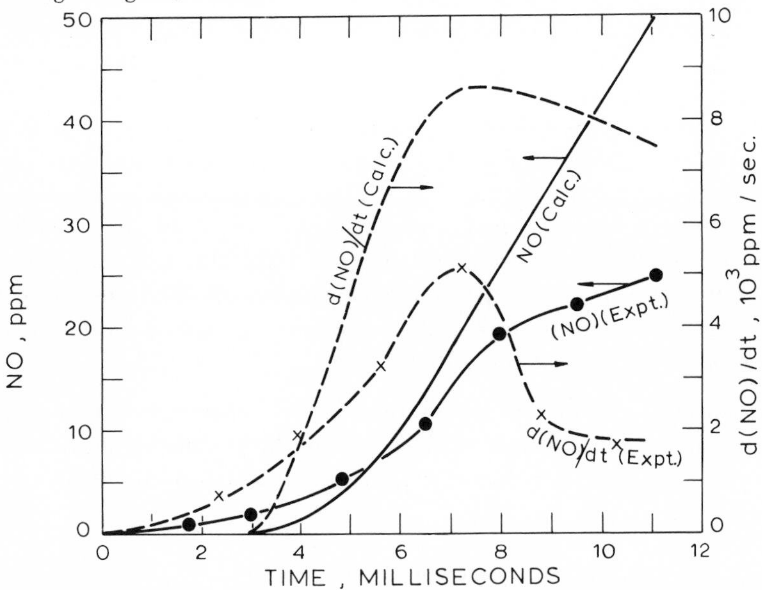

Fig. 5. Calculated and experimental concentrations and rates of formation of NO; energy release rate = 200,000 Btu/hr.

The computed and measured concentrations and rates of formation* of NO are compared in Figs. 5, 6 and 7. The calculated *peak* rates of formations and *maximum* nitric oxide concentrations are within a factor of two of the measured values. The much better agreement between experiments and theory obtained using the modified Zeldovich rate expression, Eq. 11, in place of the unmodified form, Eq. 4, provides

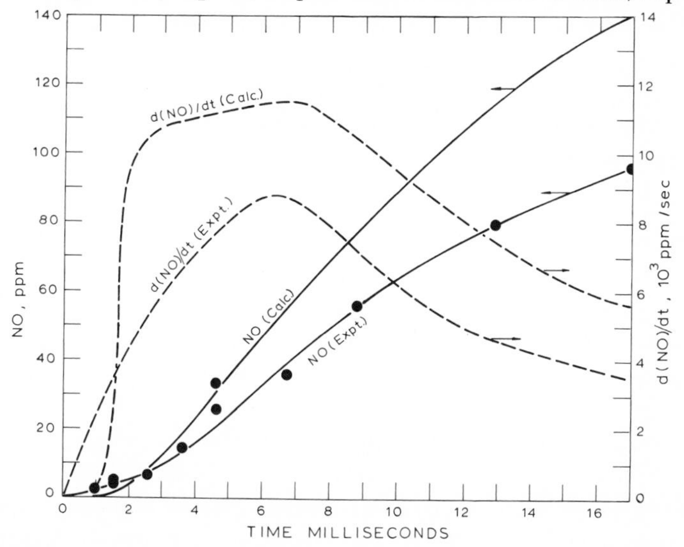

Fig. 6. Calculated and experimental concentrations and rates of formation of NO; energy release rate = 300,000 Btu/hr.

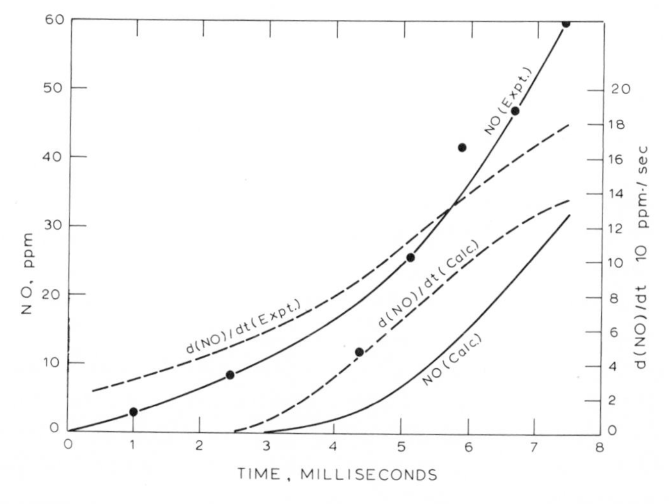

Fig. 7. Calculated and experimental concentrations and rates of formation of NO; energy release rate = 400,000 Btu/hr.

* *Rates of formation are related to local concentrations by assuming that axial diffusion is negligible for the high output system under study (local velocities 20 to 50 ft./sec.).*

References p. 154

partial support for the postulate that reactions 6 and 7 are equilibrated. The calculated rates within the first few milliseconds, however, are consistently low, partially as a consequence of the use in the calculations of a mean enthalpy termperature for the temperature of a highly nonisothermal zone containing multiple jet flames and partly due to the effect of the HC reactions on the free radical concentration. In the later stages, although there is general agreement between the gross features of the measured and predicted rates of formation, the discrepancies are sufficient to suggest that reactions other than 6 and 7 need to be considered in relating the oxygen radical concentrations to those of H_2, O_2, and H_2O. This will be discussed in greater detail below.

The modified Zeldovich mechanism provides a measure of the effect of departures from equilibrium on the rates of NO formation but does not provide a predictive tool since the non-equilibrium H_2 concentration must be provided either by experiment or an appropriate theoretical model. The H_2 concentration profile observed in this study was unexpected since previous studies of premixed hydrocarbon-air flames showed approximate coincidence of the positions of the peaks of the intermediates formed in the flame zone (6, 7), usually with the peak in CO concentration occurring subsequent to that in the H_2 concentration. A more elaborate kinetic model is required to explain the observed anomaly. It is interesting to note that the rapid increases in NO and H_2 concentrations approximately coincide, supporting the speculation that whereas high H_2 concentrations are indicative of high rates of NO formation, appreciable NO concentrations inhibit the rate of H_2 oxidation.

The present results and those by Thompson et al. both suggest that in the post-flame region the kinetics of NO formation are an order of magnitude faster than that given by Eq. 4, in apparent contradiction to Zeldovich's experimental results which were fitted by an equivalent form of Eq. 4. This is partially explained by the higher temperatures (ca. 2500°K) in Zeldovich's experiments, at which the O_2 dissociation reactions are faster, and by the long residence times in his experiments relative to the rate of reaction (leading to NO concentrations in excess of 2000 ppm). Examination of Figs. 5 and 6 shows that the effect of super-equilibrium oxygen radical concentrations peaks at the temperature level of 2000°K at about 7 milliseconds. At longer residence times the rate of NO formation begins to approach the value calculated from the unmodified Zeldovich equation (800 ppm/sec). It is, therefore, expected that in systems such as industrial furnaces where the combustion products are at peak temperatures for times significantly longer than the times in this study the use of the unmodified Zeldovich equation might yield acceptable results since the excess NO formed in the first few milliseconds would be small relative to the total amount formed.

Kinetics of Carbon Monoxide Burnout — The rate-limiting step for CO oxidation is that due to the reaction with OH radicals:

$$CO + OH \rightleftharpoons CO_2 + H \qquad (12)$$

The rate constant for reaction 12 is well known (7) but the free radical concentrations particularly in the flame and post-flame zones are not. Expressions for the over-all burning rate of CO have been derived either from experiment or assumptions concerning the equilibration of the free radical species. A common form for the global reaction rate is given by

$$\frac{d(CO_2)}{dt} = A \exp\,[-B/RT]\; f_{O_2}^{a}\, f_{CO} f_{H_2O}^{b}\, (P/RT)^{1+a+b} \qquad (13)$$

TABLE 2

Selected Rate Constants for CO Oxidation.

Investigators	A*	B	a	B	Comments
Hottel, Williams,	12x10^{10}	16,000	0.3	0.5	Expt. CO oxid.
Neirheim, Schneider (8)	2.9x10^{10}	15,000	0.35	0.4	Expt. C_3H_8 oxid.
Koslov (9)	10^{13}†	32,000	0.25	0.5	Expt. CO oxid.
Fristrom, Westenberg (6)	4x10^{13}	45,000	0.25	0.5	Assumed 20 = O_2 equilib.

* $(sec)^{-1}(mole/cm^3)^{-(a+b)}$

† Rate expression included a temperature dependence of $(T)^{-2.5}$ which has been altered to fit the $(T)^{-1.75}$ power at 1500°K, a little above the upper temperature limit of 1373°K of Koslov's experiments.

Selected values for the rate constants reported by various investigators are shown in Table 2. The two sets of data by Hottel, et al. (8) were based on stirred reactor studies, and the data could be fitted adequately by a mechanism which assumed that the rate limiting step was the forward reaction of Eq. 12, that reactions 5, 6, and 7 were equilibrated and that the free radicals production by the chain branching reaction associated with the net oxygen comsumption was balanced by a three body termination step. Fristrom and Westenberg (6), noting the higher activation energy associated with Koslov's correlation of the Russian data, suggested that a rate expression could be derived assuming that reactions 1, 6 and 7 were equilibrated, and obtained the exponents for Eq. 13 given in Table 2. It should be noted that equilibration of reaction 1 is inconsistent with the rates of NO formation reported in the previous section, and that the oxygen radical concentrations in this study appear to be an order of magnitude higher than those given by Eq. 1. One additional expression can be derived based on the assumption that Eqs. 5 and 6 are equilibrated.

Addition of Eqs. 5 and 6 yields

$$H_2 + O_2 = OH + OH \qquad (14)$$

or

$$(OH) = K_{14}^{1/2}\,(H_2)^{1/2}\,(O_2)^{1/2} \qquad (15)$$

References p. 154

and therefore, from Eq. 12 and 15,

$$\frac{d(CO_2)}{dt} = 3.87 \times 10^{13} \exp\left[-16{,}000/RT\right] f_{CO} f_{H_2}^{1/2} f_{O_2}^{1/2} (P/RT)^2 \quad (16)$$

As, in the case of the modified Zeldovich equation, Eq. 16 can be applied only for those cases in which an independent measure of the H_2 concentration is available.

The applicability of the above reactions to the calculation of the CO burnout in the post-flame zone of the compact heater used in this study is best discussed in relation to Fig. 8. The predicted values of $-d\ln P_{CO}/dt$ at 2000°K, $f_{O_2} = 0.04$, $f_{H_2O} = 0.126$ are, from Eq. 13, 1.9×10^4 sec^{-1} using the rate constants for CO oxidation from Ref. 8, 1.1×10^4 sec^{-1} using the rate constants for C_3H_8 oxidation from Ref. 8, 6.0×10^4 sec^{-1} using the rate constants from Ref. 9, 0.91×10^4 sec^{-1} using the rate constants from Ref. 6, and, from Eq. 16, 3.6×10^4 sec^{-1}. These rates are within a factor of 7 of one another but are several orders of magnitude higher than that observed towards the end of the combustion chamber where the temperatures and concentrations are very close to the values for which the calculations were made (the slope marked 2000°K in Fig. 8 corresponds to the rate predicted using the rate constant for C_3H_8 from Ref. 8). In the early stages of the combustion process the rate of CO oxidation must have been much faster than that in the later stages, assuming that all of the carbon in the propane fuel was converted to a CO intermediate prior to complete oxidation to CO_2. If the propane fuel had been converted instantaneously to CO at the burner plate, a mixture with $f_{CO} = 0.09$, $f_{O_2} = 0.0826$, $f_{H_2O} = 0.121$, and $T = 1180^{\circ}$K would have been produced. The burning rate at this hypothetical initial condition, calculated using the constants for C_3H_8 oxidation from Ref. 8, is shown on Fig. 8; it should provide, together with the rate at 2000°K, bounds for the slope of the CO-t curve in the initial phases. The first data points for CO concentration are consistent with an average initial burning rate equal to or greater than the value calculated for $T = 1180^{\circ}$K. The apparent reduction in burning rates at longer times must be due to the increasing importance of the reverse reaction. (The rate constants for reaction 12 are sufficiently fast so that estimated equilibration times are of the order of 0.2 milliseconds). A lower bound on the CO concentration is then imposed by the H/OH ratio.

If reaction 7 were equilibrated, then

$$(CO) \geq \frac{K_7}{K_{12}} \frac{(CO_2)(H_2)}{(H_2O)} \quad (17)$$

Values of this bounding value at 2000°K calculated from the equilibrium constants K_7 and K_{12}, the measured concentrations of CO_2 and H_2, and the concentration of H_2O inferred from stoichiometry are shown on Fig. 8. This is the value of CO that would be expected for the case of equilibration of reactions 7 and 12. Also shown for comparison is the value for CO at 2000°K if complete thermodynamic equilibrium was achieved. It is clear from Fig. 8 that since the CO concentrations fall below the

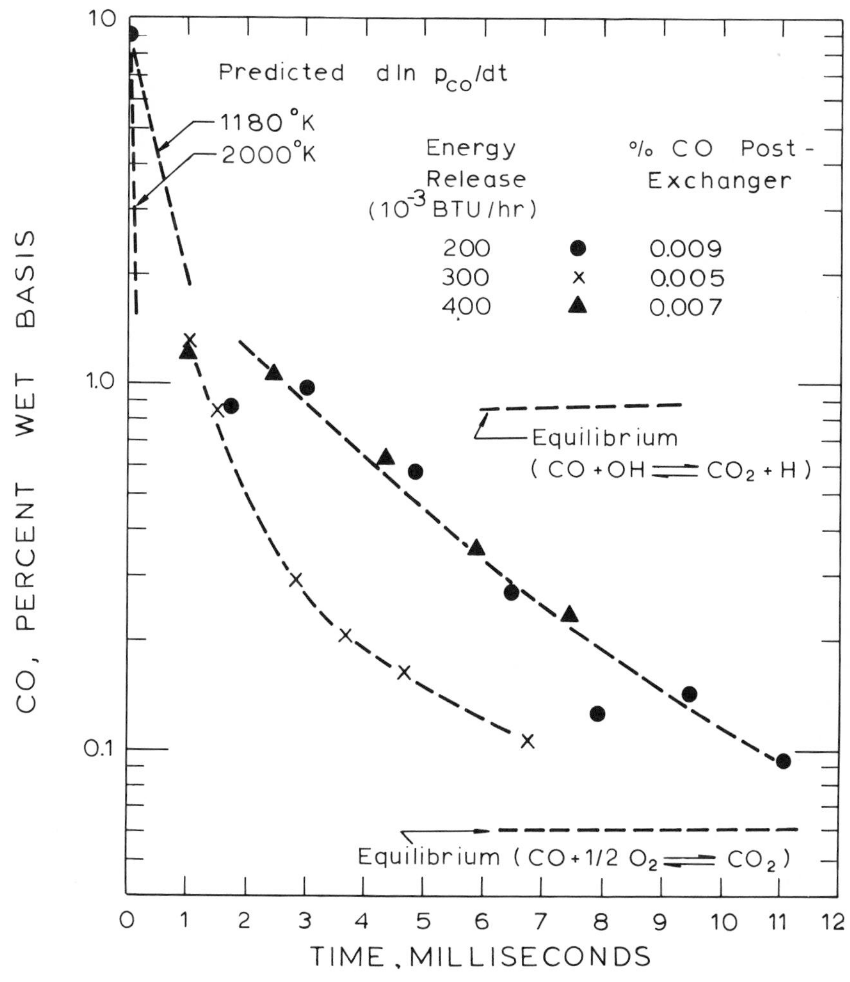

Fig. 8. CO concentration versus gas residence time.

bounding value given by Eq. 17 the concentration of H atoms must be smaller than the value expected if reaction 7 were equilibrated, but its concentration must be still sufficiently high to retard the rate of carbon monoxide burnout.*

Additional evidence for the apparent retardation of the CO burnout is provided by the reduction of the CO concentrations across the heat exchanger. As the temperature is reduced, the rate of CO_2 + H reaction falls off much more rapidly than that of the CO + OH and, consequently the apparent rate of CO oxidation increases. If the

** Some uncertainty in the measured CO concentrations is introduced by the possibility of reactions occurring within the probe. The cooling rates of 3 x 10^7°K/sec are sufficient to quench the retarded CO reaction but not that predicted from the one-way reaction using the constants in Table 2.*

References p. 154

forward reaction had been rate limiting, the CO concentration would have been frozen at a value close to that entering the heat exchanger. The results obtained here may explain other cases reported in the literature of rates of CO burning much lower than those obtained from Eq. 13 (see, for example, Ref. 10).

ACKNOWLEDGEMENTS

The authors are grateful to Steam Engine Systems Corporation for providing the burner and heat exchanger and to the Air Pollution Control Office of the Environmental Pollution Control Office for providing partial support for the study under Contract No. 68-02-0018. The authors also take pleasure in acknowledging the many useful contributions made to the experimental program by Dr. David H. Fine.

REFERENCES

1. *Steam Engine Systems Corporation, "Design and Development of an Automobile Propulsion System Utilizing a Rankine Cycle Engine (water based fluid)," 3rd Quarterly Progress Report, EPA Contract 68-04-0004, December 15, 1971.*
2. *C.P. Fenimore, "Formation of Nitric Oxide in Premixed Hydrocarbon Flames," Thirteenth Symposium (International) on Combustion, Combustion Institute, 1971, p. 373.*
3. *Y.B. Zeldovich, "The Oxidation of Nitrogen in Combustion Explosions," Acta Physicochimica U.S.S.R., Vol. 21, 1946, p. 577.*
4. *D.L. Baulch, D.D. Drysdale, D.G. Horne, and L.C. Lloyd, "Critical Evaluation of Rate Constants for Homogeneous, Gas Phase Reactions of Interest in High Temperature Systems," Department of Physical Chemistry, University of Leeds, England, Report No. 4, December, 1969.*
5. *D. Thompson, J.M. Beer, and T.D. Brown, "The Formation of Oxides of Nitrogen in a Combustion System," Preprint No. 37e, 70th National AICHE Meeting, Atlantic City, August, 1971.*
6. *R.M. Fristrom, and A.A. Westenberg, Flame, Structure, McGraw-Hill, New York, 1965.*
7. *R.M. Fristrom, A.A. Westenberg, "Flame Zone Studies. IV Microstructure and Material Transport in a Laminar Propane-Air Flame Front," Combustion and Flame, Vol. 1, 1957, p. 217.*
8. *H.C. Hottel, G.C. Williams, N.M. Nerheim, and G. Schneider, "Combustion of Carbon Monoxide and Propane," Tenth Symposium (International) on Combustion, Combustion Institute, 1965, p. 111.*
9. *G.I. Koslov, "On the High Temperature Oxidation of CH_4," Seventh Symposium (International) Combustion, Combustion Institute, 1959, p. 142.*
10. *J.M. Singer, E.B. Cook, M.E. Harris, V.R. Rowe, and J. Grumer, "Flame Characteristics Causing Air Pollution: Production of Oxides of Nitrogen and Carbon Monoxide," Bureau of Mines, RI 6958, 1967.*

DISCUSSION

J. M. Beér and D. Thompson *(University of Sheffield)*

This is an interesting and valuable paper, and the extension of the consideration of non-equilibrium radical concentrations in prediction of pollutant emission to a system closely related to commercial practice represents a welcome advance.

The improved general agreement between observed and predicted nitric oxide formation rates obtained by use of the modified Zeldovich equation is encouraging. The absence of 'instant NO' is important; in our related study of nitric oxide formation in fuel-lean methane-air combustion, concentrations of nitric oxide of the order of 20 ppm were observed above the multiport flameholder. The ports were considerably larger than those of the present system, and the early combustion zone consisted of distinct flamelets, so that it was not possible to determine whether the NO was due to local high temperatures due to near-adiabatic combustion, and normal N_2 + O reaction, or the additional unknown reaction observed by Fenimore. The results presented here indicate the former source, and therefore show that inaccessible stabilizing flamelets are a characteristic to be avoided if NO emission is to be minimized.

The observed loss of carbon monoxide rather than the expected reduction in H_2 concentration during passage through the heat exchanger is significant, since it suggests that close proximity of flame and heat exchanger, which is seen from the results to be an important factor in the reduction of NO emission, need not have a serious effect on the CO level.

It is possible that the reduction of this distance still further may be feasible, since the results show that at the two lower firing rates combustion was very near to completion at 1½" from the flameholder, and even at the highest firing rate, was near completion at 2". Quenching at these locations would have resulted in considerable reduction in NO emission. Relatively slow quenching at rates typical of heat exchangers could reduce the somewhat high CO levels to similar levels to those observed under the conditions studied. The possible need for reduced heat transfer efficiency at the upstream side of the heat exchanger which might be required to obtain sufficient CO removal need not seriously affect its effectiveness in NO emission reduction, because of the high activation energy of the rate determining step of the reaction scheme involved in formation of the latter.

As the authors indicate, loss of CO in the pre-exchanger region implies imbalance of the hydrogen combustion reactions; the required condition for the observed loss of CO is a deficit of hydrogen atoms. An order of magnitude calculation carried out on the results indicates that the excess of radicals implied by the hydrogen concentration was sufficient for the hydrogen atom concentration to be similar to that of hydroxyl radical and oxygen atoms (because of the increase in the ratio of hydrogen atom concentration present to thermodynamic equilibrium concentration as the cube of the same ratio for hydroxyl radical and 3/2 power of that ratio for oxygen atoms). The presence of this large concentration may have favoured reactions H + H + M and H + OH + M and other termolecular reactions involving H sufficiently for these to be the dominant paths of radical removal. (Although H_2 exceeds equilibrium overall, the excess is not as great as for H, consequently reaction H + H + M will 'go'). This may have caused the observed H deficit; the effect on H_2 concentration need not be as great as that on CO because of adjustment within the radical 'pool', and some hydrogen regeneration by H + H + M.

The implied imbalance is more severe than would be expected to be due to the above cause alone, and it seems possible that some loss may have occurred during sampling, possibly due to H diffusion to, and removal at the walls of the probe.

It appears that further details of the relationships between the radical concentrations and those of related stable species, under the practically encountered conditions of approach to overall equilibrium, may be required to determine pollutant emission levels with high accuracy. Nevertheless, the results presented here show that it is possible to design a practical combustor with acceptable emission levels, by quenching in two stages: first cooling the combustion products from their maximum temperature down to about 1600°K within a period of about two milliseconds – this step will reduce the rate of formation of NO sufficiently – followed by a stage of low heat loss at around 1600°K for about 20-30 millisecs to allow CO oxidation to be completed after which the cooling to the final exhaust gas temperature could follow.

R. M. Fristrom *(Johns Hopkins University)*

This interesting paper presents the study of a new steam generator design and correlates combustion volume with levels of CO and NO pollution in the exit gases. These results are compared with predictions by several partial mechanisms. Another significant and fortunate observation is that NO production is quenched by the heat exchanger while the depletion of CO by reaction continues.

Several remarks seem in order. First, this is a rather good boiler. Volumewise, it is 100 times more efficient than a typical home oil furnace and some 20 times as good as a Navy boiler. On the other hand, it is twentieth of the volume rate attained in a ram jet and a hundredth of the theoretical limit set by kinetics. (1)

The authors find no evidence of prompt NO (2) and appear to favor the attribution of non-equilibrium NO to super-equilibrium concentrations of O atoms in the flames. This is an intuitively satisfying viewpoint since it allows a discussion of the problem in terms of known reactions. It is both an experimentally and theoretically established fact that radical concentrations in excess of final thermal equilibrium values occur in flames. (3,4) In order to make reliable apriori predictions of NO and CO formation or disappearance rates in flames, it is necessary to know these radical concentration levels.

Radicals in the O-C-H flame system can be shown to be in partial equilibrium with each other even when the total radical concentration is in excess of equilibrium. (4)

This occurs because the reactions which exchange the radicals are rapid compared with the other steps in particular radical formation and fuel and oxygen disappearance. This concept of partial equilibrium offers a tool for estimating the relative concentrations of radicals in a system, but it offers no handle on the absolute concentrations. As the authors note that if one knows the oxygen and hydrogen concentrations in the reaction region, estimates can be made; but the existence of substantial concentrations of hydrogen is also a kinetic phenomena and the prediction of such a quantity requires some knowledge of the dominant kinetics.

A number of observations are pertinent to the existence of thermodynamic excess of free radicals. They occur because the branching reactions forming them are more rapid than the reverse recombination reactions, since the branching reactions will normally depend linearly on radical concentrations while the reverse reactions will depend quadratically. (5,6) At low temperatures, the recombination will dominate because of its 1/T dependence compared with negative exponential dependence of the branching step. There will be a crossover point (5) beyond which excess radicals will be generated. The generation dominance will continue until the depletion of source species for the branching and the buildup of radical concentration finally limit the branching. This limiting point will always occur despite the exponential dependence of the rate of the bimolecular step because exp (-E/RT) can never exceed unity while the ratio of branching source molecule to radical concentration can approach zero.

The amounts of radicals formed should be equivalent to the number of cycles of the branching. In oxygen flames, it is usually assumed that the $H + O_2 \rightleftarrows OH + O$ reaction is responsible for the branching. If, as appears to be the case, all of the oxygen disappears by this reaction; the upper limit for radical concentration in such flames would be attained if recombination was negligibly slow compared with branching. This would mean that the peak radical flux would be equal to the inlet oxygen flux and since radical concentrations goes through a maximum that the concentration maximum would be only slightly lower. The observed concentrations in flames are an order of magnitude below this upper limit and one must conclude that in the flame reaction zone the rates of formation and recombination of radicals must be almost balanced. (7)

With a restricted model of this type, one can in principle estimate peak radical concentrations in flames and from this derive rates of the rate and velocity controlling steps in a flame. The alternate to restricted models of this type is a complete formulation of the flame equations which for a system of this geometry and chemical complexity is at present prohibitively expensive as a mathematical exdreize especially so because certain rate constants are not known with sufficient precision for the purpose.

For the present, the approach of the authors of making a judicious combination of experiment and theory to determine the controlling factors is a wise choice.

References

1. *W. H. Avery, "Space Heating Rates and High Temperature Kinetics", Fifth Symposium (International) on Combustion, The Combustion Institute, 1955, p. 86.*
2. *C. P. Fenimore, "Formation of Nitric Oxide in Premixed Hydrocarbon Flames", Thirteenth Symposium (International) on Combustion, The Combustion Institute, 1971, p. 373.*
3. *J. O. Hirschfelder, C.F. Curtis and R. B. Byrd, Molecular Theory of Gases and Liquids, John Wiley, New York, 1954, p. 785.*
4. *R. M. Fristrom and A. A. Westenberg, Flame Structure, McGraw-Hill, New York, 1965, p. 205.*
5. *F. Weinberg, "The Significance of Low Activation Energies in Flames", Proc. Roy. Soc., Vol. A230, 1955, p. 331.*
6. *R. M. Fristrom and R. Sawyer, "Flame Inhibition Chemistry", Proceedings of AGARD Conference #84 on Aircraft Fuels, Lubricants and Fire Safety Sec. 12, Paper AGARD-CP-84-71, 1971.*
7. *R. M. Fristrom and R. F. Sawyer, "An Ultra Simple Model for Flame Chemistry" (In Preparation).*

A. F. Sarofim

I wish to thank Prof. Beér and Dr. Fristrom for their kind comments. Their prior contributions on the effects of superequilibrium radical concentrations assisted us greatly in the interpretation of our experimental results. In response to Dr. Fristrom's question relating the selection of cooling rate, we were fortunate to have selected an exchanger that cooled the combustion products at a rate that quenched the NO formation reactions while permitting the CO burnout reactions to proceed. We have set up a mathematical model to determine optimum cooling rates but we expect the results to be qualified by the uncertainty in the kinetic expression for CO burnout.

R. M. Fristrom

As far as the kinetics are concerned, may I beat the drum for chemical kinetics. After all, this is one of my fields. There is definitely an improvement in the reliability of chemical kinetics which is available. Beyond that, the data are getting into a form, more easily accessible. The Leeds work which has been mentioned prominently at this meeting is by far the most reliable. In addition to this work, Prof. Sawyer of the University of California and a number of other persons are evaluating critically kinetic constants which is really what we want for practical devices.

Two very excellent surveys of the chemical kinetics literature have been conducted which are now available. One is by Milne and Dickenson which has recently been updated in 1971, and which effectively surveys the gas phase chemical kinetics. The other survey by Prof. Kondratiev at Moscow is similar. So now we have two independent sources. In general, I feel that the literature has been well searched. The Russian text was of course published in Russia in small quantities. We have made a translation of this Russian paper at APL, and it will be published by the Bureau of Standards. This compilation, which does include incidentally some critical evaluations by the group at Moscow, should be available through the government printing office hopefully sometime toward the end of the year.

E. S. Starkman *(Environmental Activities Staff, GMC)*

I think that this work points up something that all of us should realize. The learning curve in continuous combustion systems is pretty steep today. We have much to learn even though we may know something about how you would set up a flame in a system such as a Rankine cycle with respect to the heat exchanger surfaces. But I want to go on beyond that and say, please, let's not extrapolate the emissions for a steam automobile from one data point. The internal combustion engine also looks pretty good cruising at 50 miles an hour, and the internal combustion engine has a lot of difficulty with emissions when changing from one load point to another, moreso than a boiler would have. Yes, a boiler has an easier job of it. But the simple extrapolations sure make a lot of people nervous.

G. C. Williams

In this paper, we evaluated only one operating point. It does not represent an entire driving cycle. Unfortunately, we do not have all of the data needed to calculate a driving cycle for this particular system. We make no claims as to what the effect would be of accelerating from zero to 60 MPH in 15 seconds or what a cold start is like with this system. I suspect that we should not be as bad as an I. C. engine.

W. H. Lipkea *(Research Laboratories, GMC)*

I wish to direct some comments to Dr. Sarofim concerning the comparison of their burner emission data with the Federal Standards. It should be emphasized that to reduce pollutant emissions based on gm/kgm of fuel to a gm/mi basis requires a value for the fuel economy. The value used in the paper of 7,100 Btu/mi at 50 mph (about 18 mpg) is not realistic and a value of 22,000 Btu/mi; based on GMR SE-101 steam car data is more representative. The effect of fuel economy on emission data is shown in Fig. 1, which is a replot of the paper data using 22,000 Btu/mi fuel economy. The most significant point is the NO emissions exceed the Federal Standards by a factor of 2.

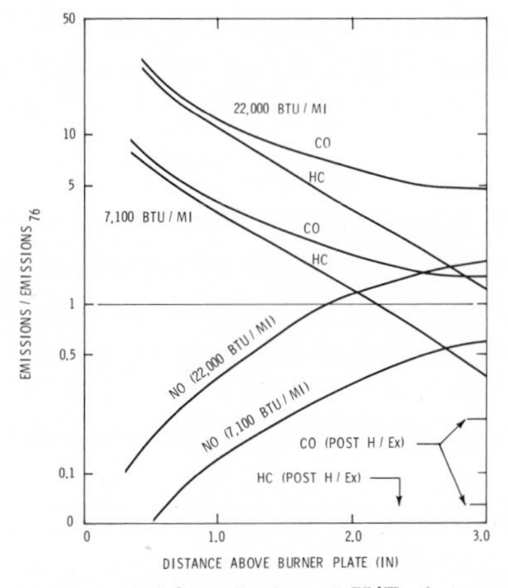

Fig. 1. Emissions as a function of 76 Standards and H/Ex burner plate gap (@ 50 mph).

A. F. Sarofim

The value of 7,100 Btu per mile is based on preliminary design considerations including allowance for the expected efficiencies of the various drive mechanisms. It is not possible to explain the differences between the value quoted here and that observed at 50 mph on the GM SE101 steam car without a detailed examination of the sources of inefficiencies. It should be noted, however, that the emission values quoted are indicative of what a laboratory burner produces under steady operation and are not presented here as values expected for either cycle operation or for a production burner. The results, however, are promising enough to warrant further study of systems of this or similar design.

R. F. Sawyer *(University of California)*

We have been performing some experiments in which the cooling is controlled fluidynamically in gas turbine type combustors and have found similar results. I am interested in two points. Firstly, how one dimensional is your experiment? It appears to resemble closely a plugged-flow reactor. Secondly, is the cooling rate a good measure of a critical cutoff for fixing carbon monoxide? In our experiments, a critical temperature describes the point beyond which the carbon monoxide reaction is so slow that no disappearance is observed. This of course depends weakly on the pressure level and probably on the type of device. This temperature is about 1500°K in our system. What is the gas temperature drop across your heat exchanger?

A. F. Sarofim

We believe that the concentration distributions were one dimensional beyond the tips of the flamelets stabilized at the burner plate. No radial traverses were made but the fuel/air ratio calculated from the traverse along the axis equaled the overall fuel/air ratio. In our experiments we were unable to observe the temperature at which the carbon monoxide reaction is effectively quenched since the temperatures at the exit of the heat exchanger (ca 1000°K) were below the effective quench temperature. For such cases the rate of cooling through the exchanger plays a dominant role.

EMISSIONS FROM GAS-TURBINE-TYPE COMBUSTORS

N. J. FRISWELL

Shell Research Limited, Thornton Research Centre, Chester, England

ABSTRACT

A combustion rig simulating gas turbine combustion has been used to study pollutant emissions, particularly those of carbon and nitrogen oxides, NO_x, as a function of air-fuel mixing pattern. The use of metallic additives to remove carbon has been investigated and mechanisms proposed to account for the action of barium and manganese in this system. The influence of combustion chamber design on NO_x emissions has shown that large variations in NO_x level may occur and some correlation with primary zone residence time has been found. For a given flame tube design, no influence of either overall or primary zone air-fuel ratio (AFR) on NO_x levels (normalized to stoichiometry) has been found. This surprising result is explained by either the existence of a combustion zone of constant AFR or diffusion-flame burning of the fuel.

INTRODUCTION

It is unnecessary to reiterate the importance of pollution today; the search for "clean" combustion has been stimulated by legislation setting limits on automotive emissions. It has frequently been claimed that the gas turbine offers the lowest pollutant emissions of any internal combustion engine. Unfortunately, the supporting evidence for this has often been distorted by the concentration units used to express exhaust pollutant values. Nevertheless, it seems clear from recent work (1) that the only pollutants in automotive turbine exhausts with mass emissions as high as, for example, those in a spark-ignition engine – and whose formation may be difficult to limit – are the nitrogen oxides, commonly referred to as NO_x. The reduction of NO_x in aircraft gas turbines is also likely to pose a difficult problem. Unlike the case with

References p. 175

aircraft gas turbines where smoke emission has been a continuing problem, with automotive turbines (carbon) smoke emission has not been claimed to be a serious problem; this is not surprising in view of the low compression ratios used. In both systems, unburnt and partially burnt hydrocarbon emissions may be a problem at idle operation but, as they result from poor fuel-air mixing at this engine condition, design changes should easily solve it.

It is envisaged that NO_x emission may well be the major problem in both automotive and aviation gas turbines. In aviation turbines, the trend is towards higher compression ratios leading to higher air inlet temperatures, which increase the rate of formation of nitric oxide (NO) in the combustion chamber. In addition, recent attempts to attain more complete combustion (primarily to reduce smoke emissions) have resulted in increased NO_X emissions (2). In the use of gas turbines for automotive purposes, the efficiency of the engine depends on the use of a regenerator to heat the incoming air – thus, again, increased air inlet temperatures will lead to higher NO_x levels. However, investigation of pollution (apart from smoke) from gas turbines has been started only very recently and, hopefully, changes in combustion chamber design may reduce NO_x emissions.

It has been mentioned earlier that smoke has been a persistent emission from aircraft gas turbines. Design changes have kept pace with the increase in smoke production with increasing compression ratio so that the overall level of smoke emission has remained fairly constant. Recent re-introductions of vaporizing fuel injectors appear (possibly only temporarily) to have reduced the problem. With the automotive gas turbine, little emission of carbon smoke has been found though even small amounts could cause problems of sooty deposits within the engine, particularly on the regenerator. The problem of smoke may, however, become more severe as changes are made to the design of the combustion chamber to achieve low levels of NO_x, e.g. it has frequently been proposed that a solution to the NO_x problem is to have a rich primary zone – but undoubtedly this could cause larger smoke emissions. Therefore, the smoke problem may become important here too and for this reason we have, in our experimental work reported here, concentrated our attention on smoke as well as NO_x emissions.

Heywood and co-workers (3, 4) have recently published theoretical considerations of NO and carbon (smoke) formation in gas turbines. Their predictions confirm those made by others that NO production in a gas-turbine combustion chamber is limited by the slow kinetics of its formation and that, additionally, the concentration attained in the combustion zone is effectively "frozen" so that no reduction in NO concentration takes place in other parts of the chamber or exhaust system. The main conclusion from this work, and one that is accepted here, is that one can only limit the rate of formation of NO and that this must be achieved by varying the method of air and fuel mixing throughout the combustion chamber. The factors controlling NO formation are temperature, free radical concentrations and residence time in the combustion zone and are, in turn, controlled by the air-fuel mixing pattern. Heywood et al (4) have predicted theoretically, from a known kinetic scheme, the formation of NO in various parts of the combustion zone as a function of air-fuel mixing. In contrast to NO emission, carbon formed in the combustion zone may be subsequently removed by oxidation in the dilution zone of the combustion chamber and Heywood et al (3) have attempted semi-quantitatively to predict this oxidation of carbon.

In the work reported here, the problems of NO_x and carbon emissions have been approached from the opposite point of view i.e. attempts have been made to relate the concentrations of NO_x and carbon measured in a combustion rig simulating gas turbine combustion to (a) the design of the combustion chamber and, in particular, to the air-fuel mixing pattern and (b) the composition of the fuel used. An integral part of this work has been the use of a water flow visualization rig to obtain, at least, qualitative information concerning the air flow distribution and the recirculation patterns within the primary zone. The composition of the fuel used has been varied in (a) the hydrocarbon type and (b) the use of metallic additives; only the latter part will be reported here. The use of metallic additives – principally anti-smoke additives – has been discussed in the past (5); their use is likely to be qualified because of their tendency to build up deposits of the metallic oxide on parts of the engine and, in some cases, because of the toxicity of the metal used. Nevertheless, the quantities required to substantially reduce smoke emission can be very small and, as a short-term expedient to the problem of smoke emission, their use can be contemplated. The work on additives reported here is an extension of previous studies in laboratory flames (6) which led to proposals of mechanisms by which various metals reduce soot. The application of these mechanisms to explain soot removal in the combustion system used here gives considerable information about the nature of the combustion process in gas turbines.

EXPERIMENTAL

The salient features of the combustion rig are illustrated in Fig. 1 and a more detailed diagram of the combustion chamber is shown in Fig. 2. The rig has been designed for operation up to a pressure of 25 bar but so far has only been run at pressures up to 3.4 bar. In the present work, unless otherwise stated, all results have been obtained at 3.4 bar with a 3.43 cm diameter nozzle at the downstream end of the chamber.

Fuel, aviation kerosene (Avtur) meeting the U.K. D.Eng.R.D.2494 specification, is supplied by a Lucas IP 60 pump, metered by a rotameter and injected into the combustion chamber by a Monarch atomizer of 3.5 flow number and 60° cone angle. Fuel additives used have been barium, manganese and iron in the forms of Lubrizol 565, Ethyl CI-2 and ferrocene respectively.

To simulate a "real" system, the combustion air can be preheated to 900 K by burning hydrogen in the gas stream before the main combustion chamber (and

References p. 175

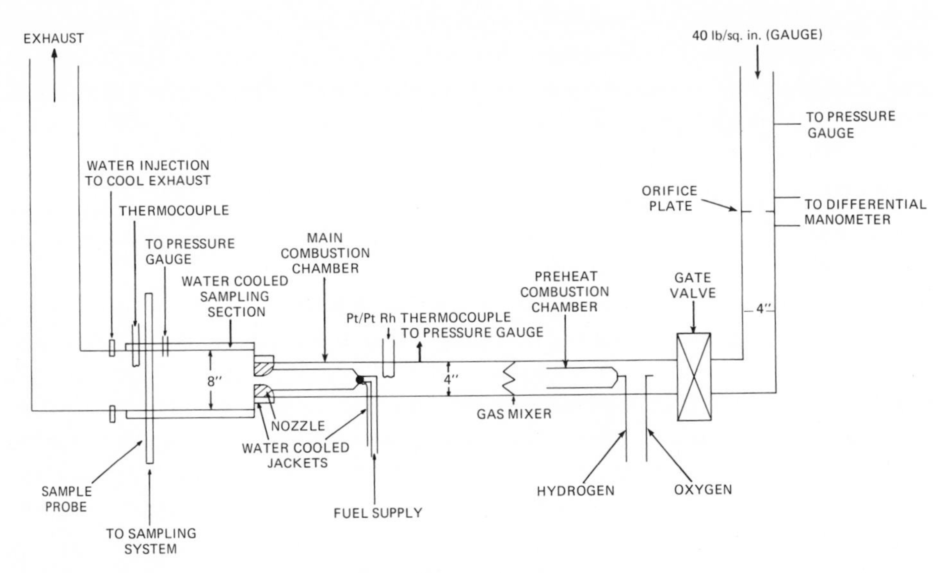

Fig. 1. The combustion rig.

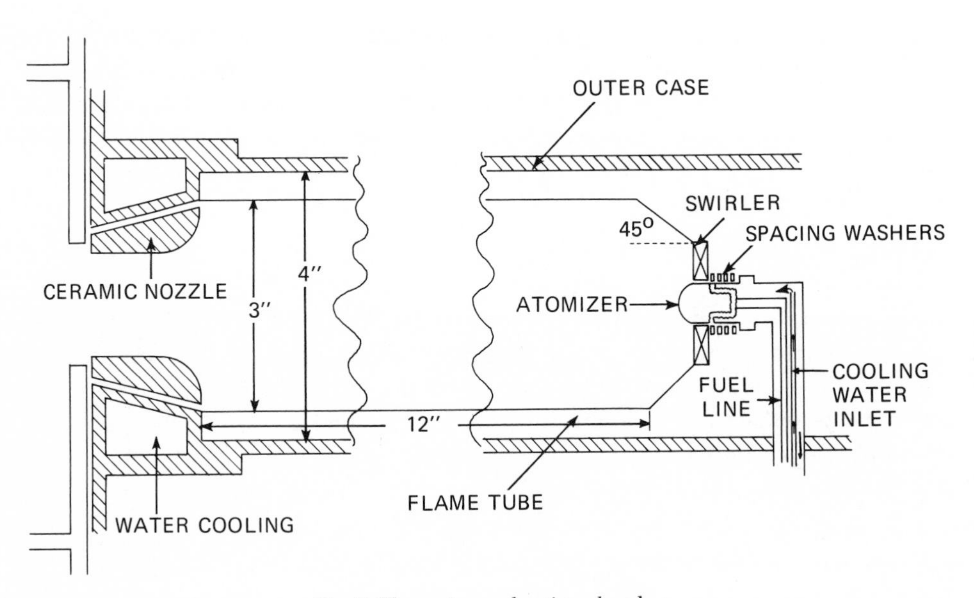

Fig. 2. The main combustion chamber.

replacing the oxygen consumed). This results in a maximum of 9% of water vapor in the air plus a small concentration of NO_x, the amount of which is measured and subtracted from that produced in the main combustion chamber.

The air flow pattern is determined by the positioning and size of the holes in the flame tube and by the design of the swirler. Three designs of tube have been used in this work in combination with a number of swirlers of various design. The basic design of flame tube is illustrated in Fig. 3 and designated as flame tube A. The other two tubes were modifications of this design and involved:

1. Replacement of the first three (upstream) rows of holes by four rows of louvred slots – tube B.

2. Lengthening the primary zone by moving the first row of dilution holes downstream by 3.8 cm – tube C.

In all cases the total area of holes in the primary zone remained constant.

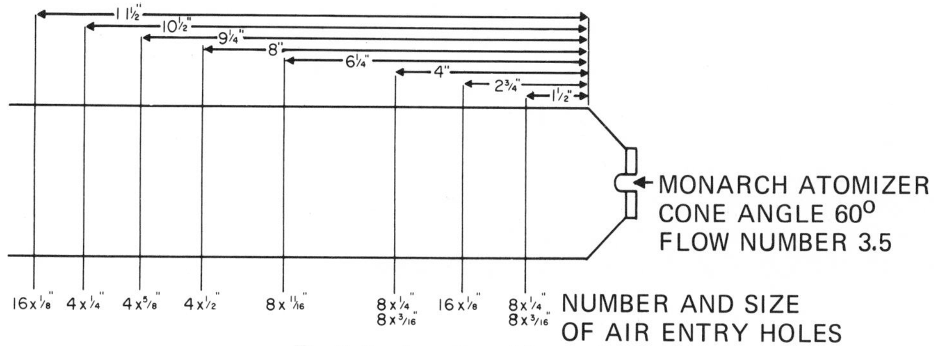

Fig. 3. Configuration of flame tube A.

The basic design of the swirler involved a disc with eight slots cut out at an angle of 60°. Variations were produced by changing the width and depth of these slots.

Analytical Procedure – Samples of exhaust gas were extracted via a water-cooled stainless-steel probe.

Carbon – A heated sampling line was used with a high gas-velocity (30 m/s) to avoid any surface deposition of carbon. Analysis was by both calibrated smokemeter and by filtration/weighing.

NO_x – A modified Saltzman apparatus (7) was used to measure both ($NO + NO_2$) and NO_2 alone (and thus NO by difference). The main features of this apparatus are oxidation of NO to NO_2 by acidified sodium dichromate supported on glass beads and reaction with the Saltzman reagent by passing the exhaust gas through a sintered thimble into the reagent. The accuracy and repeatability of this method has been estimated as ± 5% (7). The value of the "Saltzman factor" used has been 0.72.

Unburnt Hydrocarbons, CO and CO_2 – Standard analytical techniques of flame ionization and NDIR (non-dispersive infra-red spectroscopy) have been used for these constituents.

References p. 175

Water Flow Visualization Rig – This apparatus and its use have been thoroughly described previously (8). In this work polystyrene tracers have been used to record flow patterns and measure residence-time distributions in the primary zone. Residence times for individual tracers injected into the atomizer orifice were measured with a stop-watch at various water flows.

RESULTS

Water Flow Rig – Flow patterns for flame tubes A, B and C are shown in Figs. 4, 5 and 6. The swirler used to obtain these patterns was, in each case, 0.476 cm wide and

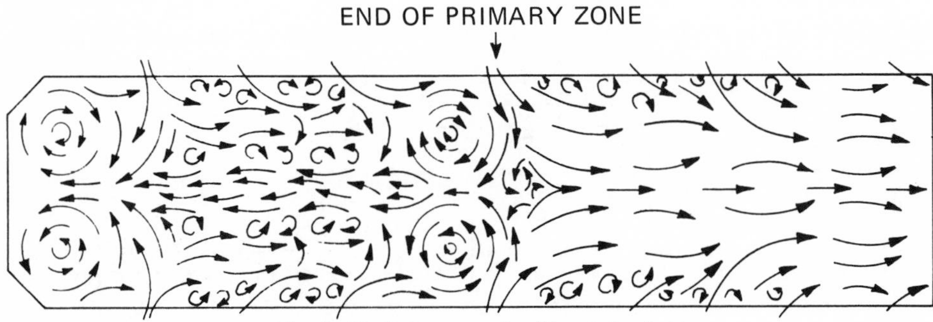

Fig. 4. Air flow pattern for flame tube A.

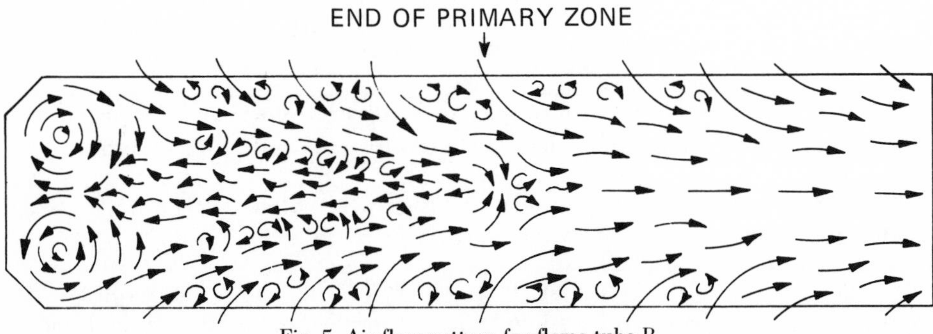

Fig. 5. Air flow pattern for flame tube B.

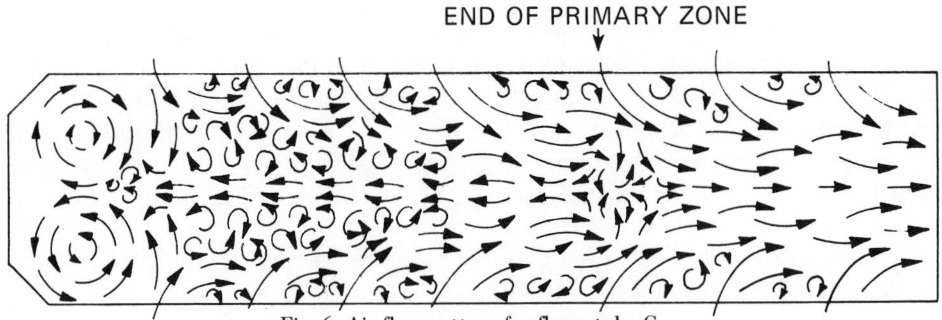

Fig. 6. Air flow pattern for flame tube C.

0.635 cm deep. The main features of these patterns are: (a) the vortices at both ends of the primary zone in tube A; (b) the elimination of the downstream vortices in tube B by the stronger flow through the slots leading to a more compact recirculation zone; and (c) the lengthening of the primary zone in tube C while the good recirculation associated with the design of tube A is still maintained.

Mean residence times in the primary zone, i.e. times for particles injected into the atomizer orifice to reach the first row of dilution holes, were determined as 1.25, 1.05 and 1.77 sec. for tubes A, B and C respectively at a water flow of 13.6 kg/sec., corresponding to a cold air flow in the combustion rig of 0.419 kg/sec.

Combustion Rig – Preliminary experiments established that manganese and iron had almost identical effects on exhaust carbon over a wide variety of conditions. A similar effect had been found in earlier work in laboratory diffusion flames (6) and therefore only barium and manganese were examined in detail in this work.

The principal results obtained are contained in Figs. 7 and 8 showing the amounts of soot removed (in μg/1) at various overall air-fuel ratios (AFR) plotted against

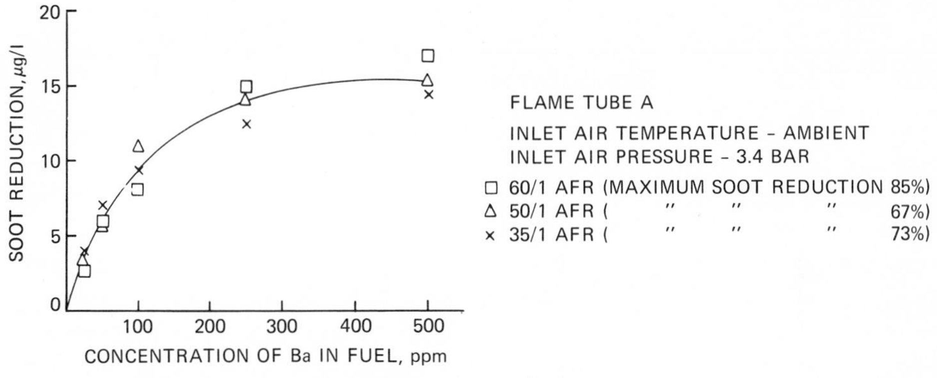

Fig. 7. Soot reduction by barium.

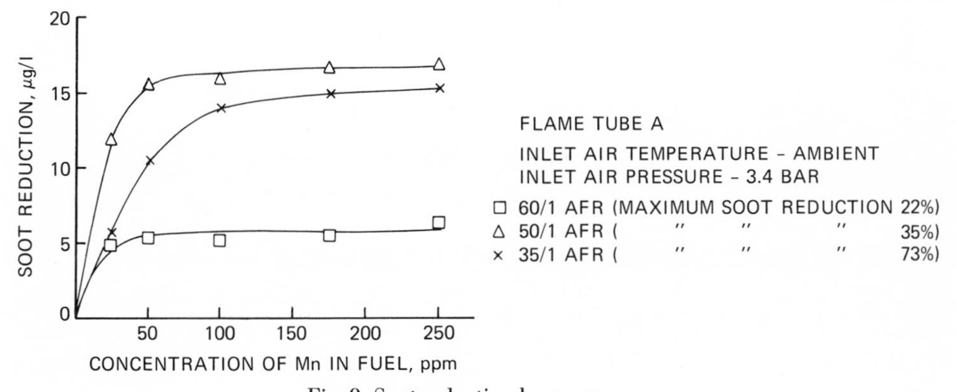

Fig. 8. Soot reduction by manganese.

References p. 175

concentration of metal in the fuel. These results are for flame tube A. Other significant results obtained with this flame tube are shown in Table 1.

TABLE 1

Effects of Ba and Mn on Carbon Formation in Flame Tube A

Air inlet temperature – ambient

Air inlet pressure – 3.4 bar

Fuel Additive	Overall AFR	Effect on Exhaust Carbon
250 ppm Ba	80/1	Complete removal
500 ppm Mn	100/1 and 70/1	No removal
250 ppm Ba	60/1	26% removal
100 ppm Mn	60/1	11% removal
250 ppm Ba + 100 ppm Mn	60/1	47% removal
250 ppm Ba	50/1	17% removal
100 ppm Mn	50/1	3% removal
250 ppm Ba + 100 ppm Mn	50/1	22% removal

Flame tube B produced much larger amounts of soot than did flame tube A – typically two to three times as much. With tube B, neither barium nor manganese was effective at any AFR at ambient air inlet temperature – though, in fact, the irreproducibility of the rig, together with the high carbon levels, could have obscured small reductions. When the inlet air was heated to 400°C, reductions of approximately 40% in carbon content were observed with both additives. During these experiments with preheated inlet air, some measurements of exhaust NO_x concentrations were taken. At an AFR of 50/1, barium increased NO_x by 30% while iron had no effect (the effects of manganese and iron on soot concentrations had been shown earlier to be virtually identical). These results are considered to be extremely significant in both confirming the proposed mechanism by which barium removes carbon and giving insight into the mechanism of NO formation.

Effect of Air-Fuel Mixing on Pollutant Emissions – Measurements of carbon, NO_x, CO_2, CO and unburnt hydrocarbons have been taken over a range of AFR from 30/1 to 120/1 at ambient air inlet temperatures for flame tubes A and B and are shown in Table 2. The values for hydrocarbons were extremely erratic and only general levels have been quoted. To take account of the different air dilutions at different overall AFR, the results have been normalized to AFR = 15 (stoichiometric) by multiplying by AFR=15, as discussed by Sawyer and Starkman (9). The normalized emissions quoted are thus proportional to the mass of pollutant per unit mass of fuel.

TABLE 2

Normalized Emissions from Flame Tubes A and B

AFR	Carbon, $\mu g/l$	NO_x, ppm	CO_2 %v	CO, %v	Hydrocarbons, ppm as Hexane
Flame Tube A					
120	24	105	12.5	0.16	
100	60	121	14.9	0.20	
80	91	119	14.7	0.30	About 15 ppm actual concentration
60	124	106	13.6	0.60	
40	115	85	13.9	0.80	
30	46	83	13.0	0.18	
Flame Tube B					
120	587	57	12.0	1.28	
100	290	74	13.3	1.40	
80	184	75	13.4	1.55	About 15 ppm actual concentration
60	200	74	12.0	1.80	
40	183	52	10.1	1.87	
35	117	–	–	–	
30	–	62	12.4	0.60	

Results from flame tube C proved to be extremely variable. Initially, measurements of NO_x only were made and the day-to-day variations were very large. (Accumulations of carbon on the walls of this tube and tube A necessitated daily cleaning of the tubes.) It is thought – though not proven – that these variations were due to the build up of deposits resulting in changes in air distribution pattern. Because of the variability of the NO_x results, detailed measurements of other pollutants were not made. Some of the NO_x results obtained are, however, incorporated in Fig. 9.

With flame tube B, the size of the swirler slots was changed considerably and also slots were added to the "dome" joining the swirler to the flame tube in an attempt to change the primary zone air-fuel ratio (and thus, presumably, NO_x levels). A few experiments with flame tube A and with slots in the dome were also carried out. However, NO_x concentrations remained virtually unchanged throughout these changes though levels of carbon and CO changed considerably.

DISCUSSION

The emissions of different exhaust pollutants from a given combustion device are necessarily interrelated to some extent. However, the emissions of carbon and NO_x will be considered separately at first and then the main conclusions will be correlated together with a picture of the combustion process in this type of system.

References p. 175

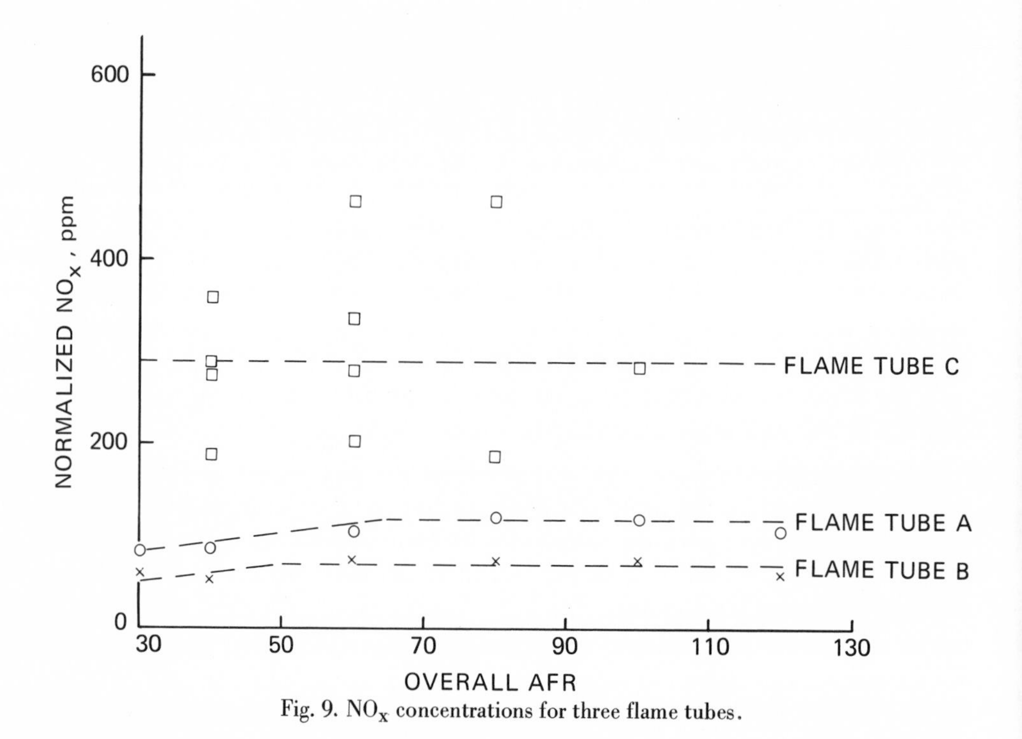

Fig. 9. NO_x concentrations for three flame tubes.

Effect of Additives on Carbon Emission – Previous work on laboratory diffusion flames (6) had examined the effects of a large number of metallic additives on soot (or carbon) production and, besides identifying the more efficient metals, proposed mechanisms by which these additives operate. The metals tested appeared to operate by at least two basic mechanisms: (a) the alkaline earth metals and also, probably, molybdenum catalyzed the decomposition of either hydrogen or water to give free radicals which rapidly removed soot; (b) other metals were probably incorporated into soot particles and subsequently catalyzed the oxidation of soot by oxygen in cooler parts of the flame.

The temperature-dependencies of these two mechanisms were very different. The alkaline earths were only effective at very high temperatures ($>$1700 K) and thus operated in the flame itself, whereas other metals were effective at lower temperatures and could operate in dilution/entrainment of flame gases with air. Therefore, it was proposed that a combination of additives from either group might be effective throughout all the temperature regimes of an engine such as a gas turbine.

In the present work, it was hoped to substantiate the above conclusions and the most effective metals operating by each mechanism – barium and manganese – were chosen for detailed study. An additional advantage of using these two metals was that they are the most widely used constituents of anti-smoke additives.

Figs. 7 and 8 show the reductions in soot at varying AFR as a function of additive concentration in the fuel for flame tube A. At AFR above 60/1, barium at 250 ppm completely eliminated soot whereas manganese had no effect on soot concentrations.

The results suggest that the effect of barium in terms of the amount of soot removed for a given concentration of metal is independent of AFR whereas that of manganese is very dependent on AFR. These results initially were extremely surprising. It had been shown (6) that the effect of barium was very temperature-sensitive, and it was expected that the temperature of the combustion zone would change considerably with changing AFR. The preliminary conclusion, therefore, might be that, in fact, the temperature of the combustion zone (as distinct from the primary zone as defined in the air flow pattern – Fig. 4) remains essentially constant as the overall AFR is changed. This point is an important one and it will be returned to later as it arises when the NO_x results are discussed.

The results for manganese suggest that this additive becomes more effective with increase in exhaust gas temperature. They support the contention that the metal is incorporated into soot particles and accelerates their rate of oxidation in the dilution zone and exhaust system. It is well-known that such impregnation of soot particles lowers their ignition temperature.

The amounts of soot removed with increasing concentrations of both manganese and barium reach a plateau at respective metal concentrations of 100 and 250 ppm. To gain support for the idea that, since the additives operate by different mechanisms and in different parts of the combustion system, a combination of the two might be effective throughout the whole system, a blend of fuel containing 100 ppm manganese and 250 ppm barium was prepared. The results are given in Table 1 and show that the amounts of soot removed by the combination of additives is approximately equal to the sum of the effects of the metals on their own. It is thought that the slight synergism which might be assumed from the 60/1 result is not real but due to experimental error.

Though all of the above results support the mechanisms proposed from previous diffusion-flame studies, they do not prove them. More convincing evidence is found in the effects of additives on NO_x concentrations. The crucial assumption in the barium mechanism (for which supporting evidence is given in Ref. 6) is that in carbon-containing parts of a flame free-radical concentrations may be reduced below their thermal equilibrium levels and that barium acts by restoring these concentrations towards equilibrium by catalysing the decomposition of water and hydrogen. The rate of NO formation in a flame is dependent on temperature, time and free-radical concentrations. Thus, if barium increases free-radical concentrations it should increase the rate of NO formation whereas metals reducing soot by different mechanisms (e.g. manganese and iron) should not. This has been confirmed by results quoted earlier namely that in flame tube B with an air inlet temperature of 400°C, barium increases NO_x by 30% while iron has no effect. Under these conditions, barium and manganese (and presumably iron) reduce soot by 40%.

References p. 175

From the foregoing we may say that the present results support the mechanism proposed (6) for the mode of action of metals in reducing soot. Therefore, the alkaline earth metals, represented in this work by barium, remove soot in the flame zone itself and their effectiveness increases markedly with flame temperature and residence time in the flame. Other metals appear to work in other regions of the combustion system i.e. the recirculation, secondary and dilution zones and the exhaust system and, with typical gas-turbine residence times, have a lower limit of operation of around 800K. There is almost certainly an upper limit to the effective concentration of additives. For barium, this is due to the fact that eventually free-radical concentrations are restored to equilibrium and any remaining exhaust smoke must be carbon that is not retained for sufficient time in the flame zone to be removed by free radicals (though of course, it could be removed, by the addition of manganese, in other parts of the system). The limiting concentration effectiveness of manganese has been noted previously (10).

NO_x Emission – Fig. 10 shows the variation in carbon, CO and NO_x concentrations (all normalized to AFR = 15) over a four-fold range in overall AFR.

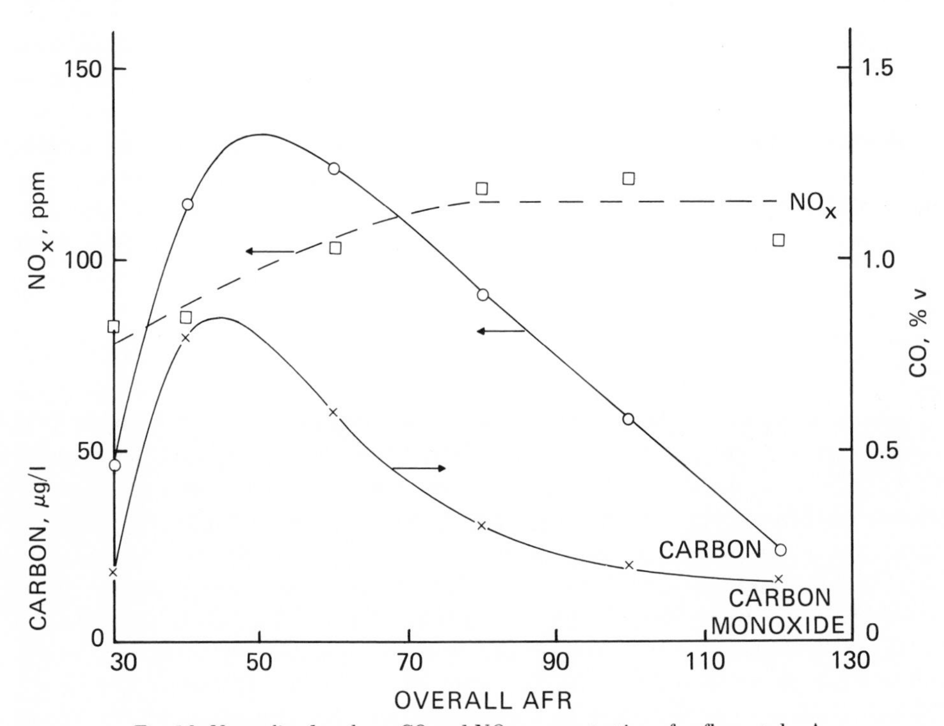

Fig. 10. Normalized carbon, CO and NO_x concentrations for flame tube A.

The striking point is that while carbon and CO concentrations vary considerably with AFR there is virtually no variation in NO_x, except for a slight reduction at low AFR. The experimental scatter is demonstrated (Table 2) by the variation in (CO_2 + CO) concentrations (unburnt hydrocarbon concentrations being negligible) and, allowing for this, normalized values of NO_x concentration are even closer to being constant. In fact, in all of the work so far with the combustion system this invariance of normalized NO_x with AFR has been found. Its significance will be discussed later. The normalized NO_x concentrations found for the three flame tubes used are shown in Fig. 9. The scatter in the results for tube C has been commented on earlier. It is believed to be due to a variable build-up of deposits on the flame tube walls which distorted the air flow pattern i.e. the detailed pattern shown in Fig. 6 may not be that which results in high levels of NO_x. Nevertheless, we are confident that the high concentrations of NO_x measured are real though we are unsure of the pattern of air-fuel mixing which has produced them. Thus, Fig. 9 shows the extent to which NO_x concentrations can be varied by changes in air-fuel mixing and also the scope which engine designers have in attempting to reduce NO_x levels in gas turbine exhausts.

Two points arising from these results need further comment, viz: (a) the invariance of NO_x with AFR, and (b) the differences in NO_x levels found for flame tubes A, B and C.

A similar invariance of normalized NO_x with AFR may perhaps be deduced from the results of Smith, Sawyer and Starkman (11) and has also been noted in unpublished measurements by the author on an industrial turbine burning gaseous fuels and running at a constant governed speed at all loads. In other turbine systems, of course, a change in load, though changing AFR, changes air inlet temperature, and sometimes pressure thus affecting the rate of formation of NO_x.

It has frequently been proposed that the best approach to limiting NO_x formation in a gas turbine is to run the primary zone at a fuel-rich condition. The results of our work in varying the size of swirler and in the addition of slots in the dome suggest that (a) changing the overall primary zone AFR does not change NO_x significantly and (b) within the primary zone there exists a "combustion zone" where NO is formed and which is not changed in composition by changes in either overall AFR or primary zone AFR. The results on the effect of barium in reducing carbon also suggested the existence of a combustion zone of constant conditions. Thus, the desired changes in NO_x may perhaps only be achieved by different air-fuel mixing patterns (including atomization changes).

The above results might be explained by the actual combustion zone remaining at a fairly constant AFR with changes in overall AFR or by the fuel droplets burning primarily as diffusion flames with the flame front at the stoichiometric "surface". The results quoted earlier from an industrial turbine, burning gaseous fuels, do not tend to support the latter view though information is lacking concerning the air-fuel mixing pattern. The slight fall in normalized NO_x concentration at low AFR could be due to a richening of the combustion zone such that there is insufficient air to burn all of the fuel droplets at the stoichiometric surface, thus lowering flame temperature

References p. 175

and rate of NO formation. The effect of more drastic changes in flame tube design and fuel injection method (e.g., air-assisted or vaporizing atomizers) is now under investigation.

The variations in normalized NO_x concentrations for the three flame tubes studied (Fig. 9) are in good qualitative agreement with the primary zone residence times found in the water flow rig though the relative increase in residence time from, say, tube B to tube A is less than that in NO_x concentration. An explanation of this might be found in Fenimore's hypothesis (12) of "prompt", or flame front, NO formation.

At present it should be stressed that the conclusions and discussions of this part of the work are preliminary and tentative. Nevertheless, the experimental results are considered to be extremely significant and merit much further investigation.

GENERAL REMARKS

From the above results and discussion, a picture of the combustion process in gas turbines emerges which is consistent with that described by Toone (13) and Heywood et al (14). A fuel-rich region exists around the fuel injector and accounts for most of the carbon formation. The fuel entrains air and, from the action of the swirler, flows and burns around the outer edge of the primary zone. Part of the burnt gas recirculates in the center of the primary zone to provide continuous ignition of the fuel-air mixture. The results presented in this paper suggest that conditions in the "combustion zone" remain fairly constant with changing overall AFR and that the NO_x measured in the exhaust gas is formed in this zone. Variation of exhaust carbon concentration with overall AFR (while normalized NO_x does not change) arises because carbon is formed in a different part of the combustion zone and is also subject to oxidation processes later in the combustion chamber, whereas NO_x concentration does not change appreciably after the combustion zone (apart from the effect of dilution).

In computing the rate of formation of NO in turbines, it is usually assumed that free-radical concentrations are at their equilibrium values (4). That this is not necessarily so is apparent from work in laboratory flames. In premixed hydrogen-air flames, very large excess radical concentrations exist in the post-reaction-zone gases (15) and this fact has been taken into account in modelling NO formation in this system and in industrial natural gas flames (16). It has also been shown (6) that in luminous-sooting-hydrocarbon flames radical concentrations can be below the equilibrium values, and this formed the basis of a proposed mechanism for the removal of soot by the alkaline earth metals. It is shown in the present work that a similar situation may arise in parts of the primary zone of a gas turbine and, thus, that the rate of NO formation is not as high as would be expected if radical concentrations were at equilibrium values.

ACKNOWLEDGEMENTS

The author wishes to acknowledge the contributions of Dr. D. H. Cotton and Mr. J. G. Dobson for the design and construction of the combustion rig, Mr. J. H. Pearson for his excellent assistance in the experimental work and his colleagues for many helpful discussions.

REFERENCES

1. *R. F. Sawyer, D. P. Teixeira and E. S. Starkman, "Air Pollution Characteristics of Gas Turbine Engines", ASME Transactions, J. Eng. Power, 1969, p. 290.*
2. *R. E. George, J. E. Verssen and R. L. Chass, "Jet Aircraft: A Growing Pollution Source", J. Air Poll. Control Assoc., Vol. 19, 1969, p. 847.*
3. *L. H. Linden and J. B. Heywood, "Smoke Emission from Jet Engines", Comb. Sci. and Tech. Vol. 2, 1971, p. 401.*
4. *R. S. Fletcher and J. B. Heywood, "A Model for Nitric Oxide Emissions from Aircraft Gas Turbine Engines", AIAA Paper No. 71-123, 1971.*
5. *M. W. Shayeson, "Reduction of Jet Engine Exhaust Smoke with Fuel Additives", SAE Paper 670866, 1967.*
6. *D. H. Cotton, N. J. Friswell and D. R. Jenkins, "The Suppression of Soot Emission from Flames by Metal Additives", Comb. & Flame, Vol. 17, 1971, p.87.*
7. *C. J. Halstead, G. H. Nation and L. Turner, "The Determination of Nitric Oxide and Nitrogen Dioxide in Flue Gases, Part I, Sampling and Colorimetric Procedure", The Analyst, in press.*
8. *E. F. Winter, "Flow Visualization Techniques Applied to Combustion Problems", J. Royal Aero. Soc., Vol. 62, 1958, p. 268.*
9. *R. F. Sawyer and E. S. Starkman, "Gas Turbine Exhaust Emissions", SAE Paper 680462, 1968.*
10. *W. G. Taylor, "Smoke Elimination in Gas Turbines Burning Distillate Oil", ASME Paper 67-PWR-3, 1967.*
11. *D. S. Smith, R. F. Sawyer and E. S. Starkman, "Oxides of Nitrogen from Gas Turbines", J. Air Poll. Control Assoc., Vol. 18, 1968, p. 30.*
12. *C. P. Fenimore, "Formation of Nitric Oxide in Premixed Hydrocarbon Flames", 13th Symposium (International) on Combustion, The Combustion Institute, 1971, p. 373.*
13. *B. Toone, "A Review of Aero Engine Smoke Emission", Cranfield Int. Prop. Symp., 1967.*
14. *J. B. Heywood, J. A. Fay and L. H. Linden, "Jet Aircraft Air Pollutant Production and Dispersion" AIAA Paper 70-115, 1970.*
15. *D. R. Jenkins and T. M. Sugden, "Radicals and Molecules in Flame Gases", Chapter 5 in "Flame Photometry", Marcel Dekker, New York, 1969.*
16. *J. B. Homer, D. R. Jenkins and M. M. Sutton, "Kinetics of Nitric Oxide Formation". Paper presented to Flame Chemistry Panel of International Flame Research Foundation at Velsen-Noord, May, 1971.*

DISCUSSION

J. B. Heywood *(Massachusetts Institute of Technology)*

In presenting his experimental data, Dr. Friswell raised several important points which show the difference between premixed systems with which the previous papers were concerned, and hetergeneous combustion systems where fuel is sprayed into the burner separately. I would like to show briefly some results from our own experiments with a hetergeneous combustion system, which illustrate several points that Dr. Friswell raised. I think that his data fit in with our explanation of how a hetergeneous combustor works – the burned gases are not uniform.

First, from the soot measurements presented, it is obvious that part of the primary zone is fuel rich even though the average is close to stoichiometric. Second, NO_x emissions show a lack of sensitivity to changes in air-fuel ratio. A similar effect has been shown in several other combustor studies; for instance, the General Motors GT-309 gas turbine data, and the Northern Research data recently become available from two different turbine combustor cans. The NO_x exhaust emissions decrease rather slowly as the overall air-fuel ratio of the burner becomes leaner. If one uses a well-stirred reactor model for the combustor primary zone, i.e., the fuel and air are mixed uniformly, one would predict a much more rapid fall off in the formation rate of NO. Another point Dr. Friswell raises is that the fuel injector characteristics play an important part in determining the NO emissions. All these points suggest that the burned gases produced in the primary zone are not uniform in fuel to air ratio, but are distributed about the mean value. Let me show you some of our own results using a simple model that illustrates this distribution effect.

We have been conducting some atmospheric pressure burner studies in the burner shown in Fig. 1. We swirl air into one end of a long tube. We burn kerosene and use

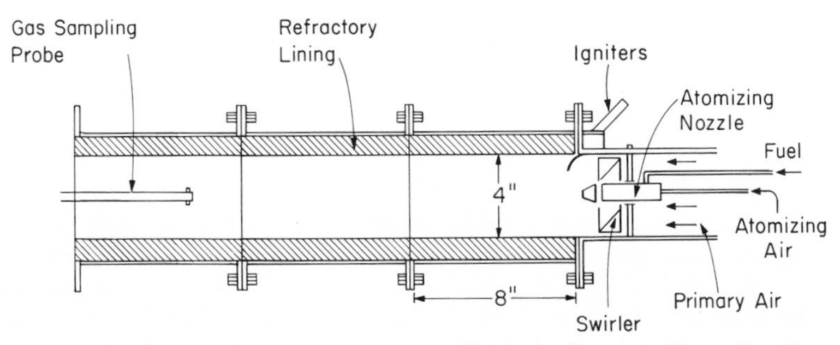

Fig. 1. Schematic of atmospheric pressure burner. The fuel is kerosene, burned at about 130,000 Btu/hr.

an air-blast atomizer to distribute the fuel. In such a burner the flow pattern is relatively simple. The way we run the experiments is to keep the fuel flow and air flow constant, and vary the atomizing air pressure. This changes the flame structure from a luminous flame at low atomizing pressures where nonuniformities are greatest to a blue flame at higher atomizing pressures where the flow is better mixed. Now, let us examine the physical model shown in Fig. 2. The assumption is that all the vaporization, mixing, and burning processes we have talked about result in burned gases with a distribution in composition about the mean. Here, the fuel-air ratio

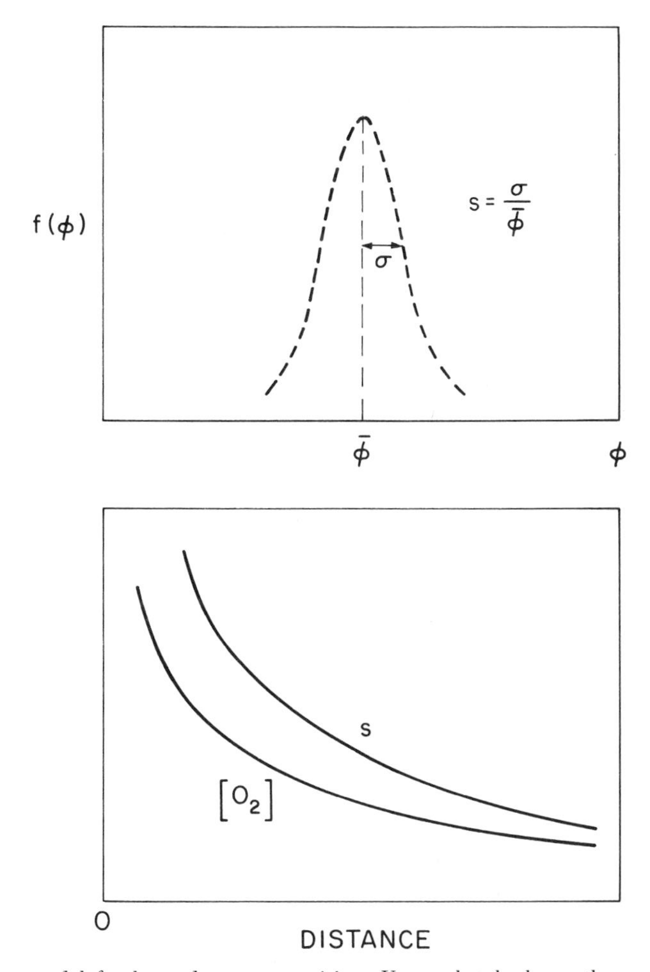

Fig. 2. Mixing model for burned gas composition. Upper sketch shows the assumed Gaussian distribution of equivalence ratio in burnt gas eddies about the mean value. The lower sketch shows the measured oxygen concentration along the burner for stoichiometric combustion, and the mixing parameter profile determined from it.

distribution is assumed to be gaussian about the mean, with a standard deviation σ. We define a mixing parameter s which is the ratio of σ to the mean equivalence ratio. This is not the same as a droplet burning model. My feeling is that the single droplet burning model is a little misleading. The droplet density in the combustor is high enough so that even though the flame around a droplet may burn at the stoichiometric fuel-air ratio, there are a lot of droplets or fuel vapor pockets around. The burnt gases from many droplets or vapor pockets end up at some fuel-air ratio that may not be the average, but could well be either richer or leaner than the mean. If one measures the oxygen concentration in a stoichiometric case after the fuel is fully burned, this distribution can be deduced – there would be no oxygen if the burned gases were uniform. We measure the kind of characteristic shown in the lower sketch in Fig. 2; combustion is complete by about one quarter the way along the burner. So we feel that this [O_2] profile corresponds to burnt gas eddies with a distribution of fuel-air ratios that narrows as mixing proceeds and as the gas flows through the burner. If one calculates a value of s corresponding to the oxygen concentration, and one checks this value by calculating the CO profile along the burner, we find that the measured CO profile is in close agreement.

Fig. 3 shows the effect of changing mixing characteristics on the exhaust CO emissions. Exhaust CO is plotted as a function of the atomizing air pressure. Each of the lines refers to a particular fuel-air ratio. There is no kinetics in the CO model. Gas cooling rates are low enough for CO to be in equilibrium. One can see that with this simple model we can predict with reasonably good agreement a wide range of CO emissions from stoichiometric to 40 per cent excess air. One can see that the mixing is worse at low atomizing pressures than at higher atomizing pressures. Now this

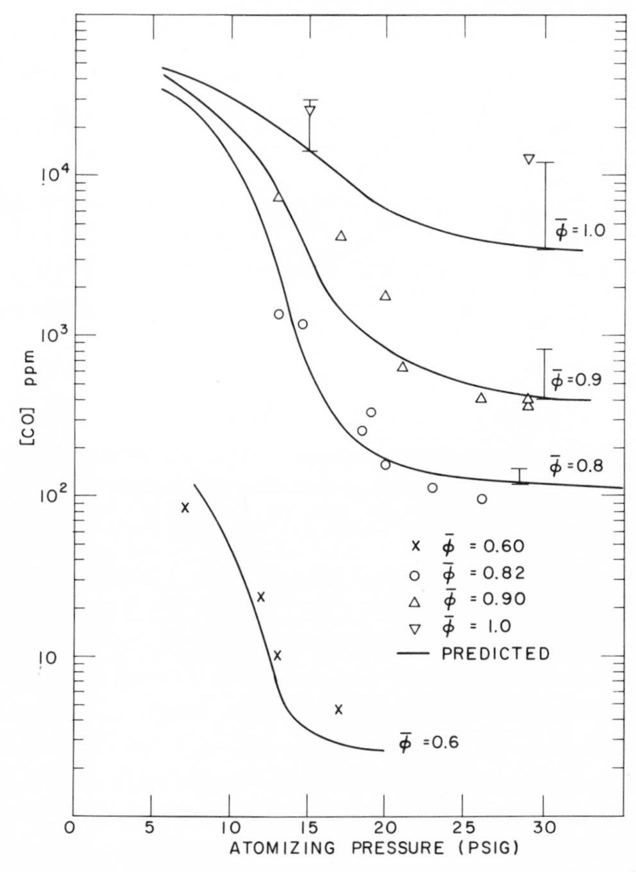

Fig. 3. Measured and predicted burner exhaust CO concentrations. CO emissions rise at low atomizing air pressures due to increasing nonuniformities in the burnt gases.

nonuniformity in the flow affects NO as well, as shown in Fig. 4. Here we are comparing measured and calculated profiles of the NO concentration along the burner. It is more or less a one-dimensional flow; the cross-sectional area is constant. The NO model allows for the distribution of fuel-air ratio in burnt gas eddies about the mean value (equivalence ratio 0.82). If we assume that all of the flow is at the average equivalence ratio, we would predict a NO level a factor of three below the measured level. This model assumes that oxygen atom concentrations are in equilibrium, but, of course, varies with the fuel-air ratio. Nonequilibrium effects may contribute, but they do not seem to dominate the NO formation process. So I feel that this lack of homogeneity is very important in modeling such heterogeneous combustion systems. As we look into the kind of results Dr. Friswell has presented, many of them can be explained with this type of model for the flow.

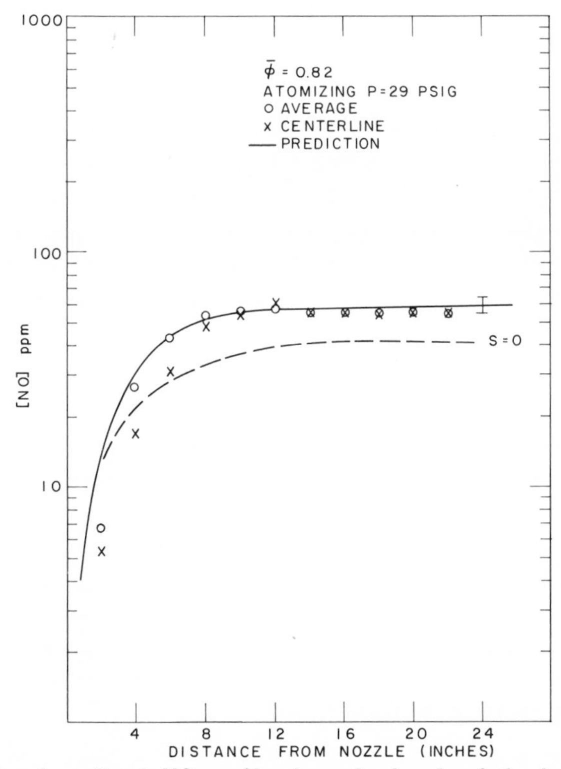

Fig. 4. Measured and predicted NO profile along the length of the burner. Solid line is calculation allowing for nonuniformities; dashed line assumes burnt gases are uniform in composition at each location.

A. M. Mellor *(Purdue University)*

First of all, I'm curious, Dr. Friswell, as to how much nitric oxide was produced by the hydrogen-oxygen pre-heater. I wonder if that varied with the pre-heater air-fuel ratio unlike in the second combustor. I would like to know what inlet temperature was used for the NO results you showed. Secondly, in flame tube C where you have the bad scatter in the data you attributed it to the possibility of a kind of random deposit in the liner. It strikes me that this explanation infers that the air flow distribution changed moreso because of the deposits in the liner than as a result of changes in the overall air-fuel ratio. Therefore, I do not think that your explanation is a very good one. Finally, you stated that the NO emissions did not increase in direct proportion to the mean primary zone residence time as you determined in the water model. You suggested prompt NO formation as a possible explanation. I want to remind you that NO concentrations would depend linearly on reaction time only if the appropriate reactant concentrations and temperature remained invariant over that time which is not the case in these systems. So you need not introduce prompt NO to explain the results.

N. J. Friswell

When the inlet air was preheated to 400°C, the NO concentration in the main combustion chamber inlet air was about 30 ppm. There is obviously a problem when using vitiated air to study NO formation (and one that I am not altogether happy about) in that one has to subtract the NO produced by the preheater from the total exhaust level to obtain the NO concentration produced in the main combustion chamber. In actual practice, in all of the work on NO formation quoted here the preheater was not used, i.e. the combustion chamber inlet air was at ambient temperatures.

The scatter in results for NO formation from flame tube C is attributed in the paper to a variable build-up of deposits changing the air-fuel mixing pattern and also the residence time in the primary zone. The air flow pattern is not expected to change drastically with changes in overall AFR as the air flow rate itself changes only slightly in the way in which the experiments were carried out and the effects of combustion on the flow pattern may not be large. I think it quite possible that two to three fold changes in NO level could be caused by changes in the air-fuel mixing pattern – it is unfortunate that we do not know the particular 'configurations' of deposits which may be responsible for them for flame tube C.

Where NO levels are substantially below equilibrium the increase of NO concentration with time is approximately linear for fixed combustion conditions. However,the relative increase in NO formation from flame tubes B to A to C are *greater* than those of the corresponding primary zone residence times. In any case the correlation between NO levels and residence times is only a qualitative one and I would not wish the results to be seen as confirmation of Fenimore's 'prompt' NO theory, rather that it might be a possible explanation for the results presented.

R. H. Essenhigh *(Pennsylvania State University)*

On the matter of smoke, there is a considerable amount of work in the literature on the effects of additives and particularly of platinum and palladium on burning carbon. In regard to the mechanism, there seems to be indications that for some reason or other platinum and palladium cause the activation energy of the reaction to drop from something like 50 or 55 kilo-calories to about 40 kilo-calories. Now exactly why, I do not know. While I presume that you have never injected platinum or palladium, what group would you associate them with? I would guess that it would be with manganese. The other thing is that you have two different groups of components. Is there any chance that one of them may in fact be preventing the formation of the smoke rather than enhancing the burning? Before you answer those questions, may I take the opportunity of commenting on the cooling rates for CO. We did some experiments and found that somewhere between 10^4 and 10^5 degrees per second seemed to be quenching that reaction.

N. J. Friswell

The effects of platinum/palladium on carbon (smoke) formation were not investigated but I would probably put them, as you suggest, with the 'manganese' group of metals. It is possible that the 'barium' group could inhibit smoke formation but I prefer the mechanism of oxidation of soot in the flame zone by free radicals. However, the 'manganese' group appears only to catalyse the soot oxidation by O_2. In our recent paper (1) on reduction of soot by metallic additives in laboratory flames we show that the 'manganese' group results correlate well with those of the considerable amount of work done on the oxidation of carbon/graphite doped with metallic compounds. In particular, it is known that phosphorus is an inhibitor of carbon oxidation and in our flame work addition of phosphorus *increased* carbon formation.

R. F. Sawyer *(University of California)*

It is a rare opportunity to be able to compare work which was done, I think, in exactly the same combustor under practically exactly the same conditions as in our laboratory. While these experiments will be discussed later in my paper, it would be appropriate to make a few comments now. First of all, it is not true that the flow pattern in this type of combustor is unaffected by the combustion. Although you didn't say this specifically, it was stated by a previous speaker. Your use of a water table analogy to describe the flow pattern neglects the effect of combustion. We can show that combustion changes air flow distribution substantially. We see changes in the nitric oxide level with the overall nitric oxide concentration decreasing as the air flow increases, not very strongly but definitely a trend in that direction. In addition,

---

1. D. H. Cotton, N. J. Friswell and D. R. Jenkins Comb. & Flame, Vol. 17, 1971, p. 87.

we have been doing some exploratory work on the effect of manganese. We see no effect whatsoever on nitric oxide emissions and a decrease in carbon monoxide when manganese is used. We are still trying to sort out what manganese does to carbon formation.

N. J. Friswell

Firstly, I would like to deny that our combustor is exactly the same as that used by Sawyer – certainly it is similar as both are based on tubular gas turbine combustion chambers, but there are significant differences in construction and method of operation (for example, in our combustor overall AFR is changed by maintaining a constant air flow and varying the fuel flow whereas Sawyer adopts the opposite procedure). The effect of combustion in our type of chamber has been shown(1) to change only slightly the flow patterns, as determined in a water flow visualization rig, and they give a good qualitative view of what is happening inside the combustion chamber.

It was found that under extreme conditions (i.e. very high soot levels) barium increased NO concentrations whereas iron had no effect. It would similarly be expected that manganese would have no effect as its mechanism of soot removal appears to be identical to that of iron.

---

1. E. F. Winter, J. Royal Aero. Soc., Vol. 62, 1958, p. 268.

GENERAL DISCUSSION

E. B. Zwick *(The Zwick Co.)*

I did some work for EPA under Contract EHS-70-125 while employed at Paxve Incorporated, Newport Beach, California on the effect of a vapor generator or a steam boiler on emissions from a burner. We made measurements over a wide range of fuel-air ratios from lean blowout to rich blowout and over a range of air flows of more than 10 to 1. When we made measurements downstream of a vapor generator, we could not detect any hydrocarbons. Measurements made from the burner exhaust showed values ranging from about less than 0.1 ppm to about 100 ppm or more as we approached lean blowout. The point here is that the hydrocarbons apparently burned out going through the vapor generator.

Data were also obtained on carbon monoxide. The unburned material in the burner exhaust was primarily carbon monoxide. Hydrocarbons were generally negligible. The point is that CO existed in the burner exhaust. It varied as a function of flow rate and mixture ratio. While CO was present at the inlet to the vapor generator, no CO was found at the outlet side of the vapor generator. Five parts per million was the sensitivity limit of our gas chromatograph. Even for fuel-air ratios which are about half of stoichiometric, the CO is burned out by the time it passes through the vapor generator.

We also experimentally determined oxides of nitrogen concentration as a function of fuel-air ratio at both the burner and vapor generator outlets. There was close similarity between the two sets of emission data. The same correspondence of NO data was shown in the paper by Williams, Sarofim and Lambert. While the oxides of nitrogen are substantially frozen by the vapor generator, the CO and the hydrocarbons burn out.

An important result of my work is the large reduction in NO_x that can be obtained by operating with more than 25 percent excess air. At 100 percent excess air, the level is 10 percent of the 1976 standards. With a vapor generator in place, the CO and the hydrocarbons, both of which were within the limits of the 1976 standards ahead of the vapor generator, disappeared during passage through the generator. From the point of view of a good combustor, it should not be difficult to meet the 1976 standards.

During this effort, NO_x and other emission data were obtained with combustion of both kerosene and propane. The air temperature and flow rates were varied over large ranges. The only parameter that seemed to correlate with the oxides of nitrogen was the fuel-air ratio, which I interpret to be a dependence on combustion temperature. When the inlet air temperature was raised from 70°F to 400°F, the NO_x values increased slightly at the same fuel-air ratio value. One important thing about my findings was that the flow rate did not seem to have any effect. We could not see any consistent influence of flow rate from less than 50 lbs/hr to over 200 lbs/hr.

It is clear that one can achieve NO_x levels of less than 10 ppm. By selecting the right fuel-air ratio and burner design, one can get low NO_x emissions without high CO or hydrocarbon emissions.

E. R. Norster *(Cranfield Institute of Technology)*

Dr. Sarofim pointed out that he sampled with a stainless steel probe of 0.016 in. internal diameter and the temperatures encountered in the sampling region were on the order of 2,000°K. Is Dr. Sarofim aware of the work done by Halstead which shows that under these conditions one can get a complete removal of NO in a stainless steel sampling probe? Did Dr. Sarofim use silicone probes as Halstead did in order to determine if different measurements are obtained with these two types of sampling probes?

A. F. Sarofim

We are familiar with Halstead's study which showed the possibility of reduction of nitric oxide within stainless steel probes under fuel-rich conditions when the probes were not cooled. Since we were operating fuel lean and with a water-cooled probe we did not encounter any reduction of NO within the probe. We have run tests with probes constructed of several materials including quartz and are satisfied that for our conditions the sampling probe does not influence the concentration of the nitric oxide in the sample.

SESSION SUMMARY

G. C. WILLIAMS
Massachusetts Institute of Technology, Cambridge, Massachusetts

The four papers comprising this session describe emissions from pre-mixed, diffusion and fuel-spray flames and NO formation in shock-induced burning. The authors and discussors present evidence supporting the theory of the effects of super equilibrium radical concentrations in enhancing the formation rate of NO and slowing the burnout of CO. A reverse theory is presented for those metallic additives presumed to aid in carbon burnout by restoring radical concentrations from below to more nearly their equilibrium concentrations in flames produced from hydrocarbon sprays. The conclusions reached from these reports, mostly of research in progress, again emphasize the great importance of the interplay between physical and chemical parameters in the performance of practical combustion systems. The effect of this interaction would appear to be of even greater importance in affecting emission levels than in the case of our previous concern over achieving high combustion efficiency and tailored temperature and velocity profiles.

The lively discussions which followed each presentation testified to the quality of and interest in the presentations.

SESSION III

EFFECTS OF OPERATING CONDITIONS AND FUEL FACTORS

Session Chairman
A. H. LEFEBVRE

The Cranfield Institute of Technology
Cranfield, Bedford, England

EFFECT OF FUEL COMPOSITION ON PARTICULATE EMISSIONS FROM GAS TURBINE ENGINES

R. M. SCHIRMER

Phillips Petroleum Company, Bartlesville, Oklahoma

ABSTRACT

A critical analysis has been made of the test methods, Smoke Point and Luminometer Number, which are used at present to evaluate the burning quality of hydrocarbon fuels for gas turbine engines. Differences in the mechanism of soot formation, between a wick lamp and a gas turbine combustor, as evidenced by the effect of hydrocarbon structure, are discussed in detail. Relevant information is reviewed on both the effect of smoke-abatement fuel additives and the morphology of particulate emissions. It is concluded that a specification based upon a fundamental fuel property, such as hydrogen content, rather than a performance test, such as sooting tendency, offers several advantages.

INTRODUCTION

No one likes soot in their environment – not even a little bit! It has a definite aesthetic nuisance, with which we all are familiar. But, perhaps more serious, the National Air Pollution Control Administration (1) has cautioned that soot may have adverse effects on man's health and welfare.

Aircraft Turbines – Soot is the particulate matter which causes the exhaust plume from an aircraft-turbine engine to be black. The Los Angeles County Air Pollution Control District (2) reported that people noticed and complained about this black smoke when commercial jet aircraft were introduced in 1959. Previously, exhaust smoke had been treated by the Military as a minor nuisance, for which they were not willing to either sacrifice aircraft performance or limit fuel availability. Never-the-less, the technology was available at that time to reduce exhaust smoke either by combustor design, as discussed by Lefebvre and Durrant (3), or by fuel selection, as

References pp. 207-208

discussed by Alquist and Schirmer (4). At a recent Conference on Aircraft and the Environment, Megonnell (5) left no doubt that its long tolerance by the airlines was a major public-relations blunder.

Automotive Turbines – The control of air pollution from motor vehicles by the Federal Government (6) has stimulated interest in alternative engines for trucks, buses, and passenger cars. Most experts agree with Starkman (7), that: "The low level of emissions shown, plus already proven applicability and acceptability to the motoring public, give promise of making the (gas) turbine a reasonable replacement for the gasoline (piston) engine."

Recent investigators, such as Cornelius and Wade (8), have sought to further reduce the gaseous pollutants from vehicle-turbine engines. Others, such as Durant (9), have cautioned that some changes in combustor design, made to reduce the emission of nitrogen oxides, can increase exhaust smoke. Experience with the aircraft-turbine engine suggests that visible soot will negate other low-pollution benefits. Therefore, it seems reasonable to conclude that the effective control of particulate emissions is a basic requirement for the development of a satisfactory vehicle-turbine engine – even when abused.

Fuel Requirement – Experimental investigations by Boegel and Wagner (10), Macaulay and Shayeson (11), and others have demonstrated that the operation of a gas turbine engine on different hydrocarbon fuels can have a pronounced effect on particulate emissions. The influence of hydrogen-carbon ratio has been clearly established, with the cleanest burning fuels being those that contain the most hydrogen. Logically, the specification of fuel burning quality should be an important consideration in the development of a gas-turbine engine to control air pollution.

We have reported (12, 13, 14, and 15) ample evidence showing that measurements of the relative burning quality of fuels by test methods based on their sooting tendency in a smoke lamp are of questionable value in their specification for gas turbine engines. Present test methods, Smoke Point (16) and Luminometer Number (17), used to evaluate the burning quality of aviation-turbine fuels should not be carried over into the specification for automotive-turbine fuels without careful consideration.

Purpose – This paper reviews some of the pertinent data which we have obtained that relate to specification of the burning quality of fuels for gas-turbine engines. Since this concerns the mechanism by which soot is formed in the flame, and its subsequent survival in the hot oxygen-rich exhaust gas, relevant information is reviewed on both the effect of smoke-abatement fuel additives and the morphology of particulate emissions.

SOOT FORMATION

Combustion processes in diffusion flames and premixed flames are quite different. These differences are known to affect the relative tendencies of various hydrocarbon fuels to form soot. This is important because the burning quality of aviation-turbine fuels are currently specified by performance test methods using diffusion flames.

Diffusion flames are characterized by separation of fuel and oxygen by a wedge of combustion products. The chemical reaction at the interface does not control the rate of the combustion process. Diffusion processes require a relatively long time for completion. The fuel is decomposed thermally while diffusing to the reaction interface. The resultant high concentration of hydrocarbon radicals favors formation of soot.

Premixed flames are essentially explosion waves traveling through intimate mixtures of fuel and oxygen. Oxygen attacks the hydrocarbon molecule to produce highly reactive fragments which normally are broken down to oxides of carbon by sustained attack. However, the combustion process may be limited by a very short residence time in the flame zone or by an insufficient supply of oxygen in the mixture.

The distinguishing features of diffusion flames and premixed flames are less apparent under turbulent flow conditions. The vortices and irregular movements cause entrainment and mixing of fuel and oxygen in turbulent diffusion flames. Such is the condition in a gas turbine combustor where the highly turbulent flame zone is fed by jets of air and fuel and is maintained by recirculating hot combustion products.

Diffusion Flames – A simple wick lamp has a diffusion flame, in which fuel and air are not premixed. Small flames may be non-luminous. In larger flames the blue reaction zone gives way to a luminous tip. In still larger flames the soot particles formed in the luminous zone can grow so large that they are not oxidized at the reaction interface and escape from the flame as smoke. The height of the flame just before it begins to smoke is the index of burning quality in the present ASTM Method of Test for Smoke Point of Aviation Turbine Fuels (16).

The ASTM Smoke Point Method evolved from work done by Kewley and Jackson (18) some 40 years ago, in which they measured the relative burning quality of illuminating oils. They observed that the candle power of the flame in a wick lamp was directly proportional to flame height. Since flame height is much easier to measure than flame radiation, the maximum height to which the flame could be adjusted without smoking was chosen as an index of burning quality. Minchin (19) used this same apparatus a few years later to demonstrate that the relative "tendency to smoke" is a function of molecular structure. He constructed a chart for homologous series of hydrocarbons showing paraffins, naphthenes, olefins, monocyclic aromatics, and polycyclic aromatics in order of increasing tendency to smoke. The Institute of Petroleum Technologists shortly standardized a test method for measuring the smoke point of kerosene.

Following a study of variables in the design of the wick lamp that affect smoke point, Rakowsky and Hunt (20) developed the Indiana Smoke Point Lamp to have maximum sensitivity in the range of aviation-turbine fuel quality requirements. They found an excellent correlation between fuel ratings with this lamp and the IPT lamp. The Indiana lamp was subsequently used by Hunt (21) to investigate the relation of smoke point to molecular structure. Its increased sensitivity was used to advantage in studying homologous series of paraffin hydrocarbons. Normal paraffins were the cleanest burning hydrocarbons in a wick lamp, and the tendency to smoke was increased by methyl branching.

The more recent Method of Test for Luminometer Number of Aviation Turbine Fuels (17) rates fuels at a constant level of flame luminosity, and presumably at a constant soot content in the luminous flame tip. The size of the flame is measured indirectly by determining the temperature rise across a modified Smoke Point lamp. Bachman (22) studied the relation of Luminometer Number to molecular structure and concluded that burning quality as measured by the Luminometer was identical with that established previously on the basis of Smoke Point.

Studies of the effect of molecular structure on the sooting tendency of hydrocarbons in diffusion flames show consistently that normal paraffins burn significantly cleaner than their branched chain isomers. Generally, the difference between normal heptane and isooctane covers about half the performance range of the apparatus, if for reference purposes its lower limit is established by toluene or tetralin. Data from the work of Hunt (21) and Bachman (22) are presented in Table 1 to illustrate this difference between paraffins.

TABLE 1

Combustion Cleanliness of Diffusion Flames

Hydrocarbon	Indiana Smoke Point, mm	Luminometer Number
Normal Heptane	147	224
Isooctane	86	100
Toluene	6	3
Tetralin	6	0

Premixed Flames – Conditions are much less favorable for the formation of soot in premixed flames, because fuel and oxygen are intimately mixed prior to reaction. Soot formation reactions must compete with the more rapid oxidation reactions for the carbon atoms. However, the deficiency of oxygen in premixed flames of rich fuel-air mixtures can leave hydrocarbon fragments to form soot.

An ordinary laboratory Bunsen burner does not exhibit a true premixed flame. Diffusion of surrounding air into the flame is substantial, and alters the mixture strength at which soot can be formed, if at all. The classical solution has been to place an outer jacket around the flame, sealing it to the barrel of the Bunsen burner to prevent air entrainment. This allows the premixed gases to burn in an atmosphere of their own combustion products. Smithells and Ingle first used this arrangement around 1892 to study the true structure of premixed hydrocarbon-air flames (23). It is commonly referred to as the Smithells' flame separator because a secondary diffusion flame can be supported from the top of the outer jacket if the primary fuel-air mixture is rich enough.

Street and Thomas (24) used a Smithells' separator to study the effect of molecular structure on the sooting tendency of premixed hydrocarbon-air flames. The critical concentration of air required to suppress flame luminosity was the index of sooting tendency. A ratio of this amount of oxygen to that required for stoichiometric combustion to carbon dioxide and water was used for comparing relative combustion cleanliness. As with diffusion flames, hydrocarbon structure was found to be important, but with premixed flames the effect was reversed with respect to isomeric differences in molecular structure. The data presented in Table 2 illustrate that the straight chain paraffins burned dirtier in premixed flames than their branched chain isomers.

TABLE 2

Combustion Cleanliness of Premixed Flames

Hydrocarbon	Air Required to Suppress Luminosity, % Stoichiometric
Normal Octane	72
Isooctane	69
Toluene	75
Tetralin	95

Combustor Flames – Performance requirements of aircraft turbine engines have necessitated the design of small combustion chambers capable of burning fuel efficiently over a wide range of operating conditions. In addition, combustion products must be diluted with excess air to uniformly lower temperature before they are fed in continuous flow to the turbine. The flame is usually maintained by a spray of fuel entering the upstream end of the combustor into which air is gradually fed along the full length of the combustor. Ideally, a fuel-air mixture near stoichiometric would be formed, and burned, in a combustion zone, followed by a dilution zone. Actually, this is never achieved. Pockets of over-rich fuel-air mixture in the combustion zone, and flames quenched before combustion reactions are complete in the dilution zone, may result in excessive soot formation.

The resulting configurations of combustors for aircraft turbine engines are characterized by turbulent diffusion combustion processes, having features of both diffusion and premixed flames. They may favor one or the other, depending upon combustor design and operating conditions. Generally, the effect of isomeric differences in molecular structure on the combustion cleanliness of hydrocarbon fuels is intermediate.

This is illustrated by the combustion cleanliness characteristics of normal and branched-chain paraffins, relative to an aromatic hydrocarbon, in a laboratory-scale combustor. Small-scale combustors, which are used because of the high cost of test facilities and fuels, have shown excellent correlations with aircraft-turbine combustors in fuel evaluation studies (12, 13, 15). A schematic of Phillips 2-Inch Combustor is shown in Fig. 1. Data obtained with this combustor, reported by Miller, Blake, Schirmer, Kittredge, and Fromm (25), are shown in Table 3. The combustor was operated at two different pressures, 5 and 15 atmospheres, with inlet air temperature of 400 F, combustor reference velocity of 100 ft/sec, and fuel-air ratio of 0.010. The

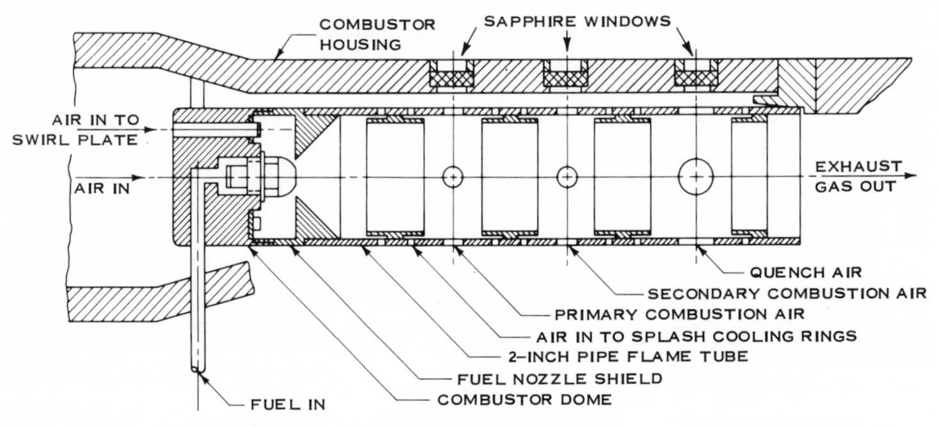

Fig. 1. Phillips 2-Inch Combustor.

TABLE 3

Combustion Cleanliness of Phillips 2-Inch Combustor

Hydrocarbon	Optical Density of Exhaust Gas, (Von Brand Smokemeter)	Transverse Flame Radiation, btu/ft_2/hr
@ 5 Atmospheres Pressure		
Normal Heptane	0.005	16,000
Isooctane	0.009	15,000
Toluene	0.070	77,000
@ 15 Atmospheres Pressure		
Normal Heptane	0.031	80,000
Isooctane	0.081	82,000
Toluene	0.347	146,000

difference in combustion cleanliness between normal heptane and isooctane is negligible when compared with toluene.

SMOKE ABATEMENT

While the mechanisms which control soot formation and soot survival in gas-turbine engines are not clearly understood, experimental investigations have shown that combustor design, operating conditions, and fuel composition are important factors for smoke abatement. Several interesting observations were made during a recent study (15) to determine the effect and interactions of operating conditions, fuel hydrogen content, organometallic additive type, and additive concentration. The Phillips 2-Inch Combustor was used to simulate the conditions for combustion found in gas-turbine engines.

The test program was designed to permit an analysis by conventional statistical procedures, and a 95 per cent confidence level was selected for determining the significance of observed differences. The four base fuels were aviation-turbine kerosenes varying in hydrogen content from 13.8 to 15.0 per cent by weight. The three additives were organometallics of manganese, barium, and calcium. Each additive was blended in each base fuel at three different concentrations; 0.012, 0.024, and 0.036 gram-atoms metal per gallon of fuel. While holding combustor reference flow velocity at 150 ft/sec, and combustor temperature rise at 1000 F, four levels of combustor pressure ranging from 7.5 to 15 atmospheres, and four levels of combustor inlet-air temperature ranging from 400 to 1000 F were used to evaluate each fuel-additive combination twice.

Operating Conditions – Equations for describing the relationship between the operating variables and smoke emissions were computed for each fuel in the program using a model derived from the results of the statistical analysis. To illustrate the trend of smoke emission for typical production JP-5, the family of curves shown in Fig. 2 was developed using the following equation:

$$\text{Optical Density} = 0.3404400 + 0.0287320\,P - 0.6194500\,T - 0.2396000\,PT + 0.2625000\,T^2. \quad (1)$$

where P is combustor pressure and T is combustor inlet-air temperature. To relate the optical density of the exhaust gas, as obtained by our operation, to the smoke levels emanating from present engine designs, an estimated scale is provided in the figure.

To illustrate the marked reduction that the manganese additive had on smoke emission for typical production JP-5, the family of curves shown in Fig. 3 was developed using the following equation:

$$\text{Optical Density} = 0.2634117 + 0.0059427\,P - 0.6817250\,T - 0.0994667\,PT + 0.4614583\,T^2. \quad (2)$$

References 207-208

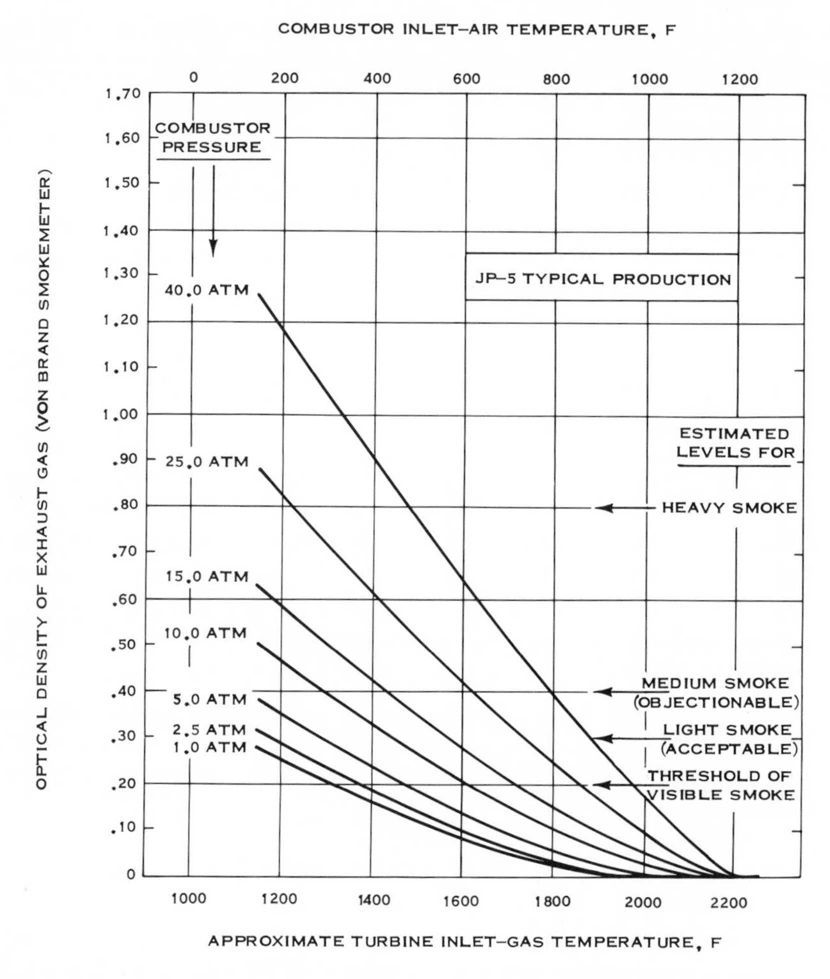

Fig. 2. Computed effect of operating conditions on exhaust smoke with typical production JP-5 fuel.

The variation in additive concentration over the range tested did not have a significant effect on smoke emissions.

Fuel Composition – The effect of fuel hydrogen content on exhaust emissions is best shown by relating the experimental data to various hypothetical gas-turbine engines operated at sea-level take-off conditions. As engine designs advance toward improved efficiency and reduced specific weight, combustor pressure and inlet air temperature will increase. Increasing both temperature and pressure tends to effect smoke emission in opposite directions; i.e., one compensates for the other. The net effect of the two variables therefore is considered of greater practical interest than the effects of holding one of the parameters constant and varying the other.

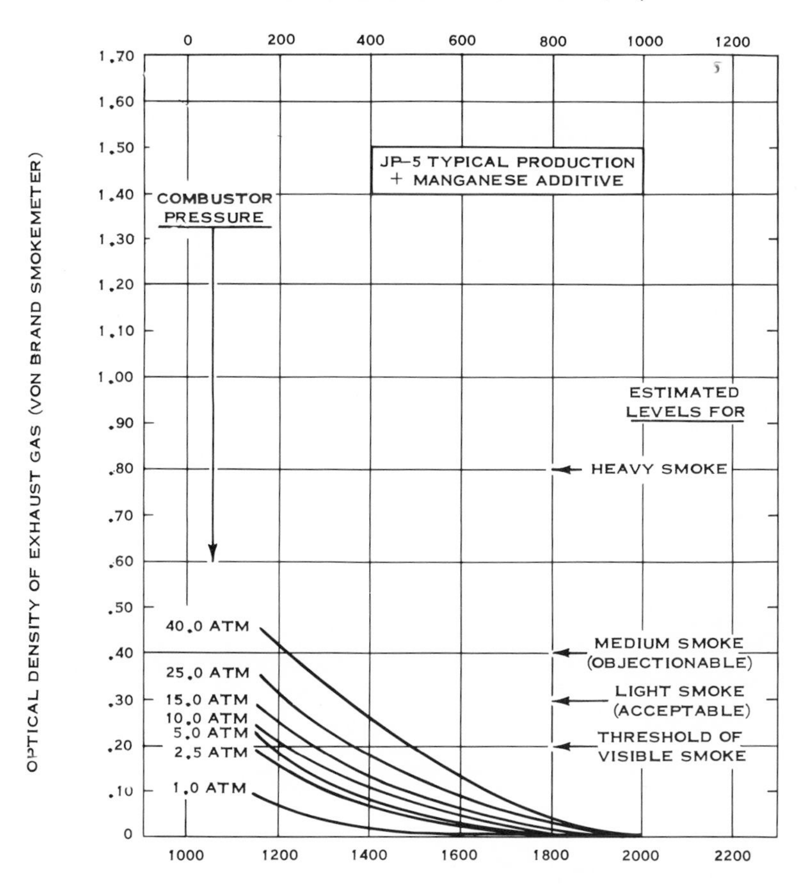

Fig. 3. Computed effect of operating variables on exhaust smoke with typical production JP-5 fuel containing manganese additive.

A statistical model which included the hydrogen content of the fuel was developed from the experimental data for the four base fuels, and the necessary coefficients were evaluated to obtain the following relationship:

$$\begin{aligned}\text{Optical Density} = {} & 5.583140 + 0.03473700\,P - 0.01111021\,T - 0.3740673\,H \\ & - 0.00002958500\,PT + 0.000005445041\,T^2 \\ & + 0.0007432044\,HT - 0.0000003648098\,HT^2. \end{aligned} \tag{3}$$

where P is combustor pressure, T is combustor inlet-air temperature, and H is hydrogen content of the fuel.

References 207-208

A graphical representation of the results of this study, based on each fuel at its respective hydrogen content, is shown in Fig. 4. The lines labeled "non-additive fuels" were computed from Eq. 3. The trends in Fig. 4 reveal that:

1. Smoke density is very sensitive to fuel quality at the mild turbine-inlet conditions of 5 atmospheres and 1300 F; i.e., smoke emission is markedly reduced as the fuel-hydrogen content increases.

2. The effect of fuel quality becomes less important as the severity of the turbine-inlet conditions increase.

3. Additives are beneficial primarily at the intermediate turbine-inlet conditions. At the mild turbine-inlet conditions only moderate improvements resulted from additives, and to obtain these slight benefits care must be exercised in the choice of additive and concentration level.

4. At severe turbine-inlet conditions of 25 atmospheres and 2100 F, the smoke problem vanishes regardless of the fuel quality and the use of additives is unnecessary.

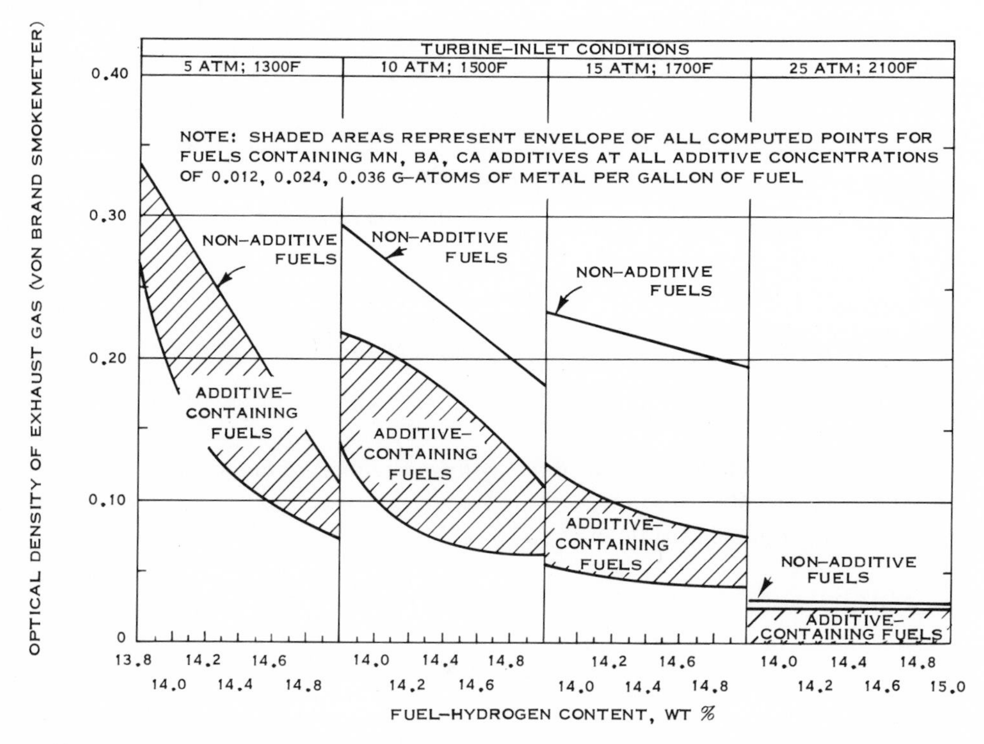

Fig. 4. Computed effect of JP-5 fuels on exhaust smoke from hypothetical engines operated at sea-level take-off conditions.

In an attempt to obtain some insight into the mechanism by which the additives effect their reduction in smoke emissions, a similar statistical model was developed using experimental data obtained on total radiant energy from the flames.

$$\text{Flame Radiance} = 315.5459 + 7.635700\,P + 0.08055969\,T - 23.81250\,H + 0.005846500\,PT - 0.00007488281\,T^2. \quad (4)$$

A similar graphical representation of the results of this study, based on each fuel at its respective hydrogen content, is shown in Fig. 5. The lines labeled "non-additive fuels" were computed from Eq. 4. The trends in Fig. 5 reveal that:

1. Flame radiance is slightly sensitive to fuel quality at the mild turbine-inlet conditions of 5 atmospheres and 1300 F; i.e., flame radiance is reduced as the fuel-hydrogen content increases. As the turbine-inlet conditions increase in severity, the beneficial effects of increasing hydrogen content are also increased.
2. The problem of flame radiance increases markedly as the severity of the turbine-inlet conditions increase. At 25 atmospheres and 2100 F, a severe flame-radiance problem is anticipated.
3. The use of smoke-abatement additives has little or no effect on flame radiance. Since flame radiation emanates primarily from soot particles produced in the primary combustion zone, and since the smoke-abatement additives have little or no effect on their concentration, it is concluded that the additives are effective by increasing their rate of oxidation.

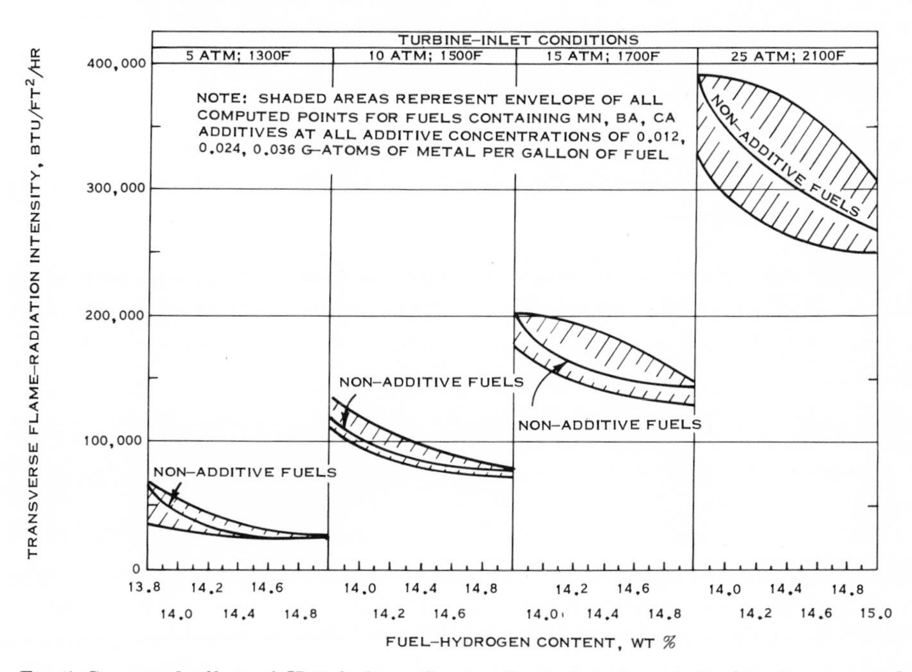

Fig. 5. Computed effect of JP-5 fuels on flame radiance from hypothetical engines operated at sea-level take-off conditions.

Although smoke-abatement fuel additives are effective in reducing exhaust smoke from gas-turbine engines, their use should be approached with extreme caution. Fiorello (26) has reported that: "As a result of continued operation on fuel containing a manganese additive, some engines showed performance losses with time of operation due to the accumulation of manganese oxides on critical areas on the turbine."

MORPHOLOGY

The morphological features of soot were first defined by electron microscopy more than 30 years ago. Transmission micrographs, such as shown in Fig. 6 at 85,000X magnification, outlined a lace-like microstructure which appeared to be chains and clusters of uniform-sized spherules. X-ray diffraction studies suggested an internal structure of small crystallites composed of several parallel planes of carbon atoms in hexagonal array. Using these methods, and chemical analysis, Clark (27) concluded that the particulate emission from a gas-turbine combustor was soot, similar to that from a wick lamp; however, he noted that "the majority of the particles are small and irregular, as if they had been partially consumed in the flame."

The soot shown in Fig. 6, and in subsequent electron micrographs, was collected from the exhaust gas of the Phillips 2-inch, fuel-atomizing, combustor operated on aviation kerosene at 15 atmospheres combustor pressure, 1000 F combustor inlet air temperature, 140 ft/sec combustor reference velocity, and 2000 F exhaust gas temperature. It also shows evidence of the ragged and shrunken microstructural appearance noted by Clark.

Chemical models of soot formation in flames, such as proposed by Street and Thomas (24), have provided for the formation of polybenzenoid fragments to build the graphite-like crystallites.

Physical models of soot formation in flames, such as proposed by Howard (28), have provided for spherule growth, by aggregation of randomly oriented crystallites, to a size sufficient for deactivation and agglomeration in a chain network with other spherules.

New Concept – Recently, it has become possible to obtain electron micrographs, such as shown in Fig. 7, with sufficient resolution to allow direct examination of the graphite layers in soot. Neither an internal structure of randomly-oriented crystallites nor the primary particles as discrete spherules are evident. Instead, the primary particles appear to be aggregates of once-individual particles which have been fused at an early stage in the flame to form a nodular structure. This structure is similar to that of the commercial carbon blacks produced in furnaces by fuel-rich combustion. The development of new concepts of morphology and microstructure for carbon blacks has been reviewed by Rivin (29).

Fig. 6. Electron micrograph of particulate emission from gas turbine type combustor.

The graphite layer network, shown at 2,000,000X magnification, appears as alternate light and dark lines, and the 3.5 Å interlayer spacing approaches that of

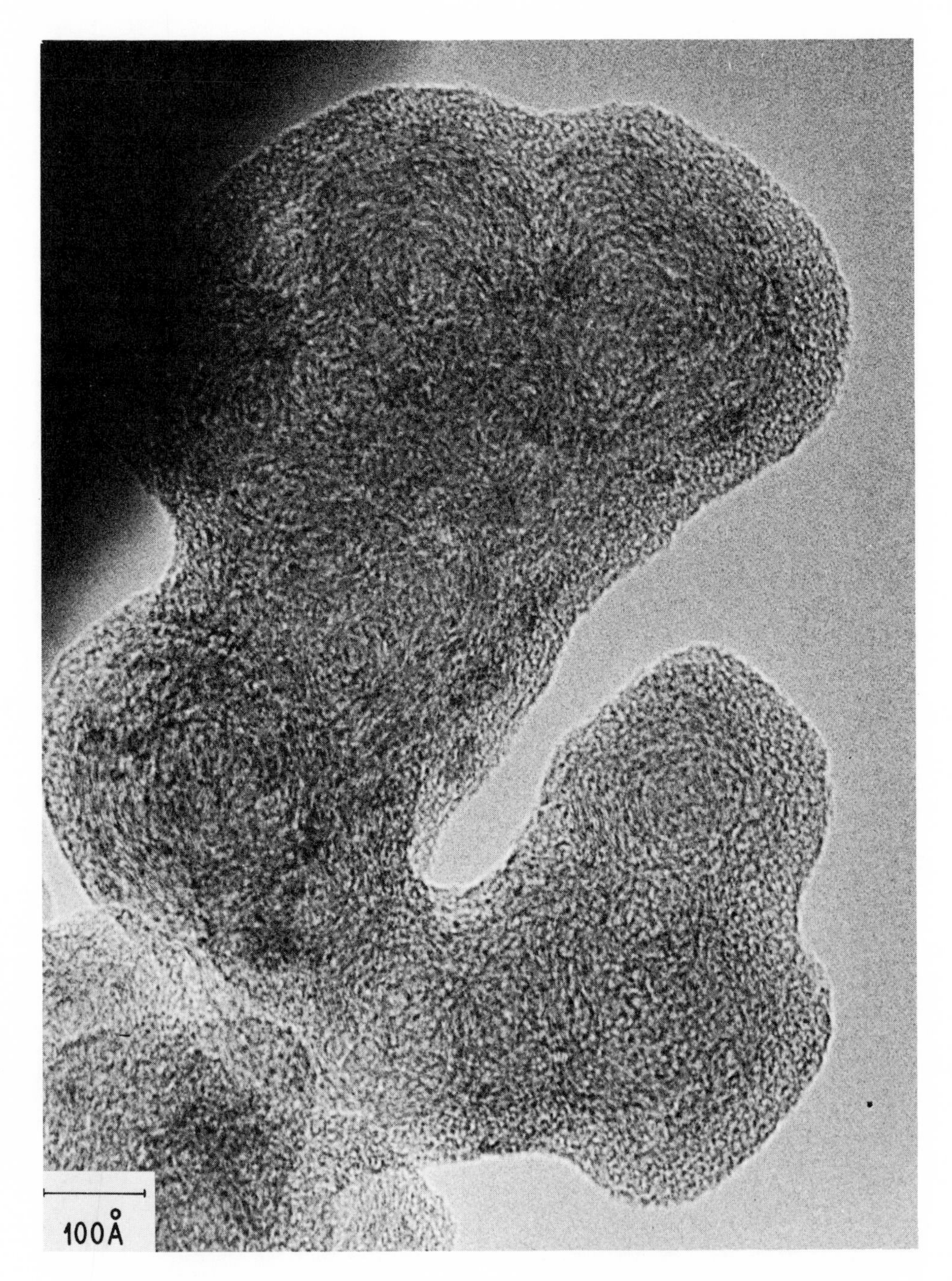

Fig. 7. High resolution electron micrograph of soot from gas turbine type combustor.

crystalline graphite. This continuous network of concentric graphite layers around several growth centers of rotation is called paracrystalline because it exhibits a degree of order somewhat between the truly crystalline and amorphous states. The layers show considerable blending and general distortion. The internal structure of randomly-oriented crystallites that was suggested by X-ray analysis apparently results from chance parallel alignment of several successive graphite layers.

Only a portion of the complete paracrystalline unit is shown in Fig. 7. Such units are extremely complex, varying greatly in size and shape. They can occur in either spheroidal or fibrous form, as shown in Fig. 8. When the structure overlaps in space it appears more dense; thus, a twisted fibrous structure, such as shown at the center of Fig. 8, might appear to be a chain with overlapping spherules when examined at lower resolution. An automated technique for measuring the key dimensions of single carbon-black units, using image-scanning procedures with suitable computer programs, has been described by Burgess, Scott, and Hess (30).

Oxidation – The most significant result of the paracrystalline structure is that a highly resistant surface of basal planes is presented to oxidative attack. Air oxidation causes this relatively smooth surface to take on a ragged appearance, such as shown in Fig. 9. There was little or no evidence of the inside-out oxidation which has been observed by Heckman and Harling (31) with furnace blacks, to form capsules of unchanged external dimensions. Rather, the units appear to burn from the outside in, thereby, causing appreciable decrease in their size as oxidation progresses.

Indications are that the paracrystalline units burn selectively, since Figs. 7, 8, and 9 simply show different locations on the same collection screen; however, it is possible that the particles which reach the exhaust survive by spending most of their time traveling in the relatively cooler areas of the combustor near the wall, as suggested by Heywood, Fay, and Linden (32).

Smoke-Abatement Additives – Comparisons were made of duplicate samples of soot collected from various locations in the Phillips 2-Inch Combustor and its exhaust gas – with and without a smoke abatement barium additive in the fuel. No differences in morphology or microstructure were evident between these duplicate samples of soot, indicating that such additives affect a reduction in particulate emissions from gas turbine engines by changing chemical processes rather than by changing the physical structure of soot.

Particle Characterization – In the absence of high shear, the soot particles flocculate under the influence of van der Waals' forces to give a multiplicity of secondary agglomerates, whose size depends in part on the collection system and the method of measurement. Without high-resolution electron microscopy it becomes impossible to state unambiguously whether a particular particle is just one somewhat irregular primary unit, or whether it is a reversibly agglomerated secondary structure; however, direct characterization by such detailed inspection normally would be impossibly tedious. This points out the difficulty and subjectivity of particle-size counting and analysis when attempting to characterize the particulate emissions from a gas turbine engine.

References pp. 207-208

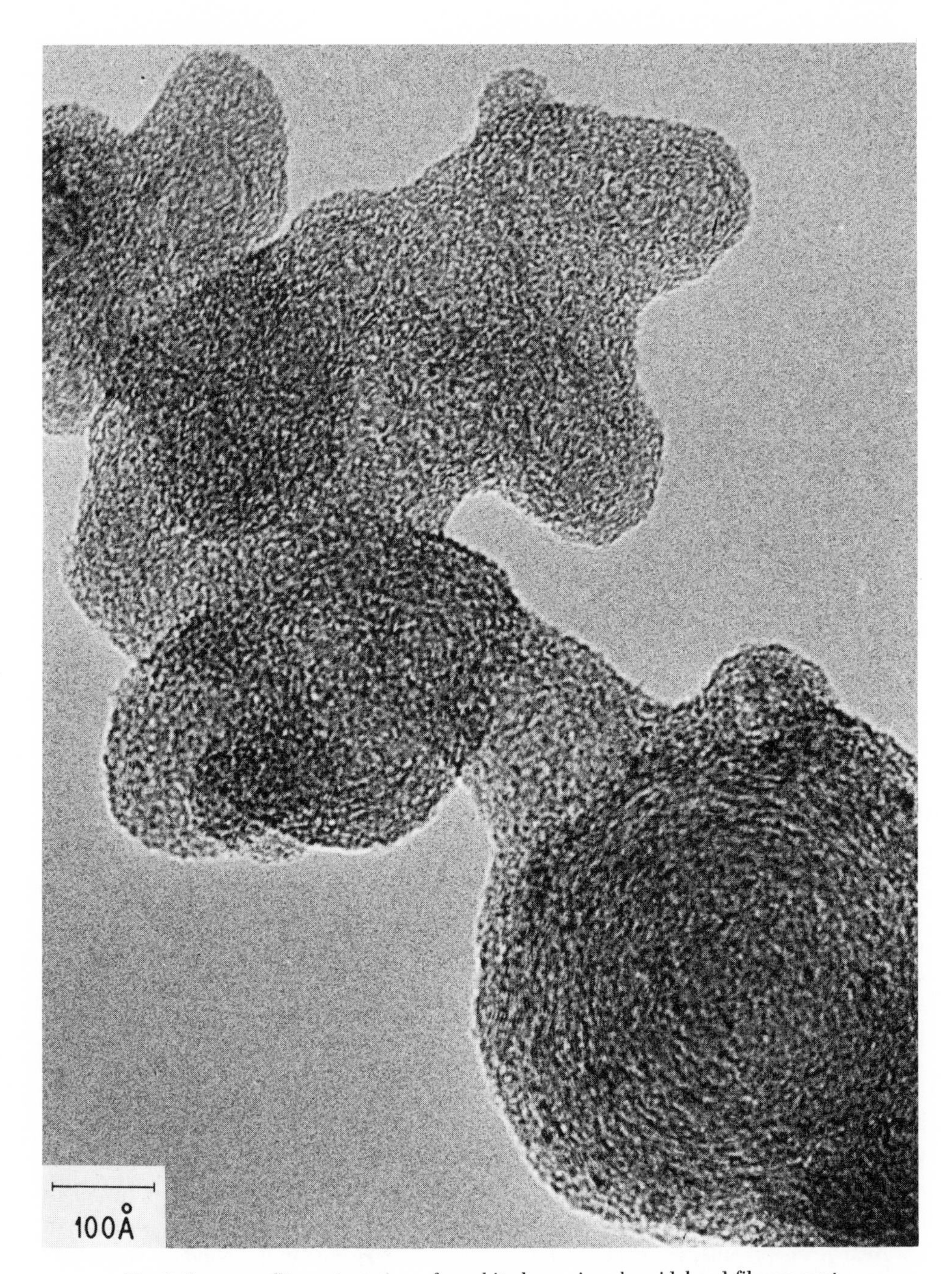

Fig. 8. Paracrystalline orientation of graphite layers in spheroidal and fibrous soot.

The preceding information on the morphology and microstructure of particulate emissions from gas turbine engines helps to explain why Dalzell, Williams, and Hottel

Fig. 9. Partial oxidation of soot while at high temperature in exhaust gas.

(33) found it necessary to use a chain-cluster model to accurately predict the growth and burn-out of soot particles in flames. It emphasizes their conclusion that the

References 207-208

validity of determinations of the size and concentration of soot particles in flames are in grave doubt when keyed to the simple model of monodisperse solid spheres.

CONCLUSIONS

It has been shown by illustration that the combustion processes in diffusion flames and premixed flames are quite different. Diffusion flames are cleanest with normal paraffin hydrocarbons, while premixed flames are cleanest with isoparaffin hydrocarbons. However, there is little or no difference between normal paraffin and isoparaffin hydrocarbons in the cleanliness of turbulent diffusion flames, such as exist in gas turbine engines.

Present methods for specification of the burning quality of hydrocarbon fuels for gas turbine engines, Smoke Point and Luminometer Number, are performance tests in which a small sample of the fuel is burned in a wick lamp to determine its relative tendency to form soot. Unfortunately, the laminar flow diffusion flame of the wick lamp differs appreciably from the highly turbulent combustion process in a modern gas turbine engine. As a result, the present methods for specification of the burning quality of hydrocarbon fuels can give a distorted picture of the actual performance of such fuels in gas turbine engines.

The hydrogen content of a pure hydrocarbon, or of a complex mixture of hydrocarbons such as in a normal aviation-turbine fuel, is a basic property of the specific fuel sample and can be determined with considerable accuracy by a variety of techniques. It has long been realized that the hydrogen content of such fuels have a direct bearing on their combustion cleanliness characteristics in gas turbine engines; in fact, correlation studies between hydrogen content and flame radiance or exhaust smoke show a near linear relationship. This characteristic is highly desirable for it prevents undue emphasis being placed upon the attainment of a fuel quality, such as a very high Luminometer Number, from which no actual benefit in terms of gas turbine engine performance or durability can reasonably be expected.

A specification based upon a fundamental fuel property, such as hydrogen content, rather than a performance test, such as sooting tendency, offers additional advantages. Hydrogen content can be determined accurately by a variety of test methods, allowing a selection from differing procedures and equipment to best meet the needs of research, process control, and product inspection. The basic guidance it provides, with respect to the requirements of refinery processes for upgrading the burning quality of fuels for gas turbine engines, is easily understood by the petroleum chemist. Therefore, the use of hydrogen content for specification of the combustion cleanliness of gas-turbine fuels is encouraged.

ACKNOWLEDGMENT

The author expresses appreciation to the U.S. Naval Air Systems Command for financial support of this work and for permission to publish the results, to Dr. J. O. Gardner for electron micrographs, to Mr. M. R. Goss for statistical analyses, and to Professor N. W. Ryan for valuable discussions.

REFERENCES

1. *"Air Quality Criteria for Particulate Matter." National Air Pollution Control Administration, U.S. Department of Health, Education, and Welfare, Washington, Publication AP-49, January 1969.*
2. *R. E. George and R. M. Burlin, "Air Pollution from Commercial Jet Aircraft in Los Angeles County." Los Angeles Air Pollution Control District, Los Angeles, April 1960.*
3. *A. H. Lefebvre and T. Durrant, "Design Characteristics Affecting Gas Turbine Combustion Performance." Paper No. 240C presented at SAE National Aeronautic Meeting, Los Angeles, October 1960.*
4. *H. E. Alquist and R. M. Schirmer, "A Critical Survey of Commercial Turbine Fuels." Paper No. 520 presented at SAE National Aeronautic Meeting, New York, April 1955.*
5. *W. H. Megonnell, "Regulation of Pollutant Emissions from Aircraft – Today and Tomorrow." Published in P-37, "SAE/DOT Conference on Aircraft and the Environment, Part 2." New York: Society of Automotive Engineers, Inc., Paper 710337, pp. 54-57.*
6. *"Control of Air Pollution from New Motor Vehicles and New Motor Vehicle Engines." Federal Register, Vol. 36, No. 128, July 1971, p. 12652.*
7. *E. S. Starkman, "The Chances for a Clean Car." Astronautics & Aeronautics, Vol. 9, No. 8, August 1971, pp. 68-75.*
8. *W. Cornelius and W. R. Wade, "The Formation and Control of Nitric Oxide in a Regenerative Gas Turbine Burner." SAE Transactions, Vol. 79, 1970, Paper 700708.*
9. *T. Durrant, "The Control of Atmospheric Pollution from Gas Turbine Engines." Paper No. 680347 presented at SAE National Air Transportation Meeting, New York, April 1968.*
10. *M. J. Boegel and J. F. Wagner, "Relation of Fuel Properties to Combustion Cleanliness in a Small Gas Turbine Engine." Paper No. 114C presented at SAE Annual Meeting, Detroit, January 1960.*
11. *R. W. Macaulay and M. W. Shayeson, "Effects of Fuel Properties on Liner Temperatures and Carbon Deposition in the CJ805 Combustor for Long-Life Applications." Paper No. 61-WA-304 presented at ASME Winter Annual Meeting, New York, November 1961.*
12. *R. M. Schirmer, L. A. McReynolds, and J. A. Daley, "Radiation from Flames in Gas Turbine Combustors." SAE Transactions, Vol. 68, 1960, pp. 554-561.*
13. *R. M. Schirmer, "Specification of Jet Fuel Hydrogen Content for Control of Combustion Cleanliness." Progress Report No. 1 to U.S. Bureau of Naval Weapons under Contract No. N600(19)-58219, June 1962. (Available from DDC as AD-282332.)*
14. *R. M. Schirmer and H. T. Quigg, "High Pressure Combustor Studies of Flame Radiation as Related to Hydrocarbon Structure." Progress Report No. 3 to U.S. Bureau of Naval Weapons under Contract No. NOw 64-0443-d, May 1965. (Available from DDC as AD-617191.)*
15. *L. Bagnetto, "Smoke Abatement in Gas Turbines. Part 2: Effects of Fuels, Additives and Operating Conditions on Smoke Emissions and Flame Radiation." Summary Report to U.S. Naval Air Systems Command under Contract No. NOO156-67-C-2351, September 1968. (Available from DDC as AD-842818.)*

16. "Standard Method of Test for Smoke Point of Aviation Turbine Fuels," ASTM Designation: D 1322-64. Published in Part 17, "1970 Annual Book of ASTM Standards." Philadelphia, American Society for Testing and Materials, 1970.
17. "Standard Method of Test for Luminometer Numbers of Aviation Turbine Fuels," ASTM Designation: D 1740-69. Published in Part 17, "1970 Annual Book of ASTM Standards." Philadelphia, American Society for Testing and Materials, 1970.
18. J. Kewley and J. S. Jackson, "The Burning of Mineral Oils in Wick-Fed Lamps." Journal of Institute of Petroleum Technologists, Vol. 13, 1927, pp. 364-382.
19. S. T. Minchin, "Luminous Stationary Flames: the Quantitative Relationship between Flame Dimensions at the Sooting Point and Chemical Composition, with Special Reference to Petroleum Hydrocarbons." Journal of Institute of Petroleum Technologists, Vol. 17, 1931, pp. 102-120.
20. F. W. Rakowsky and R. A. Hunt, Jr., "Variables in Lamp Design that Affect Smoke Point." Analytical Chemistry, Vol. 28, October 1956, pp. 1583-1586.
21. R. A. Hunt, Jr. "Relation of Smoke Point to Molecular Structure." Industrial and Engineering Chemistry, Vol. 45, 1953, pp. 602-606.
22. K. C. Bachman, "Relation of Luminometer Number to Molecular Structure." Symposium on Jet Fuels, Vol. 5, No. 4-C, pp. C39-C47. Sponsored by Division of Petroleum Chemistry of the American Chemical Society, New York, September 1960.
23. W. A. Bone and D. T. A. Townend, Flame and Combustion in Gases. London, Longmans, Green and Co. Ltd, 1927.
24. J. C. Street and A. Thomas, "Carbon Formation in Pre-Mixed Flames." Fuel, Vol. 34, 1955, pp. 4-36.
25. E. C. Miller, A. E. Blake, R. M. Schirmer, G. D. Kittredge, and E. H. Fromm, "Radiation from Laboratory Scale Jet Combustor Flames." Special Report for U.S. Bureau of Naval Weapons Contract No. NOas 52-132-c, October 1956. (Available from DDC as AD-138244.)
26. S. C. Fiorello, "The Navy's Smoke Abatement Program." Paper 680345 presented at SAE National Air Transportation Meeting, New York, April 1968.
27. T. P. Clark, "Examination of Smoke and Carbon from Turbojet Engine Combustor." Research Memorandum E52126, National Advisory Committee for Aeronautics, Washington, November 1952.
28. J. B. Howard, "On the Mechanism of Carbon Formation in Flames." Twelfth Symposium (International) on Combustion, The Combustion Institute, 1969, pp. 877-887.
29. D. Rivin, "Surface Properties of Carbon." Rubber Chemistry and Technology, Vol. 44, No. 2, April 1971, pp. 4-71.
30. K. A. Burgess, C. E. Scott, and W. M. Hess, "Carbon Black Morphology: New Techniques for Characterization." Rubber World, Vol. 164, May 1971, pp. 48-53.
31. F. A. Heckman and D. F. Harling, "Progressive Oxidation of Selected Particles of Carbon Black: Further Evidence for a New Microstructural Model." Rubber Chemistry and Technology, Vol. 39, 1966, pp. 1-13.
32. J. B. Heywood, J. A. Fay, and L. H. Linden, "Jet Aircraft Air Pollutant Production and Dispersion." Paper No. 70-115 presented at AIAA 8th Aerospace Sciences Meeting, New York, January 1970.
33. W. H. Dalzell, G. C. Williams, and H. C. Hottel, "A Light Scattering Method for Soot Concentration Measurements." Combustion & Flame, Vol. 14, 1970, pp. 161-70.

DISCUSSION

N. J. Friswell *(Thornton Research Centre)*

We have similarly studied the effect of fuel hydrocarbon formulation on soot formation from a gas turbine type chamber as discussed by Mr. Schirmer. A large number of fuels with composition and properties grouped around those of aviation kerosene were studied. The conclusion in agreement with Mr. Schirmer, was that for this range of fuels the hydrogen-carbon ratio of the fuel gave the best correlation with measured soot levels. I would like to ask Mr. Schirmer if he noticed any effect of fuel volatility on soot emission in view of the fact that soot is formed in his system in the center of the fuel cone spray.

In my paper yesterday, I gave mechanisms by which barium and manganese removed soot and that this was by accelerating soot oxidation – though in different parts of the combustion chamber. Therefore, I was pleased to see the same conclusion drawn by Mr. Schirmer from his studies. I would like to ask him if any decrease in flame radiation was produced by the addition of either barium or manganese. I would have expected a decrease for barium but not for manganese since only barium appeared to operate in the main combustion zone.

R. M. Schirmer

There can be a very large effect of fuel volatility on deposit formation in the combustors. This has been well documentated and is carried over into our present specifications for turbine fuels where we use a fuel volatility index which is a combination of the boiling point of the fuel and the smoke point. However in regard to soot formation, we are concerned more with the combustion characteristics of the vaporized hydrocarbons. The more volatile fuels generally have more hydrogen and less carbon available; and as a result, usually produce less soot emissions.

The influence of fuel volatility on soot emissions was evaluated by testing a fifth fuel, a typical production JP-4, which was more volatile than the other four kerosine-type fuels. The NASA K-Factor, described in NASA Research Memorandum E52B14, was used to characterize these fuels:

$$\text{K-Factor} = 0.7\,(t + 600)\,\frac{H/C - 0.207}{H/C - 0.259}$$

where t is volumetric average boiling point, F, and H/C is hydrogen to carbon weight ratio. The necessary coefficients were evaluated to obtain the following relationship:

$$\begin{aligned}\text{Optical Density} = {} & -0.7334728 + 0.03591360\,P + 0.0005562556\,T + 0.003078621\,K \\ & -0.00003072800\,PT + 0.0000002262500\,T^2 - 0.000003305887\,KT\end{aligned}$$

where P is combustor pressure, T is combustor inlet-air temperature, and K is NASA K-Factor.

In regard to the effect of the barium additive on flame radiation, there was a statistically significant reduction in flame radiation with the barium additive at some operating conditions. This was not apparent with the manganese and calcium additives.

MEASUREMENT OF NITRIC OXIDE FORMATION WITHIN A MULTI–FUELED TURBINE COMBUSTOR

C. W. LaPOINTE and W. L. SCHULTZ

Ford Motor Company, Dearborn, Michigan

ABSTRACT

Factors affecting the nitric oxide (NO) emission level of a regenerative turbine combustor are reviewed. Differences attributable to fuel type are discussed. Temperature and composition measurements obtained with a water-cooled choked sampling probe are presented for a turbine combustor operating on two different fuels: #2 diesel oil and methanol. Methods of averaging the discrete data are developed. Temperature is computed from measurement of the choked sample flow rate. Methanol displays lower NO emission because of decreased high temperature residence duration, lower flame temperature and diminished oxygen and nitrogen availability. It is found that the combustion and NO formation zones occupy a small portion of the regenerative combustor while the CO oxidation zone occupies roughly half of the combustor under investigation. Nitric oxide never reaches equilibrium values, whereas carbon monoxide exceeds equilibrium values throughout most of the combustor.

INTRODUCTION

In a conventional regenerative turbine combustor certain liquid fuels consistently produce lower NO emissions than others, all other controllable factors being equal. This was demonstrated by early experiments utilizing a combustor test rig in which airflow, inlet temperature and pressure were held constant as the fuel flow rate was adjusted to achieve the desired exit temperature. Identical combustors and nozzles were employed. Some typical results are shown in Fig. 1 where the NO_x ($NO + NO_2$) emission levels of diesel oil, kerosene, Indolene and methanol are plotted versus average exit temperature. For this set of conditions, the NO_x exhaust concentration

References pp. 232-233

for methanol was about one-fourth of that for the hydrocarbons. Replotting these data as a function of overall equivalence ratio leads to the same hierarchy of emission levels. Other tests showed that the NO_x emission level for ethanol was approximately 50 per cent greater than that for methanol.

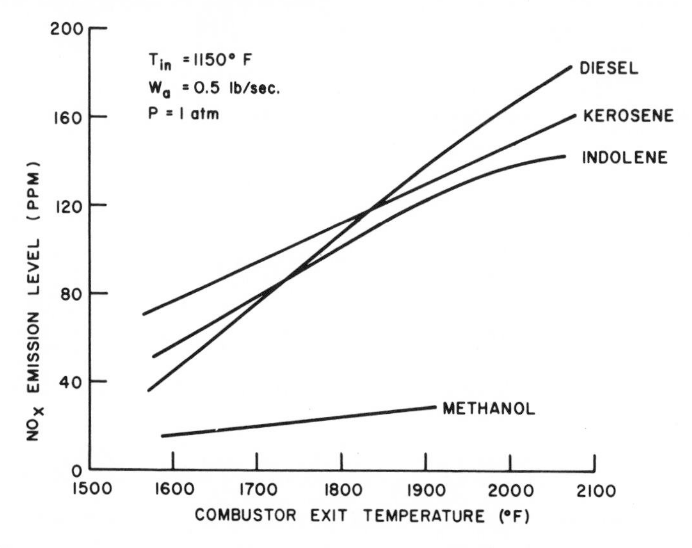

Fig. 1. NO_x emission levels vs. combustor exit temperature for several fuels.

Several factors, which differ between fuels, may be presumed to influence the ultimate level of NO_x emitted by a particular combustor. They are both chemical and physical. The distinction is more than academic; physical factors which render one fuel a lower emitter could conceivably be implemented in the injection process of another fuel whereas purely chemical factors could not. For the sake of brevity, the following discussion will be confined to chemical and physical differences between diesel fuel and methanol. An effort will be made to quantify these differences insofar as they affect relative NO emission levels. This will point out the need for detailed experimental information concerning the combustion process and lead to a description of further experiments which were performed to obtain some of this information. In particular, the combustor energy addition and air addition profiles are of crucial importance for they determine the temperature and residence time in the NO-forming region. The conditions listed below in Table 1 and referred to throughout the remainder of this work pertain to these latter experiments.

Chemical Considerations – Purely chemical differences between diesel oil and methanol include heats of combustion and vaporization, stoichiometric air-fuel ratio, and adiabatic flame temperature. (See Table 1) The difference in flame temperature is of the order of 100F. According to the Zeldovich thermal theory (1), this results in a factor of two difference in the NO formation rate for combustion in the neighborhood of 4000F. Blumberg and Kummer (2), in that part of their paper

TABLE 1

Experimental Conditions

Fuel	Diesel No. 2	Methanol
Inlet Temp. (°F)	1000	1000
Exit Temp. (°F)	1400	1400
Air flow (lb/sec)	1.6	1.6
Fuel Flow (lb/hr)	40.5	84
Fuel Temp. (°F)	220	190
Lower Heating Value (BTU/lb)	18500	8600
Latent Heat of Vaporization (BTU/lb)	155	503
Stoichiometric A/F	14.6	6.4
Overall A/F	141	68
Sauter Mean Diameter (μm)	127	55.1
Injection Velocity (fps)	65	139
Drop Reynolds Number	106	99
Pressure (atmospheres)	3	3
Combustor Pressure Drop (%)	3	3
Space Rate (BTU/hr-ft^3-atm)	1.8 x 10^6	1.8 x 10^6

dealing with the effects of fuel differences in *reciprocating* engines, do predict a factor of two decrease in NO emission between gasoline and methanol.

Consideration of the NO formation rate equation reveals another chemical factor affected by fuel type. The Zeldovich thermal theory states that the rate at which NO forms is given by:

$$\frac{dX_{NO}}{dt} \sim P^{1/2}\, X_{O_2}^{1/2}\, X_{N_2}\, T^{-1}\, e^{-122000/T} \left[1 - \left(\frac{X_{NO}}{X_{NO_e}}\right)^2\right] \qquad (1)$$

For a given combustor air addition profile, methanol will constitute a greater fraction of the air-fuel mixture than will diesel fuel at comparable equivalence ratios and extent of reaction. This means that, as a percentage of the total flow, less oxygen and nitrogen will be available to form NO. Since these constituents enter into the formation rate equation non-linearly, the NO produced will not only be fractionally less for methanol but will be less on a true mass emission basis. Quantitatively, this is expressed by

References pp. 232-233

$$X_{O_2}(t) = \frac{0.21}{1 + \frac{\varphi(t)}{(A/F)_s}} [1 - \eta(t)\varphi(t)] \tag{2}$$

and

$$X_{N_2}(t) \cong \frac{0.79}{1 + \frac{\varphi(t)}{(A/F)_s}} \tag{3}$$

At stoichiometric conditions, the true NO mass formation rate ratio (methanol/diesel) neglecting the derivative of the air flow rate would therefore be

$$\frac{d(\dot{m}_{NO})_M}{d(\dot{m}_{NO})_D} = \left(\frac{1 + \frac{1}{14.6}}{1 + \frac{1}{6.4}} \right)^{1/2} = 0.96$$

In other words, the NO formation rate for methanol on a mass basis, is 4 per cent lower than that for diesel fuel due to this dilution effect. On a parts-per-million basis, this effect could account for an 11 per cent reduction.

Physical Considerations – Physical differences concern the fuel injection process and arise mainly due to the differences in stoichiometric air-fuel ratio. They ultimately influence the energy addition rate and the residence time in the NO formation zone. For a given air addition profile, more than twice as much methanol must be injected to achieve the same local equivalence ratio as would diesel fuel. Taking into account fuel density differences, this means that a droplet of methanol enters the combustor with a velocity of 139 ft/sec as compared to 65 ft/sec for diesel fuel. The initial Sauter mean diameter (S.M.D.) of the droplet distribution was computed from

$$D_{32} = 0.815\, D_0^{1.589}\, \dot{m}_f^{-0.6}\, \sigma^{0.594}\, \mu_f^{0.22} \tag{4}$$

where the units, reading from left to right are: μm, mils, pounds/hr, dynes/cm and centipoise (3,4). Inserting the appropriate values in Eq. 4 leads to an S.M.D. equal to 127 μm for diesel fuel and 55 μm for methanol. Thus the methanol fuel spray contains smaller drops which initially have greater velocity than those contained in the diesel fuel spray.

The droplet trajectory and evaporation rate determine the energy addition profile. Adopting a simple one-dimensional point of view and ignoring recirculation, one may construct the following model of this process. Initially, the droplets move too fast relative to the primary zone gases, to support combustion on the drop itself (5). The droplet therefore evaporates or shatters into smaller droplets which evaporate. Those droplets which shatter have Weber numbers (6) greater than four. Based on their respective S.M.D., methanol has a $W_e \cong 3$ and diesel has a $W_e \cong 1$. That is, some large droplets in the methanol spray will shatter, whereas fewer large droplets will shatter in the diesel fuel spray. Assuming the combustion reaction to be diffusion limited, it is the evaporation rate which then determines the combustion completeness. The evaporation constant (7), corrected for convection (8), is given by

$$K = K_s \left[1 + 0.15\,(Re\,Sc)^{0.6}\right] \tag{5}$$

where K_s is the static evaporation rate constant

$$K_s = \frac{8k}{C_p \rho_f} \ln \left\{1 + \frac{1}{L}\left[\frac{Y_{O_2} \Delta H_c}{(O_2/F)_s} + C_p\,(T_\infty - T_v)\right]\right\} \tag{6}$$

Godsave's evaporation law (7) gives the average droplet diameter versus time as

$$D_{32}^2\,(t) = D_{32}^2\,(o) - Kt \tag{7}$$

One may, on this basis, define a representative combustion completeness as

$$\eta(t) = 1 - \frac{\dot{m}_f(t)}{\dot{m}_f(o)} = 1 - \left(1 - \frac{t}{t_b}\right)^{3/2} \tag{8}$$

where the total average burning (evaporation) time is given by

$$t_b = \frac{D_{32}^2(O)}{K} \tag{9}$$

Inserting representative values into Eq. 9 yields t_b = 3.5 msec (diesel) and t_b = 1.2 msec (methanol). The values of the transport coefficients which enter into this calculation are estimates for the presumed primary zone conditions. The conclusion that the burning time for methanol is less than that of diesel fuel is valid, however, and is due in large part to the strong dependence of t_b on particle size.

It will be shown during the discussion of experimental results that fuel variation has little effect on the air addition pattern in a combustor with a small number of discrete dilution ports. This being the case, then any air dilution jet capable of quenching NO formation must occupy the same combustor location irrespective of the fuel employed.

Since time is the fundamental variable for NO formation, whereas distance is the independent variable for dilution or quenching of NO formation, the fluid element trajectory (x vs. t) is required in order to obtain the residence time at high temperature. For the sake of discussion, if it is assumed that each fuel droplet maintains its injection velocity until it is completely consumed, then a diesel droplet (dia. = S.M.D.) will travel 2.7 inches compared to 2.0 inches for a methanol droplet before it is consumed. That is, a methanol droplet deposits its energy in less distance than a diesel droplet, with the result that methanol-heated air elements are heated earlier in *their* trajectories. They will expand, accelerate, and reach the quench location earlier than diesel-heated air elements. On the other hand, they will reach elevated, NO-producing temperatures earlier than their diesel counterparts. The incorporation of droplet drag into this discussion model does not alter the essential features: early temperature rise and early quench of methanol-heated combustor air elements. Droplet deceleration due to drag decreases the calculated burning distances and changes the shape of the $\eta[x(t)]$ curve. Thus, there are two opposing effects on the net production of NO due to physical differences in the injection of methanol and diesel fuel. Further discussion of the relative weight of each effect will be deferred until experimental data have been presented. The preceding discussion serves to illustrate the factors involved and to point out the need for data concerning the quench location and temperature distribution within the combustor.

DESCRIPTION OF THE EXPERIMENT

Test Stand – Two series of experiments, utilizing No. 2 diesel oil and methanol as fuel, were conducted in the combustion test stand shown in Fig. 2. Preheated air

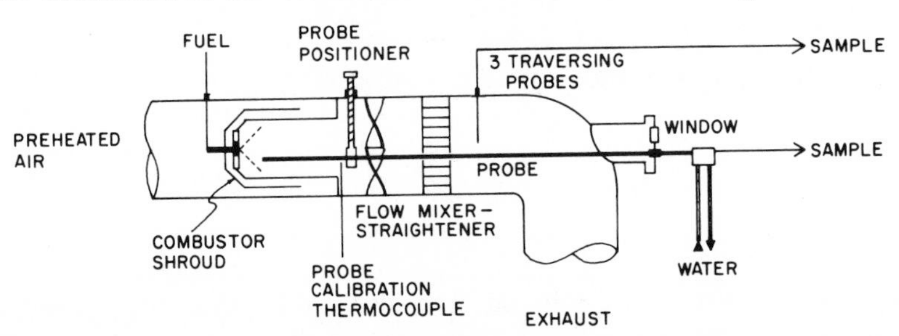

Fig. 2. Combustor test rig.

entered at a flow rate of 1.6 lb/sec as measured by a calibrated orifice, a temperature of 1000 F and an absolute pressure of 3 atmospheres (Table 1). Downstream of the shrouded combustor, there are mixing and flow straightening vanes followed by a sampling section consisting of three aspirated Chromel-Alumel thermocouples equally spaced circumferentially and mounted on radially traversing mechanisms. During burner operation, gas samples were gathered and local temperatures were measured at 13 radial locations with these probes. The measurements were then averaged as described in the Appendix. Beyond this station, the flow passes through a water-cooled elbow and ultimately through a pneumatically controlled gate valve which regulates the system pressure.

Combustor – The combustor configuration is shown in Fig. 3 along with the average air admission profile determined by cold flow, equal-Reynolds-number,

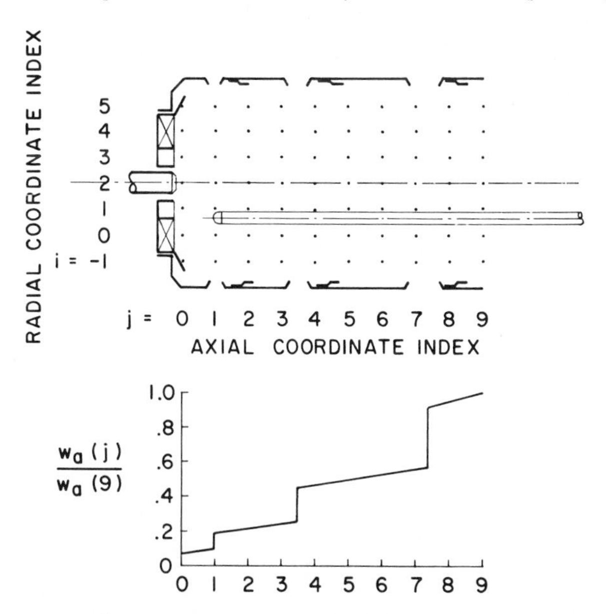

Fig. 3. Combustor sampling locations and air admission rate predicted from cold flow measurement.

measurement. It is constructed of Hastelloy with three sets of six plunged dilution ports. Wall cooling is achieved by means of splash-cooling rings located downstream of each set of ports. A vane-type annular swirler surrounds the pressure-atomizing fuel nozzle. Ignition is effected by means of a conventional spark ignitor. Table 2 contains a list of the pertinent combustor parameters.

TABLE 2

Combustor Dimensions*

Length: 9.25 Diameter: 5.75

	Dia.	Location (from nozzle)
Primary Dilution Ports:	0.5	1.2
Secondary Dilution Ports:	0.7	3.75
Tertiary Dilution Ports:	1.0	7.5

Swirler: 1.72 I.D. x 2.7 O.D.
15° air entry angle with respect to frontal plane.

Nozzle: 90° hollow cone, orifice diameter = 0.025

* *All dimensions are expressed in inches.*

During operating, a short section of the fuel line was immersed in preheated air. Diesel fuel, therefore, entered the nozzle at 220 F while methanol, with its greater flow rate, entered at 190 F. Typical nozzle differential pressures were 100 psi for diesel oil and 320 psi for methanol. Fuel flow was measured either by a rotameter or by a turbine flowmeter.

The combustor sampling coordinates are shown in Fig. 3 and in cross-section in Fig. 4 along with the associated area weights (see Appendix). With the water-cooled

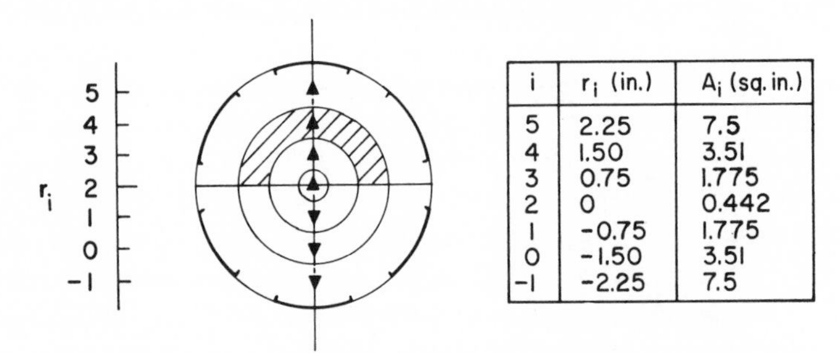

i	r_i (in.)	A_i (sq. in.)
5	2.25	7.5
4	1.50	3.51
3	0.75	1.775
2	0	0.442
1	-0.75	1.775
0	-1.50	3.51
-1	-2.25	7.5

Fig. 4. Radial sampling locations and associated area weights.

probe, positioned by means of a screw (Fig. 2), a continuous gas sample was extracted at 7 radial positions in each of 10 cross-sectional planes. A complete diagonal traverse was made. Traverses along additional diagonals would have been desirable at certain locations (plane 7, in particular) had time permitted. Temperature and concentration measurements at each individual radial location normally required between 10 and 20 minutes.

Sampling Probe – Sawyer and co-workers (9,10) have pioneered in the development of water-cooled sampling probes for turbine combustor research. They extracted and analyzed several species at a point within a combustor and were thereby able to deduce the chemical composition and local air-fuel ratio. Temperature was computed from the local equilibrium chemical composition and the combustion completeness was computed from this temperature. Attempts at measuring temperature by means of a thermocouple met with limited success. The present series of experiments draws heavily on the above techniques with the exceptions that temperature is measured in a different manner and the probe is positioned by flexure rather than by rotation.

In the present experiments, high combustion temperatures stemming from the use of pre-heated air precluded the use of thermocouples at the outset. The elevated combustion pressure (3 atm.) allows the determination of local static temperature by another method. Using this pressure to drive the sample mixture through a choked orifice allows the temperature to be computed (11) as follows:

$$\dot{m}_s = C_D \left[\frac{W}{R} \Upsilon \left(\frac{2}{\Upsilon + 1} \right)^{\frac{\Upsilon + 1}{\Upsilon - 1}} \right]^{1/2} \frac{PA^*}{T^{1/2}} \qquad (10)$$

If the molecular weight, W, and ratio of specific heats, Υ, are constant, this relation shows that the sample mass flow is inversely proportional to the square root of temperature. Since pressure varies by, at most, 3 per cent throughout the combustor, it may be assumed constant. Molecular weight (see the Appendix) varied between 29 and 30 throughout the accessible region of the combustor. Variations in the orifice area, A*, due to thermal expansion introduce a 2 per cent maximum error over the temperature range between the location, in the combustor exhaust, of a calibration thermocouple and the maximum temperature measured. In operation, the sampling probe was withdrawn to the thermocouple location and a tare reading of $\dot{m}_s$ taken by means of a rotameter operating at atmospheric pressure in the sampling system. A pressure tap immediately downstream of the orifice was used to verify that the orifice pressure differential was sufficient to insure critical flow. The Mach number, as determined by the pressure measured in the small reservoir behind the orifice, was typically 1.1. This implies a static temperature ratio of 0.8 due to expansion which aids in quenching further reaction within the probe. An advantageous feature of this technique is that temperature is measured at the same location and time as is the chemical composition. A disadvantage is encountered when sampling in the immediate vicinity of the fuel nozzle, where raw fuel clogged the orifice. This problem is not restricted to choked orifice probes; efforts at sampling in this location with a large diameter probe were prevented by sample line blockage due to condensation of fuel and water. When orifice blockage did occur, it was unmistakable and lead to no ambiguity in temperature determination.

Instrumentation – Samples from both the water-cooled probe and the three exhaust traversing probes passed through approximately 30 feet of stainless steel tubing to the instrumentation console in the control room of the test cell. The lines were unheated since any condensation would occur primarily in the 4-foot length of the cooled probe. The sample temperature as well as the water outlet temperature was typically 90 F. Sample flow rates (3-5 SCFH) through the choked probe were sufficient at all but the lowest (highest temperature) flows to allow simultaneous determinations of CO, CO_2, HC, NO and O_2. Since the pressure was low ($\sim$1 atm) throughout the sampling system, the NO $\rightarrow$ NO_2 conversion which is proportional to P^3 was minimized. (The authors are indebted to Prof. John Lenczyk of the University of Akron for this observation.)

Oxides of nitrogen were measured with a Ford-built Nitric Oxide Optical Detector (12) (chemiluminescence method) equipped with a NO_2 converter. No exhaust sample treatment other than particle filtration was required with this instrument. The data obtained indicated that, in all but a few locations sampled, NO constituted at least 95 per cent of the total observed oxides of nitrogen. It is assumed in the data presentation that all NO_2 is formed in the sample lines.

References pp. 232-233

Carbon monoxide and carbon dioxide were measured with Beckman nondispersive infrared analyzers. The CO_2 sample was dried by passage through a container filled with calcium sulfate (drierite) and subsequently filtered. The CO sample was dried, filtered and the CO_2 removed because of its interference at low CO concentrations. The CO_2 removal was accomplished by passing the sample through a container filled with asbestos granules impregnated with sodium hydroxide (ascarite). Further drying was then required to remove the water produced in this process.

The unburned hydrocarbon concentration was measured with a Beckman flame ionization detector (FID). The only sample treatment used with this device was particle filtration. Oxygen concentration was determined by a Beckman Process Oxygen Analyzer.

EXPERIMENTAL RESULTS

Radial Profiles – The radial profiles of NO mole fraction taken in several planes perpendicular to the combustor axis are shown in Figs. 5 and 6. They are

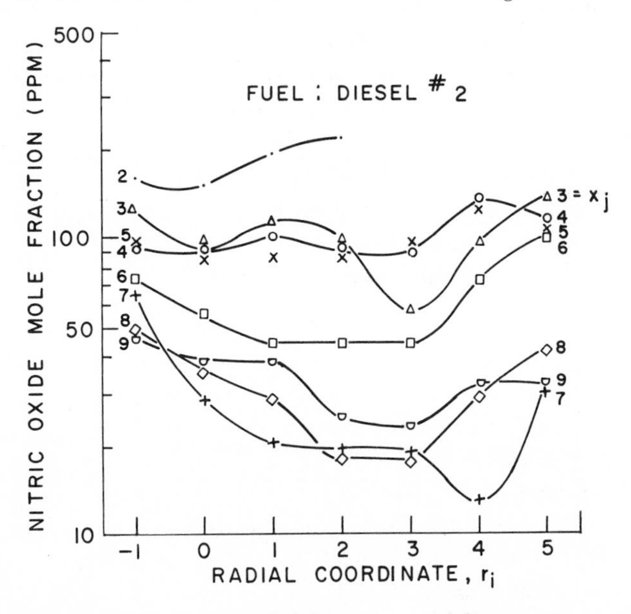

Fig. 5. NO mole fraction vs. radial distance – axial distance a parameter.

unsymmetrical but reproducible. Note that both fuels display similar profiles at each axial location with the most distortion evidenced at planes 3 and 7 which are dilution port locations. The profile at plane 2 is incomplete due to probe clogging in the vicinity of the fuel nozzle. Although no firm conclusions will be drawn regarding chemistry from these curves at this time, it is noted that the trend in NO mole fraction is downward as the distance from the nozzle is increased and that the similar character of the curves for both fuels suggests that dilution patterns are changed little by the fuel differences. This observation is reinforced by the character of the A/F and temperature profiles, which follow, and then ultimately by consideration of the cross-sectional averaged profiles.

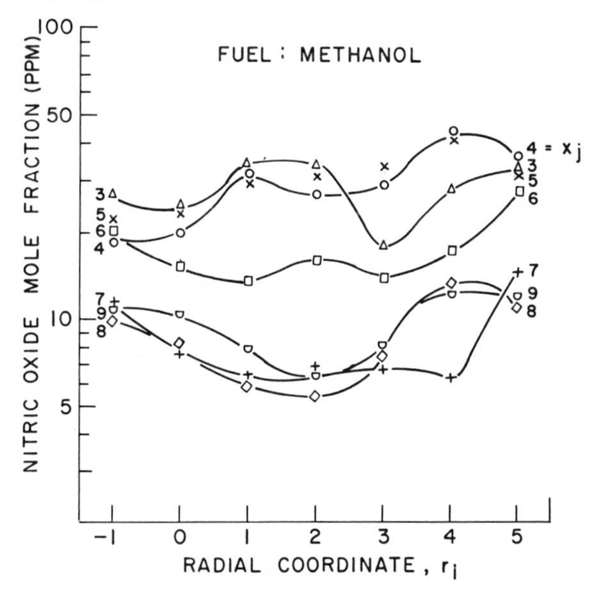

Fig. 6. NO mole fraction vs. radial distance – axial distance a parameter.

Figs. 7 and 8 illustrate the radial variation of measured air-fuel ratio as a function of axial distance for diesel and methanol. They are assymmetrical and appear to be

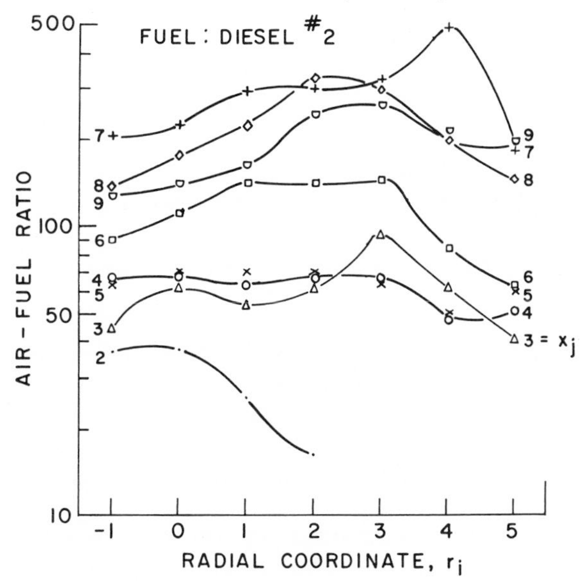

Fig. 7. Local air-fuel ratio vs. radial distance – axial distance a parameter.

References pp. 232-233

reflections of the NO profiles. Again the greatest distortion appears in planes 3 and 7, the dilution port locations. If all of the NO has been formed prior to the first measured plane (X_2), then the locations of high A/F should correspond to locations of low NO. This is indeed the situation, as will be seen later.

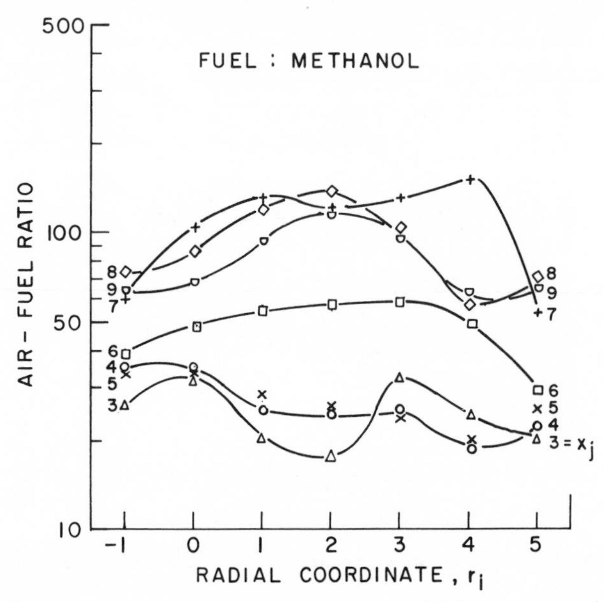

Fig. 8. Local air-fuel ratio vs. radial distance — axial distance a parameter.

The A/F patterns for both fuels differ by a constant factor, the ratio of stoichiometric A/F ratios for diesel fuel and methanol. This indicates that the air dilution pattern, for the regions sampled, is not affected greatly by the fuel differences. The profile at plane 7 is not only distorted but everwhere higher than the overall exit air-fuel ratio. This, of course, is impossible and suggests that the seven sampling points in this plane are not sufficient to provide a representative ensemble. The same remark applies to the NO profile at this location. It is lower, on the average, than the exit NO profile and could prompt the erroneous conclusion that NO is formed between planes 7 and 9. More will be said on this subject when the averages are presented.

The temperature profiles, as measured by the choked-orifice technique, are shown for diesel and methanol in Figs. 9 and 10. They are similar and of comparable magnitude for both fuels and are reminiscent of the NO profiles. This suggests that the combustor mixing processes are less than complete; gas originating in hot regions where NO is formed retains its identity throughout much of the combustor.

Cross-Sectional Averages — In the preceding section, consideration of the behavior of the radial profiles of NO, A/F and T lead to the hypothesis that all NO has already formed prior to the location of the first plane which could be reliably sampled, the

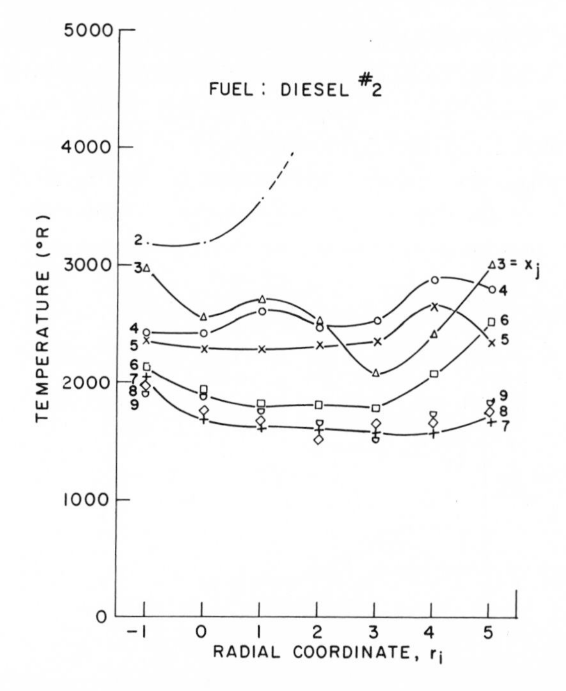

Fig. 9. Local temperature vs. radial distance – axial distance a parameter.

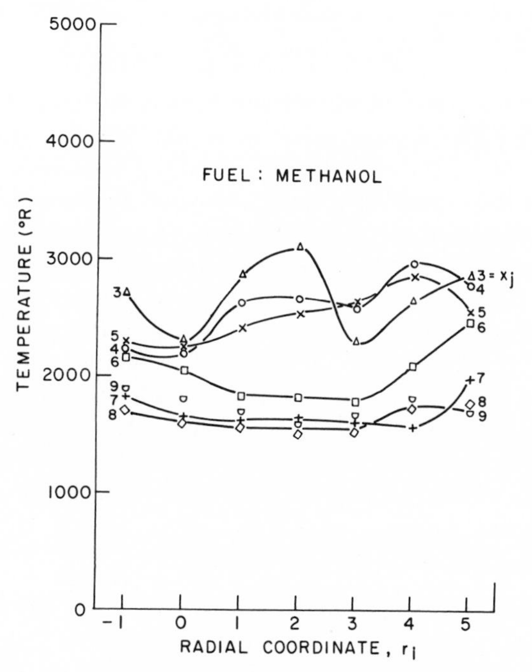

Fig. 10. Local temperature vs. radial distance – axial distance a parameter.

References pp. 232-233

plane located two inches downstream of the fuel injector. Subsequent changes in NO concentration are then due solely to dilution. Fig. 11 illustrates the axial behavior of average NO mass fraction for the two fuels. NO mass fraction and mole fraction are essentially equivalent since the molecular weight of NO and of the mixture are approximately equal. The averages were taken by the methods outlined in the Appendix. Points in the figure are experimental, whereas the solid curves are predictions based on a one-dimensional plug flow combustor model described below.

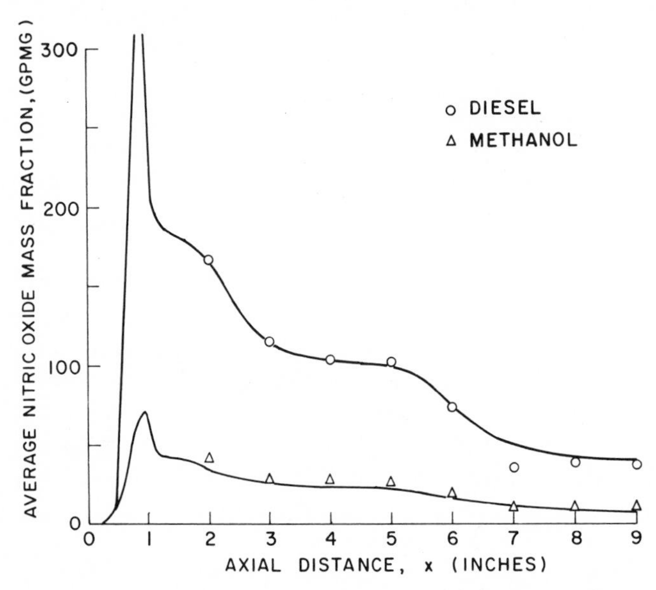

Fig. 11. Cross-sectional average NO mass fraction vs. axial distance, points are experimental, curves are computed (see text).

Consider, at first, the experimental points for diesel fuel combustion only. If it is assumed that their decreasing values as a function of axial distance are due solely to dilution, then the combustor air addition rate may be calculated from their variation. This was done and the resulting $w_a(x_j)$ profile agreed well with the air addition rate (Fig. 3) predicted from cold flow measurements. The same assumption, when made for the methanol NO values, yielded an identical air dilution rate. Further evidence that this assumption is valid comes from similar considerations involving average CO_2 and H_2O axial variations and also the average temperature variation to be discussed later. It may, therefore, be stated that all of the NO is formed in an extremely small region in the vicinity of the fuel nozzle. This region is smaller than that observed by Sawyer, no doubt due to more rapid combustion brought about by the use of preheated inlet air and elevated combustor pressure.

The foregoing interpretation not only depends upon rapid NO formation kinetics but implicitly contains a rapid quench mechanism to freeze the reaction. Such a rapid quench could be provided by the dilution jets located one inch from the front of the combustor can (Fig. 3). The predicted average air dilution profile contains a step at this location. It is evident that all of the interesting chemical processes except CO oxidation occur in a region inaccessible to direct measurement when these fuels are used. The large amount of water produced by methanol and the incomplete combustion of heavy hydrocarbons preclude the use of cooled probes within two inches of the injector. Indeed, since the NO formation region is so small, the dimensions of the probe itself become appreciable by comparison and can no longer be assumed to introduce only a small perturbation on the measurements. The use of other fuels such as methane might allow somewhat closer measurement, but since it is the original purpose of this investigation to compare diesel fuel and methanol, one must resort to analytical means to piece together the processes in the NO formation zone. The experiment, as far as it has been conducted, has provided at least some essential data with which to implement the analysis: the combustor air addition rate and an upper limit on the length of the combustion zone.

Since the air addition profile as measured by the NO dilution corresponds well to that predicted (except at plane 7 where the radial sampling points were insufficient to obtain a representative sample as previously mentioned), a hybrid air addition profile may be constructed composed of the cold flow prediction up to plane 2 and the NO dilution measurements from planes 2 through 9. This hybrid profile may now be used in conjunction with the one-dimensional combustion model mentioned previously. The combustor model requires the air addition rate and heat addition rate to be specified one-dimensionally. It incorporates the Zeldovich NO formation mechanism and employs equilibrium chemistry for the other species. It computes, as a function of axial distance, the residence time, temperature, velocity, density, nitric oxide mass fraction and the equilibrium values of NO, CO and CO_2 mole fractions. The model was programmed for digital computation by Dr. Terry Sharp of the Ford Scientific Research Staff.

It will be assumed, on the basis of observation of the measured A/F profiles for the two fuels, that the quench *location* does not vary for each fuel. Beside the air addition rate, the combustion completeness (η) must be specified. Experiment has shown that combustion is complete before plane 2 is reached. Indeed, three measures of η were computed at each data location and never indicated less than 95 per cent combustion completeness. These measures were based upon the amount of fuel consumed, the amount of oxygen consumed and the temperature attained. They were computed both locally and on an appropriate cross-sectional average. In the absence of directly measured η Eq. 8 was used to describe the shape of η and the burning time was fitted such that NO curve for diesel fuel was computed as shown in Fig. 11. The complete computer predictions are listed in Tables 3 and 4 for diesel fuel and methanol respectively. In order to predict the NO profile for methanol, the same η curve was employed but with the time to complete burning reduced by a factor of 0.73. This is in accord with the general remarks made in the section on physical

References pp. 232-233

considerations. The predicted NO profile for methanol is also shown in Fig. 11. Note that at no time do the NO production levels of either fuel attain equilibrium values.

TABLE 3

One-Dimensional Combustor Model Computer Print-Out: No. 2 Diesel Oil

COMBUSTION PROFILE

DIST % OF TOTL	TIME CUM MSEC	TIME CUM PCT	TEMP DEG R	SPEED FT/ SEC	DENS RHO/ RHOO	FUEL BURN CUM%	TOTAL AIR CUM%	NITRIC OXIDE EQ. PPM	NITRIC OXIDE ACTUAL PPM	CO EQ. PPM	CO2 EQ. MOL%
0.0	0.00	0.	1459	6.7	1.105	0.0	6.0	0.	0.	0.	0.00
0.5	0.47	4.	1938	9.3	0.825	6.7	6.2	71.	0.	0.	1.51
2.5	1.60	14.	3255	17.9	0.477	31.8	7.0	3188.	0.	24.	6.25
5.0	2.45	21.	4106	26.2	0.368	59.2	8.0	7316.	13.	2647.	9.77
7.5	3.09	26.	4427	32.3	0.333	81.5	9.0	6724.	156.	12967.	10.79
10.0	3.63	31.	4500	36.7	0.323	96.8	10.0	5753.	379.	20345.	10.81
11.5	3.85	33.	3395	53.4	0.434	100.0	20.2	3941.	199.	59.	6.87
12.5	3.99	34.	3382	53.6	0.435	100.0	20.3	3872.	197.	54.	6.82
15.0	4.34	37.	3351	54.1	0.439	100.0	20.7	3708.	193.	44.	6.70
17.5	4.69	40.	3229	56.2	0.456	100.0	22.4	3092.	179.	20.	6.21
20.0	5.02	43.	3182	57.1	0.462	100.0	23.1	2865.	174.	15.	6.03
22.5	5.33	46.	3063	59.6	0.481	100.0	25.1	2335.	161.	6.	5.56
25.0	5.63	48.	2925	62.9	0.503	100.0	27.8	1791.	145.	2.	5.03
27.5	5.92	51.	2809	66.1	0.524	100.0	30.5	1392.	133.	1.	4.60
30.0	6.19	53.	2724	68.8	0.540	100.0	32.8	1143.	124.	0.	4.28
32.5	6.46	55.	2676	70.5	0.550	100.0	34.2	1015.	118.	0.	4.10
35.0	6.71	57.	2616	72.8	0.562	100.0	36.2	870.	112.	0.	3.89
37.5	6.97	60.	2585	74.1	0.569	100.0	37.3	801.	109.	0.	3.77
40.0	7.22	62.	2561	75.2	0.575	100.0	38.2	749.	106.	0.	3.69
42.5	7.47	64.	2548	75.7	0.577	100.0	38.7	723.	105.	0.	3.64
45.0	7.71	66.	2540	76.1	0.579	100.0	39.0	707.	104.	0.	3.61
47.5	7.96	68.	2538	76.2	0.580	100.0	39.1	702.	104.	0.	3.60
50.0	8.20	70.	2534	76.4	0.581	100.0	39.2	695.	104.	0.	3.59
52.5	8.45	72.	2528	76.7	0.582	100.0	39.5	682.	103.	0.	3.57
55.0	8.69	74.	2518	77.1	0.584	100.0	39.9	663.	102.	0.	3.53
57.5	8.93	76.	2492	78.4	0.590	100.0	41.0	614.	99.	0.	3.44
60.0	9.16	78.	2448	80.7	0.601	100.0	43.0	537.	95.	0.	3.28
62.5	9.39	80.	2388	84.1	0.616	100.0	46.0	445.	89.	0.	3.07
65.0	9.60	82.	2311	89.3	0.637	100.0	50.5	343.	81.	0.	2.80
67.5	9.79	84.	2217	96.9	0.663	100.0	57.2	244.	71.	0.	2.48
70.0	9.97	85.	2131	105.7	0.690	100.0	65.0	173.	63.	0.	2.18
72.5	10.14	87.	2073	113.0	0.709	100.0	71.5	135.	57.	0.	1.99
75.0	10.30	88.	2030	119.4	0.724	100.0	77.2	111.	53.	0.	1.84
77.5	10.46	90.	1994	125.5	0.737	100.0	82.6	94.	50.	0.	1.72
80.0	10.60	91.	1972	129.8	0.746	100.0	86.5	85.	47.	0.	1.64
82.5	10.74	92.	1952	133.7	0.753	100.0	90.0	77.	46.	0.	1.58
85.0	10.88	93.	1940	136.5	0.758	100.0	92.5	72.	44.	0.	1.54
87.5	11.02	94.	1929	138.8	0.762	100.0	94.6	68.	43.	0.	1.50
90.0	11.15	96.	1923	140.4	0.765	100.0	96.0	66.	43.	0.	1.48
92.5	11.29	97.	1917	141.8	0.767	100.0	97.2	64.	42.	0.	1.46
95.0	11.42	98.	1912	143.2	0.769	100.0	98.5	62.	42.	0.	1.44
97.5	11.55	99.	1908	144.1	0.771	100.0	99.4	61.	41.	0.	1.43
100.0	11.68	100.	1905	144.8	0.772	100.0	100.0	60.	41.	0.	1.42

Measured and predicted average temperatures versus axial distance are shown in Fig. 12. The computed curves are based upon the same air dilution and combustion

TABLE 4

One-Dimensional Combustor Model Computer Print-Out: Methanol

CØMBUSTIØN PRØFILE

DIST % ØF TØTL	TIME CUM MSEC	TIME CUM PCT	TEMP DEG R	SPEED FT/ SEC	DENS RHØ/ RHØO	FUEL BURN CUM%	TØTAL AIR CUM%	NITRIC ØXIDE EQ. PPM	NITRIC ØXIDE ACTUAL PPM	CØ EQ. PPM	CØ2 EQ. MØL%
0.0	0.00	0.	1459	8.2	1.029	0.0	6.0	0.	0.	0.	0.00
0.5	0.38	3.	2038	11.8	0.731	8.2	6.2	95.	0.	0.	1.50
2.5	1.26	12.	3434	22.5	0.423	38.6	7.0	3306.	0.	75.	6.22
5.0	1.96	18.	4148	31.1	0.341	70.4	8.0	4592.	11.	4624.	9.43
7.5	2.51	23.	4287	36.3	0.323	93.2	9.0	2416.	49.	17264.	9.89
9.0	2.81	26.	4288	38.4	0.322	100.0	9.6	2333.	64.	18125.	9.95
10.0	3.00	28.	4278	39.5	0.325	100.0	10.0	3079.	76.	13208.	10.10
12.5	3.36	31.	3268	55.3	0.439	100.0	20.3	2971.	42.	27.	6.23
15.0	3.69	34.	3241	55.8	0.443	100.0	20.7	2859.	41.	23.	6.12
17.5	4.03	37.	3170	57.2	0.453	100.0	21.8	2567.	39.	14.	5.84
20.0	4.35	40.	3093	58.8	0.465	100.0	23.1	2263.	37.	8.	5.54
22.5	4.65	43.	2986	61.3	0.483	100.0	25.1	1875.	34.	3.	5.13
25.0	4.95	46.	2863	64.6	0.504	100.0	27.8	1471.	31.	1.	4.67
27.5	5.22	48.	2758	67.9	0.524	100.0	30.5	1171.	29.	1.	4.28
30.0	5.49	51.	2680	70.6	0.540	100.0	32.8	975.	27.	0.	4.00
32.5	5.75	53.	2635	72.3	0.550	100.0	34.2	873.	26.	0.	3.84
35.0	6.00	55.	2580	74.6	0.562	100.0	36.2	756.	24.	0.	3.65
37.5	6.25	57.	2552	75.9	0.568	100.0	37.3	699.	24.	0.	3.54
40.0	6.49	60.	2529	77.0	0.574	100.0	38.2	657.	23.	0.	3.46
42.5	6.74	62.	2517	77.5	0.576	100.0	38.7	635.	23.	0.	3.42
45.0	6.98	64.	2510	77.9	0.578	100.0	39.0	622.	23.	0.	3.40
47.5	7.22	66.	2507	78.0	0.579	100.0	39.1	618.	23.	0.	3.39
50.0	7.46	69.	2504	78.2	0.580	100.0	39.2	612.	23.	0.	3.38
52.5	7.70	71.	2498	78.5	0.581	100.0	39.5	602.	23.	0.	3.36
55.0	7.93	73.	2489	78.9	0.583	100.0	39.9	586.	22.	0.	3.32
57.5	8.17	75.	2465	80.2	0.589	100.0	41.0	545.	22.	0.	3.24
60.0	8.40	77.	2423	82.5	0.599	100.0	43.0	481.	21.	0.	3.10
62.5	8.61	79.	2367	86.0	0.614	100.0	46.0	402.	19.	0.	2.90
65.0	8.82	81.	2294	91.1	0.634	100.0	50.5	314.	18.	0.	2.65
67.5	9.01	83.	2205	98.8	0.661	100.0	57.2	227.	16.	0.	2.35
70.0	9.19	85.	2123	107.6	0.687	100.0	65.0	164.	14.	0.	2.08
72.5	9.35	86.	2067	114.9	0.706	100.0	71.5	129.	13.	0.	1.90
75.0	9.51	88.	2025	121.4	0.721	100.0	77.2	107.	12.	0.	1.76
77.5	9.66	89.	1991	127.4	0.734	100.0	82.6	91.	11.	0.	1.65
80.0	9.81	90.	1968	131.7	0.742	100.0	86.5	82.	11.	0.	1.58
82.5	9.95	92.	1950	135.6	0.750	100.0	90.0	75.	10.	0.	1.52
85.0	10.08	93.	1937	138.4	0.754	100.0	92.5	70.	10.	0.	1.48
87.5	10.22	94.	1927	140.8	0.759	100.0	94.6	67.	10.	0.	1.44
90.0	10.35	95.	1921	142.3	0.761	100.0	96.0	65.	10.	0.	1.42
92.5	10.48	96.	1915	143.7	0.763	100.0	97.2	63.	9.	0.	1.41
95.0	10.61	98.	1910	145.1	0.766	100.0	98.5	61.	9.	0.	1.39
97.5	10.74	99.	1906	146.1	0.767	100.0	99.4	60.	9.	0.	1.38
100.0	10.87	100.	1903	146.8	0.768	100.0	100.0	59.	9.	0.	1.37

References pp. 232-233

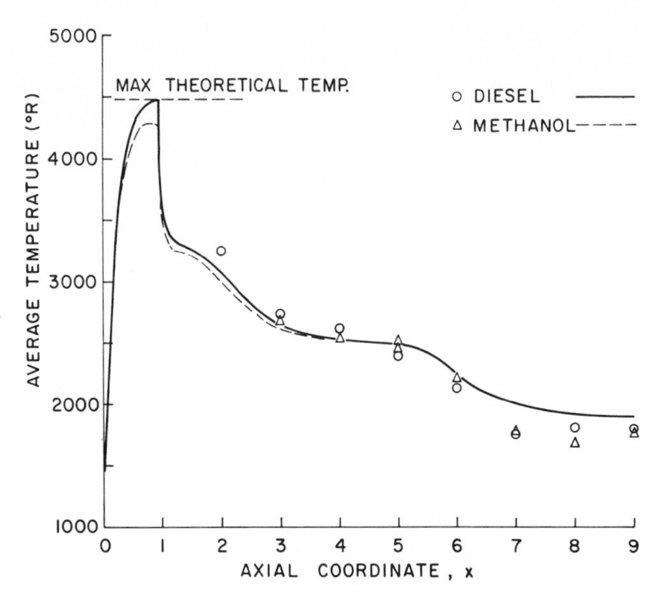

Fig. 12. Average temperature vs. axial distance, points are experimental, curves are computed.

profiles used to fit the NO data. The agreement between the experimental points as measured by the choked orifice technique and the predicted curves is quite good in view of the limited number of values which enter into the computation of the average. Note that the maximum temperature attained with methanol is about 200 F less than that achieved with diesel fuel. Identical exit temperatures are ultimately attained because there is more mass which must be cooled by dilution in the case of methanol. The lower peak temperature, of course, accounts for much of the decreased NO production with methanol.

Another reason for the decreased NO production with methanol becomes apparent when the computer predictions are replotted versus time instead of distance (Fig. 13). From the figure, it is noted that in the case of the faster burning methanol, NO begins to be produced earlier. It is produced at a lower rate, as determined by the slopes of the NO mole fraction curves, because the temperature is lower and because not as much oxygen and nitrogen is available with methanol as was mentioned earlier. The products of methanol combustion reach the quench location about 0.5 msec earlier than do those of diesel combustion due to earlier acceleration and greater mass flux and despite a somewhat lower peak temperature. One half of a millisecond is sufficient to cause a 25 percent difference in the NO produced by the two fuels. This may be seen by extrapolating the methanol NO mole fraction curve to the quench time for diesel. Instead of a peak of 76 ppm, a peak of 140 ppm would have been reached with methanol during the same period that a peak of 379 ppm was obtained with diesel. On this basis, one may say that 25 percent of the difference in the ultimate NO emission levels of the two fuels may be attributed to decrease residence time at high temperature for methanol and that 75% of the difference is due to other causes such as lower temperature and decreased oxygen and nitrogen availability.

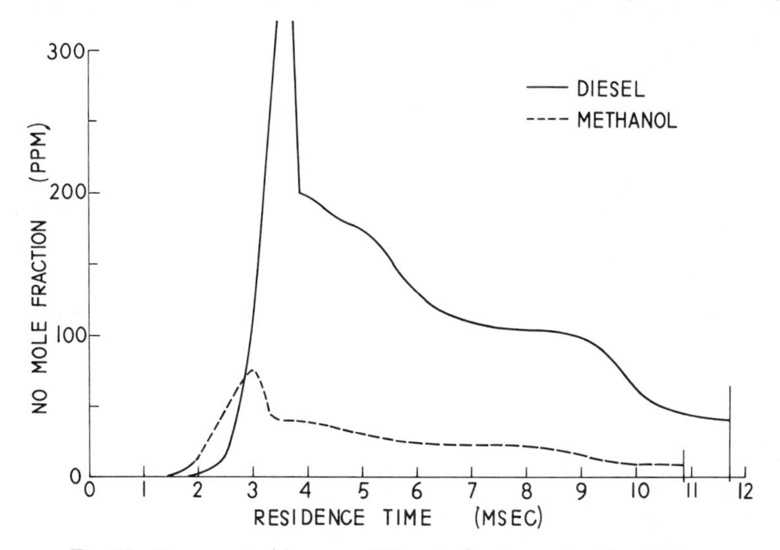

Fig. 13. Cross-sectional average NO mole fraction vs. residence time.

The average mass fractions of CO produced by diesel oil and methanol are shown plotted against axial distance in Fig. 14. In general, more CO was produced with

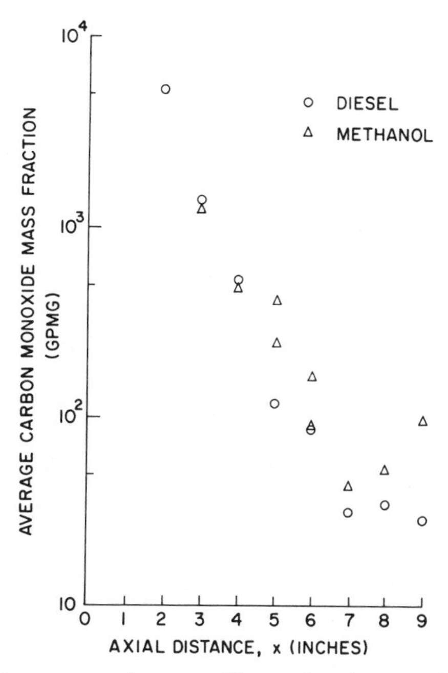

Fig. 14. Cross-sectional average CO mass fraction vs. axial distance.

References pp. 232-233

methanol and the methanol data exhibited more scatter. It is apparent from Fig. 14, that the rapid decrease of the mass fractions with axial distance is due to more than dilution. Comparison of the measured values with the equilibrium concentrations as given by Tables 3 and 4 reveals that the actual CO levels are always greater than equilibrium values except near the highest levels measured. The experimental points, if extrapolated to plane 1, approach the equilibrium values predicted at that point. Assuming these levels as a starting point for CO oxidation and subtracting the effects of dilution, the curves of Fig. 15 may be constructed. They indicate the extent of CO conversion to CO_2 for each fuel. It may be seen that conversion is essentially stopped by the time the combustor midpoint is reached. Viewed in another way, one may say that by the time the temperature is reduced to approximately 2000 F, the CO-CO_2 conversion rate is essentially zero. The conversion is not complete at this point however. The small percentage remaining unconverted is, of course, the cause of some concern itself. Since NO has already been quenched near plane 1, it would seem appropriate to position the dilution holes of plane 3 somewhat farther downstream to expedite further CO conversion.

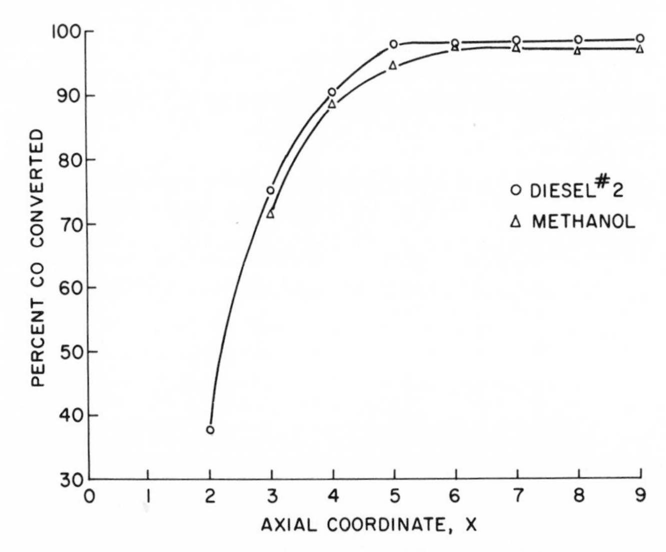

Fig. 15. Percent CO Converted to CO_2 vs. axial distance.

Hydrocarbon concentrations as measured by the FID were extremely low for both fuels. The exhaust HC concentration for diesel fuel was 4 ppm carbon for samples from both the water-cooled probe and the three uncooled traversing probes. This concentration is near the lower limit of detectivity of our system. The HC levels measured for methanol combustion were erratic but, in general, were higher than those for diesel by a factor of at least two. Nowhere in the measurable portion of the combustor did the average HC levels exceed 800 ppm carbon.

Pressure Effect – The NO formation rate, as given by Eq. 1, is proportional to the square root of pressure. For constant combustor mass flow, the overall residence time is proportional to pressure. Pressure weakly enhances the rate of combustion and inhibits dissociation. From these considerations, one would expect the ultimate level of NO produced by a given combustor, holding temperature and mass flow constant, to be proportional to pressure raised to some power greater than 1.5. This was investigated experimentally and the results are shown in Fig. 16 for the two fuels. It may be seen that the emitted level of NO does indeed increase with pressure over the range 2 - 4 atmospheres but not to the extent anticipated. The variation appears to be roughly linear with pressure for both fuels. It may be added that such an increase cannot continue indefinitely because, as equilibrium NO concentrations are approached in the primary zone, the formation rate must approach zero. It is anticipated that combustors of unconventional geometry may exhibit different pressure dependences.

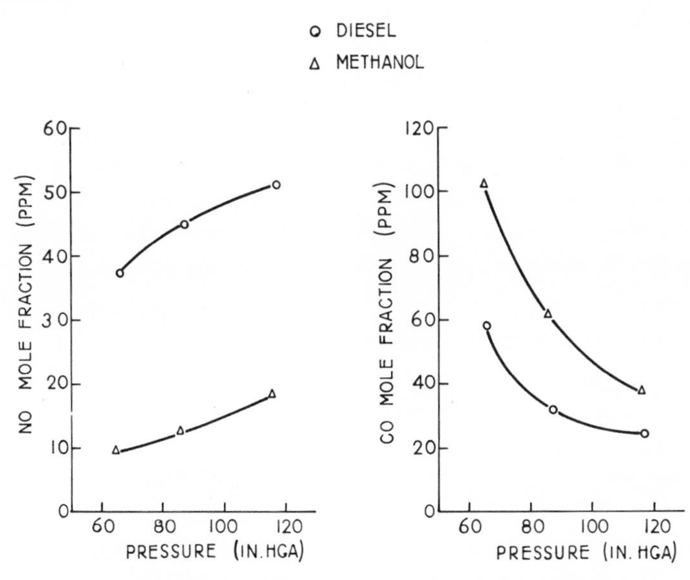

Fig. 16. Pressure dependence of NO and CO emission levels.

Also shown in Fig. 16, is the pressure dependence of the CO emission levels. They are decreasing functions of pressure due to enhanced combustion efficiency and greater residence time at high temperature.

SUMMARY AND CONCLUSIONS

The nitric oxide formation process in a regenerative gas turbine combustor has been studied to determine the causes of varying NO emission levels produced by two different fuels. Diesel #2 fuel and methanol were chosen as representative of two groups of liquid fuels which display different characteristics. A water-cooled sampling probe equipped with a sonic orifice by which temperature is measured was employed to determine combustion parameters within a turbine combustor test rig. The two

References 232-233

fuels were interchanged while the entrance and exit temperatures were maintained at 1000 F and 1400 F respectively, the chamber pressure held at 3 atmospheres and the air flow at 1.6 pounds per second. Although the probe could not be operated because of clogging in the immediate (2 in.) vicinity of the fuel nozzle, it did provide a necessary input (the combustor air addition rate and an upper limit on the length of the NO formation zone) for the mathematical description of the combustor processes. Its use allowed the following conclusions to be reached.

Nitric oxide is formed and the reaction subsequently quenched in a small region near the fuel nozzle. For this combustor, only the first 10 per cent of the air addition profile has an effect on the eventual amount of nitric oxide emitted. The small size of the NO formation zone is due to rapid combustion at elevated inlet temperature and pressure. The local actual concentrations of NO are always lower than local chemical equilibrium values.

The zone in which carbon monoxide is chemically active is considerably larger than the NO formation zone. For the conditions of the present experiment, the CO continued to oxidize until the average temperature fell below approximately 2000 F at the midpoint of the combustor. The maximum CO concentrations approached equilibrium values in the primary zone and were everywhere else higher than local chemical equilibrium indicates. Modification of the central portion of the combustor air addition profile may be expected to influence the ultimate amount of CO emitted.

The sonic orifice method of temperature measurement performed satisfactorily in a regenerative turbine combustor environment.

The difference in NO emission levels of diesel and methanol fuels is partly due to residence time differences and partly due to temperature and oxygen availability differences. The primary effect of drop size is a shorter burning time for the smaller methanol drops. This leads to earlier acceleration and diminished residence time for combustion gases in the primary zone.

Nitric oxide emission levels are approximately proportional to pressure for the conditions and geometry of the combustor considered.

ACKNOWLEDGEMENTS

The authors would like to acknowledge the assistance of Dr. T.E. Sharp who helped to develop the NO prediction computer program, Mr. N. Azelborn who programmed the data reduction analysis and Messers. N. L. Smith, G. W. Andrews, T. F. Amman, and M. D. Noldy for their assistance during the experimental phase of this study.

REFERENCES

1. *Ya. B. Zeldovich, P. Ya. Sadovnikov, and D.A. Frank-Kamenetskii, Oxidation of Nitrogen in Combustion, Publishing House of the Academy of Sciences, USSR, Moscow, 1947. (Translated by M. Shelef, Scientific Research Staff, Ford Motor Company. Copies are available from the translator).*

2. *P. Blumberg and J.T. Kummer, "Prediction of NO Formation in Spark-Ignited Engines – An Analysis of Method of Control," Combustion Sciency and Technology, 4, 1971, p. 73.*
3. *W.R. Marshall, "Atomization of Spray Drying," Chem. Eng. Prog. Series No. 2, Vol. 5, AIChE, New York, 1954.*
4. *Delavan Mfg. Co., Des Moines, Iowa, private communication with R. Slezak.*
5. *S. Way, "Combustion in the Turbojet Engine", Selected Combustion Problems II, AGARD, Butterworths, London, 1956.*
6. *A.A. Ranger and J.A. Nicholls, "Aerodynamic Shattering of Liquid Drops," AIAA Journal, 7, 2, Feb. 1969.*
7. *S.S. Penner, Chemistry Problems in Jet Propulsion, Pergamon Press, New York, 1957.*
8. *N.D. Ingebo, "Vaporization Rates and Heat Transfer Coefficients for Pure Liquid Drops," NACA TN 2368, July, 1951.*
9. *P.G. Parikh, R.F. Sawyer, and A.L. London, "Pollutants from Methane Fueled Gas Turbine Combustion," Univ. of California, College of Engineering Report No. TS-70-15, Berkeley, January, 1971.*
10. *L. Caretto, R.F. Sawyer, and E.S. Starkman, "Formation of Nitric Oxide in Combustion Processes," College of Engineering Report No. TS-68-1 Univ. of California, Berkeley, March, 1968.*
11. *A.M. Keuthe and J.D. Schetzer, Foundations of Aerodynamics, John Wiley & Sons, New York, 1959.*
12. *H. Niki, A. Warnick, and R.R. Lord, "An Ozone – NO Chemiluminescence Method for NO Analysis in Piston and Turbine Engines," SAE Paper No. 710072.*
13. *A.G. Piken and C.H. Rouf, "Chemical Composition of Automobile Exhaust and A/F Ratio," Scientific Research Staff, Ford Motor Company, June, 1968.*

NOMENCLATURE

A	Area
a	Hydrogen-carbon ratio of fuel
b	Hydrogen-carbon ratio of unburnt hydrocarbons
C	Specific heat at constant pressure
D_o	Fuel nozzle orifice diameter
D_{32}	Sauter-mean diameter in fuel spray
h	Static enthalpy
i	Radial coordinate index
j	Axial coordinate index
K	Evaporation constant
k	Thermal conductivity coefficient
L	Latent heat of vaporization
ℓ	subscript denoting ℓ^{th} chemical species

$\dot{m}$	Mass flow rate
n	Number of moles of a reactant divided by total number of moles of products
P	Pressure
R	Universal gas constant
r_i	i^{th} radial coordinate
T	Temperature, °R
t	Time
V	Velocity
W	Molecular weight
W_e	Weber Number
w_a	Air flow rate
X	Mole fraction
x	Number of carbon atoms in unburnt hydrocarbon
x_j	j^{th} axial coordinate
Y	Mass fraction
y	Number of hydrogen atoms in unburnt hydrocarbon
z	Number of oxygen atoms in methanol combustion product
γ	Ratio of specific heats
ΔH_c	Heat of combustion
η	Combustion completeness
ν	Moles of product
μ_f	Fuel viscosity coefficient
ν'	Moles of reactant
ξ	Number of carbon atoms in fuel
φ	Fuel-air equivalence ratio
ρ	Density

APPENDIX A

DATA REDUCTION – LOCAL VALUES

A gas sample is extracted with the probe at the point (r_i, x_j) within the combustor. Denoting measured mole fractions by an asterisk, the relations between measured and actual mole fractions are given by

$$3X^*_{HC} = X_{HC} \quad \text{(A-1)}$$

$$X^*_{NO} = X_{NO} \quad \text{(A-2)}$$

$$X^*_{O_2} = X_{O_2} \quad \text{(A-3)}$$

$$X^*_{CO_2} = X_{CO_2}/(1 - X_{H_2O}) \quad \text{(A-4)}$$

$$X^*_{CO} = X_{CO}/(1 - X_{H_2O} - X_{CO_2}) \quad \text{(A-5)}$$

The factor of 3 in the HC relation arises because the FID is calibrated using propane. CO_2 and CO measurements are taken on dried samples and the CO sample has CO_2 removed by the use of ascarite. At the low CO concentrations of interest, the presence of CO_2 leads to an NDIR interference comparable to the indicated CO reading.

For a hydrocarbon fuel, the combustion reaction is assumed to be

$$\nu'_1 C_\xi H_\eta + \nu'_{O_2} O_2 + 3.76\, \nu'_{O_2} N_2 \rightarrow$$
$$\nu_2 C_x H_y + \nu_{O_2} O_2 + \nu_{N_2} N_2 + \nu_{CO_2} CO_2 + \quad \text{(A-6)}$$
$$\nu_{H_2O} H_2O + \nu_{CO} CO + \nu_{NO} NO$$

This relation ignores H_2 which may be present in amounts comparable to CO. In the past (13), equilibrium of the water-gas shift reaction has been assumed and the H_2 computed from the concentrations of H_2O, CO_2 and CO. Since CO is always present in super-equilibrium amounts as was seen in the section on experimental results, this equilibrium relation cannot be employed. It is theoretically possible to use the normalization condition, $\sum_\ell x_\ell = 1$, to solve for X_{H_2} but this would entail differencing large numbers to obtain a number in the ppm range and is therefore not practical in view of the accuracy of the measurements involved. Neglecting H_2 introduces no serious error because it is a minor constituent over the major portion of the combustion chamber. The computation of the local air-fuel ratio depends mainly on O_2 and CO_2 mole fractions.

Dividing Eq. (A-6) by $\sum_{\ell} \nu_{\ell}$ expresses the products in terms of mole fractions.

$$n_1 \xi CH_a + n_{O_2} O_2 + 3.76\, n_{O_2} N_2 \rightarrow$$

$$x X_2\, CH_b + X_{O_2} O_2 + X_{N_2}\, N_2 + X_{co_2}\, CO_2 + X_{CO}\, CO + \qquad (A\text{-}7)$$

$$X_{H_2O}\, H_2O + X_{NO}\, NO$$

Here, the H/C ratios of the fuel and the unburned hydrocarbons have been designated a and b respectively. For diesel fuel, a = 1.86 and b is assumed to be equal to a/2. The quantity $xX_2 = X_{HC}$ in Eq. (A-1).

From the conservation of atomic species, H_2O and N_2 may now be computed:

$$X_{H_2O} = \frac{a}{2}\left[(1-\frac{b}{a})xX_2 + X_{CO_2} + X_{CO}\right] \qquad (A\text{-}8)$$

$$X_{N_2} = 1.88\left[2X_{O_2} + 2X_{CO_2} + X_{CO} + X_{H_2O} + 0.734X_{NO}\right] \qquad (A\text{-}9)$$

Substitution of equations (A-1) through (A-5) into these relations expresses X_{H_2O} and X_{N_2} in terms of measured quantities. This completes the chemical description of the gas mixture at a point for the case where a hydrocarbon mixture characterized by the ratio, a, reacts with air. As a check on accuracy, the sum, $\sum_{\ell} x_{\ell}$, was computed at each point. In general, it remained within 5 per cent of unity for all points. The local molecular weight was computed from

$$W(r_i, x_j) = \sum_{\ell} W_{\ell} X_{\ell}(r_i, x_j) \qquad (A\text{-}10)$$

and found to vary between 29 and 30 for all combustor locations which could be reliably sampled. (In the immediate vicinity of the fuel nozzle, there were indications that W exceeded 30). For purposes of this computation, the HC molecular weight, xW_{HC}, was assumed to be 86 (hexane).

The local air-fuel ratio is computed from Eq. (A-6) and the atom conservation equations

$$\frac{A}{F}(r_i, x_j) = \frac{138\, n_{O_2}}{\xi n_1 (12 + a)} = \frac{36.7}{12 + a}\left(\frac{X_{N_2} + X_{NO}/2}{xX_2 + X_{CO_2} + X_{CO}}\right) \qquad (A\text{-}11)$$

In the case of methanol combustion, a similar treatment may be made assuming the following combustion reaction:

$$\nu_1' CH_3OH + \nu_{O_2}' O_2 + 3.76\, \nu_{O_2}' N_2 \rightarrow$$

$$\nu_2\, C_x H_y O_z + \nu_{O_2} O_2 + \nu_{N_2} N_2 + \nu_{CO_2} CO_2 + \nu_{H_2O} H_2O \quad \text{(A-12)}$$

$$+ \nu_{CO}\, CO + \nu_{NO}\, NO$$

Here, for lack of chromatographic analysis, the unburned fuel is assumed to be methanol. The response of the FID to methanol is reduced by 25 per cent such that now

$$4X_{HC}^{*} = X_{HC} \quad \text{(A-13)}$$

while the other measurements remain unchanged. Water and nitrogen are computed from:

$$X_{H_2O} = 2\, X_{CO_2} + 2\, X_{CO} + (x - \frac{y}{4})\, X_2 \quad \text{(A-14)}$$

and

$$X_{N_2} = 1.88 \left[3\, X_{CO_2} + 2\, X_{CO} + 2\, X_{O_2} + 0.734\, X_{NO} + (z - \frac{y}{4})\, X_2 \right] \quad \text{(A-15)}$$

For pure methanol, $x = z = \frac{y}{4}$, so that X_2 does not appear in these computations.

The local air-fuel ratio for methanol combustion is given by

$$\frac{A}{F}(r_i, x_j) = 1.148 \left(\frac{X_{N_2} + X_{NO}/2}{X_{CO_2} + X_{CO}} \right) \quad \text{(A-16)}$$

These are point values. They are representative of conditions in a small volume surrounding the sampling point. The change in any quantity from point to point may be due to either chemical reaction or dilution and may not be determined unless the history of the fluid element passing through the sampling volume is known. Since the origin of all the fluid passing through the incremental sampling volume is unknown, one must resort to averages taken across the combustor to separate dilution from chemical effects and this may only be done if the average dilution profile is known as a function of axial distance. The latter may be estimated from cold flow measurement or obtained from the decrease in the cross-sectional mass-average value of a chemically inert tracer gas. Nitric oxide satisfies this criterion throughout much of the combustor. Appendix B deals with the relationship between point values and averages taken in a cross-sectional plane.

APPENDIX B

DATA REDUCTION – CROSS-SECTIONAL AVERAGES

Associated with each data point (r_i, x_j) is an area, A_i, in the x_j plane, over which the measured values are assumed constant. For the data network employed here, (Fig. 3), the magnitude of A is completely specified by the radial coordinate alone. Through A_i, the net mass flux in the x_j – direction is given by

$$\dot{m}\,(r_i, x_j) = \rho\,(r_i, x_j)\,V\,(r_i, x_j)\,A_i \tag{B-1}$$

The net mass flux of species, ℓ, through A_i is similarly

$$\dot{m}_\ell\,(r_i, x_j) = \rho_\ell\,(r_i, x_j)\,V\,(r_i, x_j)\,A_i \tag{B-2}$$

where the diffusion velocity of species, ℓ has been neglected compared to the mixture mass-averaged velocity, $V(r_i, x_j)$.

The local mass fraction of species, ℓ, is by definition

$$\begin{aligned} Y_\ell\,(r_i, x_j) &= \frac{\dot{m}_\ell\,(r_i, x_j)}{\dot{m}\,(r_i, x_j)} \\ &= \frac{W_\ell X_\ell (r_i, x_j)}{W\,(r_i, x_j)} \end{aligned} \tag{B-3}$$

The total mass flux of species, ℓ, through the entire cross-section is the sum over all the A_i. Introducing the perfect gas law and neglecting the small pressure change with r_i, this becomes

$$\dot{m}_\ell\,(x_j) = \frac{P}{R}\,\Sigma\,\frac{W_\ell\,X_\ell\,(r_i, x_j)}{T\,(r_i, x_j)}\,A_i V\,(r_i, x_j) \tag{B-4}$$

The total mass flux of all species is obtained by summing this relation over the index, ℓ.

$$\dot{m}\,(x_j) = \frac{P}{R}\,\underset{L}{\Sigma}\,\frac{W\,(r_i, x_j)}{T\,(r_i, x_j)}\,A_i V\,(r_i, x_j) \tag{B-5}$$

Velocity is not measured but we may, without loss of generality, define a cross-sectional average velocity

$$V(x_j) = \frac{R\,\dot{m}_\ell(x_j)}{P \sum_i \frac{W_\ell\, X_\ell(r_i, x_j)}{T(r_i, x_j)} A_i} \tag{B-6}$$

and a similar one in terms of the total mixture mass flux. The cross-sectional average mass fraction of species, ℓ, is now defined as the ratio of equations B-4 and B-5.

$$Y_\ell(x_j) = \frac{\sum_i \frac{W_\ell\, X_\ell(r_i, x_j)\, A_i}{T(r_i, x_j)}}{\sum_i \frac{W(r_i, x_j)\, A_i}{T(r_i, x_j)}} \tag{B-7}$$

This is the desired expression. It relates the average mass fraction to the locally measured mole fractions and the measured local temperature. It is essentially a mass-weighing of the individual point values to yield an average mass fraction which can then be used in conjunction with an independently obtained $m(x_j)$ to calculate

$$\dot{m}_\ell(x_j) \equiv Y_\ell(x_j)\, m(x_j) \tag{B-8}$$

The average cross-sectional temperature may be similarly related to the locally measured temperatures. In this case, one considers the flux of enthalpy through A_i, sums over A_i to get the total enthalpy flux, assumes that mass and energy convect with the same average velocity and arrives at an expression for cross-sectional average enthalpy in terms of local temperatures

$$h(x_j) = \frac{\sum_i \frac{W(r_i, x_j)\, A_i\, h(r_i, x_j)}{T(r_i, x_j)}}{\sum_i \frac{W(r_i, x_j)\, A_i}{T(r_i, x_j)}} \tag{B-9}$$

In the present data reduction program, the specific heat is assumed constant so that this expression reduces to

$$T(x_j) = \frac{\sum_i W(r_i, x_j)\, A_i}{\sum_i \frac{W(r_i, x_j)\, A_i}{T(r_i, x_j)}} \tag{B-10}$$

Thus far, it has been possible to average mass fractions and temperatures to their appropriately weighted locally measured values. This is not the case for the air-fuel ratio. The cross-sectional average air-fuel ratio is defined

$$1 + \frac{A}{F}(x_j) = \frac{\dot{m}\,(x_j)}{\dot{m}_f\,(x_o)} \tag{B-11}$$

where it has been assumed that all the fuel enters the combustor at the plane x_o. This may not be computed from the individually measured local air-fuel ratios without knowledge of the fuel distribution across the plane x_j. This may be seen by defining the local air-fuel ratio in terms of the individual fluxes passing through A_i and summing as with $Y_\ell\,(x_j)$. There results the following expression:

$$1 + \frac{A}{F}(x_j) = \frac{\sum_i \dfrac{W\,(r_i,x_j)\,A_i}{T\,(r_i,\,x_j)}}{\sum_i \dfrac{\dot{m}_f\,(x_o)}{\dot{m}_f\,(r_i,\,x_o)}} \sum_i \left\{ \frac{T\,(r_i,\,x_j)}{W\,(r_i,\,x_j)\,A_i} \left[1 + \frac{A}{F}(r_i,\,x_j) \right] \right\} \tag{B-12}$$

Everything in this expression relating A/F (x_j) to A/F (r_i, x_j) is known, except the sum in the denominator which may be termed the fuel distribution factor. It is the sum of the total fuel flow divided by the individual fuel fluxes passing through each A_i. Were this factor known, individually measured local A/F ratios could be summed to yield A/F (x_j) or, equivalently, the average combustor dilution rate. Since the latter is more easily obtained from Y_{NO} (x_j), for example, Eq. B-12 might be better employed to compute the fuel distribution factor as a function of x_j. The utility of such an exercise escapes the authors at the moment; we merely note that the factor for diesel and methanol, computed at the exit plane where A/F (x_j) is most accurately known, is 160 and 154, respectively, in our coordinate system. A completely uniform fuel flux density $(\rho_f\ V_f)$ distribution would yield a distribution factor equal to 107. These fuel distribution factors indicate that both fuels are similarly but not uniformly distributed at the exit plane. Assuming the fuel distribution factor to remain constant with respect to x_j results in poor agreement with A/F (x_j) computed by other means. Perhaps the greatest utility of this factor lies in providing a figure of merit with which to gauge various fuel injection methods.

DISCUSSION

A. H. Lefebvre

I should warn Dr. LaPointe that if you play around with computer programs long enough, you can usually get the answer you want. The trouble is that it may not be the answer you need.

F. V. Bracco *(Princeton University)*

In your experiments you seem to change both the fuel and the characteristics of the spray simultaneously. By changing both at the same time, it is then difficult to say whether the difference in NO production is due to the difference in the spray characteristics or to the difference in the chemical properties of the fuel such as, for instance, the ignition delay. The experiment might be useless if you cannot tell which one is responsible. Would it be possible to use one fuel and change the injector design so as to maintain a constant fuel-air ratio while producing sprays of different characteristics, or vice versa to change the fuel while keeping the same spray characteristics?

C. W. LaPointe

I think that it would be easier to change the injector to obtain different injection characteristics for a single fuel because to achieve the same exit temperature with methanol as with a hydrocarbon fuel we must put in 2.2 times as much methanol. The injection velocity is higher with methanol. I had a pre-conceived notion that since the injection velocity was higher the methanol would flow out of the primary zone quicker resulting in a leaner flame. It does not look like that happened. It appeared that the droplets of methanol were small enough so that they accommodated to the air velocity very quickly and the gas flowed out of the primary zone quickly because of its increased mass and earlier temperature rise.

Trajectories were computed for a methanol and diesel droplet having diameters equal to their respective SMD's and injected into a quiescent primary zone environment. They were assumed to evaporate according to Godsave's law modified for convection and to experience drag according to the intermediate Reynold's number law, $C_D = 18.5/R_e^{0.6}$, appropriate to the range $2 \leq R_e \leq 500$. The results are shown in Fig. 1A. Methanol is completely evaporated within one inch of the nozzle, whereas diesel travels considerably farther.

The combustion completeness for these trajectories is shown in Fig. 1B. Bearing in mind that the fuel enters at a 45° angle with respect to the combustor axis, it is seen

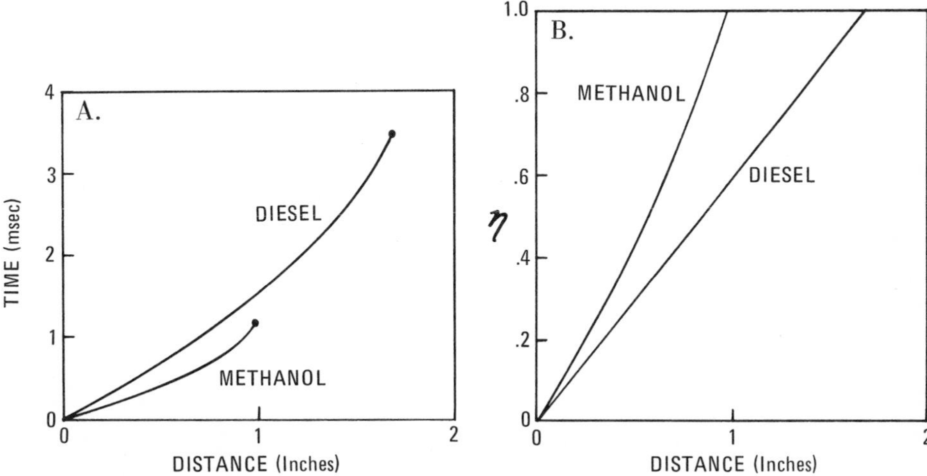

Fig. 1. Trajectory and associated combustion completeness of diesel and methanol droplets injected into quiescent primary zone environment.

that combustion is completed for both fuels in a very small region of the can. Methanol burns faster than diesel due to its small drop size and, in spite of its greater injection velocity, deposits its energy in a smaller volume. Drag quickly decelerates the smaller particles.

F. V. Bracco

Do you plan on separating the two effects in the future?

C. W. LaPointe

Yes, we hope to.

EXPERIMENTAL STUDIES OF CHEMICAL PROCESSES IN A MODEL GAS TURBINE COMBUSTOR

R. F. SAWYER

University of California, Berkeley, California

ABSTRACT

Composition and temperature within a model gas turbine combustor have been measured in a series of experimental investigations. Species measured include CO_2, CO, O_2, total HC, NO, NO_2, and particulates. The primary effort was devoted to the understanding of the process controlling the emissions of CO, HC, and NO. CO and HC levels are fixed by the thermal quenching of their oxidation reactions in the dilution zone of the combustor. NO levels are determined by the kinetics of formation and are strongly influenced by the maximum local temperature. The chemical kinetics of the controlling reactions should allow tailoring of combustor temperature-time characteristics in a manner to control the emission of all three species simultaneously.

INTRODUCTION

During the past five years a series of experimental investigations on gas turbine combustion with particular emphasis on the formation of pollutants have been conducted. Some of this work has been reported previously in the literature (1-4), a study of methane combustion recently has appeared in report form (5) and some investigations including the current continuing efforts have not been published. The present paper summarizes this work.

Although initial studies were simply measurements of combustor exhaust composition, more recent work has focused upon the determination of composition and temperature within the combustor. The one common feature of these studies has been that the same cylindrical can-type model gas turbine combustor has been employed in all of the experiments. Species measured include CO_2, CO, O_2, total HC,

References p. 252

NO, NO_2, and particulates with the primary effort upon understanding the mechanism of formation of CO, HC, and NO. As a part of this work, the necessary sampling and analysis techniques have been developed.

The combustion processes within the gas turbine combustor which control pollutant formation are a complex combination of fluid mechanics and chemical kinetics. In the present studies the complexities of the flow field often have been ignored. Such an attitude in no way implies that fluid mechanics is not an important part of the problem but simply reflects an effort to determine if the gross features of pollutant formation could be described in terms of primarily chemical phenomena. It seems likely that techniques for treating at least simple cases of combined fluid mechanics and chemistry are approaching practical application. For the present, however, the main justification for our approach to these studies is that it is difficult to do otherwise. Knowledge of the distribution of chemical species within the combustor should prove a valuable tool in efforts to minimize pollutant emissions.

EXPERIMENTAL STUDIES

The experimental equipment employed in this study has evolved over the period of the experiments so that the recent work benefits from more refined techniques and reflects more precise measurements than the earlier work. The basic combustor apparatus actually predates the current air pollutant studies by more than fifteen years. (6)

Combustor – The combustor employed in these studies bears perhaps only an historical resemblance to gas turbine combustors. The cylindrical geometry and uniform spacing of air holes, Fig. 1, with the resulting approximately linear introduction of air, are features which promote axisymmetry and minimize the importance of fluid mechanical variables. The division between a primary and secondary zone is fixed by the fuel and air flow rates, not the combustor geometry. Typical combustor parameters are given in Table 1.

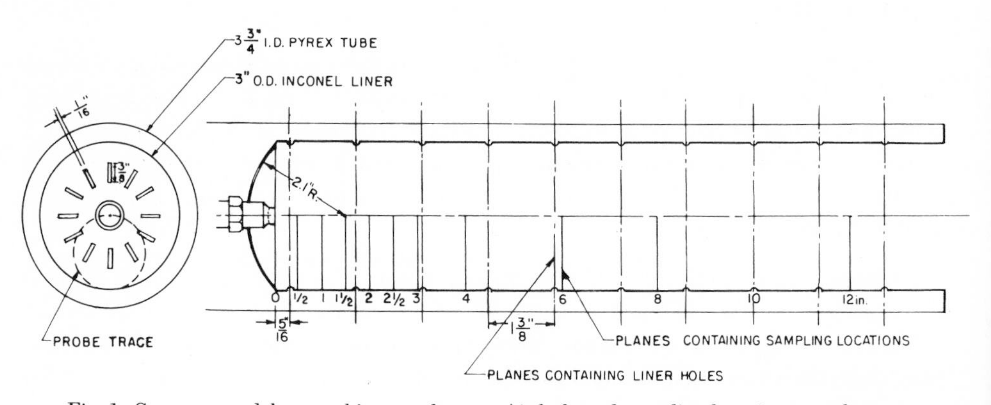

Fig. 1. Can-type model gas turbine combustor. Air hole and sampling locations are shown.

TABLE 1

Combustor Characteristics

Type	Cylindrical (3 in dia by 14 in)
Holes	80 3/16 in dia., uniformly spaced, 10 rows, 8 holes per row
Dome	Hemispherical, no swirl
Loading	2 x 10^6 Btu/ft^3-hr-atm (total volume) 10^7 Btu/ft^3-hr-atm (primary zone)
Mean Velocity	50 ft/sec
Residence Time	20 msec (total volume) 3 msec (primary zone)
Fuel Flow	5 lb/hr

Sample Probe – Water cooled sample probes, the latest version of which is shown in Fig. 2, were employed in all experiments. Although the probe shown includes a thermocouple for temperature measurement, the determination of temperature from the chemical composition has proved to be a better technique. The probe is positioned axially in the combustor through translational movement and radially through rotation. Typical sample flow rates are 1200 cc/min (STP) and initial cooling rates are estimated at 5 x 10^6 °C/sec. A larger, isokinetic probe is used for the sampling of particulates.

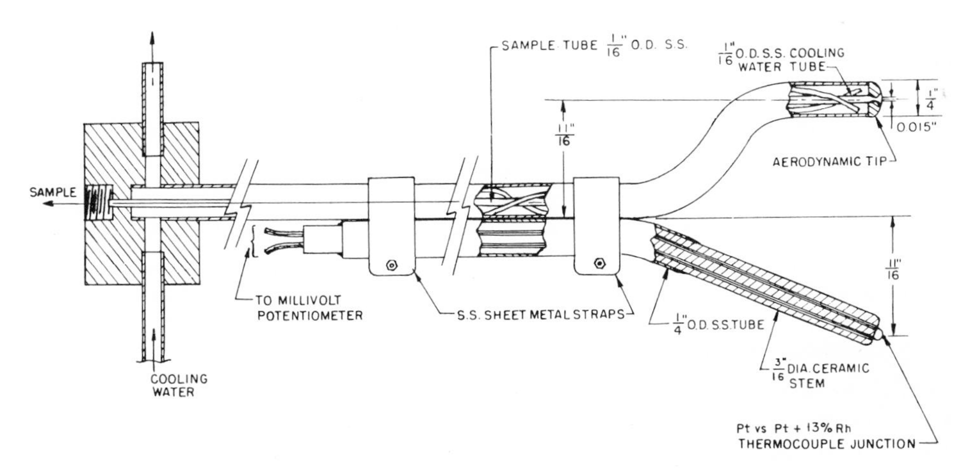

Fig. 2. Water-cooled sampling probe and thermocouple.

Sample Analysis – The problem of sample analysis is complicated by both the nature of the species to be detected and the low concentrations of interest, in some cases on the order of one part per million by volume. Analysis techniques are

References p. 252

summarized in Table 2. Where several different techniques are listed, a historical evolution generally is described. In some cases the techniques are still under development. Species not measured directly include H_2O and H_2. Continuous as opposed to batch analysis techniques have been sought wherever possible. The development of the chemiluminescent detector for NO has been reported previously. (7) The pressure at which the sample is transferred from the combustor to the analyzer generally is determined by the analyzer, for example, the nondispersive infrared analyzers require samples at or above atmospheric pressure; the chemiluminescent analyzer accepts samples at pressures of a few millimeters of mercury. Low pressure sampling and analysis generally is preferred because of reduction of sample size requirements, condensation difficulties, and transit times.

TABLE 2

Sample Analysis

Species	Technique
CO_2	nondispersive infrared
CO	nondispersive infrared
O_2	electrochemical
HC	nondispersive infrared heated flame ionization detector gas chromatography
NO	nondispersive ultraviolet (batch) nondispersive ultraviolet (continuous) nondispersive infrared chemiluminescent
NO_2	nondispersive ultraviolet
Particulate	filter (for mass) impaction (for microscopy) Neutron activation (for chemical analysis)
Aldehydes	precipitation and gas chromatography

Experimental Parameters – In most investigations the primary experimental variable was the overall equivalence ratio. Variation usually was produced by holding fuel flow or loading constant and varying the air flow. All studies have been of HC fuels and air oxidizer. Studies to date have all been with the reactants initially at ambient temperatures, i.e. no studies with air preheat have been conducted. A summary of the experimental parameters is presented in Table 3.

Experimental Results – Composition measurements have been examined in three different forms. First, combustor centerline measurements are the most easily

TABLE 3

Experimental Parameters

Fuels	diesel, jet, heptane, methane
Oxidizer	air
Pressure	1-4 atm
Equivalence ratio	0.2 - 0.6
Fuel injection	conical spray (liquids) high and low velocity jets (gaseous)

obtained and are particularly interesting because of generally axisymmetric character of the flow. Although the flow is not one dimensional, centerline composition profiles, especially downstream of the primary zone, provide a time history of at least part of the total combustor flow. Secondly, some of the compositions have been analyzed as concentrations averaged across the combustor diameter. Third, and perhaps most interesting (and complex) are the concentration profiles presented in radial and longitudinal coordinates. A few measurements of azimuthal variations indicate that the combustor is only approximately axisymmetric (two dimensional). Efforts are directed toward improvement of axisymmetry rather than measurement and presentation of concentrations in three dimensions. Two dimensional combustor maps have been constructed through least squares curve fitting techniques to provide concentration contours.

INTERPRETATION OF SELECTED EXPERIMENTAL RESULTS

Experimental measurements have been selected for their ability to demonstrate the phenomena controlling pollutant concentrations. These results are organized by species so that each subsection contains information obtained in different investigations and with different fuels.

Temperature – Chemical composition has been previously employed to successfully determine combustor gas temperatures. (8) The present work has confirmed the advantages of this approach over thermocouple measurements which suffer difficulties of probe integrity in the combustor environment and radiation correction. Temperature maps for three different air flows are shown in Fig. 3.

Carbon Dioxide – CO_2 in itself is not of great interest. Measurement, however, is essential to the determination of local equivalence ratio and temperature (if calculated from the chemical composition). If O_2 is not measured, the CO_2 provides useful information on the dilution air distribution.

Carbon Monoxide – Averaged CO concentration profiles are shown in Fig. 4. The phenomenon referred to as "thermal quenching" results from a cooling of the gases

References p. 252

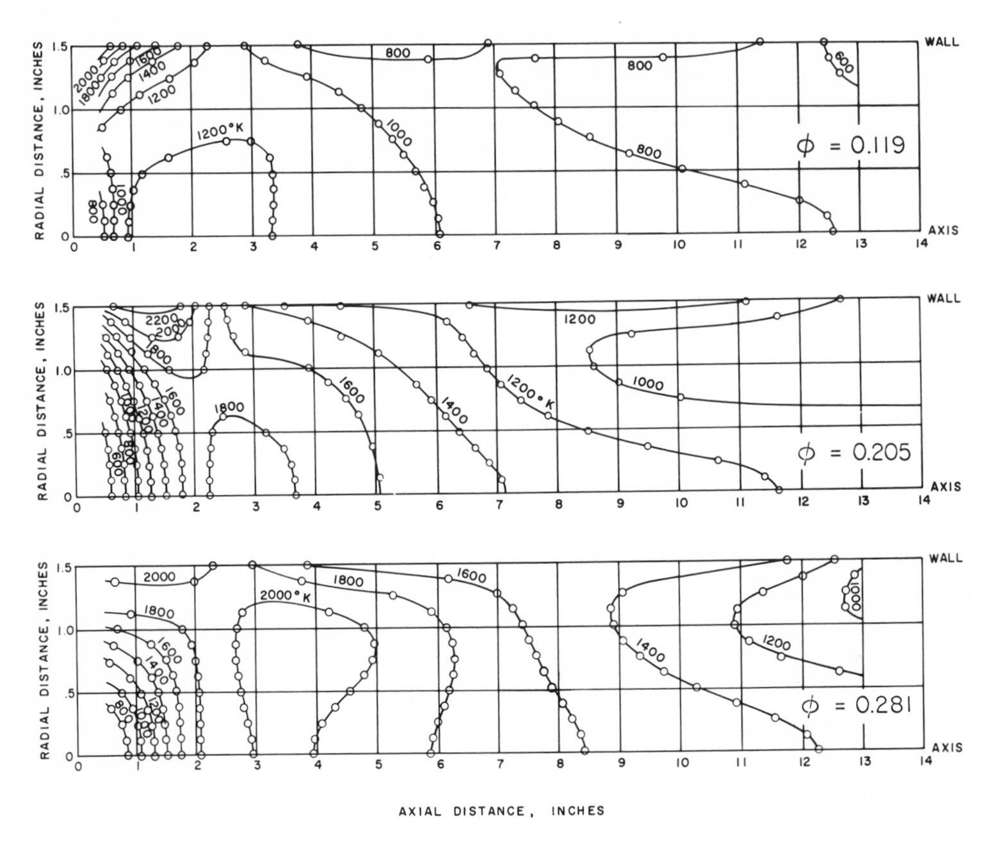

Fig. 3. Combustor temperature contours for overall equivalence ratios of ϕ = 0.119, 0.205, 0.281. Heptane/air, 1.5 atm pressure.

below the temperature at which CO oxidation occurs. Experimentally, this temperature appears to be approximately 1500°K. Note that the higher equivalence ratio case has a larger initial (primary zone) CO level but results in a smaller CO level at the exit. CO is approximately at its equilibrium concentration at the beginning of the secondary zone. A decreased air flow in the $\phi_o = 0.27$ case delays the time of thermal quench and allows greater CO oxidation to CO_2. High injection velocities for methane promote quenching and result in higher exit CO levels.

Oxygen – The measurement of both CO_2 and O_2 generally is redundant. Note that it is not necessary to measure all species since some may be calculated from conservation of elemental mass. H_2O is particularly difficult to measure directly and generally is determined from either O_2 or H_2 conservation. Again, O_2 in itself is not of particular interest to the problem of pollutants.

Hydrocarbons – Averaged HC are presented as a function of combustor length in Fig. 4. The repeatedly observed phenomenon of thermal quenching of the HC oxidation is evident. With greater air flow (lower equivalence ratio), quenching occurs

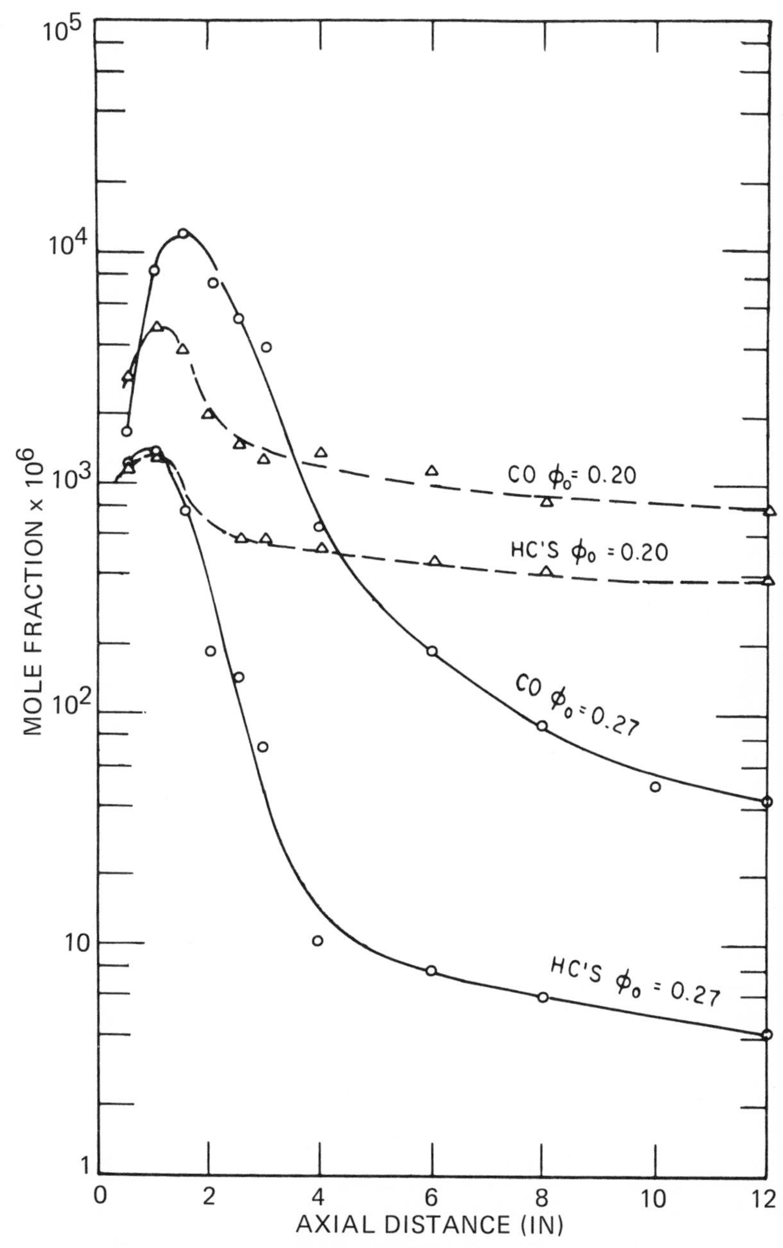

Fig. 4. Axial distribution of averaged CO and HC for overall equivalence ratios of ϕ = 0.20, 0.27. Methane/air, 1.5 atm pressure, Methane injection velocity 1100 ft/sec.

sooner and results in a higher exit level of the HC. The quench temperature for HC is approximately 1500-1700°K, usually higher than for CO. Heavier HC appear to be quenched at a higher temperature than CH_4. High fuel injection velocities increase the problem of HC quenching. In most experiments, the quenching of the HC has been observed to precede the quenching of CO. In all experiments, the cessation of HC oxidation in the secondary zone as opposed to the initial HC level was the dominant factor controlling the HC exit concentration.

Nitric Oxide – Unlike CO and HC, NO levels are fixed by the kinetics of formation rather than the kinetics of destruction. Peak levels are reached just downstream of peak temperatures, i.e. roughly at the beginning of the secondary zone of the combustor (see Fig. 5). These peak values are well below the local equilibrium concentrations of NO.

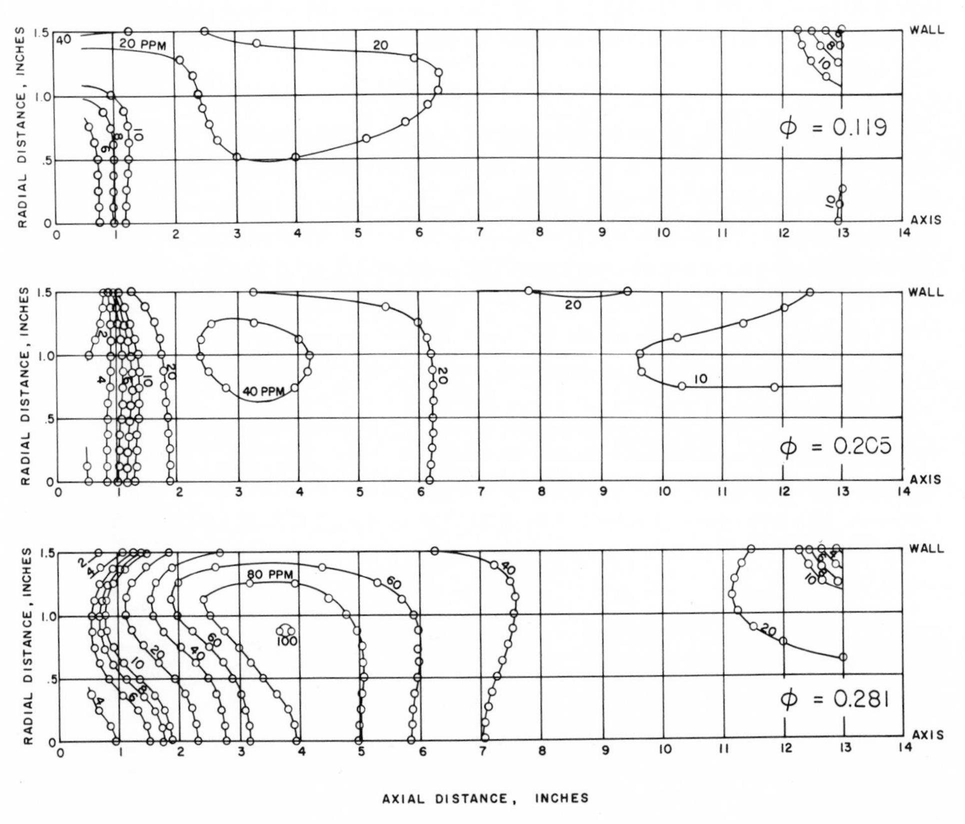

Fig. 5. Combustor NO contours for overall equivalence ratios of ϕ = 0.119, 0.205, 0.281. Heptane/air, 1.5 atm pressure.

No evidence of destruction of NO has been observed. This is consistent with theoretical considerations. NO only exceeds equilibrium levels at temperatures well below those at which NO equilibration would be rapid enough to be observed. It is concluded that NO once formed does not go away. The formation of NO is strongly temperature dependent with the critical temperature for the onset of formation falling in the range of 1900–2000^{0}K.

NO levels from the combustion of CH_4 were less than one-half for liquid HC at comparable combustor loading. The difference results from mixing characteristics rather than as a consequence of differences in either thermochemistry or chemical kinetics.

Primary zone temperatures for liquid HC were near the stoichiometric adiabatic flame temperature. One is tempted to say that the primary zone equivalence ratio is near one. The concept of a constant primary zone equivalence ratio with a liquid fuel, however, is of questionable validity. With CH_4, the average primary zone equivalence ratio was about 0.5 or quite lean. Again, the primary zone is not homogeneous and a distribution of mixture ratios exists. In this case even the richest regions are still on the lean side of stoichiometric, at least on the size and time scale resolvable by the sampling system.

Nitrogen Dioxide – Measurable quantities, greater than about 5 parts per million, of NO_2 were never detected. At high NO concentrations, greater than 100 ppm, it is therefore reasonable to say that nearly all oxides of nitrogen appear as NO. Although one is tempted to generalize this statement to all of the present experimental observations, the direct confirmation of the predominance of NO over NO_2 is lacking. Since local NO concentrations as low as one ppm were measured (our chemiluminescent detector is sensitive to about 0.1 ppm NO), it is possible that comparable levels of NO_2 could be present but undetected.

Particulate – Only preliminary measurements of carbon particulates have been made. Centerline particulate mass is constant over the last two-thirds of the combustor (see Fig. 6). The apparent decrease is due to dilution rather than chemical reaction. Electron micrographs reveal particles of about 0.5 - 0.6μ diameter formed from the aggregation of smaller particles of about 0.06 - 0.07μ diameter.

CONCLUSIONS

The levels of CO and HC produced in the model gas turbine combustor are determined by the chemical kinetics of their oxidation reactions. For both species, a temperature is reached in the combustor at which reaction effectively stops and exit emissions are fixed. The phenomenon is referred to as "thermal quenching" and may be delayed through control of the rate at which dilution air is added. The level of nitric oxide produced is determined by the chemical kinetics of its formation. The reduction of maximum local temperatures is the most promising way to limit NO formation.

While the thermal quench temperature for HC is in general higher than for CO, both temperatures are lower than the critical temperature for the formation of NO. Hence, the possibility exists for tailoring the combustor temperature-time characteristics to reduce simultaneously all three species. The present measurements also

References p. 252

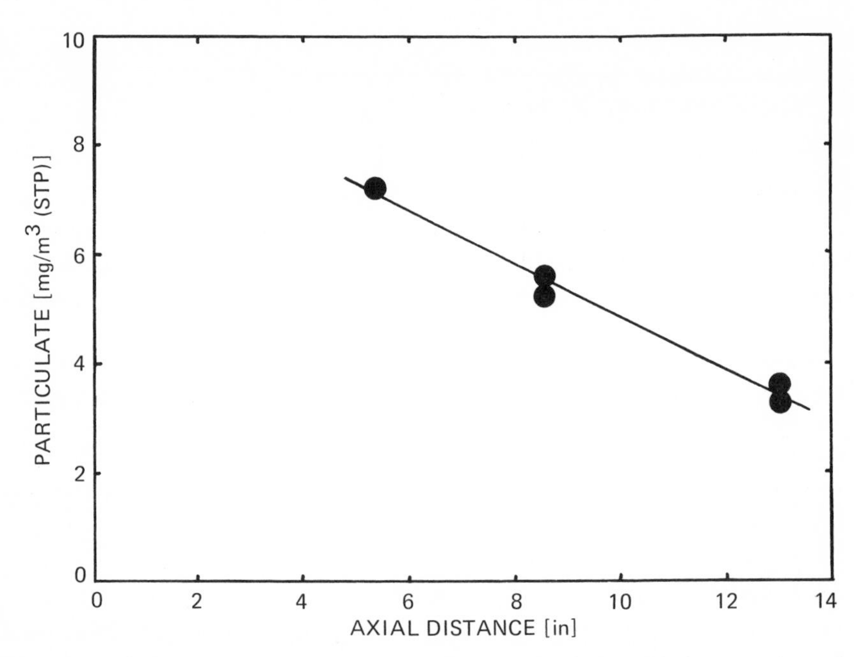

Fig. 6. Axial distribution of centerline particulates for overall equivalence ratio of ϕ = 0.275. JP-4/air, 2.2 atm pressure.

emphasize the importance of local variations in composition and the gains to be realized from increased mixture homogeneity. The use of a homogeneous, lean primary zone, possibly obtained through premixed, prevaporized fuel injection, is suggested as a means to minimize combustor emissions.

REFERENCES

1. *D. S. Smith, R. F. Sawyer and E. S. Starkman, "Oxides of Nitrogen from Gas Turbines," J. of APCA, Vol. 18, January 1968, p. 30.*
2. *R. F. Sawyer, D. P. Teixeira and E. S. Starkman, "Air Pollution Characteristics of Gas Turbine Engines," Trans. ASME, of Eng. for Power, Vol. 91A, October 1969, p. 290.*
3. *R. F. Sawyer, "Reducing Jet Pollution before it becomes serious," Astro. and Aero., Vol. 8, April 1970, p. 62.*
4. *E. S. Starkman, Y. Mizutani, R. F. Sawyer, and D. P. Teixeira, "The Role of Chemistry in Gas Turbine Emissions," Trans. ASME, J. of Eng. for Power, Vol. 93A, July 1971, p. 333.*
5. *P. G. Parikh, R. F. Sawyer and A. L. London, "Pollutants from Methane Fueled Gas Turbine Combustion," College of Eng., Univ. of Calif., Berkeley, Report No. TS-70-15, January 1971.*
6. *E. S. Starkman, A. G. Cattaneo and S. H. McAllister, "Carbon Formation in Gas Turbine Combustion Chambers," Ind. Eng. and Chem., Vol. 43, December 1951, p. 2282.*
7. *J. A. Paterson, M. W. McElroy, R. F. Sawyer and T. Singh, "A Prototype Chemiluminescent NO Analyzer," College of Eng., Univ. of Calif., Berkeley, Report No. TS-70-9, September 1970.*
8. *R. C. Williamson and C. M. Stanforth, "Measurement of Jet Engine Combustion Temperature by the use of Thermocouples and Gas Analysis," SAE Paper No. 690433, Nat. Air Trans. Meeting, New York, 1969.*

DISCUSSION

J. M. Beér *(University of Sheffield)*

This is very impressive work and the general interest it has created shows how good experimental information is welcomed by engineers and theoreticians alike. Similar studies have been performed on furnace flames at atmospheric pressure conditions for some time. I believe that this is the first time that a detailed mapping of flame properties in a gas turbine combustor has been published. I was particularly interested in what Prof. Sawyer said about the inhomogeneity of the combustion region. The significance of this to the performance of the combustor and to the emission of pollutants is generally appreciated. While I agree that the combustion region should be homogeneous I do not think that this necessarily has to lead to the adoption of prevapourisation.

Relatively little attention has been given in gas turbine combustor design to the possibilities of appropriately combining swirling jet flows with hollow cone fuel sprays. Recent studies at the International Flame Research Foundation and at the University of Sheffield have shown that high degree of homogeneity of the combustion region results when the regions of high fuel concentrations match those of high turbulent shear stresses in a swirling jet. It was demonstrated that burning fuel at the same input rate and with the same overall air-fuel ratio the flame length could be reduced to one fifth by correctly overlapping concentration and flow patterns in swirling jets. For this to be achieved, detailed information on the flow and concentration patterns in the gas turbine combustor is necessary together with the variation of these properties with design and operating variables. Considerable amount of information on swirling jet flames burning under atmospheric pressure conditions is available and it would be of great interest to see how this information might be used in gas turbine combustor design.

Perhaps the next step could be to learn more about the flow pattern in the gas turbine combustor. At the present stage it would perhaps be too ambitious to seek experimental data on the fluctuating components of velocities and concentrations but tools developed for the measurement of time mean velocity distributions in the flame might be used together with sampling for determining gas and soot concentrations in the combustor. It would be of particular interest to compare results so obtained with those of isothermal model experiments, with the objective of further developing modeling techniques. There is a specially high premium on the successful modeling of a gas turbine combustor because of the great difficulties in prototype experimentation.

R. F. Sawyer

I think that the best way to avoid this problem is to burn gaseous fuels. With our sintered filters, we did not retain the liquid component. Since we collected samples at high temperature and dessicated the filter, we examined strictly the solid phase. I certainly agree with you that it would be advantageous to combine this work with velocity measurements inside the combustor. This has been done and is described in an Egyptian paper in which ball-type pitot tubes were used to make these measurements. When I say pre-vaporized pre-mixed, I mean simply that these processes be accomplished before combustion takes place. It is fine if this can be accomplished inside the combustor. I do not mean to imply that it has to be done outside the combustor. The problem is to get the fuel into the vapor phase and well-mixed with the air at a lean mixture before the combustion process takes place. While I have heard some suggestions that a rich mixture would suffice, I disagree.

EFFECTS OF FUEL INJECTION METHOD ON GAS TURBINE COMBUSTOR EMISSIONS

E. R. NORSTER and A. H. LEFEBVRE
Cranfield Institute of Technology, Bedford, England

ABSTRACT

A series of tests have been carried out using a single segment of a tuboannular, aircraft-type, gas turbine combustion chamber to investigate the influences of operating conditions and fuel injection method on exhaust emissions. The test range included pressures up to 200 psia and inlet air temperatures from 380 to 900°K (684 to 1620°R). Concentrations of nitric oxide, carbon monoxide, unburned hydrocarbons and smoke were determined from gas samples obtained in the combustor exit plane.

The results indicate that nitric oxide concentration is strongly dependent on inlet air temperature and, to a lesser extent, on pressure. In all tests, an increase in inlet temperature or pressure produced an increase in nitric oxide emission, accompanied by reductions in the concentrations of carbon monoxide and unburned hydrocarbons. The separate effects of air pressure and temperature on exhaust smoke were examined and it was found that exhaust smoke increases with pressure but diminishes as the inlet temperature is raised.

Comparative tests were carried out on a standard dual-orifice atomizer and an "air spray" atomizer in which the fuel was atomized by the action of high-velocity air created by the pressure drop across the liner. It was found that the air-spray atomizer offers significant advantages over conventional swirl atomizers in terms of reduced emissions of all types, but especially in regard to smoke. Tests have shown that when an air-spray atomizer is used in conjunction with a small pilot fuel spray, the burning range is considerably enlarged at the price of a small increase in emission levels.

References p. 278

INTRODUCTION

The main pollutants produced in gas turbine combustion are smoke, carbon monoxide (CO), unburned hydrocarbons (HC) and nitric oxide (NO). The relative importance of these emissions varies with engine operating conditions. Smoke and NO are normally associated with operation at high pressures, whereas CO and HC only become significant at low pressure conditions, such as idling. Although the basic reactions governing the exhaust concentrations of these pollutants are highly complex, from an engineering standpoint considerable simplifications seem feasible. For example, it is now clear that the rate of NO formation is primarily a function of temperature and residence time. This gives encouragement to the development of correlating parameters and modelling techniques that could be of great value to the combustion engineer, leading to better control of emissions from conventional combustors, and to simplifications in the analysis and design of new concepts, such as staged combustion and variable geometry.

A major obstacle to the development of suitable modelling techniques is a lack of experimental data against which to check the accuracy of the predictions. Although the results of measurements carried out on engine exhausts are becoming increasingly available, much of the data is of limited value owing to doubts concerning the validity of the instrumentation and/or the sampling procedures employed. Another difficulty is that published data are usually confined to a few specific power settings such as take off, cruise and idle, representing wide variations in all three main parameters of combustor pressure, inlet air temperature and air-fuel ratio.

There is clearly a need for emissions data obtained under conditions where each of the main variables are studied separately in turn, and this was a main incentive for the present study. A further objective was to examine the effect on emissions of the method of fuel injection. Two types of fuel injector were investigated; (a) dual-orifice pressure atomizers of the kind widely used in present day aircraft and industrial engines, and (b) "airblast" or "airspray" atomizers which appear to show most promise for engine applications in the future.

EXPERIMENTAL

The entire test program was carried out on a single, tubular, aircraft-type combustor, featuring air admission through an upstream axial swirler, secondary holes, tertiary holes and film-cooling slots. The main dimensions and flow proportions are given in Table 1.

Air compressors, used in conjunction with a pebble-bed heater, enabled exhaust emission surveys to be conducted at pressures up to 200 psia and inlet air temperatures from 384 to 900°K. The fuel employed was standard aviation kerosene.

Measurement of Exhaust Smoke – Smoke samples were taken using a single-point probe, which was located at the center of a water-cooled exhaust pipe at a point six

TABLE 1

Combustor Details

Liner length	– 14 in.
Liner diameter	– 5.3 in.
Primary zone volume	– 39 $in.^3$
Total combustion volume	– 139 $in.^3$
Total liner volume	– 254 $in.^3$
Primary zone air	– 25%
Total combustion air	– 60%

feet downstream of the combustor, immediately after the pressurizing valve. Smoke measurements were obtained on a Hartridge meter which had been modified slightly to improve accuracy at the low smoke levels associated with airblast atomizers. Instead of reading the meter directly, a digital voltmeter was used to monitor the output from the light-detecting photocell inside the instrument. Comparison was made between the voltmeter reading and actual Hartridge scale readings, using the smoke from a smoke lamp, suitably diluted with clean air. The calibration graph for the range 0 to 10 units is reproduced in Fig. 1.

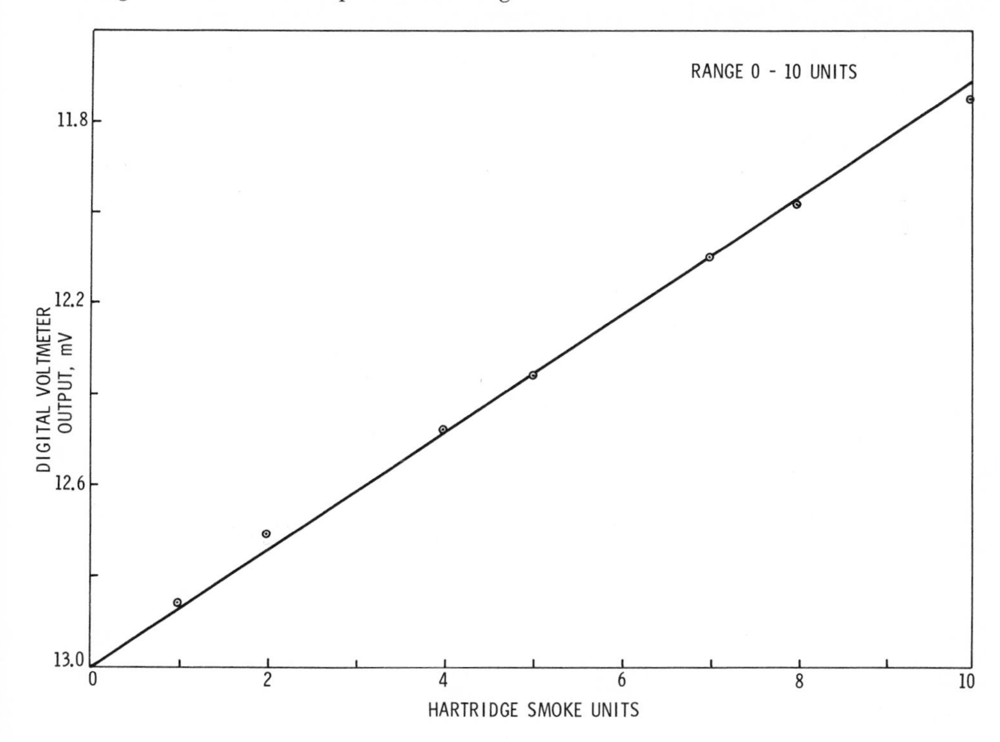

Fig. 1. Response characteristics of Hartridge smoke meter.

References p. 278

Measurement of Gaseous Emissions – Exhaust gases were extracted at the combustor exit through a diametral probe containing nine holes of 0.040 in. diameter, spaced to give equal area sampling. Although the probe itself was water cooled, the line between the test rig and the adjoining control room was electrically heated to between 180° and 200°C. The instrumentation employed is shown diagrammatically in Fig. 2. A small positive pressure was maintained on all instruments to overcome possible problems of leakage. A freezing trap and spray trap were included to cool the sample and remove excess water. Additional drying was accomplished by passing the sample through calcium sulphate. This drying agent was replenished before the start of each test run, since it was found that NO readings were

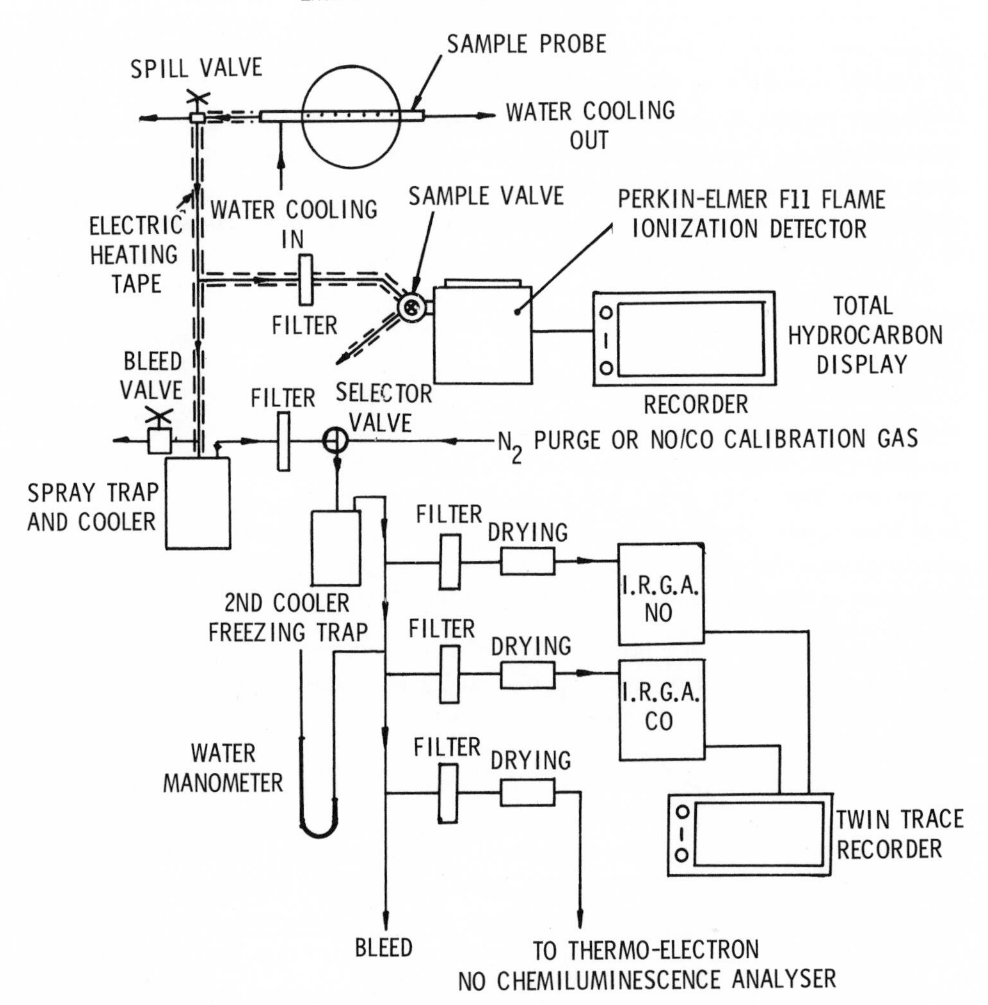

Fig. 2. Schematic diagram of equipment employed in emissions sampling and analysis.

sensitive to slight traces of water. It was also found necessary to replace the first filters after each test run in order to minimize errors arising from the presence of particulate matter.

Good agreement was obtained between NO readings taken on the infra-red and chemiluminescence analyzers for concentrations above 10 ppm. At lower concentrations the chemiluminescence analyzer gave more consistent readings.

No special problems were encountered in the measurement of unburned HC which were evaluated as equivalent hexane. Collection of fuel in the sample line and valve was avoided by heating all components between the sampling probe and the flame ionization detector. The overall response time of the sampling and instrumentation system was typically around ten seconds.

EXHAUST SMOKE

Exhaust smoke is caused by the production of finely-divided soot particles in fuel-rich regions of the flame. Soot may be generated in any part of the combustion zone where mixing is inadequate, but with pressure atomizers the main soot-forming region lies inside the fuel spray at the center of the liner. This is the region in which the recirculating burned products move upstream towards the fuel spray, and where local pockets of fuel and fuel vapor are enveloped in oxygen-deficient gases at high temperature. In these fuel-rich regions, soot may be produced in considerable quantities.

Most of the soot produced in the primary zone is consumed in the high temperature regions downstream. Thus from a smoke viewpoint, a combustor may be considered to comprise two separate zones. The primary zone governs the rate of soot formation. The other zone, which includes the secondary zone and on modern high-temperature engines the dilution zone also, determines the rate of soot consumption. The soot concentration actually observed in the exhaust gases is an indication of the dominance of one zone over the other.

Recognition of the roles played by these two regions helps to explain many of the apparent anomalies that occasionally arise in regard to the effects of various factors on exhaust smoke. For example, when injection of water into the combustion chamber was employed as a means of augmenting thrust, it vastly increased the level of exhaust smoke in some cases while in other cases it eliminated smoke entirely. This conflicting experience arose because of differences in the method of water injection. When well atomized water is injected directly into the primary zone, soot generation is drastically reduced and the exhaust gases are virtually smoke-free. However, if water is injected downstream of the primary zone, the soot-consuming reactions are quenched and the smoke output may be considerably higher than the "dry" value. (1)

The influence of temperature on exhaust smoke should be considered with regard to its effect on both soot-forming and soot eliminating processes. So far no tests have

References p. 278

been reported in which the primary-zone temperature has been varied without changing temperatures in other regions of the combustor. Indirect evidence suggests that soot formation is enhanced by an increase of temperature in this zone. The effect of an increase in inlet air temperature is to accelerate both soot-forming and soot-consuming processes; the net result is to reduce smoke, as illustrated in Fig. 3.

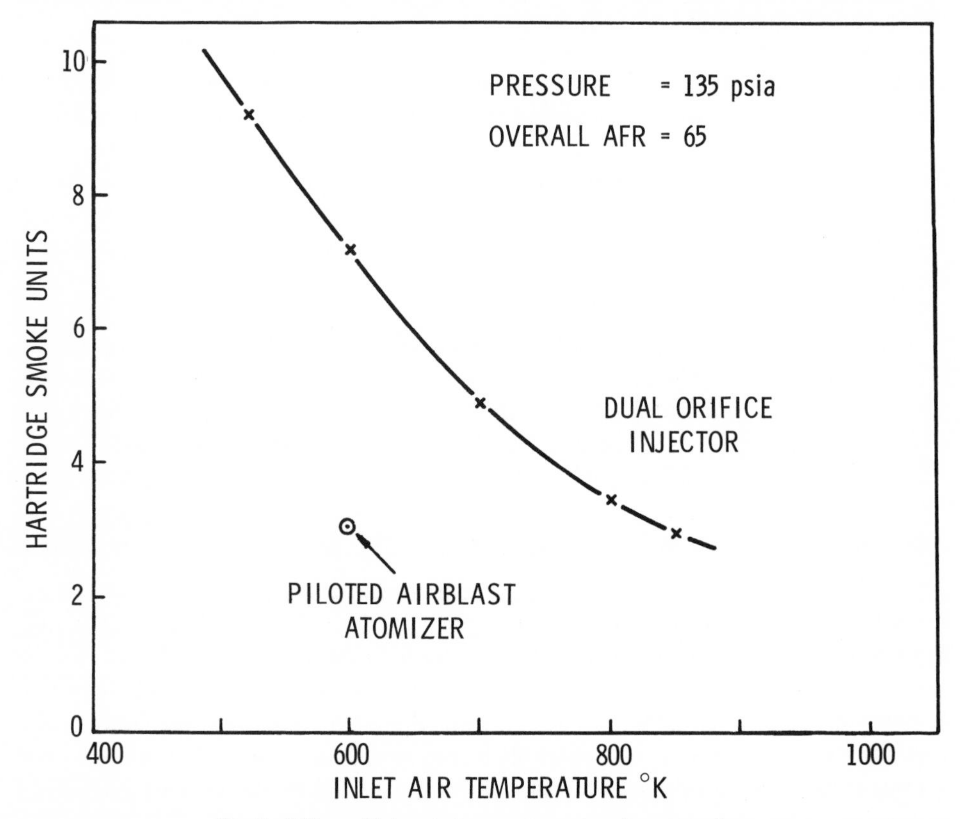

Fig. 3. Effect of inlet air temperature on exhaust smoke.

Smoke problems are usually most severe at high pressures. There are several reasons for this, some of which derive from chemical effects, while others stem from physical factors which affect spray characteristics and hence also the distribution of mixture strength in the soot-generating regions of the flame.

One important consideration is that burning limits widen with pressure. Thus at high pressures, soot is produced in regions which, at lower pressures, would be too rich to burn. (2) Another effect of an increase in pressure is to delay the evaporation of fuel drops and so provide more time for soot formation in the liquid phase. Increase in pressure also accelerates chemical reaction rates, so that combustion is initiated earlier and a larger proportion of the fuel is burned in the fuel-rich regions adjacent to the spray.

These considerations affect all types of atomizers, but with pressure atomizers the main effects of pressure are manifested through its influence on fuel spray characteristics. For any given maximum fuel pump pressure, an increase in combustor pressure implies a reduction in available fuel injection pressure and hence an increase in mean drop size. This extends the drop lifetime causing higher heat absorption and enhanced soot formation. Data illustrating the manner in which soot formation increases with fuel drop size have been reported by Durrant. (3) Somewhat in contrast Faitani (4) found that major changes in fuel injection pressure, corresponding to wide variations in mean drop size, had only a slight effect on smoke.

Conflicting evidence of this kind often arises with pressure atomizers due to inherent complexities in the mechanisms of fuel atomization and distribution. In general exhaust smoke decreases with mean drop size, but if improved atomization is accompanied by reduced spray penetration the smoke output may actually increase.

Reduced spray penetration is one of the main causes of smoke on high-pressure ratio engines. Calculations indicate that with a given fuel nozzle an increase in combustor pressure from one to twenty-seven atmospheres (with velocity and overall air-fuel ratio (AFR) held constant) causes a three-fold reduction in spray penetration. Thus whereas at low pressures the fuel is distributed across the entire combustion zone, at high pressures it tends to concentrate in the soot-forming region just downstream of the nozzle.

Another adverse effect of an increase in pressure is to reduce the cone angle of the spray. (5) This encourages soot formation, partly by increasing the mean drop size, but mainly by raising the mixture strength in the soot-forming zone.

The total effect of all these factors is that smoke emission increases steeply with pressure, as shown in Fig. 4. In practice, high combustor pressures are normally accompanied by high inlet air temperatures and, in the interests of optimum engine performance, by high turbine inlet temperatures also. The temperature effects partly offset those caused by pressure, but the usual outcome of an increase in engine pressure ratio is an increase in the level of exhaust smoke. (3)

Since fuel-rich combustion zones are a prime requisite for the production of soot, it is only to be expected that fuel-air ratios, both local and overall, should have a significant effect on soot formation and smoke. In an early paper on the subject of exhaust smoke, Lefebvre and Durrant (1) advocated the use of leaner primary zones and quoted results obtained on a Conway engine which demonstrated how a modest increase in primary air flow effected a four-fold reduction in smoke output. Since then, this technique has been successfully applied to other engines. (4,6) It is now normal practice to employ the maximum possible amount of air in primary combustion, the upper limit being set by the important but conflicting requirements of stability and high-altitude ignition performance.

References p. 278

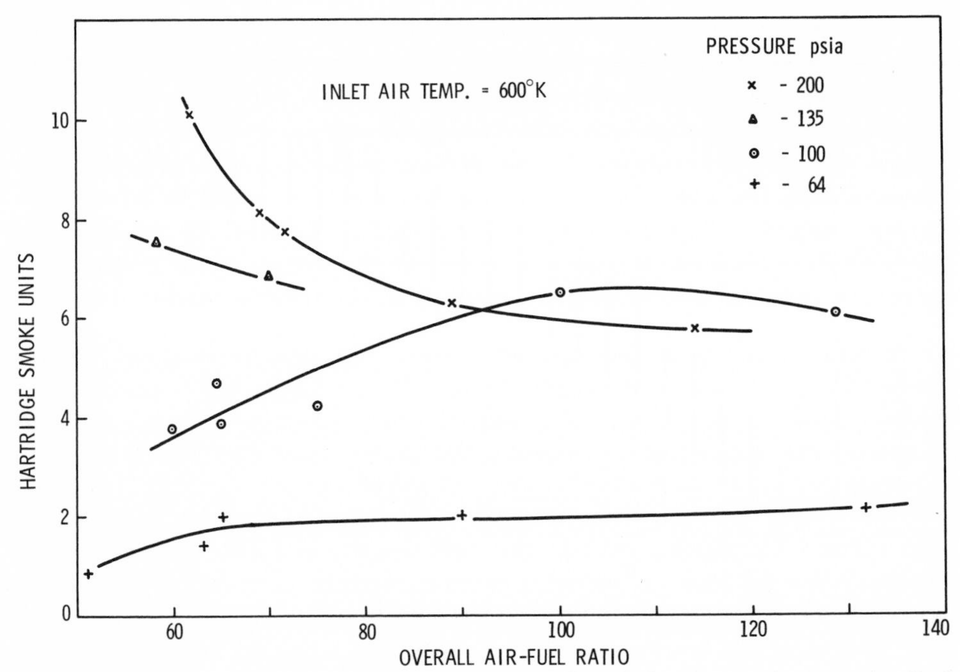

Fig. 4. Influence of combustion pressure and overall air-fuel ratio on exhaust smoke (dual orifice atomizer).

The manner in which exhaust smoke varies with overall air-fuel ratio, in a combustor fitted with a dual-orifice atomizer, is illustrated in Fig. 4. In this figure, it is of interest to note that the curves corresponding to higher pressures show an opposite trend with air-fuel ratio to those obtained at lower pressures. This occurs because reduction in fuel flow at lower combustor pressures causes a deterioration in atomization quality and consequently an increase in exhaust smoke. At high combustion pressures, corresponding to high fuel injection pressures, atomization quality is no longer a limiting factor, and exhaust smoke increases with fuel flow in a manner that is consistent with accepted notions on soot formation.

CARBON MONOXIDE

The emission of CO is not only undesirable from a pollution aspect but its presence in the exhaust gas also indicates combustion inefficiency. The close relationship that exists between combustion inefficiency and exhaust concentrations of CO is illustrated in Fig. 5, which is based on engine data.

Fig. 6 shows how CO concentrations vary with overall air-fuel ratio for conditions of constant pressure and inlet temperature. All three curves indicate high values of CO at high air-fuel ratios, presumably because of low combustion efficiencies and low temperatures in the primary zone. Increase in fuel flow elevates the flame

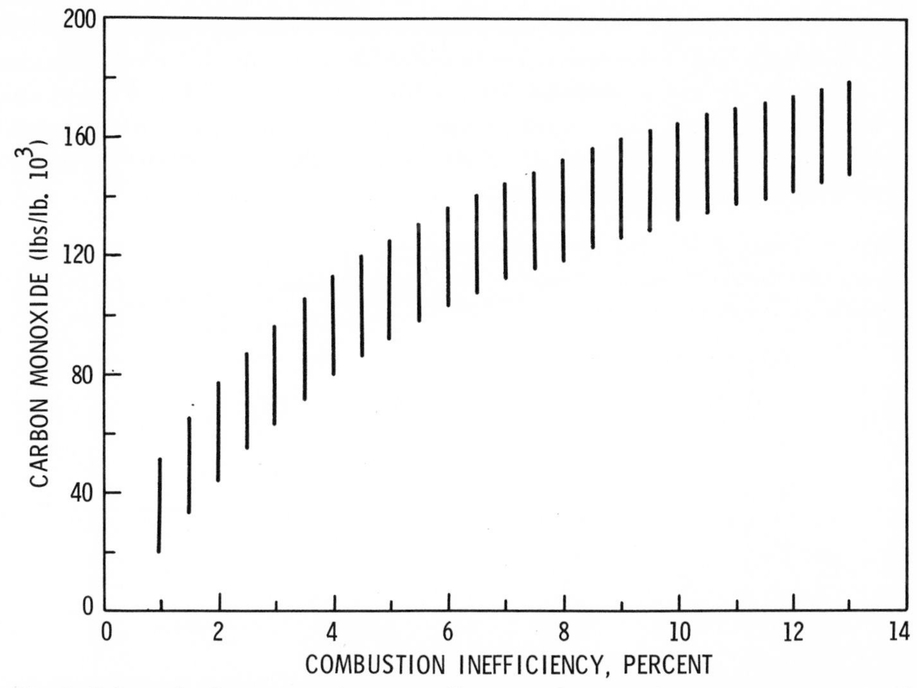

Fig. 5. Relationship between combustion inefficiency and engine exhaust concentration of CO.

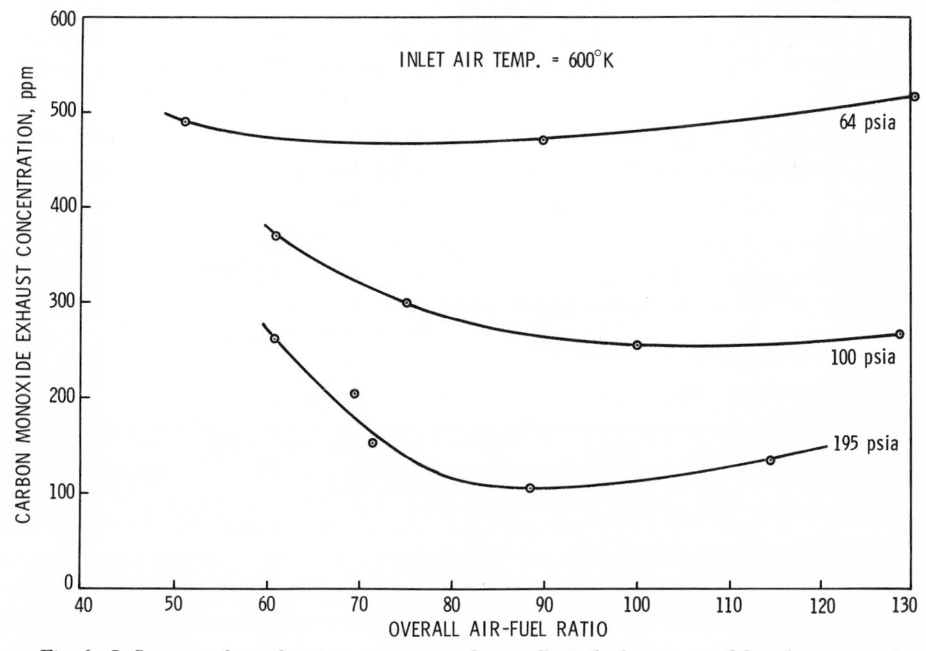

Fig. 6. Influence of combustion pressure and overall air-fuel ratio on CO exhaust emissions (dual orifice atomizer).

References p. 278

temperature and thereby reduces the production of CO but, as the primary-zone mixture strength starts to become richer than stoichiometric, the output of CO begins to rise again due to lack of available oxygen. Thus the general picture is one of high CO levels at high and low overall air-fuel ratios, with minimum concentrations occurring at an intermediate value of air-fuel ratio that corresponds roughly to slightly weaker than stoichiometric conditions in the primary zone.

Figs. 6, 7 and 8 illustrate the detrimental effects of reductions in pressure and temperature, both of which increase the concentration of CO by lowering the primary-zone combustion efficiency.

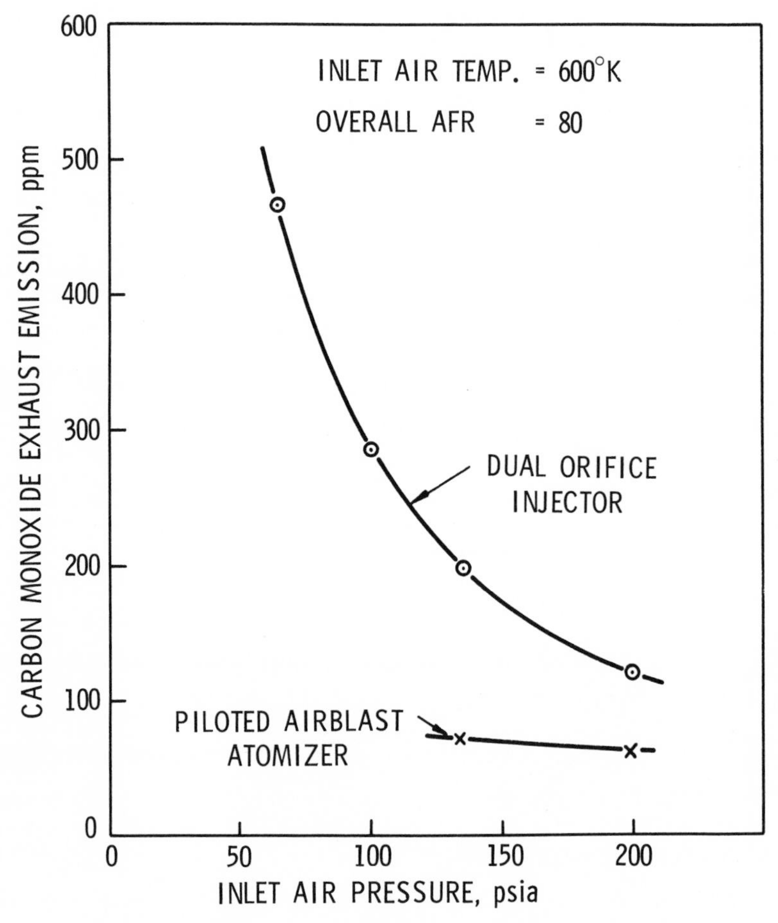

Fig. 7. Effect of combustion pressure on CO exhaust emissions.

From a design viewpoint, effective means for minimizing CO emissions include the following:

- Improved atomization and aeration of the fuel spray

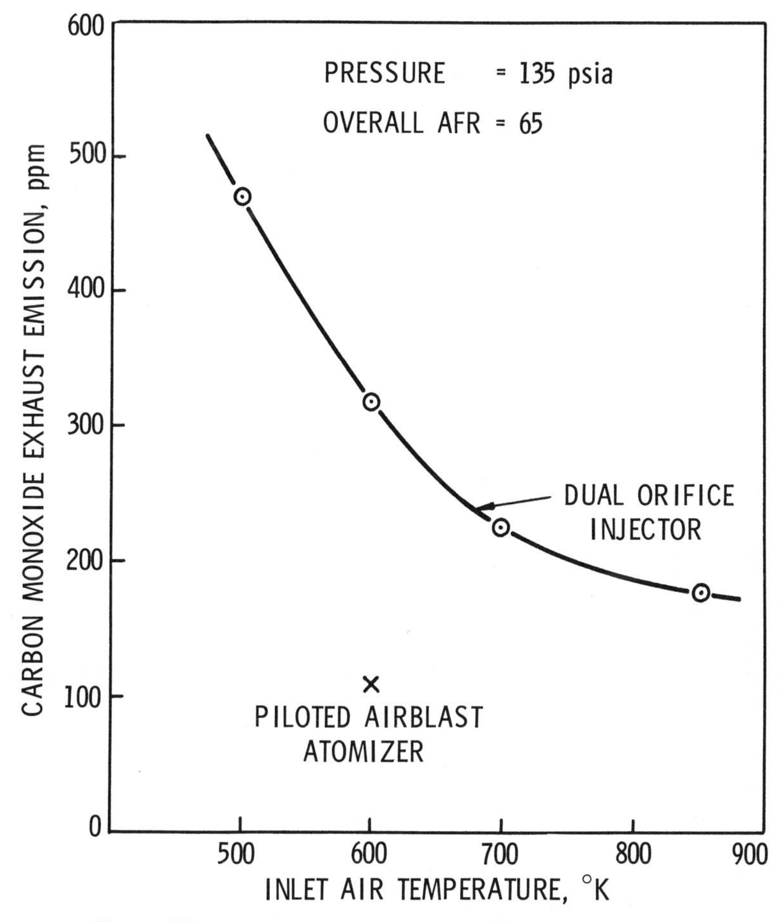

Fig. 8. Effect of inlet air temperature on CO exhaust emissions.

- Redistribution of the air flow to bring the primary-zone mixture strength closer to the stoichiometric value
- Increase in primary-zone volume
- Reduction of film-cooling air

UNBURNED HYDROCARBONS

Unburned HC include fuel which emerges at the combustor exit in the form of droplets or vapor, and also the products of thermal degradation of the parent fuel into species of lower molecular weight, such as methane and acetylene. They are normally associated with poor atomization, inadequate burning rates, the chilling effects of film-cooling air, or any combination of these. A relationship between unburned HC and combustion efficiency, based on measurements carried out on a number of current engines, is shown in Fig. 9.

References p. 278

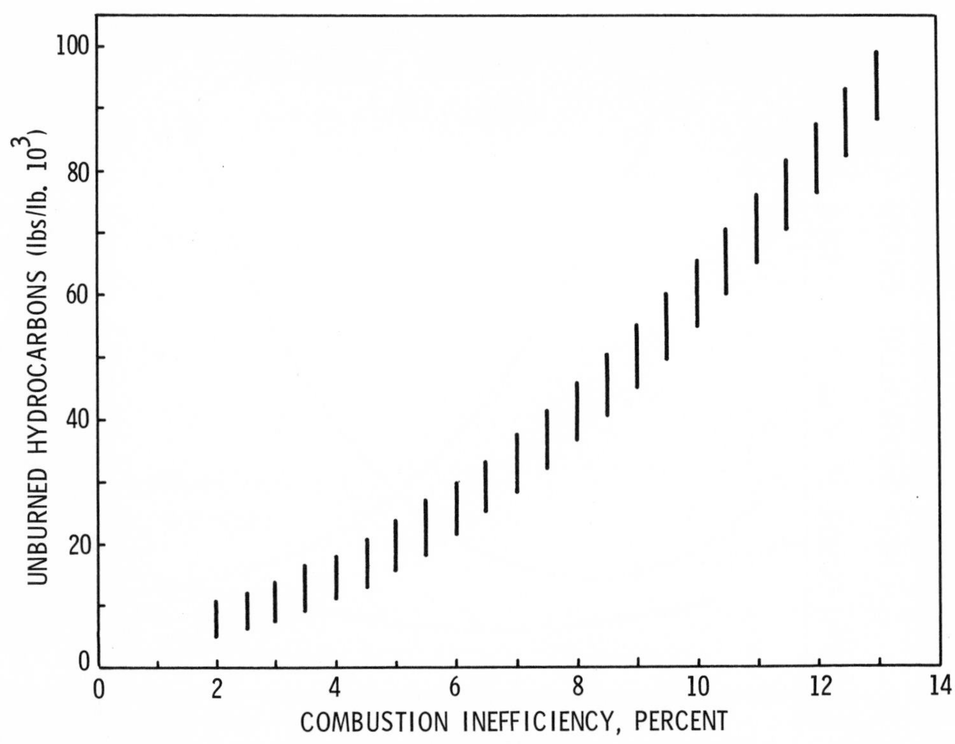

Fig. 9. Relationship between combustion inefficiency and engine exhaust concentration of unburned HC.

Figure 10 illustrates the variation of unburned HC plotted as equivalent hexane, as a function of pressure and overall air-fuel ratio for a constant inlet air temperature of 600°K. At high pressures, the level of unburned HC is low and appears to be insensitive to variations in fuel flow. At low pressures, where emission levels are higher, the variation with air-fuel ratio follows a similar pattern to that obtained with CO, i.e. reaching a minimum point at a primary-zone fuel-air ratio of around stoichiometric.

Tests carried out at lower values of pressure and inlet temperature gave much higher values of HC. At an overall air-fuel ratio of 75:1 and a pressure of 60 psia, lowering the inlet air temperature from 600 to 384°K raised the exhaust HC concentration from 4 to 120 ppm. At the same temperature of 384°K, a reduction in pressure from 60 to 35 psia produced a further increase from 120 to 180 ppm.

The methods advocated for alleviating CO emissions are also recommended for the treatment of unburned HC, with slightly more emphasis on improved atomization. Reduction in film-cooling air can also be very effective. (3)

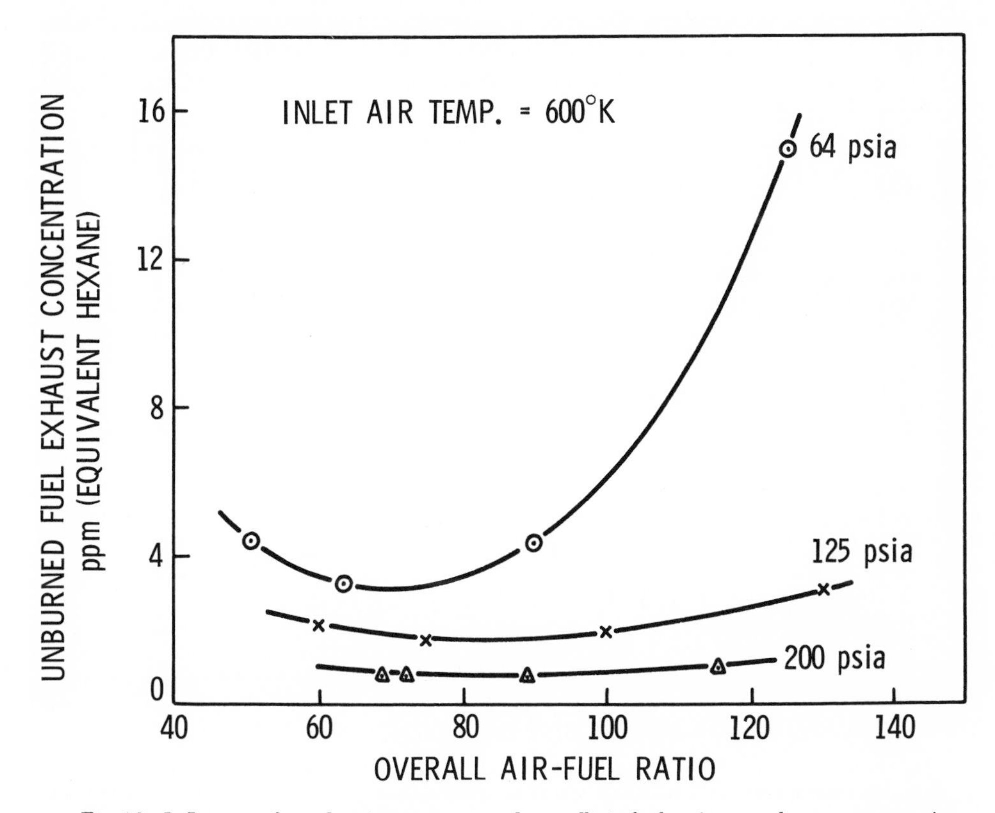

Fig. 10. Influence of combustion pressure and overall air-fuel ratio on exhaust concentrations of unburned fuel (dual orifice atomizer).

OXIDES OF NITROGEN

Oxides of nitrogen (NO_X), of which the predominant compound is NO, are produced by the oxidation of atmospheric nitrogen in high-temperature regions of the flame. Measured NO concentrations in combustor and engine exhausts are appreciably lower than calculated values based on thermal equilibrium in the primary zone. This indicates that the production of NO is limited by the kinetics of formation, and that for any given combustor the exhaust concentration will depend on the pressure, temperature and composition of the combustion zone, and on the time available for reaction. (7)

Data confirming the rate dependence of NO formation have been obtained by Cornelius and Wade (8), who used a standard engine combustor to investigate the effects of operating conditions and various design changes on exhaust emissions of CO, HC and NO. The results obtained in the present investigation are in broad agreement with their findings.

References p. 278

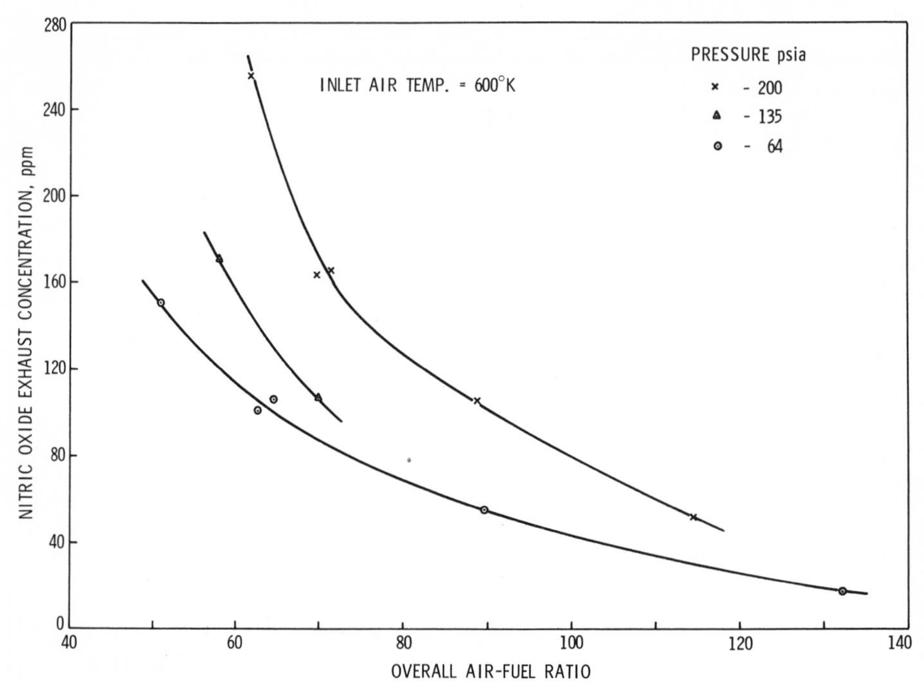

Fig. 11. Influence of combustion pressure and overall air-fuel ratio on NO exhaust concentrations (dual orifice atomizer).

Fig. 11 shows how NO exhaust concentrations increase with pressure and fuel-air ratio. In this figure, it is of interest to note that peak values were not obtained when the primary-zone mixture strength was stoichiometric, which for this combustor corresponds to an overall air-fuel ratio of around 60:1, but that emissions continued to rise with fuel flow up to the highest values tested. This occurs because when the primary-zone fuel-air ratio reaches stoichiometric proportions, any additional fuel is consumed at high temperature in the secondary zone where ample oxygen is available.

Of the variables tested, the one having the most pronounced effect was inlet air temperature. Fig. 12 is typical of the results obtained. From inspection of Fig. 11 and 12, it is obvious why advanced technology engines, featuring high temperatures and pressures, are characterized by high NO exhaust concentrations.

From the foregoing, it is clear that NO emissions may be effectively reduced by lowering the reaction temperature. This is most readily accomplished by increasing the flow of air into the primary zone. Unfortunately, any reduction in flame temperature, while lowering NO, also leads to an increased output of CO and HC. This conflict applies to factors other than flame temperature. In general any change in operating conditions or combustor configuration that reduces NO tends also to accentuate the problems of CO and HC, and vice versa. (8) This implies that conventional combustors are fundamentally incapable of providing a low-pollution exhaust over the entire range of engine operation, and adds emphasis to the need for active development of new design concepts such as variable geometry and staged combustion.

INFLUENCE OF FUEL INJECTION PROCESS

Almost all aircraft engines in service today employ dual-orifice spray atomizers. This type of injector comprises a central pilot orifice, which supplies all the fuel

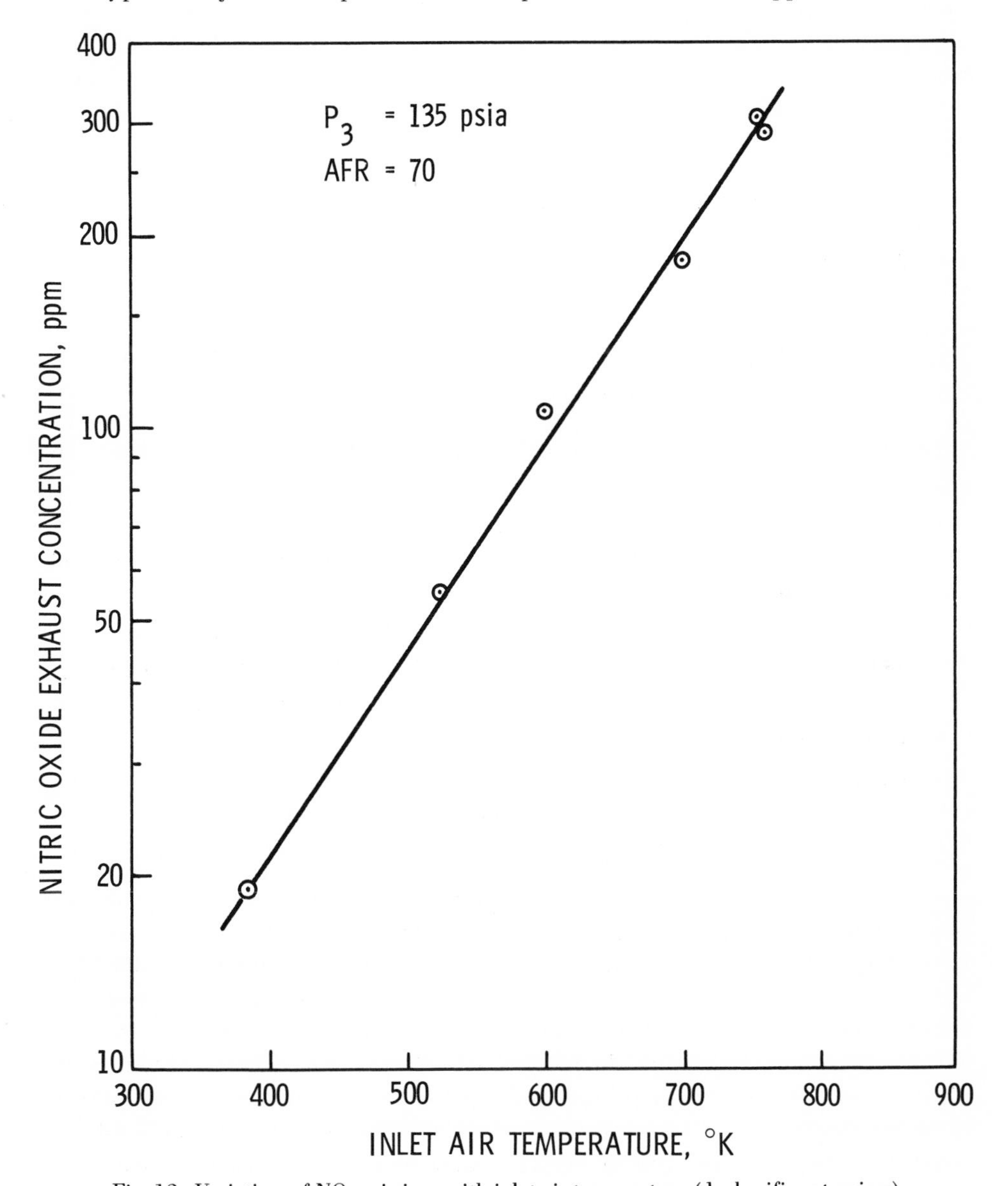

Fig. 12. Variation of NO emissions with inlet air temperature (dual orifice atomizer).

References p. 278

required at low fuel flow conditions, such as light-up, surrounded coaxially by an annular main orifice which supplies most of the fuel at normal operating conditions.

As discussed earlier, with dual-orifice atomizers the main factor contributing to high smoke output at high pressures is the reduction in cone angle and penetration of the spray, which creates large concentrations of fuel in the soot-forming regions of the flame. Some improvement can be gained by increasing the air flow into the primary zone, but only at the expense of stability and ignition performance. Thus, since primary-zone air must be used sparingly, it should be introduced only where it is likely to prove most effective, i.e. at the point of fuel injection. This is a key feature of the so-called "airblast" injector and is the chief reason for its success in alleviating soot formation and smoke.

The airblast atomizer employs a simple concept whereby fuel is exposed to the atomizing action of high velocity air which then enters the combustion zone carrying the atomized fuel along with it. Tests have shown that airblast atomizers, when operating at the natural air velocities available in gas turbine combustors (a liner pressure drop of 3 per cent produces an air injection velocity of between 240 and 370 feet/sec., depending on air temperature) are capable of producing fuel drops of comparable size to those obtained from pressure atomizers. (9)

In comparison with dual-orifice injectors, airblast atomizers have the following advantages:

- Because the atomization process ensures good mixing of fuel and air prior to combustion, soot formation is drastically reduced and problems of flame radiation are much less severe.
- Fuel distribution is dictated mainly by the air-flow pattern, which for any given liner configuration, remains constant. Thus the outlet temperature traverse is fairly insensitive to changes in operating conditions and fuel flow.
- High fuel pump pressures are not required.
- Atomization improves with increasing chamber pressure, thereby promoting more rapid fuel vaporization, mixing and combustion. (10)

The main drawback to the airblast atomizer is that, owing to the high degree of mixing achieved between fuel and air, it suffers from narrow stability limits, the lean blow-out point being in the region of 300:1 air-fuel ratio. Moreover, at conditions where the chamber velocity is low, for example, at start-up, atomization quality is poor. These problems can be readily overcome by incorporating a pilot nozzle, as shown in Fig. 13. At low fuel flows, all of the fuel is supplied from the pilot nozzle, and a well atomized spray is obtained giving efficient combustion at start-up and idling. At higher power settings, fuel is supplied to both the airblast atomizer and the spray nozzle. The relative amounts are such that at the highest fuel flow conditions most of the fuel is supplied to the airblast atomizer. By this means, the performance requirements of wide burning range and minimum exhaust smoke are both realized.

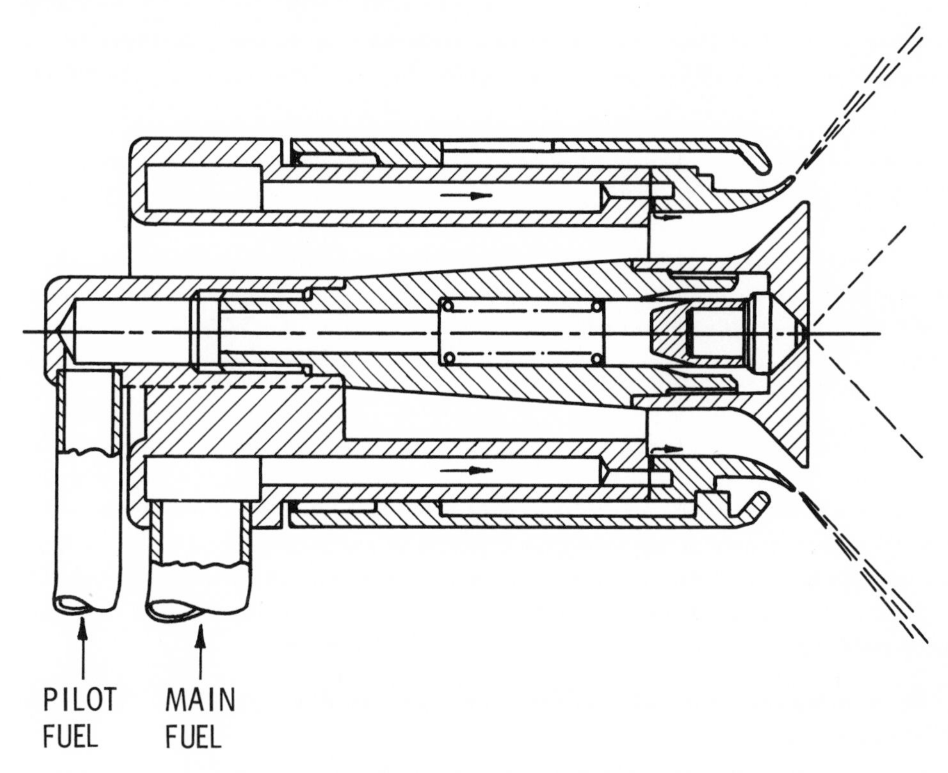

Fig. 13. Basic design features of piloted airblast atomizer.

From an emissions viewpoint, another important advantage of the piloted airblast atomizer is that atomization quality is high over the entire fuel flow range. With a dual orifice injector, owing to the interaction of the pilot and main sprays, there is always a range of fuel flows, starting at the point where the main fuel is first admitted, over which atomization quality is poor. Since atomization quality has a direct bearing on CO and HC emissions, it follows that the use of dual-orifice atomizers must inevitably aggravate the emission problem over a significant range of engine operation. With the piloted airblast injector, there is no physical interference between the pilot and main sprays and, because no fuel is introduced into the airblast section until the air velocity through it has attained its normal operating level, atomization quality is high throughout the entire range of engine operation. The atomization characteristics of dual-orifice and piloted airblast atomizers are shown qualitatively in Fig. 14.

The marked improvement in exhaust smoke that can be achieved with airblast atomizers is illustrated in Fig. 15, in which smoke levels are plotted against chamber reference velocity for various values of overall air-fuel ratio. Comparison is made between a dual orifice atomizer (main flow number = 3.5, pilot flow number = 0.25) and an airblast atomizer of the type shown in Fig. 13 (main flow number = 6.4).

References p. 278

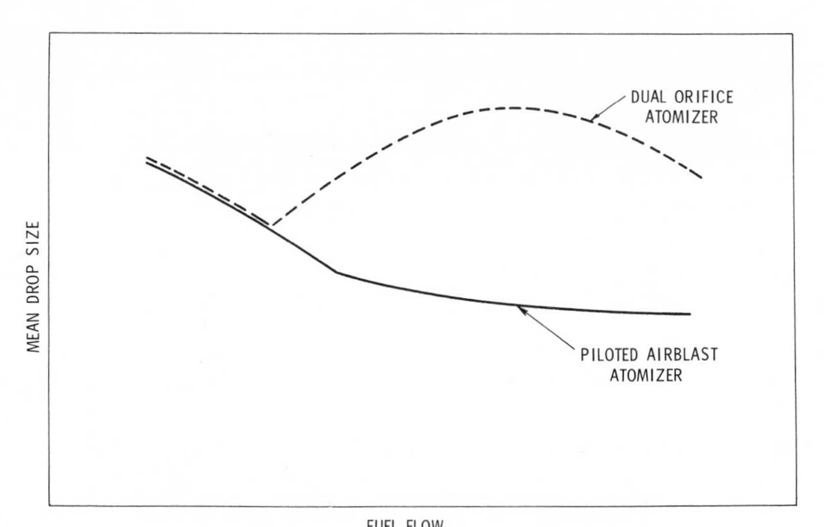

Fig. 14. Atomizing characteristic of two types of fuel injectors.

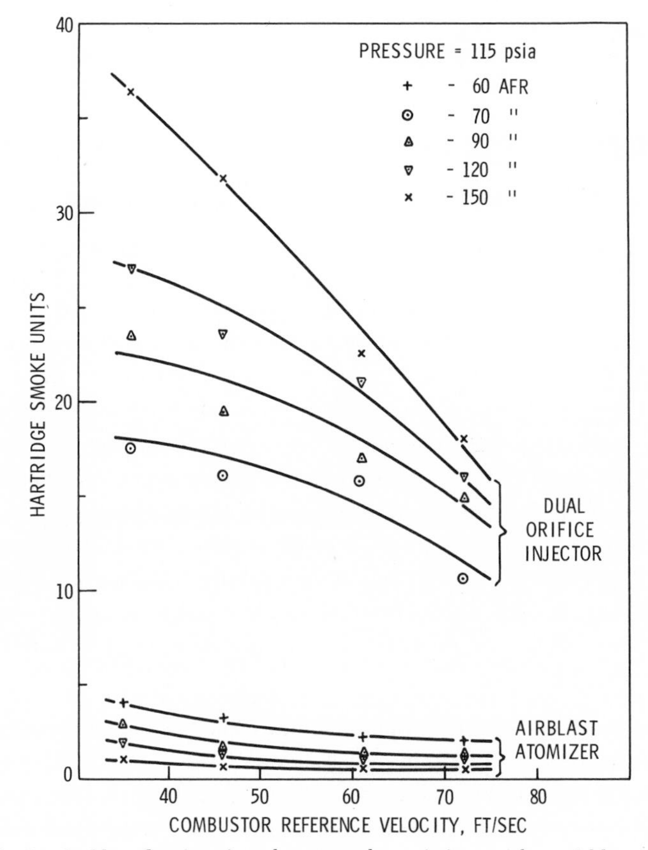

Fig. 15. Attainable reductions in exhaust smoke emissions with an airblast atomizer.

Fig. 16 is based on data extracted from Fig. 15 at the design reference velocity and also includes results obtained from the same airblast atomizer with the pilot spray in operation. The lean blow-out characteristics of all three injectors are shown in Fig. 17, in which the unsatisfactory performance of the pure airblast atomizer is very apparent as is also the considerable improvement obtained by the use of a pilot spray. The extent of this improvement is unnecessarily great as a weak extinction requirement of 1600:1 air-fuel ratio is unrealistic.

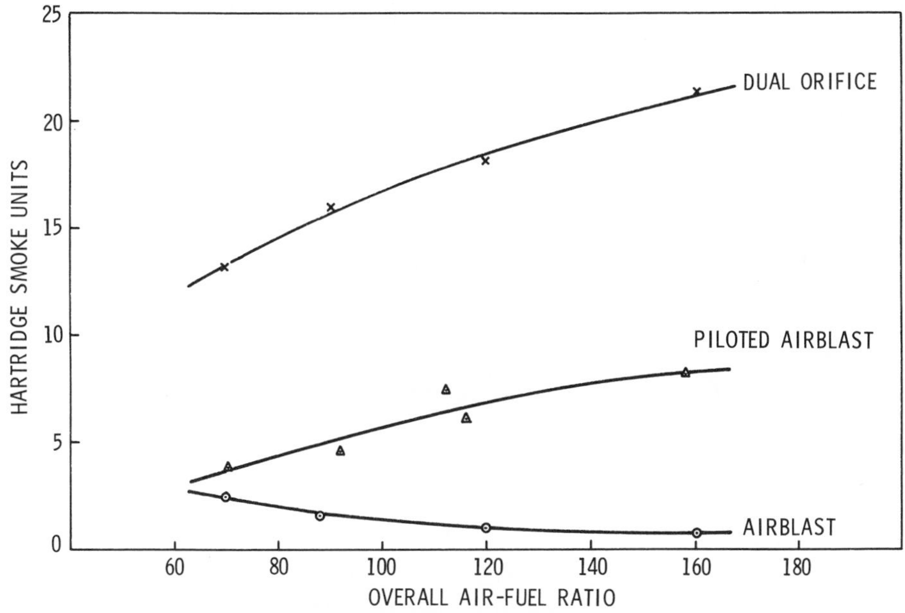

Fig. 16. Influence of fuel spray characteristics on exhaust smoke.

In order to minimize undesirable emissions, the pilot spray angle should be made as wide as the lean blow-out requirement will permit. In the tests reported here, the pilot spray angles were 45° for the airblast atomizers and 60° for the dual orifice atomizer. In view of the well known dependence of exhaust smoke on spray angle (1), it is clear that an increase in pilot spray angle from 45° to 60° would produce a further reduction in smoke for the piloted airblast atomizer, below the values shown in Fig. 16.

In Figs. 15 and 16, it is of interest to note that the variation of smoke output with air-fuel ratio shows an opposite trend to that obtained with the dual-orifice injector. This occurs because the superior atomization and mixing performance of the airblast atomizer ensures that burning is virtually complete within the primary zone. Thus any increase in overall air-fuel ratio leads automatically to improved aeration of the combustion process and hence to a decrease in smoke. This is in contrast to the pressure atomizer where, as discussed earlier, decrease in fuel flow at relatively low

References p. 278

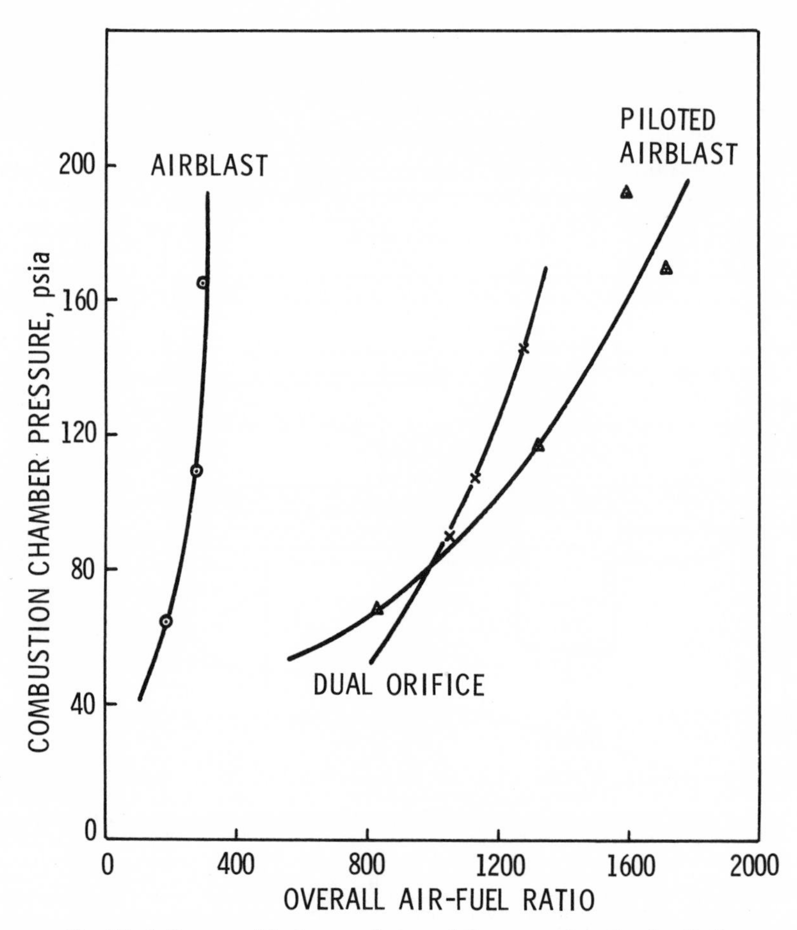

Fig. 17. Influence of fuel spray characteristics on weak extinction limits.

pressures tends to increase soot formation by virtue of a reduction in atomization quality.

Tests were also carried out on double-swirler, piloted airblast atomizer (main flow number = 3.15, pilot flow number = 0.25, pilot spray angle = 60°) as illustrated in Fig. 18. This atomizer is similar to that shown in Fig. 13, except that the central pintle is replaced by a small air swirler whose swirl component is opposite in direction to that of the main air swirler. The three main advantages claimed for the double swirler design are:—

- High burning rates are achieved by injecting the atomized main fuel into the region of high shear produced at the junction of the two contra-rotating swirler flows.

- Problems of overheating and carbon deposition on the pintle face are avoided.

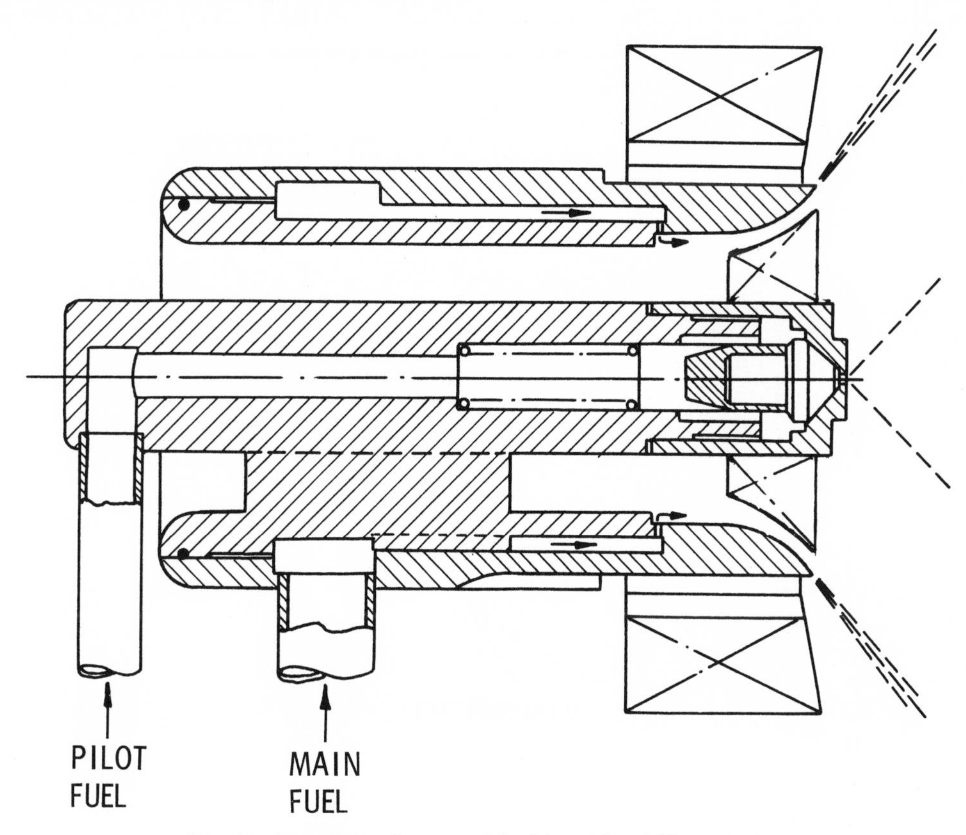

Fig. 18. Basic design features of double-swirler airblast atomizer.

- When used in conjunction with a pilot fuel nozzle, the flow recirculation induced by the inner air swirler can be matched to the pilot spray characteristics to give optimum combustion performance at low fuel flow conditions.

The results obtained with the double-swirler injector show significant reductions in exhaust concentrations of smoke, CO and HC, as illustrated in Fig. 3, 7, 8, 19 and 20. These improvements are due to better atomization and aeration of the fuel spray, resulting in more rapid evaporation, mixing and reaction, and hence higher levels of combustion efficiency. At the same time the economies in film-cooling air, made possible by the attendant reductions in flame radiation, should produce further lowering of CO and HC emissions.

Although airblast atomizers were not expected to possess any advantages over pressure atomizers in regard to NO emissions, the limited tests carried out so far indicate that the airblast atomizer has potential for small but worthwhile reductions in NO emissions. Fig. 21 shows the reduction obtained with the double swirler airblast atomizer at high fuel flows, due presumably to the beneficial effect of premixing in avoiding hot spots in the combustion zone.

References p. 278

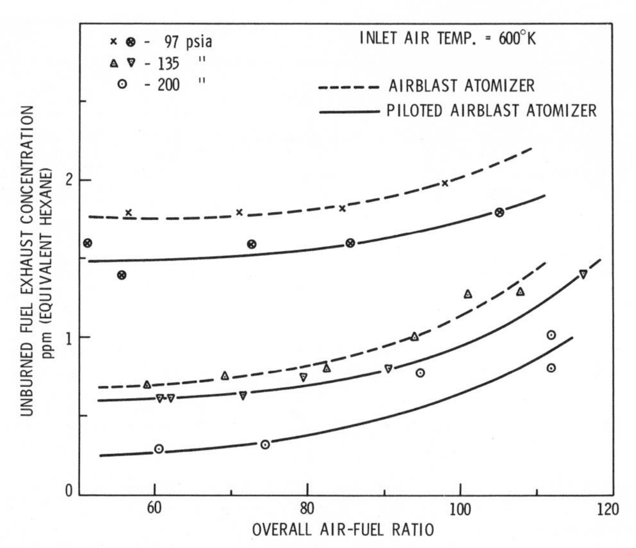

Fig. 19. Influence of pressure and overall air-fuel ratio on unburned HC emissions (double-swirler airblast atomizer).

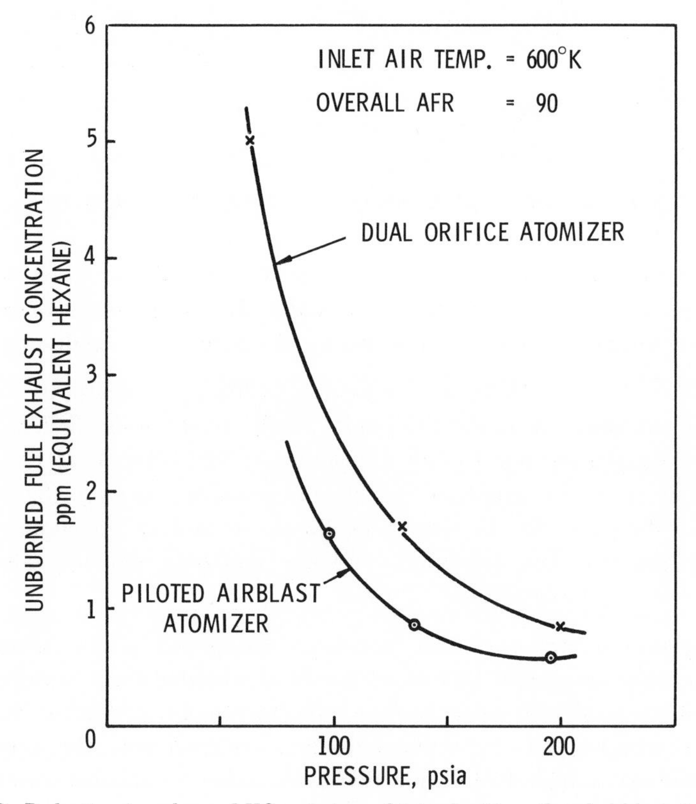

Fig. 20. Reduction in unburned HC emissions obtained with a piloted airblast atomizer.

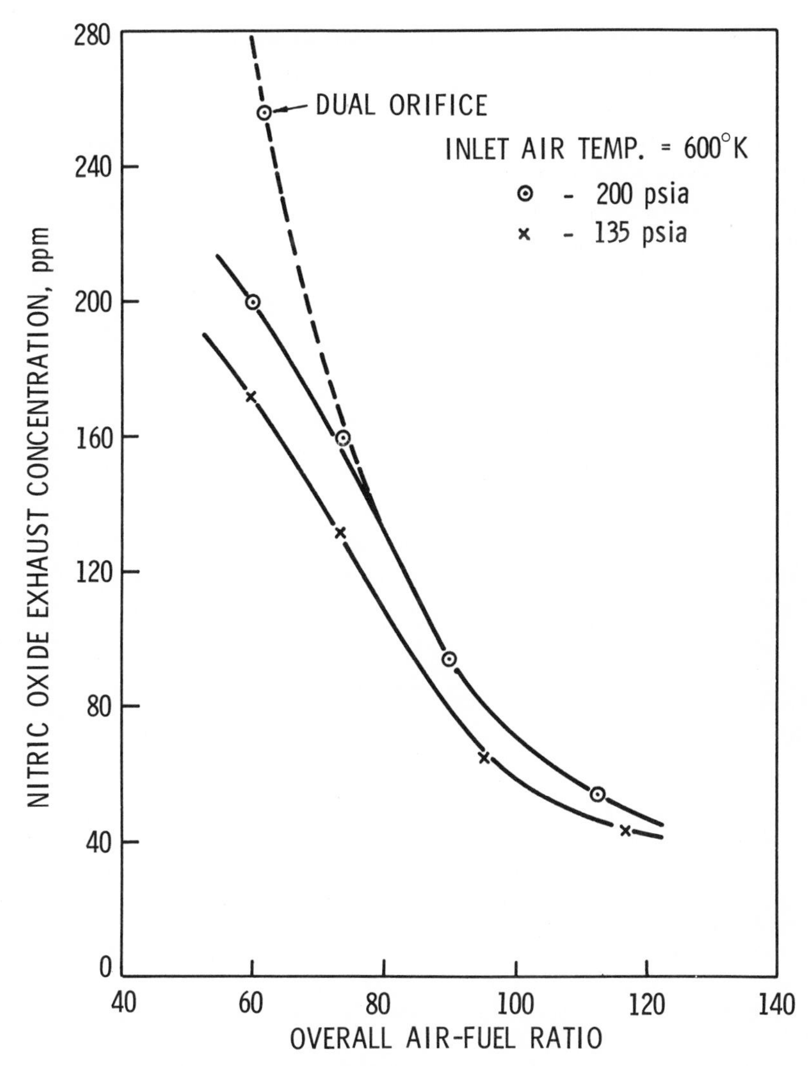

Fig. 21. Effects of pressure and overall air-fuel ratio on NO exhaust emissions (double swirler atomizer).

CONCLUSIONS

Exhaust smoke increases with operating pressure and diminishes with increase in inlet temperature. With combustion chambers featuring dual orifice injectors, at high pressures, corresponding to high fuel flows, exhaust smoke increases with decline in overall air-fuel ratio due to fuel-enrichment of the primary zone. However, at lower combustion pressures, where the atomizing pressure at the main orifice is relatively small, an opposite trend of increase in smoke with air-fuel ratio is observed, due to deterioration in atomization quality with reduction in fuel flow.

References p. 278

Emissions of CO and unburned HC are caused by low reaction temperature, poor atomization and the quenching effects of film-cooling air. In general any change in operating conditions, fuel injection method or liner configuration that elevates primary-zone combustion efficiency, will reduce CO and HC exhaust emissions. Thus raising the pressure and temperature of the inlet air, adjusting the combustor airflow proportions to bring the primary-zone fuel-air ratio closer to stoichiometric, enlarging the primary-zone volume, improving fuel atomization, premixing of fuel and air and reducing the film-cooling air, all of which tend to improve primary-zone combustion efficiency, are also effective in reducing the output of these emissions.

Exhaust emissions of NO increase with fuel flow and also with inlet air pressure and temperature in a manner that indicates a very strong dependence on reaction temperature. The influence of inlet air temperature is especially marked.

The airblast atomizer has significant advantages over conventional pressure nozzles in terms of reduced emissions of CO, HC and, to a lesser extent, NO. It is particularly effective in reducing exhaust smoke. Tests have shown that the addition of a small pilot spray nozzle to an airblast atomizer allows it to retain substantially all of its advantages in regard to low emissions and, at the same time, considerably extends the stability and relight performance.

REFERENCES

1. *A. H. Lefebvre and T. Durrant, "Design Characteristics Affecting Gas Turbine Combustion Performance". Paper 240C presented at SAE National Aeronautic Meeting, Los Angeles, October 1960, and published in Esso Air World, Vol. 13, No. 3, November/December, 1960.*
2. *J. J. Macfarlane, F. H. Holderness, and F. S. Whitcher, "Soot Formation Rates in Premixed* C_5 *and* C_6 *Hydrocarbon Flames at Pressures up to 20 Atmospheres", Combustion and Flame, Vol. 8, No. 3, September 1964.*
3. *T. Durrant, "The Control of Atmospheric Pollution from Gas Turbine Engines", Esso Air World, Vol. 21, No. 3, November/December 1968.*
4. *J. J. Faitani, "Smoke Reduction in Jet Engines Through Burner Design", Esso Air World, Vol. 21, No. 2, September/October 1968.*
5. *S. M. DeCorso, "Effects of Ambient and Fuel Pressure on Spray Drop Size", ASME Gas Turbine Conference (Scientific Paper 8-0524-P12), ASME, 1959.*
6. *D. W. Bahr, J. R. Smith and M. J. Kenworthy, "Development of Low Smoke Emission Combustors for Large Aircraft Turbine Engines", AIAA Paper No. 69-493, 1969.*
7. *R. F. Sawyer, D. P. Texeira, and E. S. Starkman, "Air Pollution Characteristics of Gas Turbine Engines", Trans. ASME, J. of Eng. for Power, p. 290, 1969.*
8. *W. Cornelius, and W. R. Wade, "The Formation and Control of Nitric Oxide in a Regenerative Gas Turbine Burner", SAE Paper No. 700708, September 1970.*
9. *A. H. Lefebvre and D. Miller, "The Development of an Airblast Atomizer for Gas Turbine Application", Cranfield Inst. of Tech., College of Aeronautics Report, Aero No. 193, June 1966.*
10. *R. Bryan, P. S. Godbole and E. R. Norster, "Some Observations of the Atomizing Characteristics of Air Blast Atomizers", Cranfield International Symposium Series, Vol. 11, Pergamon Press, Edited by E. R. Norster, 1971.*

EFFECT OF OPERATING VARIABLES ON POLLUTANT EMISSIONS FROM AIRCRAFT TURBINE ENGINE COMBUSTORS

J. S. GROBMAN

NASA – Lewis Research Center, Cleveland, Ohio

ABSTRACT

The purpose of this paper is to review NASA-Lewis combustor research aimed at reducing exhaust emissions from jet aircraft engines. Experimental results of tests performed on both conventional and experimental combustors over a range of inlet total pressure, inlet total temperature, reference velocity, and fuel-air ratio are presented to demonstrate the effect of operating variables on pollutant emissions. Combustor design techniques to reduce emissions are discussed. Improving fuel atomization by using an air-assist fuel nozzle has been shown to significantly reduce hydrocarbon (HC) and carbon monoxide (CO) emissions during idle. A short-length annular swirl-can combustor has demonstrated a significant reduction in nitric oxide (NO) emissions compared to a conventional combustor operating at similar conditions. The use of diffuser wall bleed to provide variable control of combustor airflow distribution may enable the achievement of reduced emissions without compromising combustor performance.

SUMMARY

The purpose of this paper is to review recent NASA-Lewis combustor research aimed at reducing or eliminating undesirable exhaust emissions from jet aircraft engines. Emission tests have been performed on both conventional and experimental combustors over a range of inlet total pressure, inlet total temperature, reference velocity, and fuel-air ratio. Experimental results are presented that demonstrate the effect of operating conditions on pollutant emissions from jet aircraft. Total HC and CO emissions are shown to increase markedly as fuel-air ratio is reduced below a value of about 0.01. This reduction is partly due to poor fuel atomization and partly due to

References pp. 302-303

the formation of fuel-air mixtures in the primary zone that are below the flammability limit. Total HC and CO emissions are shown to increase rapidly as a correlating parameter, P_3T_3/V_R, (in which P_3 is combustor inlet total pressure, T_3 is combustor inlet total temperature, and V_R is combustor reference velocity) is reduced below a value of about 10^5 lbs-sec-°R/ft^3. The emission index for CO is shown to be particularly sensitive to reference velocity. The emission index for the oxides of nitrogen (NO_x) is shown to increase with increasing inlet total temperature and decreasing reference velocity. For a given primary zone airflow distribution, smoke number is shown to increase with increases in combustor inlet total pressure and with decreases in combustor inlet total temperature.

Experimental tests have shown that improving fuel atomization by using an air-assist fuel nozzle can significantly reduce HC and CO emissions during idle. Additional experimental designs are being investigated that optimize the local fuel-air ratio in the primary zone during idle by either 1) using diffuser wall bleed to control primary zone airflow or 2) by using staggered fuel nozzles. A short length annular swirl-can combustor has demonstrated a significant reduction in NO emissions compared to a conventional combustor operating at similar conditions. This reduction may be attributed to reduced reaction dwell time as a result of both reduced burning length and rapid mixing of combustion gases and dilution air. The premixing and carbureting of fuel and air in the swirl-can may also play a part in this NO reduction. An experimental combustor segment designed with increased primary zone airflow and increased mixing intensity has been tested that demonstrates extremely low smoke numbers at elevated pressures; however, the altitude relight capability of this configuration has not been satisfactory. Further research is being pursued to determine if techniques such as diffuser wall bleed may be used to control primary zone airflow so that lean primary zone fuel-air ratios can be obtained during take-off to reduce smoke formation and so that richer primary zone fuel-air ratios may be obtained during idle and altitude relight.

INTRODUCTION

Exhaust emission data (1) for a typical commercial jet engine are plotted against engine power setting in Fig. 1. The products of inefficient combustion, CO, and total HC, are shown to be highest at a power setting of engine idle while NO_X is shown to be highest during take-off. Lower combustion efficiency at idle is caused by 1) poor fuel atomization at low fuel flow rates, 2) lower fuel-air ratios, and 3) lower combustor inlet total pressure and temperature. Higher NO emissions during take-off are known to be caused by a higher combustor inlet total temperature which affects its rate of formation (2). The data shown in Fig. 1 were obtained from a JT8D engine which has a compressor pressure ratio of 16:1. In general, it would be expected that engines with lower compressor pressure ratios would exhibit higher CO and HC emissions during idle while engines with higher compressor pressure ratios would exhibit higher NO emissions.

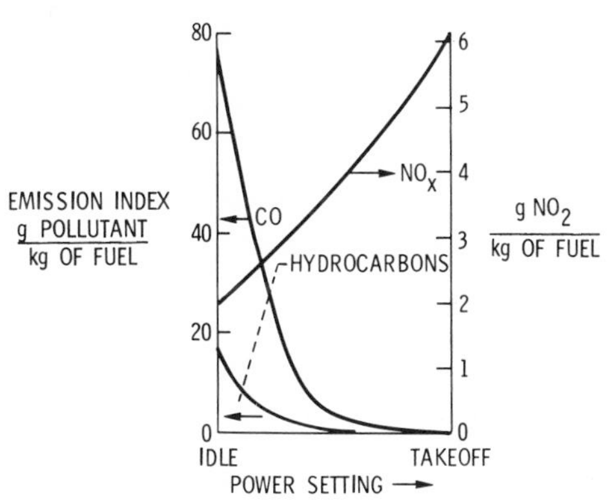

Fig. 1. Effect of engine operating conditions on exhaust emissions of JT8D engine.

Smoke number usually tends to maximize at the take-off power setting. A recent effort was made by the airlines industry to retrofit an existing commercial engine with a combustor redesigned with increased primary zone airflow in order to reduce exhaust smoke density to a value below the threshold of visibility (3). The newer gas turbine engines are being designed with smoke numbers that are below the visible threshold of smoke.

Previous combustor design and development effort has been concentrated on the optimization of combustion efficiency, total-pressure loss, durability, exit temperature profile, and altitude relight. Recent research has been devoted to the development of short length combustors for advanced high-temperature gas turbine engines (4). The additional requirement for both low gaseous emissions and low smoke number raises the question as to the difficulty or even feasibility of attaining all of the prescribed emission limits without seriously compromising the other combustor performance criteria. In addition, the problem arises that a certain design change to minimize a given pollutant may inadvertently lead to an increase in some other pollutant. Combustor research is required to gain a better understanding of the design trade-offs that are necessary for minimizing pollutant emissions at various operating conditions and to devise new combustor design techniques to obtain reduced emissions without sacrificing engine performance. Furthermore, it is necessary to consider how future engine design requirements might be affected by a restraint on pollutant emissions.

The purpose of this paper is to review recent NASA Lewis combustor research (4 - 13) aimed at reducing or eliminating undesirable exhaust emissions from jet aircraft engines. Emission tests were performed on both conventional and experimental combustors over a range of inlet total pressure, inlet total temperature, reference velocity, and fuel-air ratio. Total HC, CO, and NO concentrations and smoke number were determined from gas samples obtained at the exhaust plane of these test combustors. Experimental results are presented that demonstrate the effects of operating variables on both combustion efficiency and exhaust emissions.

References pp. 302-303

The relative influence of the various operating variables on the formation of each pollutant is discussed. Exhaust concentrations of total HC, CO, and NO expressed in terms of grams of pollutant per kilogram of fuel burned are plotted against the various combustor operating variables and against a correlating parameter that has previously been used to analyze combustion efficiency data. Results are also presented from several experimental combustor configurations that have demonstrated reduced emission levels.

Most of the experimental data presented in this report that relates emissions with operating variables were obtained from a program (5 - 7) in which a single J-57 combustor was tested in a 12-inch diameter housing. Tests were conducted at the following conditions: inlet total pressure, 1 - 20 atm; inlet total temperature, 100° - 600°F; reference velocity, 25 - 150 ft/sec; and fuel-air ratio, 0.004 - 0.015. Photographs of the test combustor used are shown in Fig. 2. Additional emission data

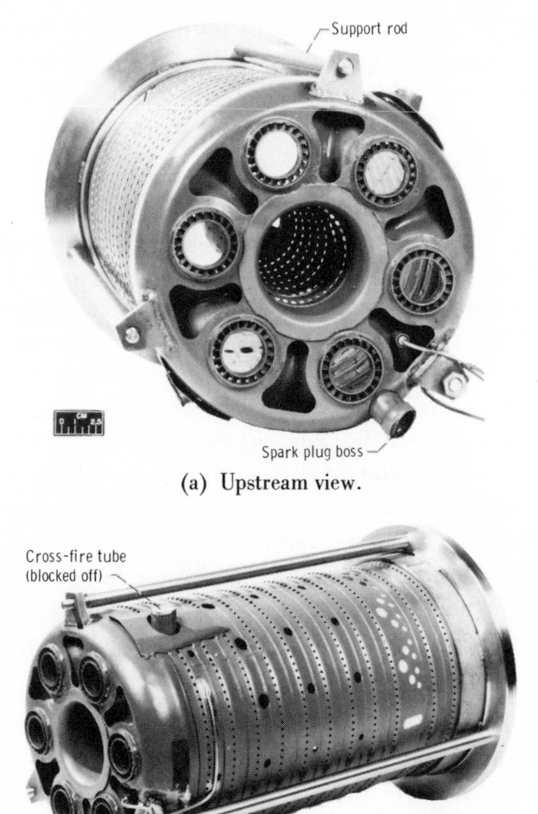

(a) Upstream view.

(b) Side view.

Fig. 2. J-57 combustor liner.

are presented for an experimental annular swirl-can combustor that was designed for elevated combustor exit temperatures (8, 9). Design details of the swirl-can combustor are illustrated in Fig. 3. The swirl-can combustor was tested at the following operating conditions: inlet total pressure, 4 - 6 atm; inlet total temperature, 600° - 1050°F; reference velocity, 70 - 120 ft/sec; and exit total temperature, 1500° - 3500°F (fuel-air ratio, 0.015 - 0.062).

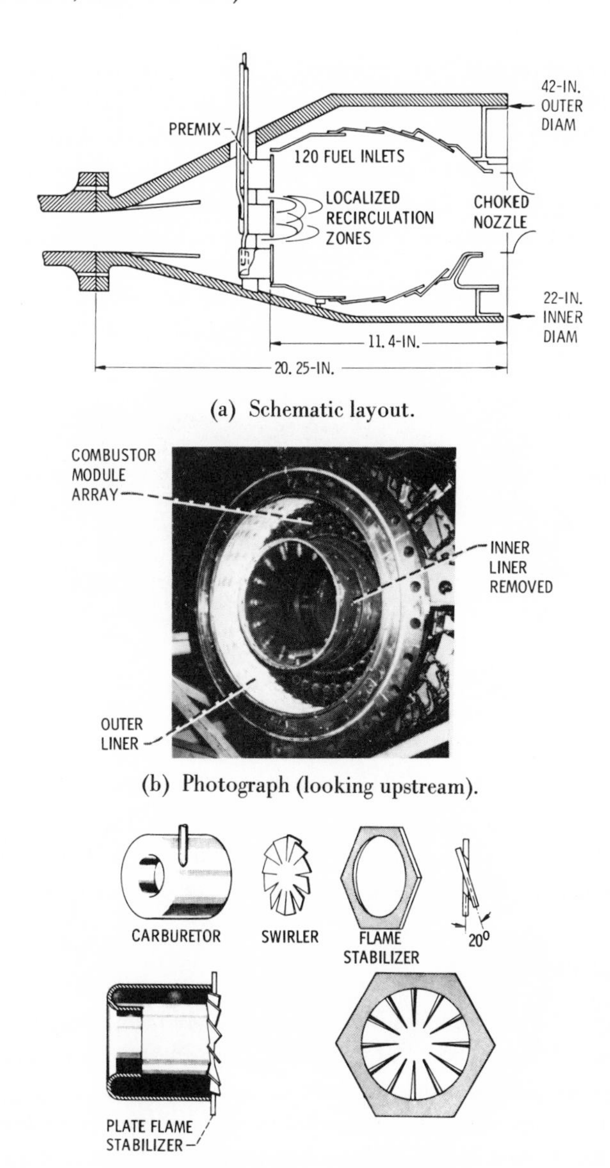

(a) Schematic layout.

(b) Photograph (looking upstream).

(c) Combustor module details.

Fig. 3. High temperature swirl-can combustor.

Smoke data, only, are presented from an experimental program (10, 11) in which a high pressure combustor was investigated. Design details of the experimental test segment are shown in Fig. 4. Smoke number data for this combustor are presented for the following range of test conditions: inlet total pressure, 10 - 27 atm; inlet total temperature, 400^0 - 900^0 F; reference velocity, 70 - 90 ft/sec; and fuel-air ratio, 0.007-0.013.

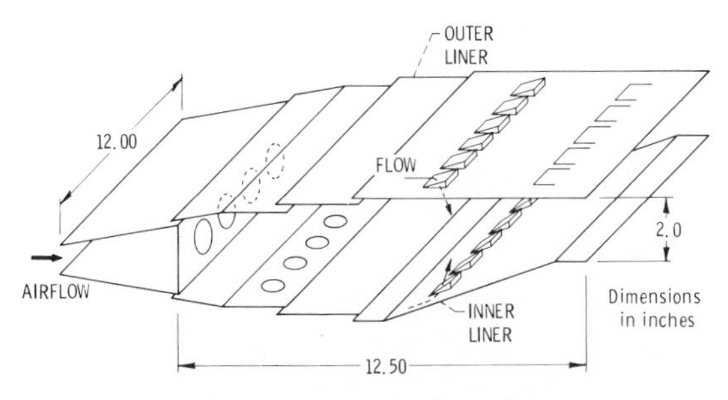

(a) Schematic sketch of combustor.

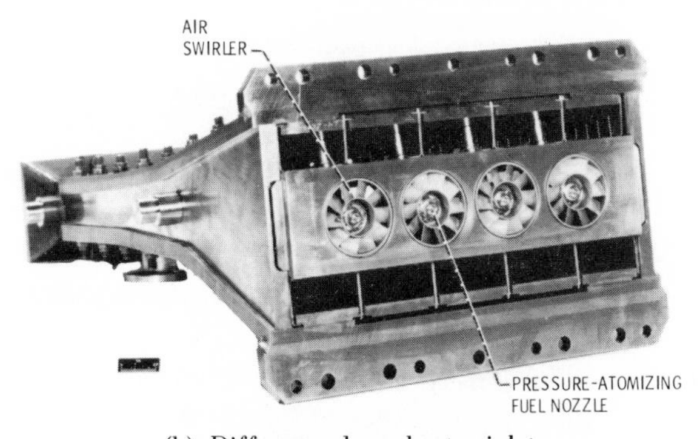

(b) Diffuser and combustor inlet.

Fig. 4. Experimental high pressure combustor test segment.

All test data presented in this report were obtained with ASTM A-1 fuel. The effect of other fuels or fuel additives on emissions will not be discussed in this report.

SAMPLING PROCEDURE

Specific details of the sampling procedures that were used in each research program are described in references (5 - 11). Practical considerations generally limited the number of gas sampling probes to one or two circumferential positions at the combustor exhaust plane at which a number of radial samples on the centers of equal areas were obtained, and then collected together as a single average sample. These probes were water-cooled to enable quenching of the reaction, and the samples were transferred through stainless steel sampling lines that were generally heated to above

300°F. Some of the gaseous emission samples were analyzed for total HC by on-line equipment; however, most of the samples were collected in sampling vessels and analyzed at a later time. Total HC were analyzed by means of a Beckman Model 106E flame ionization detector. CO was analyzed by a Beckman GC4 gas chromatograph, and NO_X was determined by a modified Saltzman method (14). Smoke number was determined from the reflectivity of smoke traces collected on filter paper (15) with the exception that a continuous moving filter tape was used. A limited number of grab samples were analyzed for carbon dioxide and hydrogen. In the case of the research program (9), a fluidic oscillator was also used to determine the fuel-air ratio of the sample gas (16, 17).

The validity of the gas sample data was checked in specific instances by comparing the combustion efficiency calculated from exhaust gas concentrations with combustion efficiency determined from thermocouple measurements, or by comparing the local fuel-air ratio deduced from exhaust gas concentrations with the fuel-air ratio determined by measured airflows and fuel flows. In general, the exhaust emission data checked rather well for test runs with combustion efficiencies greater than about 90 percent; however, as combustion efficiency measurements determined by thermocouples fell below 90 percent, the error in the exhaust sample increased greatly. Specifically, the total HC concentration in the exhaust sample appeared to be much lower than that predicted from calculated combustion efficiencies from thermocouple measurements. The sampling error at low combustion efficiencies may be attributed to the following: 1) at low combustion efficiencies, especially at low combustor inlet total temperatures, liquid fuel may pass by the combustor exhaust plane and be undetected by the gas sample probes if the liquid is centrifuged onto the walls of the duct; 2) the use of only several circumferential sampling positions may in many cases not provide a representative exhaust sample, and 3) isokinetic sampling was not used. Despite the fact that the absolute accuracy of the exhaust emission data presented for low combustion efficiencies is in error, the emission trends with operating conditions should still be of relative significance.

All gaseous emission results presented herein are expressed in terms of an emission index in grams of pollutant per kilogram of fuel burned. The NO_x present in the exhaust sample is believed to consist mainly of NO with lesser amounts of nitrogen dioxide (NO_2); nevertheless, by convention, the emission index for the NO_x is expressed in terms of grams of NO_2 per kilogram of fuel burned. In the text, this quantity will be referred to as either NO (nitric oxide) or NO_x (oxides of nitrogen).

PRODUCTS OF INEFFICIENT COMBUSTION

Effect of Fuel-Air Ratio and Fuel Atomization – Combustors for aircraft gas turbine engines are required to operate over a relatively wide range of fuel flows. Conventional combustors are, therefore, normally equipped with a dual range fuel nozzle consisting of a primary orifice to cover low fuel flows and a secondary orifice to handle high fuel flows that cuts in at higher pressure differentials across the fuel nozzle. Fig. 5a shows a plot of combustion efficiency against fuel-air ratio for the J-57 test combustor. The combustion efficiency data presented in this report were determined from thermodynamic calculations using thermocouple measurements. The accuracy of the combustion efficiency data determined in this manner is estimated to be within about ±3 percent. These data were obtained at an inlet total pressure of 2 atm; an inlet total temperature of 300°F; and a reference velocity of 50 ft/sec. Data

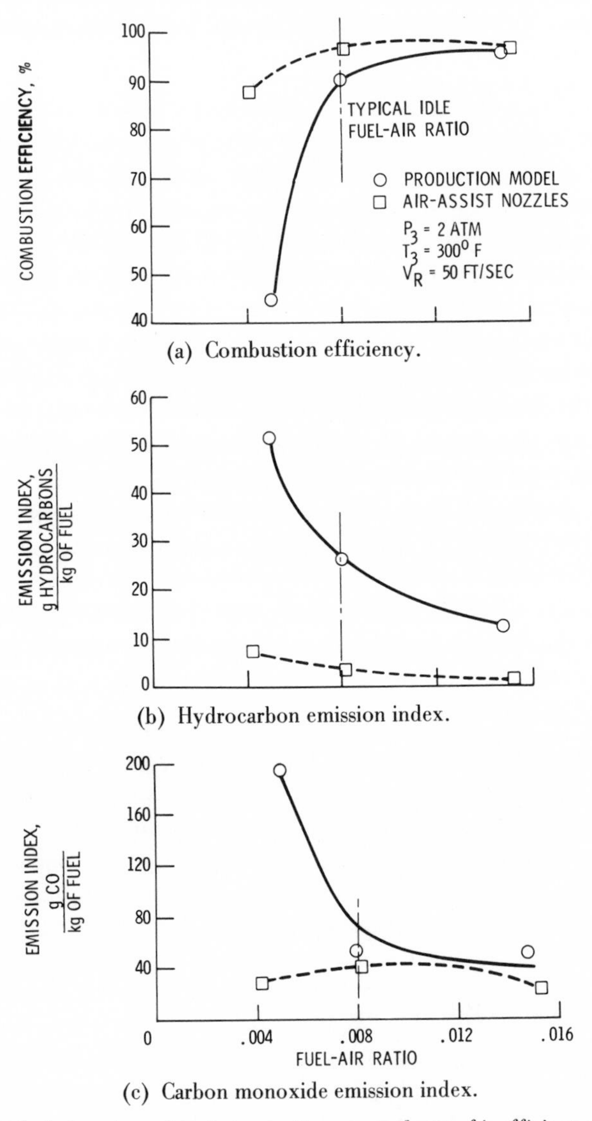

(c) Carbon monoxide emission index.

Fig. 5. Effect of fuel-air ratio and fuel atomization on products of inefficient combustion for the J-57 combustor.

are presented for the production J-57 fuel system which contains six dual orifice nozzles and a modification thereof which incorporates air atomization. For the range of fuel-air ratio shown, only the primary orifice of the nozzle is used for the production model. The pressure drop across the production fuel nozzle is about 50 psi at a fuel-air ratio of 0.008 for this operating condition. Combustion efficiency is shown to fall off rapidly as fuel-air ratio is lowered below a value of about 0.01. This effect is attributed to poor fuel atomization and reduced fuel penetration because, at low fuel flows, the pressure drop across the fuel nozzle is too low to provide effective swirl atomization. Poor fuel atomization causes larger fuel droplets to form which increases the time required for vaporization while reduced fuel penetration results in ineffective mixing of fuel and air.

The air-assist nozzle data shown in the same plot were obtained by connecting a source of high pressure air to the secondary orifice of the production model. A marked improvement in combustion efficiency is shown to be obtained by the introduction of the high pressure air which essentially improved fuel atomization of the production combustor. The improvement in combustion efficiency is especially large at the lowest fuel-air ratio where fuel atomization was the poorest for the production combustor. However, even with improved fuel atomization, there is still some reduction in combustion efficiency as fuel-air ratio is decreased. This may be attributed to the formation of fuel-air mixtures in the primary zone that are below the lean flammability limit. The data presented in Fig. 5 were obtained at a combustor inlet total pressure and temperature that are typical of the engine idle power setting. Additional data (7) were obtained for higher combustor inlet total pressures and temperature for the same range of fuel-air ratio. A similar but less steep reduction in combustion efficiency with decreasing fuel-air ratio was observed. A higher combustion efficiency is, of course, expected at higher combustor inlet total pressures and temperatures, in addition, fuel atomization is greatly improved at higher fuel flows.

Fig. 5b shows the corresponding hydrocarbon emission index plotted against fuel-air ratio for the same data. At a typical idle fuel-air ratio of 0.008 the HC emission index was reduced from a value of 26.6 to a value of 3.3 by the improved fuel atomization of the air-assist nozzle configuration. Similarly Fig. 5c shows the emission index for CO plotted against fuel-air ratio for the same data. At the typical idle fuel-air ratio of 0.008, the CO emission index was reduced from a value of about 60 to a value of about 50. At fuel-air ratios below 0.008, the reduction in HC and CO emissions is quite pronounced when the air-assist nozzle configuration is used. The application of an air-assist fuel nozzle to reduce HC and CO emissions during idle is discussed in a later section of this paper.

Effect of Combustor Inlet Total Pressure – The effect of combustor inlet total pressure on combustion efficiency of the J-57 combustor is shown in Fig. 6a. These data were obtained at a reference velocity of 50 ft/sec; fuel-air ratios of 0.0075 and

References pp. 302-303

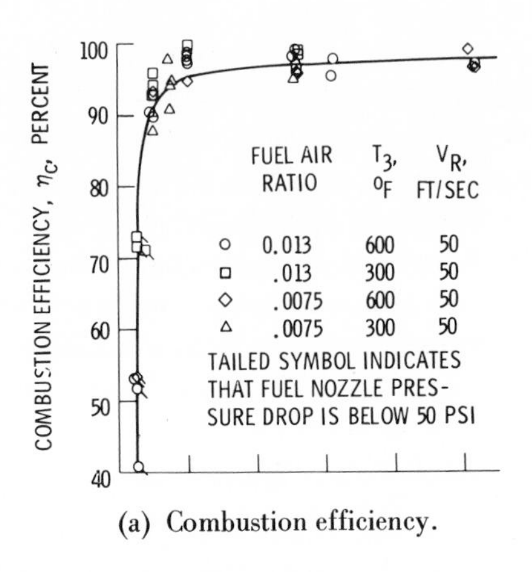

(a) Combustion efficiency.

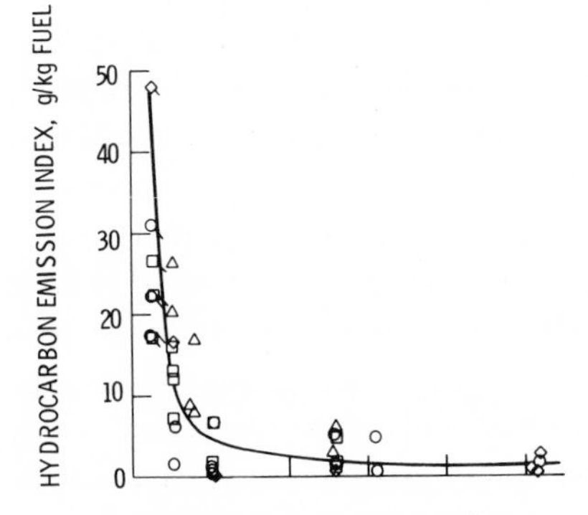

(b) Hydrocarbon emission index.

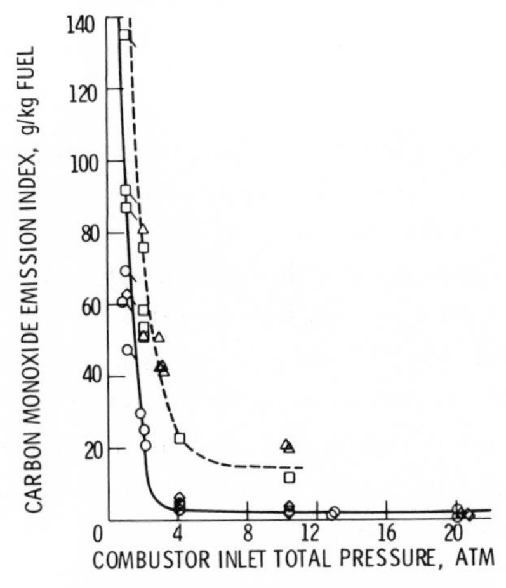

(c) Carbon monoxide emission index.

Fig. 6. Effect of combustor inlet total pressure on products of inefficient combustion for the J-57 combustor.

0.013; and combustor inlet total temperatures of 300° and 600°F. Combustion efficiency is shown to decrease rapidly as combustor inlet total pressure is lowered below about 4 atm. The decrease in combustion efficiency is partly attributed to the reduction in inlet total pressure but is considered to be predominately caused by poor fuel atomization at the lowest combustor pressures as the result of low nozzle pressure drop. The tailed symbols shown on the plot indicate data points with nozzle pressure drops lower than 50 psi. Figs. 6b and 6c show a corresponding increase in emission index for total HC and for CO as the combustor inlet total pressure is reduced below a value of about 4 atm.

Effect of Combustor Inlet Total Temperature – Fig. 7a shows a plot of combustion efficiency against combustor inlet total temperature for values of fuel-air ratio of 0.013, inlet total pressures of 2 and 10 atm, and reference velocities of 50 and 100 ft/sec. Within the degree of accuracy of the data, combustion efficiency is not shown to be strongly affected by inlet total temperature for the range of data shown except at the higher reference velocity.

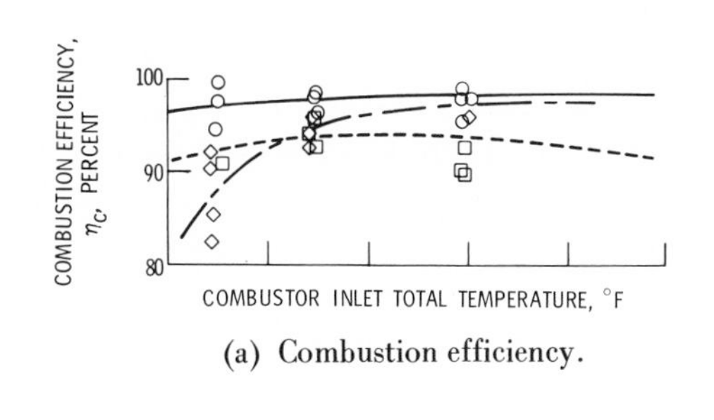

(a) Combustion efficiency.

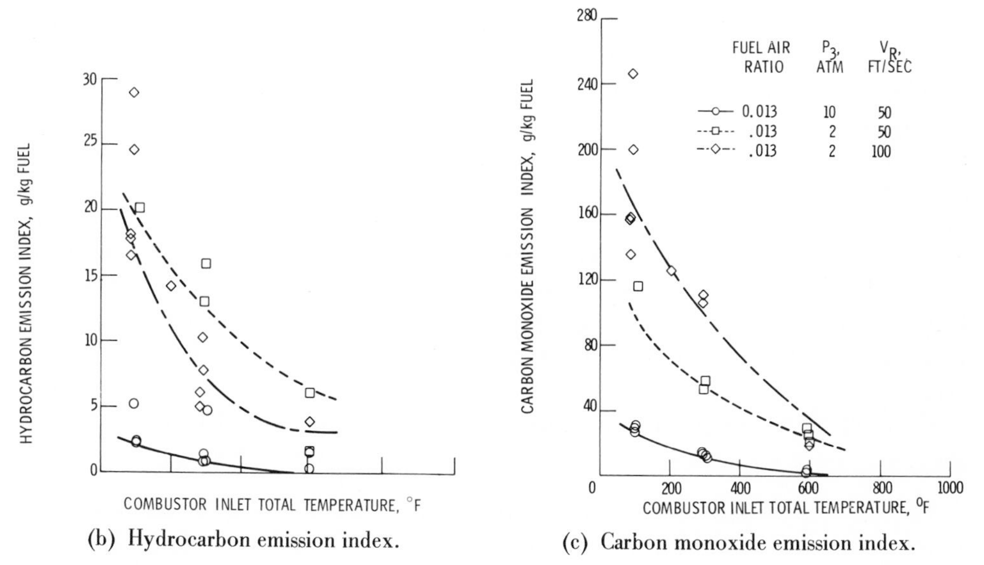

(b) Hydrocarbon emission index.

(c) Carbon monoxide emission index.

Fig. 7. Effect of combustor inlet total temperature on products of inefficient combustion for the J-57 combustor.

References pp. 302-303

The HC emission index results for the same data are shown plotted in Fig. 7b. The rate of increase of HC emissions with reduced combustor inlet total temperatures is shown to be greater for the data at the lower inlet total pressure. The corresponding increase in the CO emission index with decreasing combustor inlet total temperature is shown in Fig. 7c. The rate of increase in the CO emission index with decreasing combustor inlet total temperature appears to be greatest for the data obtained at the lower inlet total pressure and higher reference velocity. It is interesting to note that the HC and CO emission indices show a significant increase with decreasing inlet total temperature even though combustion efficiency appeared rather insensitive to varying inlet total temperature. This is attributed to the fact that the over-all variation in emission level that was observed for some of the data corresponds to a change in combustion efficiency that is less than the experimental error in combustion efficiency determined from thermocouple measurements.

Effect of Combustor Reference Velocity – Reference velocity is defined as the total combustor airflow divided by the product of combustor inlet density and maximum cross-sectional area. Fig. 8a shows a plot of combustion efficiency against reference velocity for the J-57 combustor at an inlet total pressure of 2 atm; fuel-air ratios of 0.0075 and 0.013; and a range of inlet total temperature from 100° to 600°F. Limited data for the annular swirl-can combustor at an inlet total pressure of 4 atm; inlet total temperature of 600°F; and a fuel-air ratio of 0.023 - 0.024 are also presented. No significant effect of reference velocity on combustion efficiency is apparent for the data obtained from either combustor at an inlet total temperature of 300° and 600°F. However, at an inlet total temperature of 100°F, there appears to be a relatively strong effect of reference velocity on combustion efficiency for the J-57 combustor. Similar low inlet total temperature data for the annular swirl-can combustor were not available. Combustion efficiency decreases from a value of about 90 to about 80 percent as reference velocity is increased from about 75 to 150 ft/sec at an inlet total temperature of 100°F and a fuel-air ratio of 0.013. The data obtained at inlet total temperature of 100°F at both a fuel-air ratio of 0.0075 and 0.013 indicate that combustion efficiency tends to fall off slightly as reference velocity is reduced from a value of about 75 ft/sec, to a value of about 50 ft/sec. Previous results have shown, that for most combustors, combustion efficiency decreases with increasing reference velocity; however, for some combustors, a maximum value of combustion efficiency occurs at a specific value of reference velocity, and will then decrease as reference velocity is either increased or decreased. A reduction in combustion efficiency with increasing reference velocity may be attributed to a reduction in flame stability and dwell time; while a reduction in combustion efficiency with decreasing reference velocity may be attributed to either poor mixing as the result of a lowering in combustor pressure drop or to poor fuel atomization as the result of a lowering in fuel nozzle pressure drop as fuel flow is lowered.

The data for the HC emission index shown in Fig. 8b tend to follow the trend that would be expected from Fig. 8a. The highest HC emission indices occur for the data at a combustor inlet total temperature of 100°F. The HC emission index for these data tend to reach a minimum at a reference velocity of 75 ft/sec. Similarly, Fig. 8a indicated a peak combustion efficiency for these data at 75 ft/sec. The CO emission index for these data are shown plotted in Fig. 8c. These results show a continuous increase in CO emissions as reference velocity is increased. This may be attributed to the strong effect that combustor dwell time has on the oxidation of CO formed in the primary zone. Apparently as dwell time is reduced, lesser amounts of CO are oxidized to carbon dioxide. These data for the J-57 combustor indicate that the increase in CO

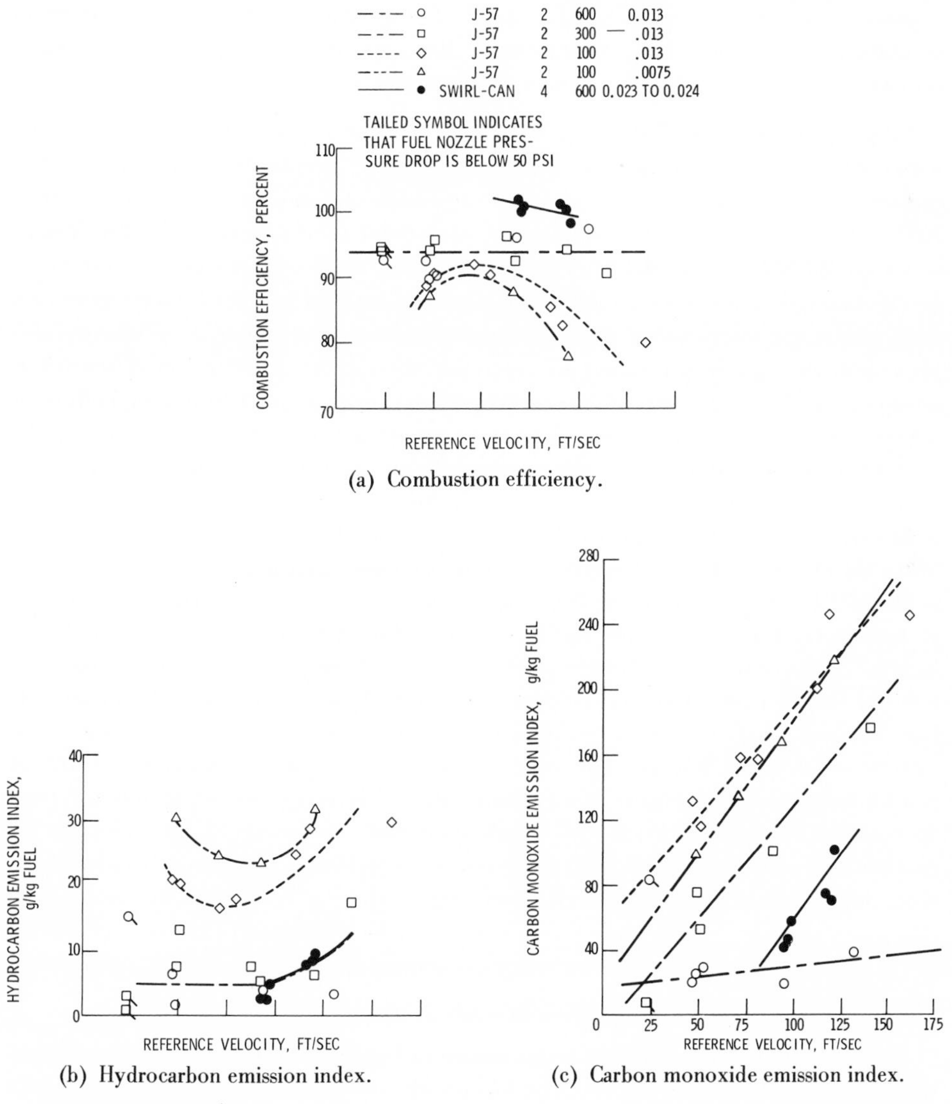

(a) Combustion efficiency.

(b) Hydrocarbon emission index.

(c) Carbon monoxide emission index.

Fig. 8. Effect of combustor reference velocity on products of inefficient combustion for the J-57 and swirl can combustors.

References pp. 302-303

emissions with increasing reference velocity is strong even at an inlet total temperature as high as 300°F. At similar operating conditions, the effect of reference velocity on the HC emission index appears to be the same for both combustors shown; however, the rate of increase in the CO emission index appears to be greater for the swirl-can combustor than for the J-57 combustor when compared at the same inlet total temperature of 600°F.

Effect of Correlating Parameter – Previous studies have correlated combustion efficiency against a combustion parameter composed of inlet total pressure P_3,

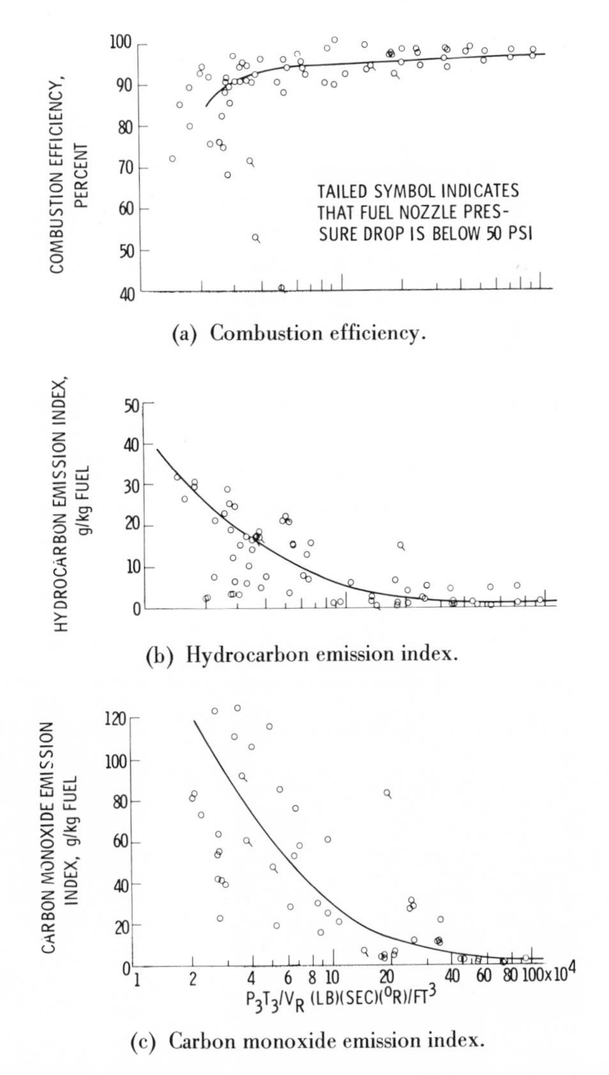

(a) Combustion efficiency.

(b) Hydrocarbon emission index.

(c) Carbon monoxide emission index.

Fig. 9. Effect of correlating parameter on products of inefficient combustion for the J-57 combustor. Fuel-air ratio, 0.013.

multiplied by inlet total temperature, T_3, and divided by reference velocity, V_R (18). In general, this technique of correlating combustion efficiency data provides a single correlating curve for only a single value of fuel-air ratio for a given combustor or combustor type. Combustion efficiency is plotted against the correlating parameter P_3T_3/V_R in Fig. 9a. These data were obtained for the J-57 combustor at a fuel-air ratio of 0.013. All of the data shown tend to correlate with the P_3T_3/V_R parameter except for the data with a fuel nozzle pressure drop below 50 psi. The data obtained at a low nozzle pressure drop would not be expected to follow the correlation because of the overriding effect of poor fuel atomization. Corresponding HC and CO emission indices are plotted for these same data against the correlating parameter in Fig. 9b and 9c. Total HC and CO emissions are shown to increase rapidly as the correlating parameter, P_3T_3/V_R is reduced below a value of about 10^5 lb-sec-°R/ft^3.

OXIDES OF NITROGEN

Effect of Combustor Inlet Total Temperature – The effect of combustor inlet total temperature on the emission index for NO_x for both the J-57 combustor and swirl-can combustor is shown plotted in Fig. 10. The J-57 data were obtained at a reference velocity of 50 ft/sec, fuel-air ratio of 0.013, and inlet total pressure of 2 atm, while the swirl-can data were obtained at a reference velocity of 99-109 ft/sec, fuel-air ratio of 0.016, and inlet total pressure of 4 to 6 atm. The emission index for NO_x is shown to increase quite rapidly as inlet total temperature is increased beyond a value of about 600°F. Similar results were previously reported (19). The increase in the emission index for NO_x with increasing inlet temperature is attributed to an increase in formation rate because of increasing flame temperature.

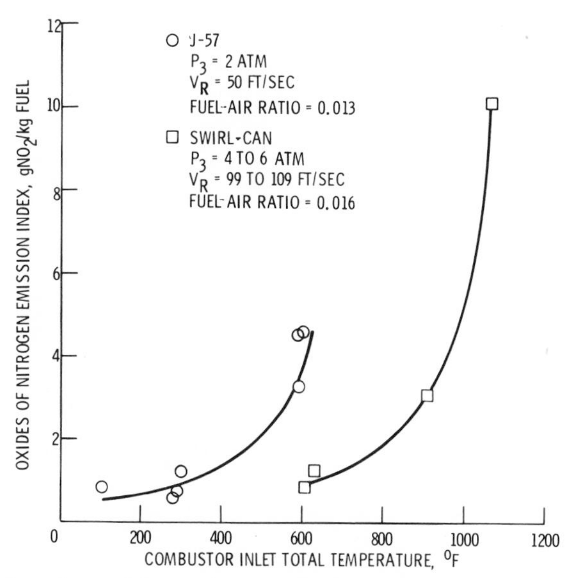

Fig. 10. Effect of combustor inlet total temperature on emission index for oxides of nitrogen.

References pp. 302-303

Facility capabilities limited the maximum inlet total temperature to 600°F for the J-57 combustor tests. Other J-57 combustor data (7) over a range of inlet total temperature of 100° - 600°F for other operating conditions display a great deal of scatter; nevertheless the general trend of these results is similar to that shown in Fig. 10. Part of this data scatter may be indicative of the difficulty in obtaining accurate NO_x samples at exhaust concentrations below 50 ppm. Since the Saltzman analysis technique used was estimated to be accurate to within about ±1 ppm, the bulk of the error may be attributed to either 1) nonrepresentative sampling or 2) absorption of a portion of the NO_x in the sampling system.

Effect of Combustor Reference Velocity – The effect of combustor reference velocity on emission index of the NO_x for both the J-57 combustor and swirl-can combustor is shown in Fig. 11. The J-57 data were obtained at an inlet total temperature of 600°F, fuel-air ratio of 0.013, and an inlet total pressure of 2 atm, while the swirl-can data were obtained at an inlet total temperature of 600°F, fuel-air ratio of 0.023-0.024, and an inlet total pressure of 4 atm. The results from both combustors indicate a rather significant increase in NO emissions with decreasing reference velocity at an inlet total temperature of 600°F. Since the formation rate of NO is reaction rate controlled, the quantity formed would be expected to be proportional to the dwell time in the reaction zone. It is reasonable to expect that dwell time would be inversely proportional with combustor reference velocity. Similar data obtained at other values of inlet total temperature for both combustors indicate that the rate of increase in the emission index for NO_x with decreasing reference velocity becomes greater as the inlet total temperature is increased.

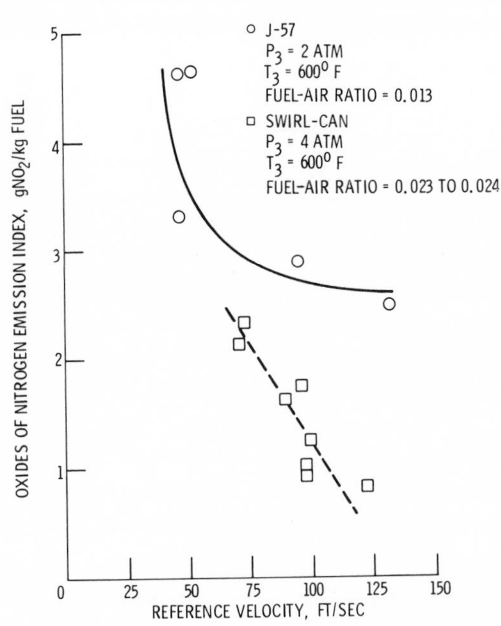

Fig. 11. Effect of combustor reference velocity on emission index for oxides of nitrogen.

Effect of Fuel-Air Ratio – The effect of fuel-air ratio on emission index for NO_x for both the annular swirl-can combustor and J-57 combustor is shown plotted in Fig. 12. The annular swirl-can combustor has been designed for high exit temperature operation and has been tested up to a combustor exit temperature of about 3600°F (8, 9). These data for the swirl-can combustor were obtained at an inlet total temperature of 600°F, inlet total pressure of 4-5 atm, and reference velocity of 67-74 and 81-88 ft/sec. The J-57 data were obtained at an inlet total temperature of 600°F, inlet total pressure of 10-12 atm, and reference velocity of 50 ft/sec. The emission index for NO_x for the swirl-can combustor increases with increasing fuel-air ratio and then reaches a peak value at a fuel-air ratio of about 0.03 before leveling off again. The NO emissions for the J-57 combustor appear to increase with increasing fuel-air ratio; however, the scatter in these data and other similar data (7) makes it difficult to precisely define the effect of fuel-air ratio for the range of data studied.

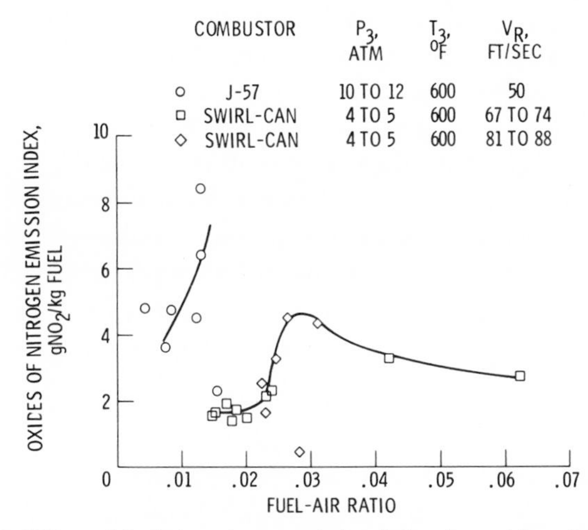

Fig. 12. Effect of fuel-air ratio on emission index for oxides of nitrogen.

Effect of Combustor Inlet Total Pressure – Exhaust emission tests have been conducted on the J-57 combustor over a range of inlet total pressures of 1-20 atm (7). At the lower end of this range in inlet total pressure (under 4 atm), where inlet total pressure has an influence on combustion efficiency, the emission index for NO_x tends to increase with increasing inlet total pressure. This effect may be attributed to an increase in the bulk temperature of the primary zone because of increased combustion efficiency. However, variations in any other operating variables that increase combustion efficiency have also been observed to increase NO emissions. Above 4 atm, NO emissions appeared to be insensitive to variations in inlet total pressure; however, there is considerable scatter in these data thus leaving some question as to the validity of this conclusion.

Effect of Correlating Parameter – The emission index for NO_x is shown plotted against the correlating parameter, $P_3 T_3 / V_R$, in Fig. 13. These data from the J-57 test combustor were obtained at a fuel-air ratio of 0.013. Despite a great deal of data scatter, there is a general tendency for NO emissions to increase with increasing values of P_3T_3/V_R. This effect is mainly attributed to increasing flame temperature and increasing dwell time which are effected by increasing inlet total temperature and decreasing reference velocity, respectively, as has been shown in the previous sections of this report. Inlet total pressure has been observed to have no effect on the NO emission level except at low values of inlet total pressure where increasing total pressure has a strong effect on improving combustion efficiency. In all cases studied, increasing combustion efficiency has been shown to increase the quantity of NO that is formed.

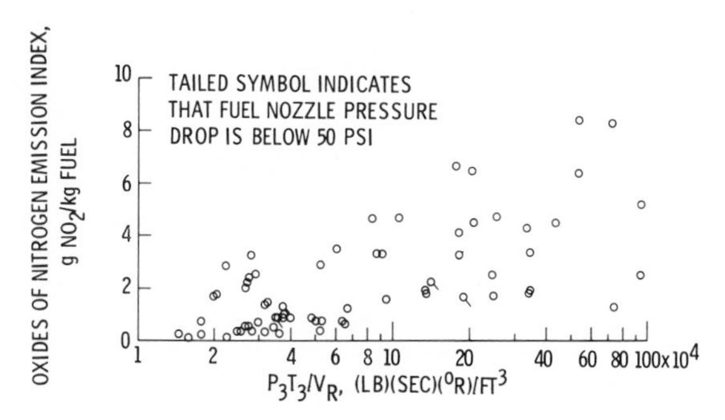

Fig. 13. Variation of emission index for oxides of nitrogen with correlating parameter for the J-57 combustor. Fuel-air ratio, 0.013.

No attempt has been made to improve the correlation shown by this figure. The correlating parameter, P_3T_3/V_R probably has limited applicability considering the fact that the data presented in Fig. 13 were limited to a maximum value of inlet total temperature of 600°F and previous results shown herein for the swirl-can combustor and the results (19) indicate a rapid increase in NO emissions as T_3 is increased above a value of 600°F. However, refinements in sampling procedure to obtain more accurate NO emission data are required prior to seeking improvements in data correlation.

SMOKE NUMBER

The present criterion used to determine acceptable smoke number values for jet aircraft is that the smoke plume be invisible. Tentative correlations between smoke number and smoke plume visibility (20) have indicated that for typical jet aircraft engines a smoke number value of 25 or less is acceptable for eliminating smoke visibility. Techniques for minimizing smoke number by means such as increasing primary zone airflow and improving fuel-air mixing in the primary zone have been discussed previously (21 - 28). The oxidation of carbon formed in the primary zone has also been recognized as an important step in controlling exhaust smoke level. Fig. 14 compares the smoke intensity (expressed as a carbon emission index) in the primary zone to that at the combustor exhaust for the high pressure combustor (Fig. 4) (10). Primary zone smoke concentrations were determined from infra-red spectral radiance measurements. These results were obtained for a combustor configuration designed with a lean, intensely mixed primary zone that had a smoke number of 32 at a fuel-air ratio of 0.013. Even in a combustor designed for low smoke formation, a significant quantity of soot is formed in the primary zone that is later oxidized before leaving the combustor. Similar results were reported previously (26). Despite the apparent importance of the soot oxidation step, results to date from short-length experimental combustor tests have not indicated any effect of combustor length on smoke number.

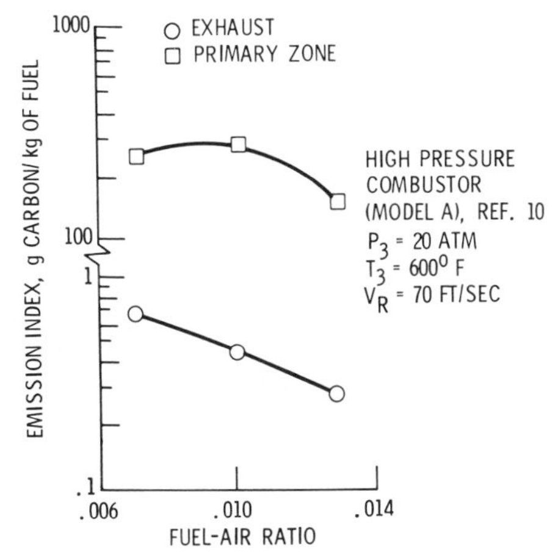

Fig. 14. Effect of fuel-air ratio on smoke formation.

Effect of Combustor Inlet Total Pressure – The effect of combustor inlet total pressure on smoke number for several experimental combustors is shown in Fig. 15. The data presented for the J-57, swirl-can, and high pressure combustors (Models A and C) (10) were obtained at a combustor inlet total temperature of 600°F. Data for several production engines including the JT8D (with and without retrofitted combustor to reduce smoke), JT9D, and CF6 are shown for comparison. Smoke number tends to increase with combustor inlet total pressure to varying degrees for different combustor configurations depending upon airflow distribution and mixing intensity. The experimental combustor configurations with the lower smoke numbers are known to have serious altitude relight limitations. Techniques for obtaining satisfactory altitude relight capabilities with combustors that have low smoke numbers will be discussed in a later section.

Effect of Combustor Inlet Total Temperature – The effect of combustor inlet total temperature on smoke number for a standard and modified J-57 combustor (5) and an experimental high pressure combustor (10) is shown in Fig. 16. The J-57 combustor data were obtained at an inlet total pressure of 12.3 atm, a fuel-air ratio of

0.013, and a reference velocity of 54 ft/sec while the experimental high pressure combustor data were obtained at an inlet total pressure of 10 atm, a fuel-air ratio of 0.013, and a reference velocity of 60 - 90 ft/sec. The reduction in smoke number with increasing inlet total temperature has also been observed (29). This effect may be beneficial in mitigating the effect of increased inlet total pressure in engines with higher compressor pressure ratios.

Effect of Other Variables – In general, over-all fuel-air ratio has not been observed to have a strong influence on smoke number. Increasing fuel-air ratio increases smoke number for some configurations and decreases it for others. In tests performed on the high temperature swirl-can combustor over a range of fuel-air ratio of 0.017 to 0.059

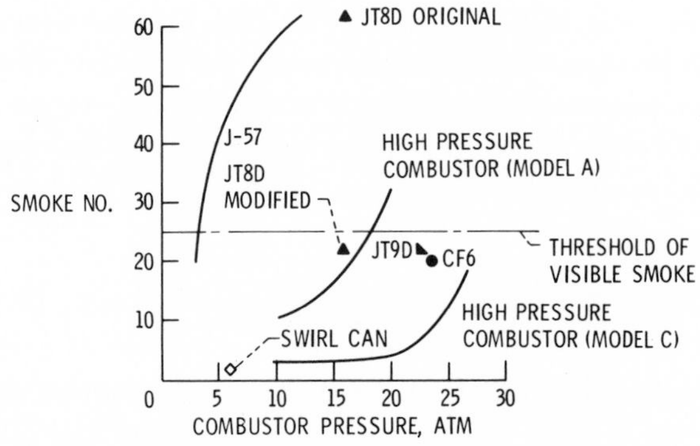

Fig. 15. Effect of combustor pressure on smoke.

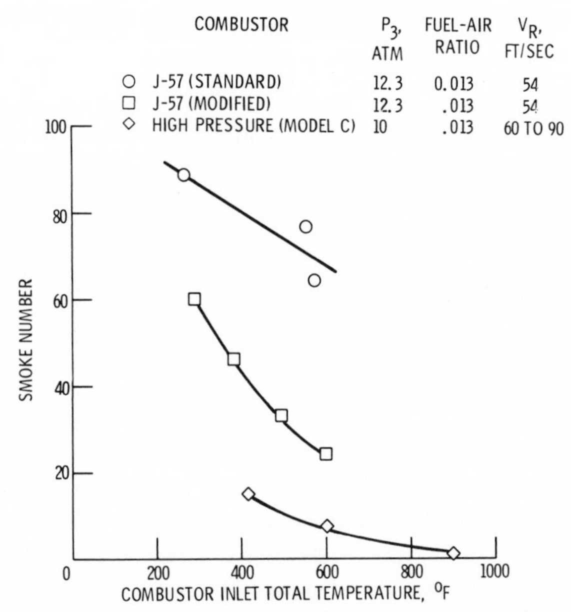

Fig. 16. Effect of combustor inlet total temperature on smoke number.

at an inlet total temperature of 600°F, inlet total pressure of 3 - 4 atm, and reference velocity of 70 - 100 ft/sec, smoke number was near zero for values of fuel-air ratio below 0.042 and approached a value of 24 at a fuel-air ratio of 0.059. The effect of fuel-air ratio on smoke number for the J-57 combustor was observed to be negligible at inlet total pressure of 12.3 atm, inlet total temperature of 600°F, and reference velocity of 54 ft/sec for a range of fuel-air ratio of about 0.007 to 0.013. Results similar to those shown in Fig. 14 for the high pressure experimental combustor have shown a small increase in smoke number with decreasing fuel-air ratio. This effect may be attributed to the formation of local fuel-rich zones because of poor atomization at low fuel flows.

Data presented for the Phillips 2-inch combustor indicated a negligible effect of combustor reference velocity on smoke number (29). Experimental data for the J-57, swirl-can and high pressure combustor to demonstrate the effect of reference velocity on smoke number have not been obtained systematically. An increase in reference velocity could conceivably decrease smoke number for a specific combustor geometry if the resulting increase in pressure loss tended to improve the primary zone mixing intensity.

No precise relationship exists between combustion efficiency and smoke number. The existence of smoke in the combustor exhaust represents only a fraction of a percent loss in combustion efficiency. A higher smoke number due to a fuel-rich primary zone design is often characteristic of a combustor configuration with both good combustion stability and good combustion efficiency. For a given combustor geometry, modifications made to reduce smoke number will generally reduce altitude relight capability.

TECHNIQUES TO REDUCE EXHAUST EMISSIONS

Total Hydrocarbons and Carbon Monoxide – Results for the J-57 combustor that were described in a previous section (Fig. 5) have shown that improving fuel atomization by using an air-assist fuel nozzle can significantly reduce unburned hydrocarbon and carbon monoxide emissions at idle operating conditions (6). The effect of atomizer air pressure drop on combustion efficiency and emissions at the same conditions as the data in Fig. 5 at a fuel-air ratio of 0.008 is shown in Fig. 17. A significant reduction in total HC and CO was obtained using an atomizer air pressure drop as low as 100 psi. The quantity of compressed air required to achieve the reductions in emissions shown in Fig. 17 amounts to less than 0.5 percent of the total combustor airflow at idle. This might be obtained from engine compressor bleed by using an auxiliary supercharger with a compression ratio of about 4.3 The secondary orifice of the fuel nozzle could be used for this purpose during idle operation, but for all other conditions, the secondary orifice could be used to handle the higher fuel flow as done conventionally in a dual orifice fuel nozzle.

References pp. 302-303

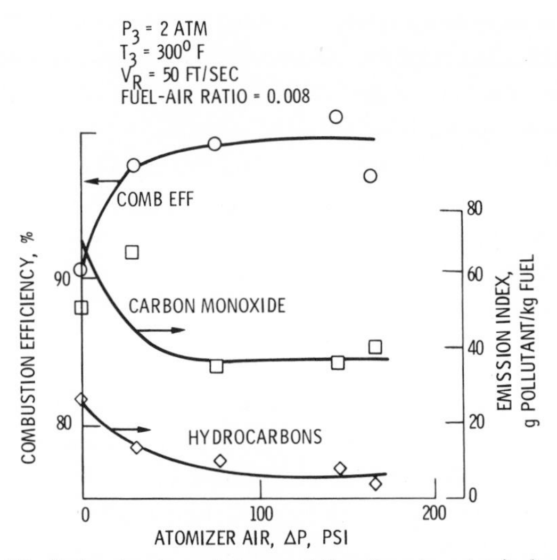

Fig. 17. Reduction in emissions at idle using air-assist fuel nozzle.

Another method being studied to reduce emissions at idle is fuel staging with the objective of operating at locally higher fuel-air ratios in the primary zone during idle to achieve higher combustion efficiency without increasing overall engine fuel flow. This might be accomplished by providing separate fuel manifolds and controls either 1) to adjacent fuel nozzles in a conventional annular combustor or 2) to separate radial zones of fuel injection as in the swirl-can combustor (Fig. 3) or the double-annular ram induction combustor (4).

An additional approach for improving combustion efficiency at idle would be to shift the combustor airflow distribution so that less air is introduced into the primary zone, thereby increasing the local fuel-air ratio in the primary zone in addition to lowering local velocities. To perform this function by means of a mechanical flow splitter or variable-area air entry port could be quite difficult and complicated because of the high temperature environment. NASA/Lewis is examining an approach for varying primary zone airflow by equipping the diffuser with wall bleed. Tests on a small-scale wide-angle annular diffuser using wall bleed have demonstrated that the radial velocity profile may be shifted to either the hub or tip by independently controlling the quantity of bleed on the inner and outer wall of the diffuser (12). This technique could be used as shown in Fig. 18. During idle when combustor pressure drop is low, wall bleed would not be used; however, the combustor inlet would purposely be designed unsymmetrical to the combustor liner in order to allow most of the air to bypass the primary zone. During takeoff and cruise when the combustor pressure drop is higher, wall bleed would be used to adjust the inlet radial velocity profile to introduce more air into the primary zone. Engine cycle efficiency would not be sacrificed because this bleed air could be used for turbine cooling. The same technique could be applied to improve altitude windmill relight capabilities.

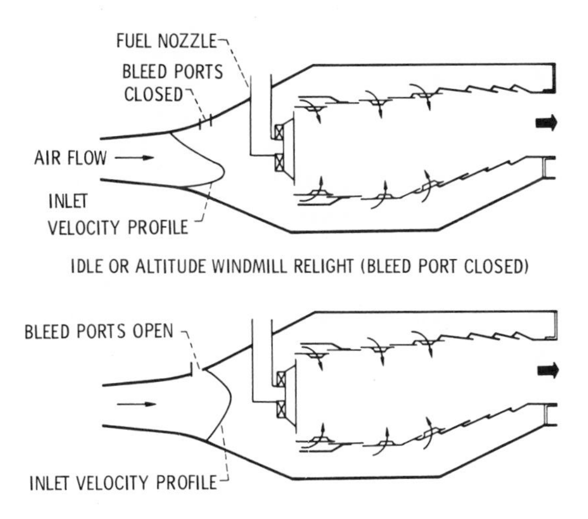

Fig. 18. Use of diffuser wall bleed to control combustor airflow distribution.

Oxides of Nitrogen – Over the past several years, NASA/Lewis has been conducting research aimed at reducing combustor length in high temperature turbine engines in order both to minimize combustor wall coolant requirements and to minimize overall engine weight (4). Fortuitously, the short-length combustor technology developed for these purposes provided us with a combustor with a short dwell time that has been shown to be effective in reducing the formation of NO_x. Tests performed on the swirl-can combustor (Fig. 3) described herein have demonstrated a significant reduction in NO emissions compared to a conventional combustor operating at similar conditions. The use of fuel prevaporization or fuel-air pre-mixing may also be useful in reducing NO emissions. Nevertheless, the reduction of NO_x in high compressor pressure ratio engines or regenerative engines which operate at combustor inlet total temperatures of 1000°F or greater may necessitate the use of water injection. Water injection reduces the quantity of NO_x formed by decreasing flame temperature.

Smoke Number – As indicated in the previous section, techniques for eliminating visible smoke plumes by increasing primary zone airflow and mixing intensity are already being utilized in advanced gas turbine engine designs. However, altitude relight capabilities may be seriously jeopardized in future higher compressor pressure ratio engines when increased primary zone airflow is used to reduce smoke number below a value of 25 (threshold of visible smoke). The technique of diffuser bleed described previously could help alleviate this design problem.

CONCLUDING REMARKS

The results presented herein have demonstrated that the degree of difficulty in obtaining low emissions is partly dependent on engine compressor pressure ratio. HC

References pp. 302-303

and CO emissions at idle are more difficult to control for low compressor pressure ratio engines, while NO emissions and smoke at takeoff are more difficult to control for high compressor pressure ratio engines. The trend in large jet engines to higher compressor pressure ratios suggests that in the future the control of both NO_x and smoke may become a more challenging problem. It is difficult at this time to speculate on the possible influence of emission regulations on future aircraft engine designs. Much will depend on how methods required to reduce emissions affect other performance criteria. Several techniques that could be used to control emissions are discussed herein. These approaches are not necessarily final solutions, but are indicative of possible directions to take in reducing emissions without penalizing combustor performance.

REFERENCES

1. *Bristol, C. W., Jr., "Gas Turbine Engine Emission Characteristics and Future Outlook." Proceedings of the SAE-DOT Conference on Aircraft and the Environment. Part 1. SAE, 1971, pp. 84-92.*
2. *Sawyer, R. F., "Fundamental Processes Controlling the Air Pollution Emissions from Turbojet Engines." Paper 69-1040, AIAA, Oct. 1969.*
3. *Anon., "Smoke Emission Control." ESSO Air World, vol. 23, no. 1, 1970.*
4. *Grobman, Jack; Jones, Robert E.; Marek, Cecil J.; and Niedzwiecki, Richard W., "Combustion." Aircraft Propulsion. NASA SP-259, 1971, pp. 97-134.*
5. *Grobman, Jack and Papathakos, Leonidas C., "Smoke Evaluation of a Modified J-57 Combustor." NASA TM X-2236, 1971.*
6. *Briehl, Daniel and Papathakos, Leonidas, "Use of an Air-Assist Fuel Nozzle to Reduce Exhaust Emissions from a Gas-Turbine Combustor at Simulated Idle Conditions." NASA TN D-6404, 1971.*
7. *Briehl, Daniel; Papathakos, Leonidas; and Strancar, Richard, "Comparison of Exhaust Emission Measurements from a Gas Turbine Combustor at Varying Operating Conditions." Proposed NASA Technical Note.*
8. *Niedzwiecki, Richard W.; Juhasz, Albert J.; and Anderson, David N., "Performance of a Swirl-Can Primary Combustor to Outlet Temperatures of 3600°F (2256 K)." NASA TM X-52902, 1970.*
9. *Niedzwiecki, Richard W.; Trout, Arthur M.; and Gustke, Eric T., "Exhaust Emissions of a Swirl-Can Primary Combustor." Proposed NASA Technical Memorandum.*
10. *Norgren, Carl T., "Determination of Primary Zone Smoke Concentrations from Spectral Radiance Measurements in Gas Turbine Combustors." NASA TN D-6410, 1971.*
11. *Ingebo, Robert; Doskocil, Albert; and Norgren, Carl T., "High Pressure Performance of Combustor Segments Utilizing Pressure-Atomizing Fuel Nozzles and Air Swirlers for Primary-Zone Mixing." Proposed NASA TN.*
12. *Juhasz, Albert; and Holdeman, James, "Preliminary Investigation of Diffuser Wall Bleed to Control Combustor Inlet Airflow Distribution." NASA TN D-6435, 1971.*
13. *Butze, Helmut F. and Grobman, Jack, "Progress in Reducing Exhaust Pollutants from Jet Aircraft." Presented at NASA Aircraft Safety and Operating Problems Conference, Langley Research Center, May 4-6, 1971.*
14. *Saltzman, Bernard E., "Colorimetric Microdetermination of Nitrogen Dioxide in the Atmosphere." Anal. Chem., vol. 26, no. 12, Dec. 1954, pp. 1949-1955.*

15. Anon., "Aircraft Gas Turbine Exhaust Smoke Measurement." Aerospace Recommended Practice 1179, SAE, May 1970.
16. LeRoy, Milton J., Jr., "Evaluation of a Fluidic Oscillator as a Molecular-Weight Sensor for Gas Mixtures." NASA TM X-1698, 1968.
17. LeRoy, Milton J., Jr., and Gorland, Sol H., "Sensing Molecular Weights of Gases with a Fluidic Oscillator." NASA TM X-1939, 1970.
18. Childs, J. Howard; Reynolds, Thaine W.; and Graves, Charles C., "Relation of Turbojet and Ramjet Combustion Efficiency to Second-Order Reaction Kinetics and Fundamental Flame Speed." NACA Rep. 1334, 1957.
19. Cornelius, Walter and Wade, Wallace R., "The Formation and Control of Nitric Oxide in a Regenerative Gas Turbine Burner." Paper 700708, SAE, Sept. 1970.
20. Champagne, D. L., "Standard Measurement of Aircraft Turbine Engine Exhaust Smoke." Paper 71-GT-88, ASME, Mar. 1971.
21. Durrant, T., "The Control of Atmospheric Pollution from Gas Turbine Engines." Rolls-Royce J., no. 2, 1968, pp. 12-18.
22. Bahr, D. W.; Smith, J. R.; and Kenworthy, M. J., "Development of Low Smoke Emission Combustors for Large Aircraft Turbine Engines." Paper 69-493, AIAA, June 1969.
23. Durrant, T., "The Reduction of Smoke from Gas Turbine Engines." Aircraft Eng., vol. 41, no. 7, July 1969, pp. 28-31.
24. Faitani, J. J., "Smoke Reduction in Jet Engines Through Burner Design." Esso Air World, vol. 21, Sept.-Oct. 1968, pp. 34-41.
25. Gleason, J. G. and Faitani, J. J., "Smoke Abatement in Gas Turbine Engines Through Combustor Design." Paper 670200, SAE, Feb. 1967.
26. Toone, B., "A Review of Aero Engine Smoke Emission. Combustion in Advanced Gas Turbine Systems." I. E. Smith, ed., Pergamon Press, 1968, pp. 271-296.
27. Taylor, W. G.; Davis, F. F., Jr.; Decorso, S. M.; Hussey, C. E.; and Ambrose, M. J., "Reducing Smoke from Gas Turbines." Mech. Eng., vol. 90, no. 7, July 1968, pp. 29-35.
28. Linden, Lawrence H. and Heywood, John B., "Smoke Emissions from Jet Engines." Rep. 70-12, Massachusetts Inst. Tech., Oct. 1970.
29. Bagnetto, Lucien, "Smoke Abatement in Gas Turbines." Part II: Effects of Fuels, Additives, and Operating Conditions on Smoke Emissions and Flame Radiation. Rep. 5127-68, pt. 2, Phillips Petroleum Co., Sept. 1968. (Available from DDC as AD-842818.)

GENERAL DISCUSSION

A. H. Lefebvre

As far as aircraft engines are concerned, quite apart from the emissions aspect, the combustion situation has been getting more difficult all the time. The so-called "advanced technology" engines pose formidable problems due to the very high pressures and temperatures at which they operate. Since high pressures and temperatures are both conducive to combustion it might be argued that life should be getting easier. However, high pressures increase the liner buckling load and also enhance flame radiation leading to higher wall temperatures. The problem of wall cooling is further aggravated by the corresponding increase in inlet air temperature, so that more air is required for film-cooling. This in turn has an adverse effect on temperature traverse quality and gives rise to combustion inefficiency at altitude cruise. At the same time the problems associated with low pressures, such as altitude relighting capability are still as important as they have always been. Another significant factor is that compressor developments have led to higher and less uniform combustor inlet velocities, which have created difficult problems of air flow distribution and control, expecially in annular chambers. All things considered it is clear that emissions is a problem that we do not really need.

It is of interest to reflect that so far the combustion engineer has been able to cope with increasingly arduous performance requirements without introducing any radical change in combustor design. Superficially, present day combustion chambers do not differ very much from their ancestors of twenty years ago. I know there are many people, including myself, who feel that existing types have reached their useful limit and we should now look for and accept new concepts in combustor design.

Until recently, when we thought or talked about emissions, we were really mainly concerned with smoke. In this area our record of success has not been outstanding. For the last ten or twelve years we have claimed to know how to get rid of smoke, but visits to airports can still prove embarrassing. The truth is that although we know how to eliminate smoke the cure is sometimes worse than the disease. That is why the subject of exhaust smoke has figured prominently in our discussions this morning. But we have also addressed ourselves to the newer and more fashionable emissions such as NO, and I feel we have achieved a satisfactory balance between the various effects of fuel type, method of fuel injection, operating conditions and design variables on the emission levels produced by gas turbine engines.

G. L. Dugger *(Johns Hopkins University)*

During Session II, Mr. Zwick remarked that if you could operate very lean, you could reduce emissions. Prof. Sawyer has referred to this possibility also. At the Applied Physics Laboratory, we have performed some experiments in which turbulent bunsen-type flames were operated at equivalence ratios as low as 0.35. We did this by pre-heating the air and using the pre-heated air to vaporize the kerosine fuel so that we had a pre-vaporized pre-mixed mixture entering the bunsen-type flame. Unfortunately we made no emissions measurements. I believe this is an approach that deserves some serious consideration.

C. A. Amann *(Research Laboratories, GMC)*

Sawyer varied overall equivalence ratio by holding burner air pressure and fuel flow constant while varying airflow. In contrast, LaPointe and Schultz varied overall equivalence ratio by holding airflow constant while varying fuel flow. Minimizing combustor residence time was their stated objective. Arbitrarily selected operating schedules such as these may be desirable for research investigations, but it is important to recognize that in an actual engine neither the combustor pressure and fuel flow nor the residence time are necessarily constant.

Gas turbine engines fall broadly into two classes – those run at constant speed, typically single-shaft machines, and those run at variable compressor speed, typical of the two-shaft automotive type. The part-load operating conditions within the combustor differ markedly between these two types.

A conceptual expression for residence time, based on a plug-flow model, is represented by the following:

$$t_r = \int^{L_p} A \left(\frac{\bar{\rho}}{w}\right) dL$$

where: t_r = residence time

A = cross-sectional area of combustor

$\bar{\rho}$ = average density over the cross-section

w = mass flow rate

L = axial length coordinate

L_p = length of primary reaction zone

The curves in Fig. 1 illustrate calculated variations in residence time with load for non-regenerative versions of the two types of engines, each designed around the same full-load operating point. The calculations are based on the method of (1).

In both engines, equivalence ratio falls as engine output is reduced. In the constant speed engine, airflow rate stays approximately constant. Average density falls with decreasing load. The primary reaction zone shortens, and residence time falls off as shown by the lower curve.

In the variable speed engine, mass flow rate falls along with density as output is reduced. At idle the low burner inlet temperature has an adverse effect on

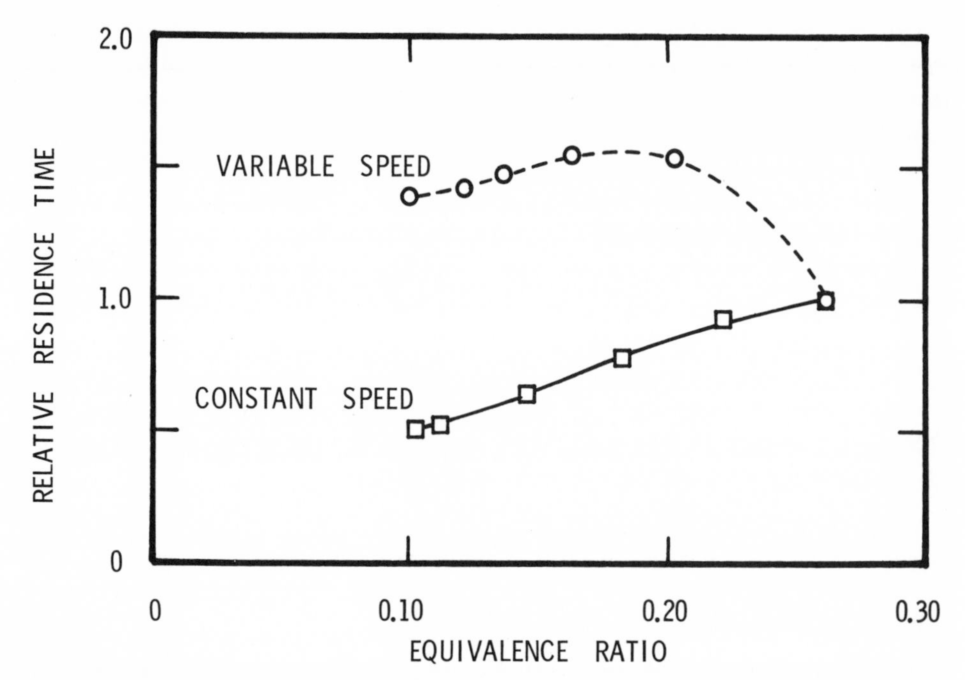

Fig. 1. Variation in residence time with load for non-regenerative versions of two types of gas turbines.

combustion, and the reaction zone lengthens. As a result, residence time increases as output is reduced.

The shapes of these curves are not universally representative. The combustor design details, turbomachinery characteristics, and design pressure ratio all have an influence. The addition of a heat exchanger to the cycle would have a major effect. The point is that as combustor investigations progress beyond the research stage and head toward engine application, it is important that combustor conditions be controlled to match those in the ultimate combustor environment.

1. *C. A. Amann, W. R. Wade and M. K. Yu, "Some Factors Affecting Gas Turbine Passenger Car Emissions," SAE Paper 720237, January 1971.*

W. R. Wade *(Research Laboratories, GMC)*

I would like to make several comments concerning the NASA swirl-can combustor described by J. Grobman. He indicated in his paper that the swirl can combustor showed significant reductions in nitric oxide emissions compared to a conventional combustor operating at similar conditions. This observation is particularly interesting to those of us who are involved in the development of a low emission burner for a vehicular powerplant.

This afternoon, I will report of General Motors' experience with conventional direct spray turbulent diffusion flame burners for vehicular gas turbine, steam, and Stirling engines. The nitric oxide emissions of these burners are significantly higher than required by the 1976 Federal emission standards. The diffusion flame burner contains local near stoichiometric regions in which the NO formation rate is high. On the other hand, a premixed burner should operate at a lean homogeneous fuel air mixture with low NO emissions.

Several years ago, the NASA Lewis Research Center began to report their work on carbureting swirl-can combustors designed to burn a premixed fuel and air mixture. At the GM Research Labs, we thought that their concepts had potential for low nitric oxide emissions. Since NASA did not report any emission measurements at that time, we fabricated and tested an experimental carbureted radial swirl can combustor, modeled closely after a NASA design for which high combustion efficiencies had been reported.

Shown in Fig. 1 is a photograph of this burner during operation at an inlet temperature of 100 F and a fuel air ratio of .013. Blue flame combustion was observed at most of the stable operating conditions. We are burning diesel No. 2 fuel here.

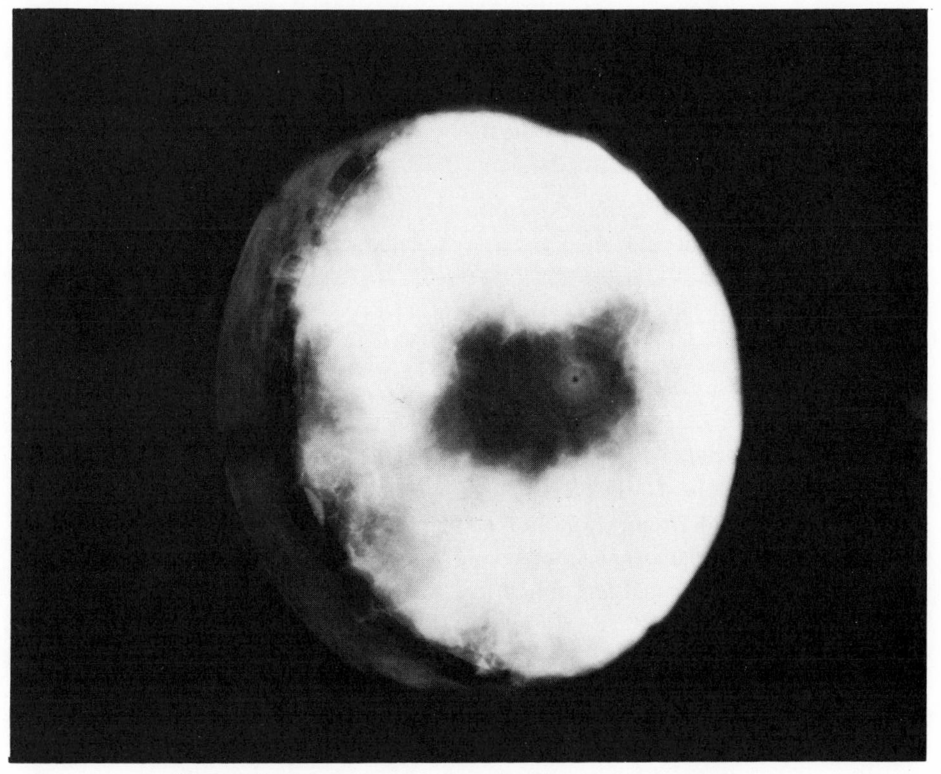

Fig. 1. Combustion in a carbureted radial swirl-can combustor.

In Fig. 2, I have shown the oxides of nitrogen emission index data which were measured from our carbureted radial swirl combustor. During operation at an inlet temperature of 600 F, our data agree fairly well with the NASA data for 120 modules of a slightly different design. I have also shown our NO emission measurements during operation at an inlet temperature of 100 F. Note, that by operating at sufficiently lean mixtures at an inlet temperature of 600 F, an NO emission index value equivalent to that obtained at 100 F can be realized.

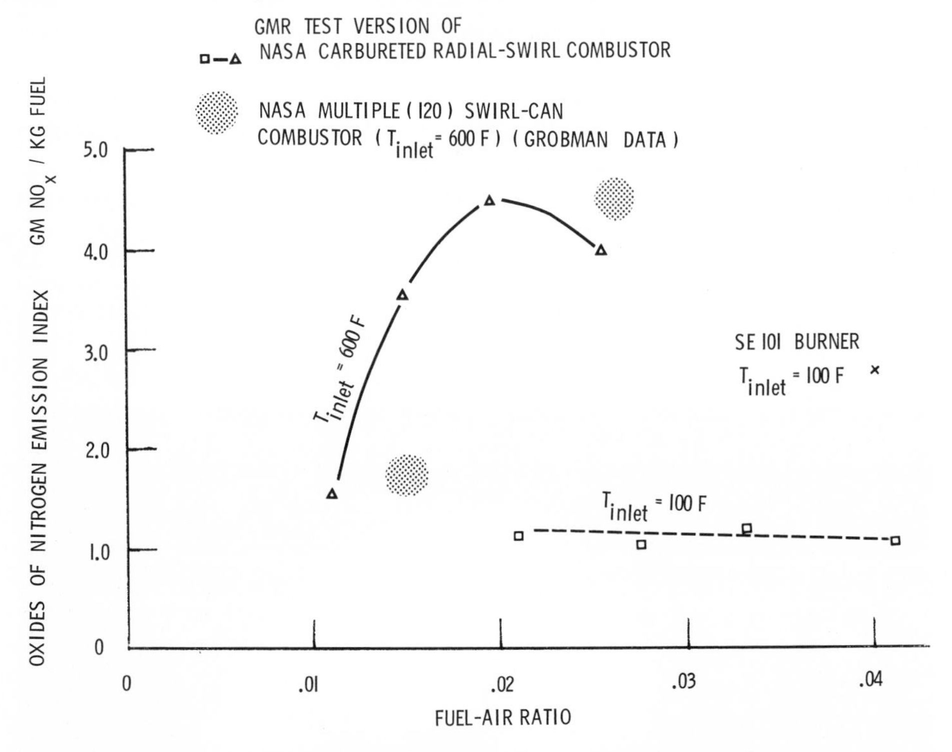

Fig. 2. Oxides of nitrogen emission indices for several swirl-can combustors.

To compare the data for the premixed swirl can combustors, I have shown the NO emission index of the GMR SE101 steam burner during operation at its design condition with an inlet temperature of 100 F. Nitric oxide emissions of the swirl can burner are 60 percent lower than for the diffusion flame burner. However, the NASA burner which we fabricated and tested had a much lower heat release rate than the SE101 burner.

In conclusion, it appears that a premixed burner may offer potential for low nitric oxide emissions. The NASA swirl can combustor is one practical method for achieving premixing. We would encourage continued emission studies of the NASA type swirl can combustor.

G. Opdyke, Jr. *(AVCO Lycoming Division)*

I just want to make a few comments about the correlation of pollutant concentrations with combustor efficiency that Dr. Lefebvre presented. In Fig. 1, I have taken his data and have plotted it against combustor efficiency – showing proportions of unburned hydrocarbons on the right and CO on the left. Lefebvre's data is the line in the center. I have compared it with the data given by Toone and also with the data by Odgers. We have a little data at Lycoming both from a can combustor and from an annular combustor which is compared as well. In general, all data seem to indicate that unburned hydrocarbons decrease as combustion efficiency increases, with fair correlation. In the combustion efficiency area of real interest – 97% and above, there is a good deal more scatter. However, the data there are all from engine operation, whereas much of the data on the left are from combustor laboratory tests.

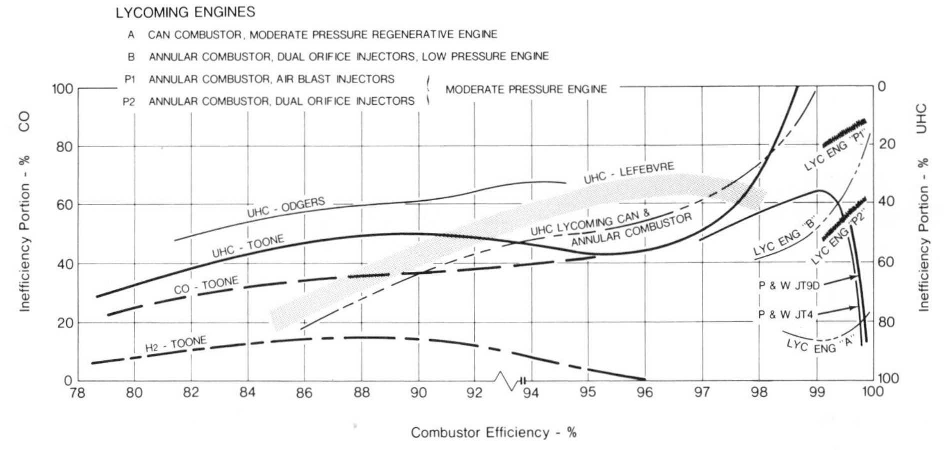

Fig. 1. Split of combustor inefficiency between carbon monoxide and unburned hydrocarbon emissions.

Here we have two different types of injection in the same annular combustor. Lycoming Engine P1 has an air blast atomizer and Engine P2 has the same combustor, but fitted with a dual orifice injection system. Lycoming Engine B is another dual orifice annular combustor which operates at somewhat lower pressures – in the order of 8 atmospheres. Engine P operates at pressure levels of 12 to 15 atmospheres.

Lycoming Engine A is a regenerative engine with a can combustor, and appears to have too much unburned hydrocarbon at high efficiency. The same thing is true of the Pratt and Whitney JT4 and JT9D engine data taken by the Los Angeles Air Pollution Control District.

There are explanations for some of the data scatter. In the regenerative engine for example, we know that CO is still being consumed at the end of the can. When the can is installed in the engine, the hot gas flows through a scroll before it reaches the turbine, and we assume that there is more CO consumption. Also, in this particular engine test, there was a minor amount of oil leaking from one of the bearing seals. Since this is a low air flow engine, about 10 lbs/sec, that can give a small increase in UHC. However, any oil leakage would have had little influence on the P. and W. data since these engines consume such large quantities of air.

So I suspect that while there is a fair correlation between combustor efficiency and chemical species found in the exhaust, differences in engines and combustor design will cause considerable scatter at very high efficiency levels.

References

1. *B. Toone, and F. Arkless, "The Application of Gas Analysis to Combustion Chamber Development", Seventh Symposium (International) on Combustion Butterworths, 1959, p. 929.*
2. *J. Odgers, "Air Pollution by Gas Turbines – Is Control Possible?", Canadian Aeronautics and Space Journal, October 1970.*
3. *R. E. George, et al, "Study of Jet Aircraft Emissions and Air Quality in the Vicinity of the Los Angeles International Airport", April 1971.*

T. Durrant *(Rolls Royce, Ltd.)*

I was very interested to see Mr. Opdyke's slide because I wondered if you, Prof. Lefebvre, had looked at a correlation between these two factors. You have expressed a correlation between the emission index, which we are all interested in, for both CO and possibly the fuel. I wondered if you had looked at this aspect and also the effect of the air-blast atomizer. I know that a later paper did in fact mention that it was mainly the unburned hydrocarbons that were attacked by the air spray burner. That was the one point I wanted to make – whether you have really looked at this factor. Having now looked at this curve I think I would tend to be rather surprised if there was a correlation because if you have a regenerative system you are operating at quite different conditions to obtain the inefficiency. In one case, one engine lies at one condition and another engine lies at another, and I wouldn't be at all surprised if you could not get a correlation.

I got the impression from one of Mr. Schirmer's curves that carbon production decreased with increasing pressure ratio. I wonder if he would elaborate on that.

A. H. Lefebvre

In the curves that we showed to illustrate the correlation between combustion inefficiency and CO and hydrocarbons we had simply plotted all the available engine data and, somewhat to our surprise, a very good correlation was obtained. Now we wouldn't want to press this too far because the observed relationship between combustion efficiency at exit and the concentrations of CO and hydrocarbons is to some extent coincidental. From an emissions viewpoint it is the primary-zone combustion efficiency that is relevant not the overall combustion efficiency.

It is appreciated that at altitude conditions any increase in overall combustion efficiency is accompanied by an increase in primary-zone efficiency, so it does not matter which efficiency is used in correlation. However, as the pressure is raised a range of conditions is encountered where the overall combustion efficiency remains close to a hundred percent, but the primary-zone combustion efficiency continues to increase. This is an area where one has to be specific in attempting to correlate emissions with combustion efficiency, and it must be the primary-zone combustion efficiency that is used. Our curves are based on overall combustion efficiency and are presented for interest and not as a near truth.

T. Durrant

Can we predice what is occurring in a combustor from measurements taken at the combustor exit? In other words, can we get some clues as to what is going on in the primary zone by looking at the proportions of CO and unburned hydrocarbon in the simple correlation that you have provided? Can we take it a little bit further, to give us a clue as to the mechanism?

E. R. Norster

Thank you very much, Mr. Opdyke, for your comments on the inefficiency problem. When one is trying to correlate the inefficiency against these measurements, one is often faced with a serious problem of inconsistent data. You are probably aware of the fact that measurements made in the pre-dilution zone or further upstream in the combustor vary from test to test and day to day. It is quite possible that the time average values in the post dilution zone will also be inconsistent. I am not surprised to see the lack of correlation in the high efficiency corner of your slide. To answer Mr. Durrant's question, I will emphasize that we took the data and from the very simple correlations shown in Figs. 5 and 9 we were agreeably surprised. CO measurements gave good correlation at high efficiency, and unburnt hydrocarbons correlated well at lower efficiency. These correlations are inherent irrespective of mechanism, combustor exit measurements can give little, if any, indication of mechanism.

R. M. Schirmer *(Phillips Research Center)*

The data in Fig. 2 of my paper show the effect of increasing combustor pressure on soot emissions, over a broad range of turbine inlet-gas temperature. It must be remembered that we are showing the soot content of the exhaust gas in this Figure, which is soot survival, and that it may not be a suitable index of soot formation. I have plotted the data on the basis of turbine inlet gas temperature because I feel this temperature is the more important controlling parameter for the survival of the soot in the exhaust gas. When you get above 2000-2600 F, you no longer have much soot emission. Fig. 4 of my paper shows the curves that we calculated to illustrate the performance of engines. The implication, of course, is that from a low-severity engine, there is quite a bit of soot. However as you go to very high pressure, and along with it to very high temperature, then soot emissions are reduced. What it really says is that perhaps the people who are now taking so much credit for producing a new series of engines that are smoke-free which are operating at very high pressures and temperatures – may be taking bows for a sort of technological accident.

E. K. Bastress *(Northern Research and Engineering Corp.)*

I will describe briefly the status of a program which we are just completing on controlling nitrogen oxides from aircraft gas turbines. The objective of our program is somewhat more worldly than some of those discussed here. It is concerned with criteria for designing combustors while minimizing nitrogen oxide emissions and yet accomplishing all of the other performance criteria which are required in aircraft engines.

The program consisted of three parts. One was the development of a model, which was described by Prof. Mellor. It was the model he referenced in the paper by Fletcher and Heywood. It is, in Prof. Mellor's terminology, a "statistical model" which entails one parameter, which is necessary to describe the distribution of fuel in the primary zone. The second part of the program had to do with experimental verification of the model which has been completed through a test program conducted at Cranfield Institute of Technology. As a partial answer to the question that Dr. Bartok asked yesterday, "Can these models be used to predict nitrogen oxide emissions?", we can say that the model can be used to successfully correlate emissions data from individual combustors. What is necessary of course is to empirically evaluate the one parameter which we need for describing fuel-air distribution. So, it is not possible at this point to predict emissions from new combustors unless some insight is obtained as to what the value of that parameter should be. Nevertheless, the model can be used quite successfully to correlate data from any given combustion chamber, and can be used to estimate emissions from new combustors by using values of the fuel distribution parameter derived from experience with existing combustors. Finally, the last part of the program dealt directly with the matter of design criteria. Here the first job was to relate the prediction of nitrogen oxides to design parameters which the combustion chamber designer has at his disposal. That is, how do you relate nitrogen oxide emissions to the design variables of the combustor such as size, pressure drop, and so on. Here again, we run into a problem of trying to relate the fuel distribution in the primary zone to things that we know about the combustor. The second part of this problem was to related other performance parameters such as combustion efficiency, pressure loss, ignitability and cooling requirements of the combustor to these same parameters. What we've been able to do in a somewhat qualitative way is to illustrate these relationships so that one can see what penalties he must be prepared to pay if he is going to reduce nitrogen oxide emissions from his combustor.

Our conclusion from this program is that if one is concerned about combustors of current design, the conventional design which Prof. Lefebvre says we have been using for 20 years, the outlook for reducing emissions is not very good because, in spite of the attractiveness of the lean, well-mixed primary zone approach, it is not practical in a system which must not only run at design conditions but has to idle, accelerate and decelerate. It is not possible to maintain those conditions (lean primary zone) throughout the entire operating range with combustors of current design. This work will be published sometime in about the next two months.

E. B. Zwick *(The Zwick Company)*

During the past year, all of our early measurements turned out to be wrong and had to be done over again. One of the things that came out this way was some of our hydrocarbon measurements. You can probe a burner with an individual probe and think that you know what your hydrocarbons are. Then when you flow the gas sample through a boiler or other heat exchanger which stirs the exhaust, and then measure the hydrocarbon concentration, you may obtain entirely different values. In determining why this was so, we found two things. One was that maldistribution of the fuel-air mixture could result in hydrocarbons existing at some point away from the locations you were probing. That is why you thought everything was so nice. The other thing which I think has considerable influence on burner hydrocarbon emissions is the temperature of the burner wall. With walls that were cold, we found hydrocarbon concentrations to be relatively high very close to the walls, falling off rapidly as the sampling probe was moved away from the wall.

Since it was difficult to sample close to the wall, I wondered if maybe the lack of correlation between the burner and engine emissions in the information just presented was the result of the stirring of the exhaust flow by the turbine. Pollutants near the burner walls that were not measured when the burner was probed would then show up in the engine exhaust. The only solution that I know of is to measure downstream of an actual engine.

H. R. Hazard *(Battelle Memorial Institute)*

At the Battelle Columbus Laboratories, we are developing a low-emission burner for automotive Rankine-cycle engines. Objectives include a turndown ratio of 100:1, a burning rate of 2,000,000 Btu per hr, total auxiliary power below 2 hp, and pollutant emission levels meeting 1976 Federal standards for light vehicles. This research is supported by the EPA Division of Advanced Automotive Power Systems Development under Contract EHS 70-117.

The design concept upon which the program was based was to provide nearly homogeneous combustion conditions by using a strong vortex flow pattern, ultrafine fuel atomization, and adequate pressure drop. With this approach, all requirements were met except that for NO_x emissions, which far exceeded research goals.

Accordingly, other approaches were then investigated, including the introduction of water-cooled coils and the use of staged-air admission with a rich primary zone and a lean secondary zone. Cooled, rich-primary-zone burners met NO_x requirements, although we have not yet completed the development.

An important aspect of the experimental data is that the real-world emission characteristics of these combustors, with their complex flow and mixing conditions, are not the same as those for the simple pre-mixed or diffusion laboratory burners discussed in earlier presentations at this Symposium. Thus, the mathematical modeling of complex combustion systems based upon the characteristics of small laboratory burners must be carried out with considerable caution to assure correct predictions.

Fig. 1 shows Configuration A, a "lean-primary-zone" burner, in which all fuel and air are well mixed in a single combustion zone. Air enters with a high level of swirl from a scroll, and superfine fuel atomization is provided by a sonic air atomizer which is a modified Hartman whistle. This configuration operated with a clear blue flame and met all emission and turndown goals except that for NO_x.

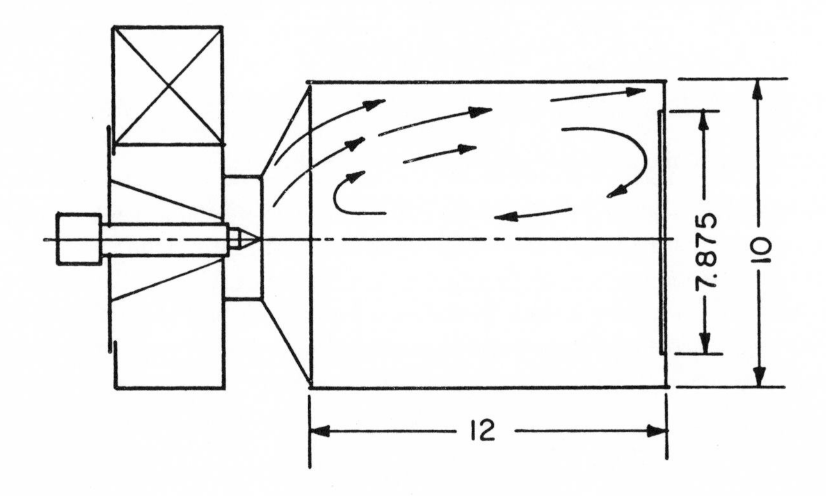

Fig. 1. Burner configuration A.

Fig. 2 shows NO_x emission for Configuration A plotted against percent of stoichiometric air, as measured with a constant fuel rate of 50 lb per hr and a variable air rate. The lowest NO_x emission levels occurred under fuel-rich conditions.

Fig. 3 shows NO_x emission data for Configurations A and B. Configuration B was similar to Configuration A except that water-cooled coils were used to remove about 10 percent of the heat released. The reduction of flame temperature resulted in a significant reduction of NO_x emission level. The NO_x emission goal, shown as a curve in Figure 3, meets 1976 Federal Standards. It is calculated on the basis of a vehicle making 10 miles per gal of fuel.

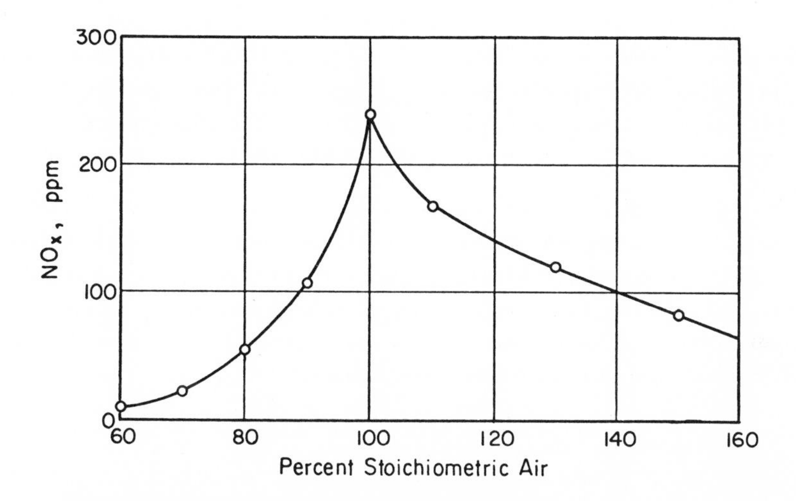

Fig. 2. Variation of NO_x with percent of stoichiometric air for configuration A.

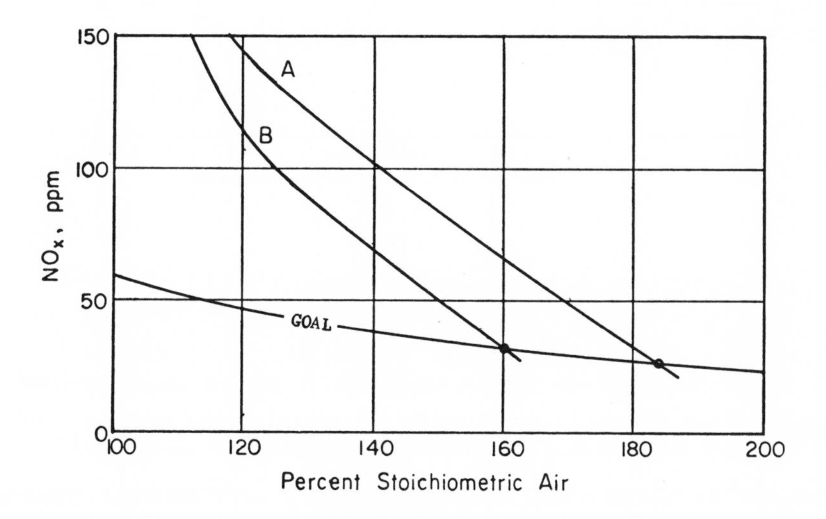

Fig. 3. Variation of NO_x with percent of stoichiometric air for configurations A and B.

Fig. 4 shows Configuration C-2, which incorporates staged air admission. A rich primary zone, in which little NO is formed, is followed by a lean secondary zone where air is added to complete the combustion and dilute the combustion products to a moderate temperature within a few millisec., thus limiting kinetically the amount of NO formed in the secondary zone. In the course of investigating configurations of this type the proportions of air to the primary and secondary zones were varied and it was found that optimum proportioning was necessary to minimize NO_x emission.

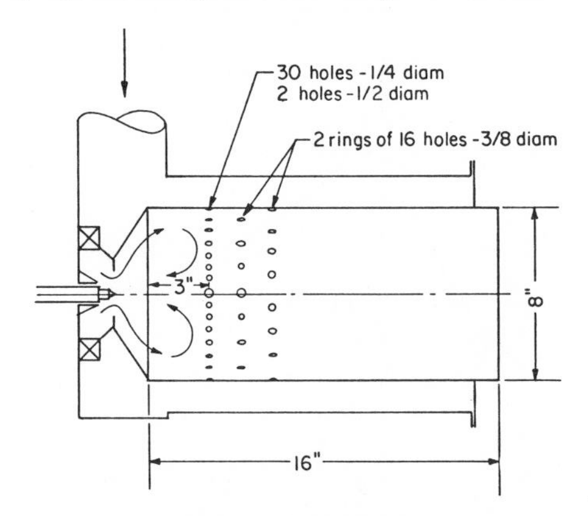

Fig. 4. Burner configuration C-2.

Fig. 5 shows NO_x emission characteristics of several rich-primary burners. In Configurations C-1 and C-2 about 60 percent of stoichiometric air entered the primary zone, resulting in less variation of NO_x with fuel-air ratio than for the lean-primary burner, Configuration A. However, with only 40 percent of the air entering the primary zone, in Configuration C-3, the emission characteristic approached that for Configuration A. Configuration D-1, also shown, was different in size and shape than Configuration C, and about 67 percent of the air entered the primary zone.

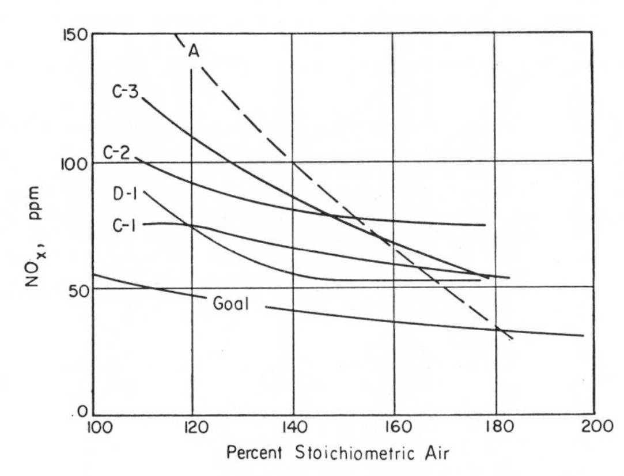

Fig. 5. Variation of NO_x with percent of stoichiometric air for configuration C.

Fig. 6 shows Configuration D, with both staged air admission and various degrees of water cooling. Configuration D-1 was uncooled, D-2 included only Coil Y, a cylindrical coil on the outer surface, D-3 included only Coil X, a conical coil on the burner end cone, and D-4 included both Coil Y and Coil Z. Coil Z included a cooled disk separating the primary and secondary zones and a 5-turn coil in the center of the primary zone.

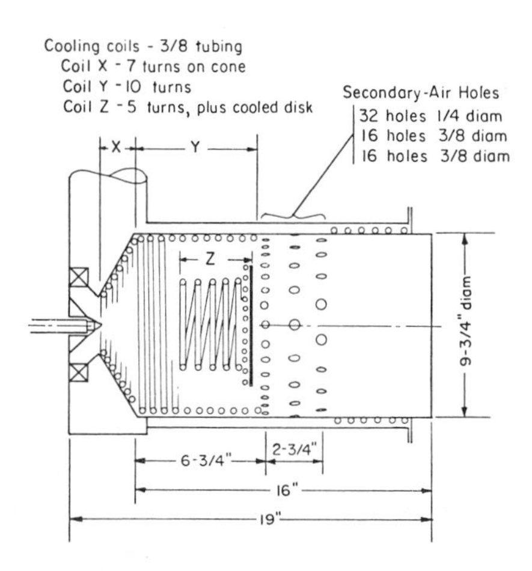

Fig. 6. Burner configuration D.

Fig. 7 shows NO_x emission data for Configuration D. All of the rich-primary burners emitted less NO_x than the lean-primary burners did, the level varying with the amount of cooling. It is of special interest that the smallest coil, Coil X, placed on the burner end cone, was more effective than the larger Coil Y in reducing NO_x emission. It is believed that the NO is formed in a small local region near the fuel nozzle, and that Coil X is most effective in cooling this region.

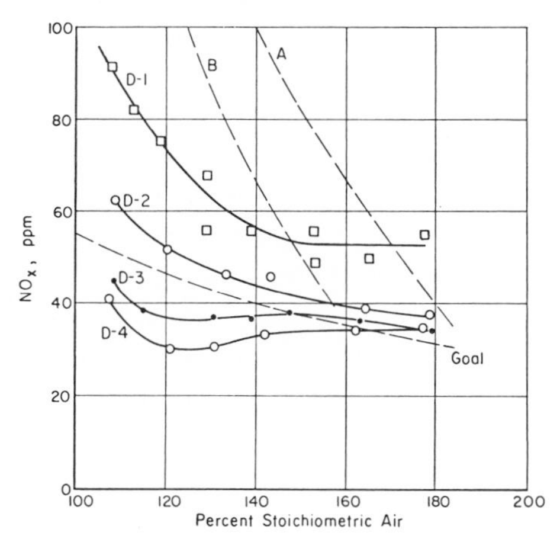

Fig. 7. Variation of NO_x with percent of stoichiometric air for configuration D.

Fig. 8 shows the variation of NO_x emission with fuel rate for three configurations. Data for these curves were obtained with nearly constant fuel-air ratio at fuel rates from 10 to 110 lb/hr, and residence time varied from about 10 to about 100 millisec. It is obvious that, under the conditions of test, the NO_x emission level was not influenced greatly by residence time.

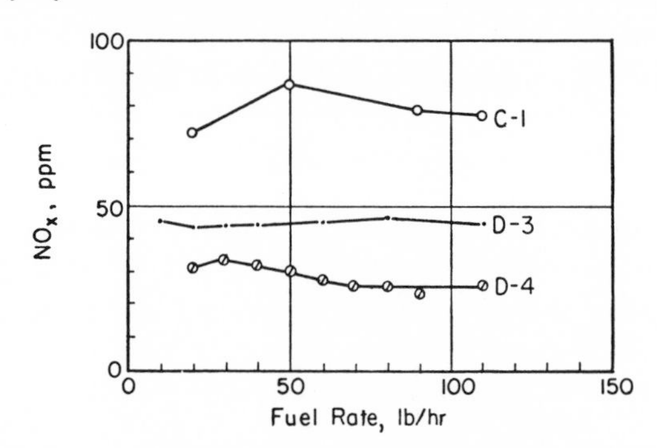

Fig. 8. Variation of NO_x with fuel rate.

From the data presented, the relative influence of flame temperature, residence time, and stoichiometry can be assessed to a certain extent. The effect of water cooling was to reduce NO_x emission in all burners. The rich-primary designs emitted less NO_x than the lean-primary designs when operated with mixtures with between 100 and 150 percent of stoichiometric air. For leaner mixtures, however, the lean-primary burners emitted less NO_x.

The effect of residence time upon NO_x emission level was similar for all experimental burners, with little variation over a range of residence times from 10 to 100 millisec. This characteristic does not agree with results from simple burners often used for fundamental kinetics studies, which show a relation between residence time and NO_x level. A possible explanation of the characteristics of the experimental burners is that the critical residence time for NO formation may be associated with a small region of high temperature and high oxygen-atom concentration, and that residence time within this region may not really change very much with firing rate. Fenimore(1) has shown that NO_x levels similar to those shown in Fig. 8 can be reached in less than 1 millisec. It appears quite probable that any level of turbulence, time, and temperature that will result in complete combustion can also provide the conditions necessary for significant NO_x formation. Thus, NO_x emission cannot be minimized by merely increasing combustion intensity to reduce residence time. Instead, it appears necessary to change the nature of combustion conditions to minimize a critical temperature, the oxygen-atom concentration, or the residence time within a very specific small region.

---

1. C. P. Fenimore, "Formation of Nitric Oxide in Premixed Hydrocarbon Flames", 13th Symposium (International) on Combustion, The Combustion Institute, 1971, p. 373.

J. P. Longwell *(Esso Research and Engineering Co.)*

Mr. Hazard, we really should take into account the effects and the possibilities of wall cooling. Our own experience, based on a more academic plane with well-stirred reactors, indicates that heat abstraction from the walls is a very important variable. The data that I showed in Session 1 included a substantial heat loss amounting to about 200°F. which reduces nitric oxide formation a great deal as we know. I think that with a clever design some advantage can be taken of this in practical systems.

A. H. Lefebvre

I recall an experiment by Mr. Durrant in which he demonstrated a twenty-fold reduction in hydrocarbon emissions by blanking off a small portion of the film-cooling air. I consider this to be highly significant and by no means a second order effect.

J. Moore *(General Electric Co.)*

I was interested in Mr. Grobman's remark that there was a lot of scatter in his nitric oxide data. Most of the flames considered at this Symposium burn in air, and I think it worthwhile to point out that scatter of nitric oxide data can result between flames burning in dry air or wet air. You can get a large reduction in nitric oxide formation by using moist air. In fact, you can get a large effect by injecting water if you do not inject too much into the fire. On an 80°F day, the air can contain as much as 2% by weight of water and according to my estimates(1) this amount of water is sufficient to reduce nitric oxide production by about 50%. Now, such a variation in the humidity of the air can contribute to the scatter in Mr. Grobman's data. I am not suggesting to Mr. Grobman that this is the reason, but it is a possible reason and when you're analyzing data it is important to include the effect of moisture in the air.

---

1. J. Moore, "The Effects of Atmospheric Moisture on Nitric Oxide Production", Combustion and Flame, Vol. 17, 1971, p. 265.

SESSION SUMMARY

A. H. LEFEBVRE
Cranfield Institute of Technology, Cranfield, England

An important conclusion from our discussions this morning is that undesirable exhaust emissions from gas turbine combustors can be effectively reduced by changes in fuel type, fuel injection method and liner geometry, and by increased primary zone homogenity. However, it is clear that a new approach to combustor design is essential if further substantial reductions in emission levels are to be attained without attendant penalties in other important aspects of combustion performance.

One possible approach is to use variable geometry as a means of regulating the amount of air entering the combustion zone. At high pressure conditions, large quantities of air would be injected to give low combustion temperatures and hence freedom from smoke and NO. At low pressure, this air flow would be partially blanked off to decrease the velocity and increase the fuel-air ratio in the primary zone; the net result being to raise the combustion efficiency and thereby reduce the exhaust concentrations of CO and unburnt hydrocarbons. An added advantage on aircraft systems is that the reduction in primary-zone velocity would also improve the altitude relighting performance. The main drawback of variable geometry devices is, of course, the complexity involved and for this reason they are not favoured by designers.

An alternative solution would be to replace the normal single combustion zone by two separate zones, each designed specifically to meet various performance requirements. The primary zone would be lightly loaded to ensure easy ignition and low emission levels at idling. The main combustion zone would be located downstream and have its own separate supply of fuel, which would preferably be pre-mixed.

Both variable geometry and multi-stage combustors appear to offer practical solutions to most of the combustion problems now being encountered, especially in regard to the control of undesirable emissions. They both involve increases in cost and complexity at a time when the combustion engineer is being pressed to produce systems that are simpler, more reliable and cost less.

Only time will show what innovations will be made to achieve these contrasting objectives, but there is considerable scope for ingenuity and we can look forward with interest to a new and exciting phase in gas turbine combustor design.

SESSION IV

POWERPLANT EMISSIONS

Session Chairman
P. S. MYERS

University of Wisconsin
Madison, Wisconsin

OPERATIONAL CONTROL OF THE NITRIC OXIDE EMISSIONS FROM STATIONARY BOILERS

B. P. BREEN

KVB Engineering, Inc., Tustin, California

ABSTRACT

Stationary boilers, such as those used in the utility industry, are a major combustor of fuel in any industrial nation. Besides being such an important part of industry, these boilers have a unique characteristic which is not available to other combustion devices; they are designed for heat removal or in more technical terms, they are inherently nonadiabatic within their combustion zone. This characteristic can be used to control the temperature history during combustion and thus to control pollutant emissions. In practice, such emission control involves understanding the combustion and pollutant formation mechanisms plus the ability to work on large scale heavy equipment. Plant operating changes have been accomplished with five major utility companies in California and New York and have led to 70 – 80% reductions in nitric oxide emission.

This paper will first develop an understanding of nitric oxide formation from the point of chemical kinetics, turbulent mixing, and the statistical nature of ignition/ combustion phenomena. Next, this understanding will lead to an exploration of the practical ways in which the combustion related temperature histories and resultant nitric oxide may be controlled, particularly by use of the nonadiabatic characteristic of boilers. The means of temperature and nitric oxide kinetic control which have been investigated include off-stoichiometric combustion, product gas recirculation, lowering air preheat, and water/steam injection.

Each method of combustion profile control will be then discussed from the point-of-view of application to large boiler systems, the limits of application, and the results which have been obtained. Within these results, a wide range of system parameters have been investigated and the effects of boiler size, firing pattern, excess

References p. 340

air operation, oil characteristics, and gas injection characteristics are explained. Basically, boiler construction requires the design of a radiant section for the loss of heat so that gases are below approximately 2400°F before entering convective tube sections. Because of this basic fact, parameters may be adjusted over a wide range of different boiler and fuel configurations so that essentially the same minimum nitric oxide operation may be achieved boiler-by-boiler with any of the available control techniques. Each nitric oxide control technique must be demonstrated to be long-term operationally feasible, to produce reduced combustible emission and to not interfere with plant efficiency.

A large amount of our work has been directed toward the extension of current plant operating data to predict new plant nitric oxide control. Our predictions are discussed in light of current and proposed pollution control regulations. The effects of different regulations in terms of ppm, lb./hr., and lb./MW or lb./Btu are shown in specific applications and the control techniques required to meet particular situations are discussed. Specifically, the way in which the wide variety of different regulations apply to actual plants is demonstrated.

Finally, the atmospheric nonadiabatic combustor is demonstrated to provide low emission characteristics in actual large scale operation through techniques which have wide range application. Future analytical modeling of the combustion processes and nitric oxide formation will point out further reductions and the equipment design direction which the nonadiabatic combustor must take. Such analytical modeling will strongly depend on an understanding of turbulence and flame holding/ignition processes.

INTRODUCTION

Nitric oxide (NO) forms at high temperature in an excess of air. At high temperature, the usually stable oxygen molecule, O_2, dissociates to the unstable oxygen atom, O, which is very reactive and attacks the otherwise stable nitrogen molecule, N_2. However, even at high temperature, if the mixture is fuel-rich the nitrogen can not compete with fuel for the scarce oxygen.

Also, the longer the mixture is at high temperature the more NO will be formed. Time is particularly important because the NO formation is slow compared to the residence time in high temperature gas zones, (i.e., the mixture is far from equilibrium). This is shown in Fig. 1. Therefore, the NO is directly proportional to the time in the primary combustion zone and this explains the usual increase of NO with increased heat loading of walls and with combustor size. This increase can be reversed by moving the combustion out of the primary zone.

In order to operate at low NO, one must first understand and control where the NO is formed; i.e., by understanding the time history of the turbulent gas eddies and by controlling the oxygen in these eddies. Then one must make this NO control technique fully compatible with good combustor operation by maintaining uniformly low CO levels and minimizing excess air requirements for the modified operation.

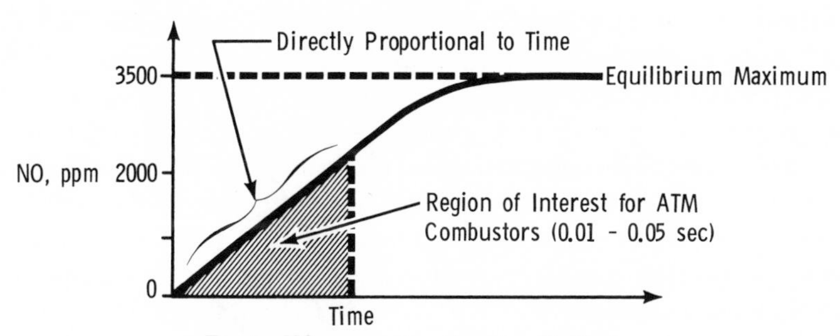

Fig. 1. NO vs. reaction time (qualitative).

NITRIC OXIDE KINETICS

Nitric oxide is formed at high temperatures in the presence of N_2 and O_2 by the following chemical reactions:

$$N_2 + O \rightleftarrows NO + N \tag{1}$$

$$N + O_2 \rightleftarrows NO + O \tag{2}$$

These reactions and the related reactions controlling temperature and O_2 and O species concentrations have been studied using thermochemical equilibrium and chemical kinetic digital computer programs (1, 2, 3). The analyses characterized the controlling influences of combustion stoichiometry and temperature and provided a basis for understanding NO formation in utility boilers. The most important kinetic analysis results are summarized in Figs. 2 and 3 which show respectively the effects of temperature and stoichiometry on NO formation. The rate of NO formation is a maximum for adiabatic, stoichiometric combustion and rapidly decreases with decreasing flame temperature and as the fuel/air ratio varies from stoichiometric.

TIME AND TEMPERATURE HISTORY OF A GAS EDDY

Many combustors operate with turbulent mixing combustion flames at the burner nozzle. (The characteristics of this type flame are vastly different from premixed flame characteristics). Turbulent eddies of air and fuel mix and are then ignited by the diffusion of excited atoms to form a primary combustion zone. The ignition process for hydrocarbons is almost instantaneous in comparison to the eddy life time.

The life time of an eddy within the primary combustion zone has been measured photographically to be between 0.01 and 0.05 seconds in typical atmospheric combustors. In this zone, little heat is lost because the lower temperature bulk gases (3000°F) have not had an opportunity to mix into the zone. (See Fig. 4). Thus, extremely high temperatures (adiabatic) are reached and this is where the vast majority of the NO is formed. (See Fig. 5). If more than 50 ppm of NO is formed, then the reaction will stop when bulk gas mixing occurs because this NO

References p. 340

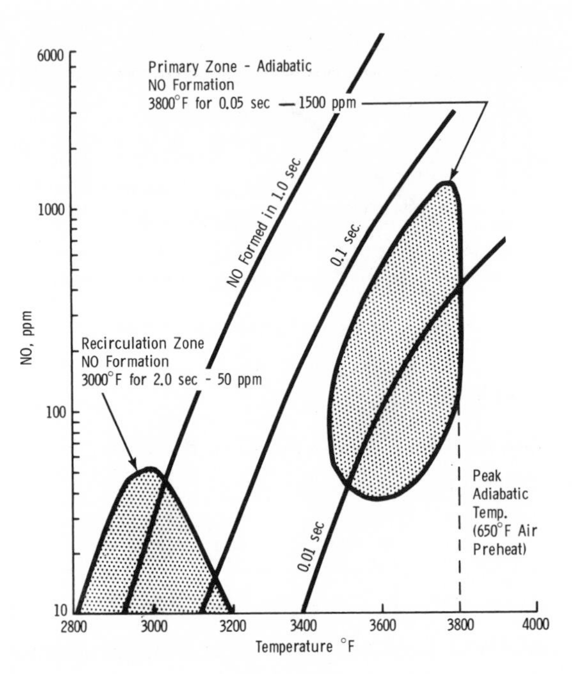

Fig. 2. Kinetic NO formation for combustion of natural gas at stoichiometric mixture ratio-atmospheric pressure.

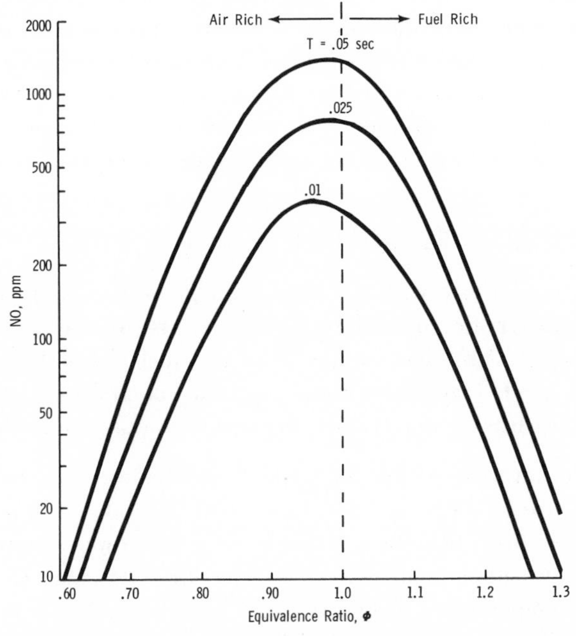

Fig. 3. Effect of equivalence ratio on kinetic NO concentration for various characteristic residence times, A/F stoichiometric = 16.3, air preheat = 650^0F, natural gas fuel.

concentration is generally greater than the equilibrium value at the bulk gas temperature.

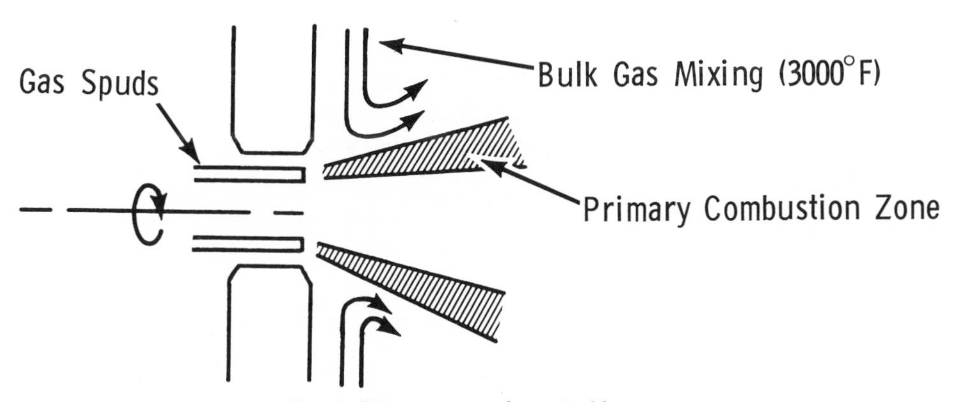

Fig. 4. Primary zone of a typical burner.

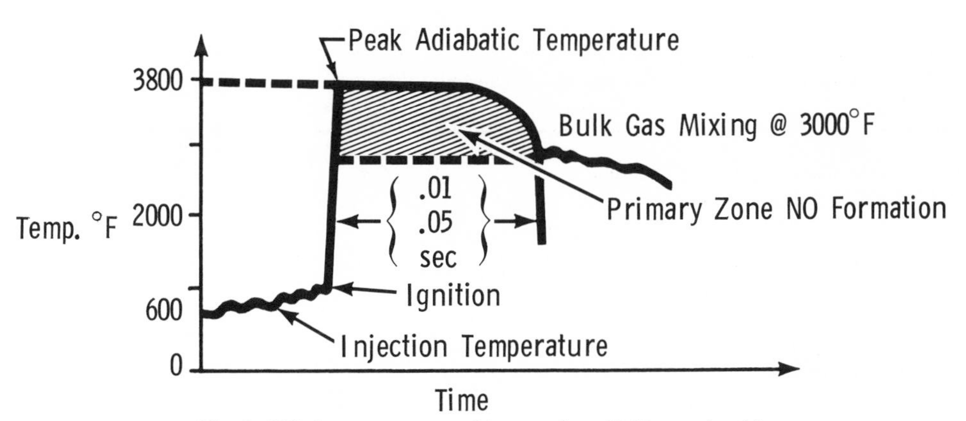

Fig. 5. Lifetime temperature history of a stoichiometric eddy.

In order for a turbulent eddy to ignite in free space, it must be very close to the stoichiometric air/fuel ratio. Thus the ignition eddy reaches a temperature of 3800°F and forms NO in the range of 1500 ppm. .This is shown in Fig. 2 and was derived through the kinetic analyses of (1, 2, 3). If the overall air/fuel ratio of the burner is within the flammability limits, all of the surrounding fuel will burn fuel-rich without going through this peak temperature excursion. Thus the eddies which ignite the surrounding fuel-rich mixture are the ones which generate NO.

These turbulent mixing and ignition phenomena which control NO are revealed in high-speed movies of gas, oil, and coal combustion. The movies show that in the primary combustion zone where adiabatic conditions can exist, *the combustion process takes place by gas/gas mixing.* Thus, droplet or particle combustion mechanisms are not the controlling influence in NO formation in the primary zone.

References p. 340

Models based on this physical picture of NO formation have been developed from the statistical turbulence point of view and from the classical point of view and are presented in Table 1.

TABLE 1.

Statistical and Classical Models of Nitric Oxide Formation in Recirculating Flow Burners

MODEL	CALCULATION	USEFULNESS
Statistical Ignition Eddies	$NO = \tau \int_0^\infty f(\phi_e, \phi_b)\, R(\phi_e)\, d\phi_e$	Determine effects of ϕ_b, ϕ_i, R. Correlate turbulent scale and ϕ_b
Bulk recirculating Combustion analysis with nitric oxide kinetics	Flow Field: $\phi, T, \bar{v} \Rightarrow NO(x)$	Correlate injection geometry, turbulence, nonadiabatic mixing, internal radiation.
Improved physical model of NO formation in bulk recirculation flow.	Flow field with correct physical turbulent/kinetic processes	More accurate prediction.

τ is the characteristic eddy lifetime
f is the number distribution of eddies, where
ϕ_e is the local stoichiometry of each turbulent eddy
ϕ_b is the average stoichiometry of the burner
ϕ_i is the fuel rich ignition limit
R is the rate of NO formation
T is the temperature
$\bar{v}$ is the velocity.

OPERATIONAL CONTROL OF NITRIC OXIDE

Utility boilers are inherently nonadiabatic in overall operation because their design is controlled by heat transfer rather than by combustion kinetics. This characteristic can be exploited by forcing the combustion process to occur at lower temperatures, in contrast to piston and turbine combustion devices; which are inherently adiabatic since combustion occurs at peak temperatures. The kinetics for the minimization of NO dictate that hydrocarbon combustion should proceed to completion only after substantial heat loss from the reacting fuel and combustion air.

With this understanding, it is possible to identify modified combustion techniques which will reduce NO but not compromise boiler operation. Two approaches, embracing several combustion control techniques, have been developed for reducing NO emissions. One approach, off-stoichiometric combustion, involves operating burners at fuel-rich mixture ratios. In this technique, the proportion of fuel burned at peak temperatures is reduced and consequently NO emissions are lowered. The remaining air required to maintain the overall furnace stoichiometry is introduced to complete combustion within the bulk gas. The second approach involves taking advantage of the strong temperature dependency of NO formation by reducing combustion temperatures. This can be accomplished in practice by several different techniques, all of which serve to lower the enthalpy per unit volume of the reactants. One technique involves recirculating product gases and mixing these gases with the combustion air. The product gases act as an inert absorbing a part of the energy released in combustion and thereby reduce the peak temperatures achieved. A second technique of lowering combustion temperatures is to simply reduce the combustion air preheat. The results obtained employing these techniques, the method of implementation and limitations are discussed below.

Off-Stoichiometric Combustion – Fuel-rich burner operation can be achieved in several ways. One technique, called "two-stage combustion" developed in the late 1950's by the Babcock & Wilcox Company and the Southern California Edison Company, involves diverting a portion of the air flow through air ports located above the burner pattern. The bypass air results in a deficiency of air at the burners and consequently fuel-rich burner operation is achieved. Two-stage combustion can be effective in NO control but it has several disadvantages. The size and number of ports fix the burner fuel-rich operation limit, and implementation requires modification of existing units and added cost of new units. Most importantly, the air ports do not provide flexibility in controlling carbon monoxide emissions as will be discussed.

A second technique is to terminate the fuel flow to selected burners and to utilize those burners as air ports while correspondingly increasing the fuel flow to the remaining burners. The same result of burner fuel-rich operation is achieved but without the disadvantages inherent in two-stage combustion. This latter approach has been most successful in controlling NO emissions.

Typical results obtained employing off-stoichiometric combustion with natural gas firing are shown in Fig. 6 (4). Fuel-rich burner operation was achieved utilizing burners out-of-service, (i.e., terminating fuel flow to selected burners) and in some cases "NO" ports. Measured NO emission levels are plotted against burner equivalence ratio (ratio of stoichiometric air/fuel ratio to actual air/fuel ratio) since this is the controlling parameter. The equivalence ratios for the data points were calculated based on the two parameters controlling the burner mixture ratio – furnace (overall) excess O_2 level and fraction of burners out-of-service. Normal operational NO emissions are indicated by data at equivalence ratios near 0.9. The data point at an equivalence ratio of 0.7 was obtained at a higher than normal excess O_2 level. The typical results shown illustrate the very significant reductions in NO emissions obtainable employing the technique of off-stoichiometric combustion.

Although NO emission levels with normal operating conditions cover a wide range, the data at maximum fuel-rich burner operation converge into a relatively narrow

References p. 340

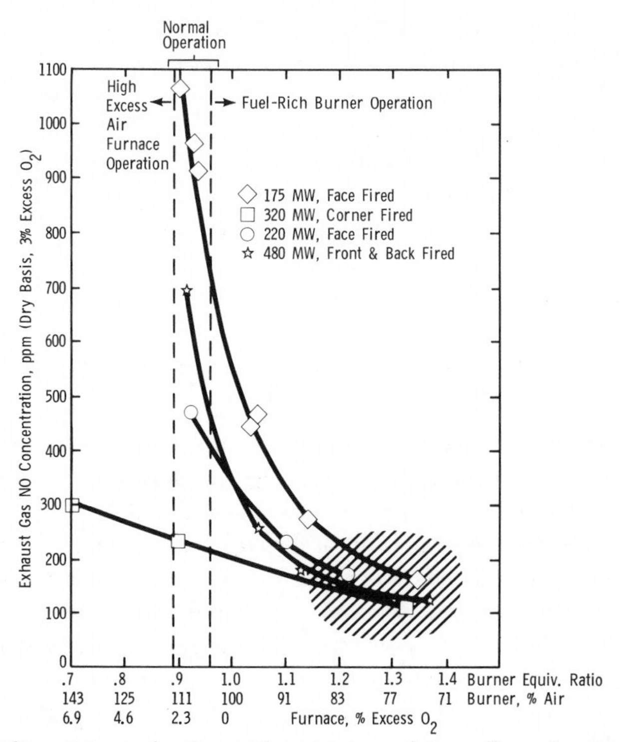

Fig. 6. Off-stoichiometric combustion with natural gas firing, effect of active burner fuel/air mixture ratio on exhaust gas NO concentration.

range of NO concentrations shown in the shaded zone in Fig. 6. The reason for this is that an increasing fraction of the NO is formed in the bulk gas as the fraction of fuel burned in the bulk gas increases. Therefore, NO concentration levels approach the bulk gas rate of NO formation of approximately 50 ppm as indicated in Fig. 2. The difference between this level and the concentrations achieved is due to NO still being formed in the flame holding region of the primary combustion zone.

The low NO emission data shown for the 320 MW corner-fired unit is characteristic of the type of burner utilized. The unit is designed to burn most of the fuel in the center of the furnace. Hence, the fraction of fuel burned and amount of NO formed in the primary combustion zone are relatively low.

These results, which are also applicable to oil firing, demonstrate the essence of the technique. The burner mixture ratio, which is dependent on both the burner excess O_2 level and fraction of out-of-service burners, controls NO formation. NO emissions are minimized for that combination of burner excess air and out-of-service burners which permits maximum fuel-rich burner operation. In practice, this operational point is achieved at minimum overall excess O_2 levels. Therefore, low overall excess air operation and two-stage combustion, which are sometimes thought of as independent techniques, are simply part of the off-stoichiometric combustion technique of operating burners at the maximum fuel-rich limit consistent with acceptable boiler operation.

The maximum degree of fuel-richness at which the burners can be operated, and hence the minimum level of NO achievable, is limited in practice by CO emissions with natural gas firing and by smoke with oil firing. If the burners are operated too fuel-rich on gas fuel, the CO concentration increases rapidly. When firing oil fuel, smoking is usually the operating limit. CO concentration and smoke can be minimized either by selecting out-of-service burner patterns which promote good bulk gas mixing, by adjusting burner air registers, or by increasing the overall excess O_2. The best patterns for controlling CO and smoke provide for maximum spacing between the air-only burners. This approach promotes maximum mixing in the bulk gas between oxygen from the air-only burners and unburned fuel from the fuel-rich burners. However, burners in the top row tend to form more NO than lower row burners and therefore the air-only burners must usually be located in the upper rows. As an example, the optimum burner pattern for low NO and CO determined during testing on a 215 MW unit was a zig-zag pattern of air-only burners in the top two burner rows. Increasing the excess O_2 reduces CO and smoke but is self-defeating since the degree of fuel-rich burner operation is reduced. The work of implementing off-stoichiometric combustion involves establishing for each unit that singular operational mode which permits maximum fuel-rich burner operation while still maintaining minimum CO and/or smoke concentrations consistent with safe, satisfactory furnace operation.

Flame Temperature Reduction Techniques – Primary zone combustion temperatures can be reduced to achieve a reduction in NO formation by several different techniques. Two of these techniques are product gas recirculation and reduced air preheat. Product gas recirculation has been demonstrated in power plant tests to be very effective. It is currently being included in the design of several new plants and is being considered for implementation on existing plants. The effectiveness of reduced air preheat has not been adequately demonstrated, but the test data available support kinetic analysis predictions which show that large reductions in NO emissions are obtainable employing this technique. The technique may be utilized in the future for NO control on particular types of plants where the economics favor this technique over gas recirculation.

Flue Gas Recirculation – Utility boilers often have the capability of recirculating product gases. This capability is included as an aid in controlling heat transfer to the boiler walls for oil fuel firing at reduced loads. In most boiler designs, the recirculated product gases enter through the bottom of the furnace. The effect is to reduce bulk gas temperatures but the influence on NO formation has been demonstrated in tests to be small. NO is formed in the primary combustion zone at peak temperatures and therefore, to be most effective, the product gases must be introduced into this zone.

References p. 340

Southern California Edison has several tangential corner-fired boilers which are unique in having the capability of mixing the recirculated product gases with a portion of the combustion air. Tests were performed on one such unit with natural gas fuel to demonstrate the reduction in NO emissions obtainable employing this technique. The test results obtained at full load operation are shown in Fig. 7. The data show a very substantial reduction in NO as the rate of gas recirculation is increased. Similar results were obtained at reduced load operation.

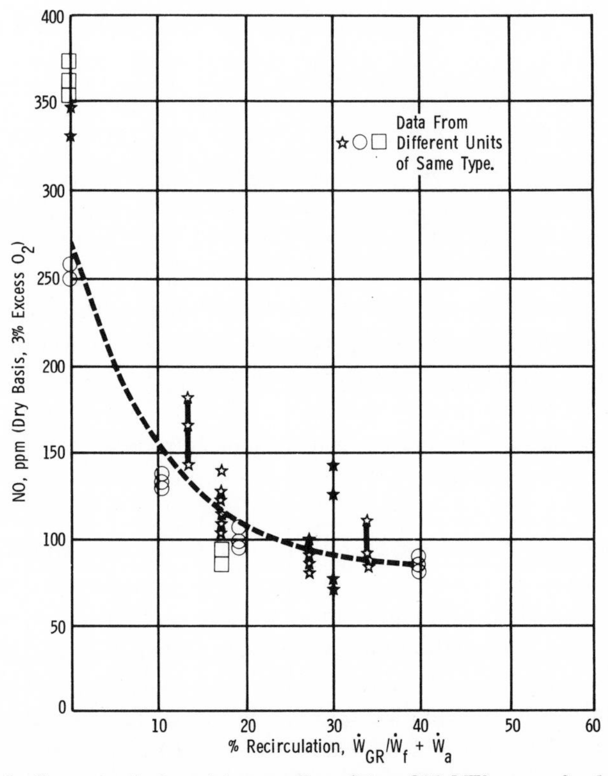

Fig. 7. Gas recirculation with natural gas firing; 320 MW corner-fired unit.

An NO reduction greater than that measured in the tests has been predicted and can be expected with optimum implementation of this technique. The fact that the recirculated gas products were mixed with only part of the combustion air means that some fraction of the fuel was still burned at the higher temperature characteristic of normal combustion. Maximum reductions would be obtained by mixing the gas products with the fuel although in practice this may not be possible to accomplish. Mixing the product gases with the combustion air is practical, however, although it will always require a modification to existing units. The most significant limitation of this technique is the resulting change in the distribution of heat transfer. Gas recirculation reduces heat transfer in the furnace and increases the energy transferred in the convective sections of the boiler. For a given boiler design, superheat and reheat steam temperature can increase excessively, as product gas recirculation is increased. This limitation can be extended somewhat by various operational procedures, but with existing units modified to mix product gases with the combustion air maximum recirculation rates will generally be limited to the 10 – 15% range. However, recirculation rates of this magnitude will result in NO reductions of 50 to 70% when this technique is utilized singly. Modification of the convection sections of existing units and designs incorporating gas recirculation in new units can potentially extend the gas recirculation limit for even greater reductions in NO. Reductions obtainable on oil firing utilizing this technique have been demonstrated, and similar results have been obtained.

Reduced Air Preheat – Utility boilers generally incorporate a heat exchanger to preheat the combustion air and improve unit thermal efficiency. The heat exchanger extracts heat from the boiler exhaust gas and preheats the combustion air to a temperature range of 400-700°F. Reducing the air preheat is a direct means of lowering primary combustion zone peak temperatures and NO formation.

Limited testing has been performed to demonstrate obtainable reductions in NO utilizing this technique. The test results are shown in Fig. 8. Although the data show a

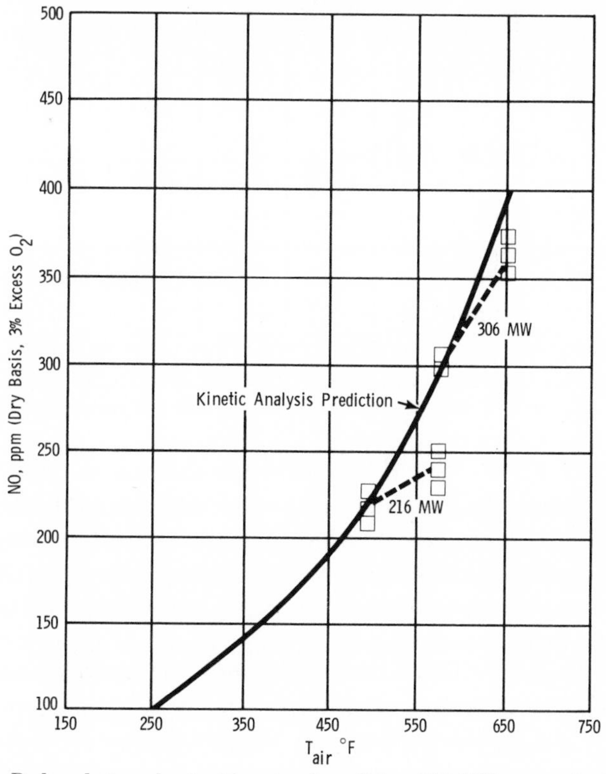

Fig. 8. Reduced air preheat with natural gas firing, 320 MW corner-fired unit.

References p. 340

decrease in NO concentration at reduced air preheat, the results are too limited to characterize the trend. However, the data are relatively consistent with the kinetic analysis prediction indicated by the solid curve and show that a 200°F reduction in air preheat will decrease NO concentration by 50%.

Operational limitations with this technique have not been established, but there is a basic disadvantage in that unit thermal efficiency is reduced. This is undesirable from an operational viewpoint and also detracts from the effectiveness of the technique since more fuel is consumed to generate the same electrical power. In future units, this limitation may be offset by careful design of the water-steam cycle.

Combination of Techniques – The techniques of off-stoichiometric combustion, gas recirculation, and reduced air preheat can be effectively utilized in combination. The reductions obtainable are complimentary since the respective methods reduce NO by different mechanisms. Off-stoichiometric combustion reduces the fraction of fuel burned at peak temperatures while gas recirculation and reduced air preheat simply reduce the level of peak temperatures.

Limited testing has been performed combining off-stoichiometric combustion with gas recirculation on a 320 MW corner-fired unit. Off-stoichiometric combustion (75% theoretical air) reduced NO emissions from 330 ppm to 110 ppm. When combined with 30-35% gas recirculation, the NO emissions were further reduced 50%. Unfortunately, the conditions necessary to achieve this NO level are incompatible with operational procedures.

The combined NO reductions were not additive combinations of the reductions obtainable individually, but the data are too limited to draw any conclusion on this point. However, the test results demonstrated that the techniques can be combined to achieve lower NO emission than is obtainable with the individual techniques. From an operational standpoint, the combination of techniques probably does not introduce additional limitations to those existent utilizing the techniques singly. However, much more testing is required to characterize the NO reductions obtainable employing the techniques in combination and to define operational and safety limitations.

IMPORTANCE OF BURNER UNIFORMITY

In light of the extra importance of excess O_2 at each fuel-rich burner, when air is being by-passed through "NO Ports" or out-of-service burners; the uniformity of the air/fuel ratio at each burner becomes predominant in the control of nitric oxide. One bad burner throat operating with a lack of air will cause the over-all excess O_2 to be increased to the point where this burner is within acceptable CO limits. However, the increased air supply in *each and every other burner* will now be producing excessive NO based on the CO production of this one bad burner.

This fact, that a single bad burner can control excess O_2 and thus the air and the subsequent NO product of every other burner can be seen from typical data shown in

Fig. 9 for a 24 burner combustor. The burners in the central right-hand side of the wind box received a poor distribution of air. These burners (10, 11, 16 & 17) produced a peak in CO measured in the exhaust duct. Although cross-mixing throughout the heat exchangers and duct work took place, this CO peak survived and is shown by the projected curve above the CO data points. To get air to these burners and to maintain CO within acceptable limits, the air throughout the complete wind box was increased. As a result, all of the units on the lefthand side received too much air; thus, causing the peak in the NO curve. These peaks would be much sharper if sampled in the flame zone. Although considerable mixing has taken place at the duct sampling point, these curves present a vivid description of how the CO at one burner can control the excess O_2 in the wind box as a whole. Thus, the most nonuniform burner becomes the controlling influence in the formation of nitric oxide. The situation is greatly exaggerated when a certain portion of burners are purposely removed from service.

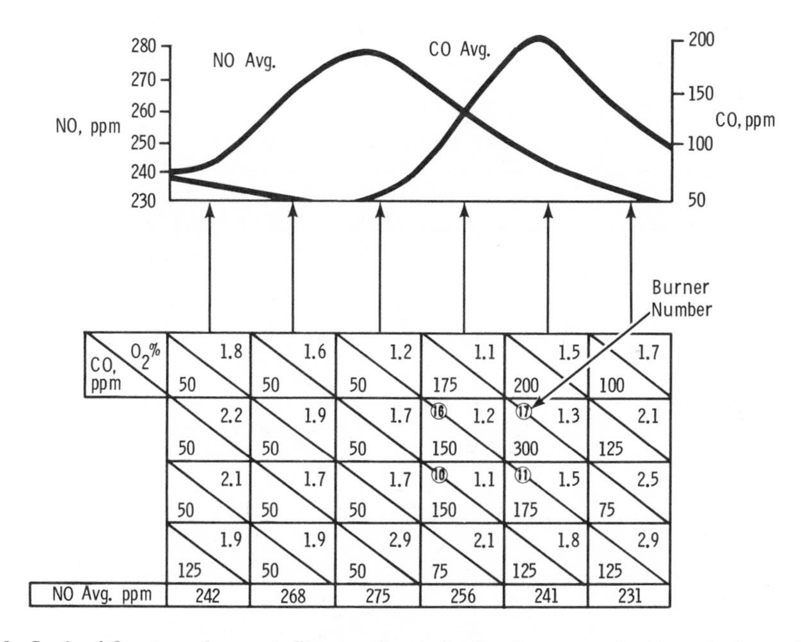

Fig. 9. Lack of O_2 at one burner influences O_2 at all other burners and ultimately determines the final NO.

APPLICATION OF NITRIC OXIDE REDUCTION TECHNIQUES TO NEW UNITS

Scattergood Steam Plant Unit 3 is being constructed by the Los Angeles City Department of Water and Power (Department). It will be the first large fossil-fuel installation to be constructed and operated under Los Angeles County Air Pollution Control District (APCD) Rule 67. This rule limits oxides of nitrogen (NO_x) emission

References p. 340

to 140 pounds-per-hour, calculated as nitrogen dioxide (NO_2). Since the permit to construct limits the fuel to natural gas, the 200 pound-per-hour limit on sulfur dioxide (SO_2) of Rule 67 is not a design factor.

Construction of Scattergood 3 was begun prior to adoption of Rule 67. However, to date, it has been held through court action that this unit does come under the requirements of the rule. Following this determination, a study was conducted based on recent test data from similar existing boilers. This study concluded that Scattergood 3 could operate under Rule 67 at 315 MW utilizing off-stoichiometric combustion, overfire air, and flue gas recirculation through the burner zone (adjacent to combustion air).

Estimated NO emissions versus load are presented in Fig. 10 and are compared with Rule 67 requirements. In predicting the NO emissions of Scattergood 3, extensive use was made of test data obtained on units very similar in construction (Southern California Edison Co. (SCE) 320 MW units), (5). The figure shows the individual effects of gas recirculation and off-stoichiometric (or overfire air) combustion, and

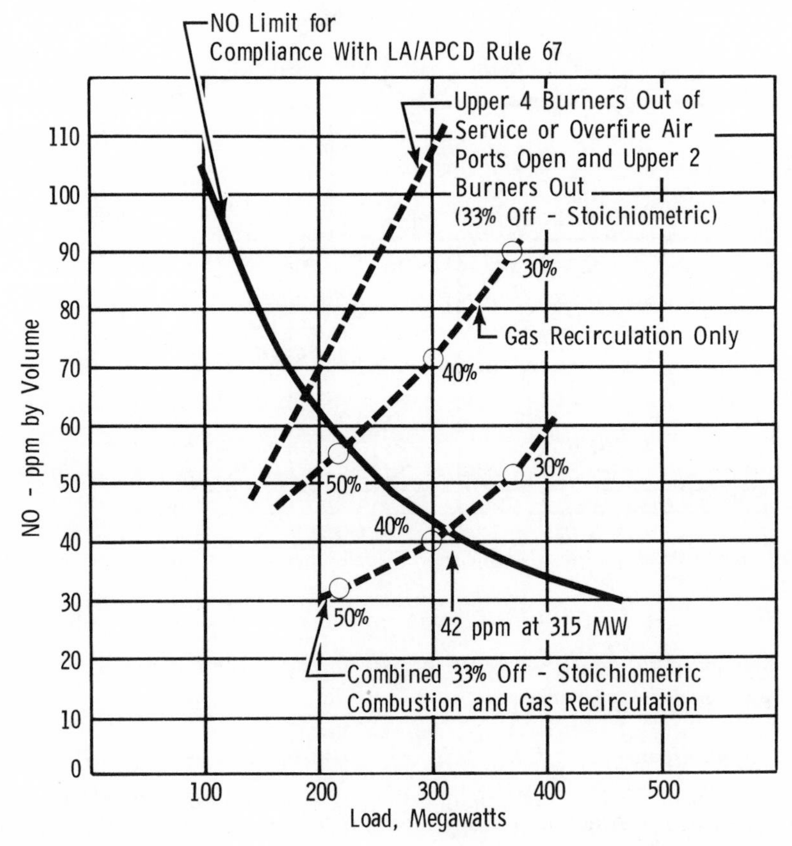

Fig. 10. Predicted NO emission vs. mode of operation as compared to Rule 67.

their combined effects, all compared against the requirements of Rule 67. From this, it was concluded that the Department's Scattergood 3 could operate within Rule 67 requirements

a. at 190 MW with overfire air only,

b. at 225 MW with gas recirculation only, and

c. at 315 MW with combined gas recirculation, overfire air, and a selected pattern of burners out-of-service in off-stoichiometric operation. The emission level at this load would be 42 ppm.

Objectives are to operate Scattergood 3 at 315 MW efficiently and in compliance with Rule 67. An extensive test program will be required in which the criteria are combustion stability, low NO concentration, high boiler efficiency, and ease of operation. These criteria must be continuously evaluated as the boiler is put in a condition of more extensive off-stoichiometric firing (percentage of burners operating on air only, plus overfire air) and a higher percentage of flue gas recirculation.

Similar tests have been conducted on Department, SCE, and San Diego Gas and Electric Company utility boilers. These tests have demonstrated both the combustion safety, and operational acceptance of methods hitherto unique to utility boilers. The implementation of these techniques has required that boiler operators be retrained to accept as normal very diffuse flames, stretching up toward and sometimes into the convective pass, as compared to the sharp, bright, short flames described in burner operating manuals.

Extensive test data on over forty utility boilers have been interpreted and applied to predict the NO emission of Scattergood 3 in light of operation within the Los Angeles County APCD Rule 67. The most pertinent data from six corner-fired 320 MW units led to the conclusion that with combined gas recirculation, overfire air, and burners out-of-service (off-stoichiometric operation), as shown in Fig. 10 Scattergood 3 at 315 MW will operate at 42 ppm of NO within Rule 67 requirements.

SUMMARY

The data presented demonstrate the effectiveness of off-stoichiometric combustion and reduced combustion temperature as techniques for controlling NO emissions from utility boilers. Off-stoichiometric combustion has been implemented on approximately 50 boilers and product gas recirculation has been implemented on units having a capability of recirculating the gas in the combustion air. Implementation of each of these techniques has resulted in a substantial reduction in total NO emissions.

References p. 340

ACKNOWLEDGEMENTS

Most of the work leading to the present paper was sponsored by the Southern California Edison Company. Also, I wish to acknowledge the support of the Los Angeles City Department of Water and Power, San Diego Gas & Electric Company, Pacific Gas And Electric Company, and the Consolidated Edison Company of New York.

REFERENCES

1. *A. W. Bell, N. Bayard de Volo, and B. P. Breen, "Nitric Oxide Reduction by Controlled Combustion Processes," Western States Section/Combustion Institute, April 1970.*
2. *F. A. Bagwell, K. E. Rosenthal, B. P. Breen, N. Bayard de Volo, and A. W. Bell, "Oxides of Nitrogen Emission Reduction Program for Oil and Gas Fired Utility Boilers," American Power Conference, April 1970.*
3. *A. W. Bell, N. Bayard de Volo, B. P. Breen, F. A. Bagwell, and K. E. Rosenthal, "Combustion Control for Elimination of Nitric Oxide Emission from Fossil Fuel Power Plants," Thirteenth Symposium (International) on Combustion, The Combustion Institute, 1971, p. 391.*
4. *F. A. Bagwell, K. E. Rosenthal, D. P. Teixeira, B. P. Breen, N. Bayard de Volo, and S. Kerho, "Utility Boiler Operating Modes for Reduced Nitric Oxide Emissions," Journal of the Air Pollution Control Association, Vol. 21, No. 11, November 1971.*
5. *H. H. Sonderling, W. W. Pepper, B. P. Breen, and A. W. Bell, "Operation of Scattergood Steam Plant Unit 3 Under Los Angeles County Air Pollution Control District Rule 67 for Nitrogen Oxides Emissions," 1971 Intersociety Energy Conversion Engineering Conference, Boston, Mass, August 1971.*

DISCUSSION

R. H. Essenhigh *(Pennsylvania State University)*

This paper by Dr. Breen is rather impressive although I think he has been a little provocative in some of his statements. For instance, he states, "Models based on the gas mixing picture of nitric oxide formation have been developed from the statistical turbulence point of view and from the practical point of view". That is all that he says about the very detailed fundamentals that the paper is based on, and most of the members of this audience would have appended many amplifying comments to such statements. This sort of comment will also frighten a lot of the plant engineers before the paper is read; but afterwards they will be pleased. I do not think that a person who has never worked in a power generation station will appreciate how nervous a plant superintendent can become when you try experiments with his boiler.

One can consider this paper on two levels. The first is the operational level, and the second is the interpretational level. We must not forget the individual who is operating a boiler that is producing NO_x in substantial amounts. He cannot wait for 10 years, or even until next year to reduce its NO_x emissions, but there are things that one can do operationally already. This paper actually represents just the tip of an informational iceberg while underneath it is a vast amount of published information describing at least three mathematical models, and these mathematical models are fairly detailed. However, there are some qualitative concepts now starting to emerge, and this present investigation shows that, if you have a situation to model that is not typical, at least you have the qualitative concept that you can start applying to the situation, with some chances of success in reducing the NO_x.

Now let me show you the information that is contained in this paper. Essentially, we find that NO_x decreases with 1) low loads, 2) fuel-riching firing, 3) staged combustion – essentially points 2 and 3 go together, 4) gas recycle (that is, taking the gas out and putting it back in again), 5) reduced air pre-heat – in a sense points 4 and 5 act together to reduce the temperature, 6) low excess air (and this may be a bit of a question mark, and 7) burner configuration. These all result in NO_x decrease.

Now, if you look at any boiler there are essentially only two types of firing systems that Dr. Breen has been investigating. The first is a type that incorporates a so-called self-stabilizing burner with a sort of small ball of flame close to the burner. This is where one can start arguing whether or not the ball really consists of pockets of flame. Or else you have another type of combustion system where in essence you have a large flame ball out in the boiler, where "the boiler is the burner".

Now, as the amounts of fuel and air are reduced for low load (item 1), in general the combustion intensity and the temperatures start to drop, and the result is a reduction in NO_x. It turns out that what you are really doing (with all seven parameters) is controlling the temperature and the time. With reduced firing there is a proportionately increased thermal load on the flame ball that is causing the temperature to decrease. In parallel, the reaction time increases, which makes the argument about time a little more difficult to work out. Now on the matter of fuel rich (point 2), if you're going to operate fuel rich, you have to stage the combustion process (point 3). Again, this is essentially a matter of dropping the temperature by effectively reduced firing rate into the primary flame zone. At the same time, particularly with coal as Dr. Longwell pointed out, as a consequence of fuel rich firing you are then more likely to separate out the nitrogen in the fuel as nitrogen and, if you can keep the subsequent temperatures down, this nitrogen will not be oxidized as it flows through the boiler. Now again, on points 4 and 5, gas recycle and reduced air pre-heat act together to reduce the temperature.

Low excess air is an interesting point because it is usually thought that NO_x decreases as the excess air is reduced. In an excellent paper that was presented by Dr. Bartok a month ago at the Atlantic City A.I. Chem. E. meeting, he showed that NO_x usually peaks near the stoichiometric on the lean side (as also indicated by Dr. Breen), but yet in boilers the NO_x was reduced by low excess air firing. However, Dr. Sarofim also pointed out at that same meeting that NO_x peaks at around stoichiometric in the case of a pre-mixed flame while he showed that for a diffusion flame the NO_x peak occurs possibly about 50% excess air. It is therefore important to know whether, in a boiler, you have a pre-mixed flame or a diffusion flame. I will summarize Dr. Sarofim's argument. (He is talking about a laminar diffusion flame, not a turbulent diffusion flame.) In the case of the diffusion flame, regarding it as a "thin flame" system, the temperatures at the flame surface are about the same as for a stoichiometric pre-mixed system, but the flame surface area is larger. In essence, the NO_x formation increased with increasing surface area of the flame. Now this concept may be applied to Dr. Breen's system. Does he have a turbulent diffusion flame? The author says yes. In that case, you don't have a thin flame surface because of the turbulence, but you could have a fairly narrow volume region of·stoichiometric "pockets" (if you like), so you could expect the NO_x to peak at high excess air, as Sarofim showed, not near stoichiometric.

However, another way to look at the problem is in regard to burner configuration. This is the difference between a single jet that must rely largely on turbulence from shear due to velocity gradients for nearly all the mixing that is needed, on the one hand; and multiple interacting jets that disperse most of the fresh fuel and oxidant throughout a "flame ball" region by forced convecting streams, with turbulence playing a secondary role to promote final cross-mixing between streams, on the other hand. Boilers generally have multiple streams and burners. This is also where one can start to argue about the real existence of the flame pockets. One can certainly always see something that can be interpreted as a pocket. However, this is in the visible region and it may be only in the eye of the beholder. If you looked at a large flame in the infrared region, you might get a different answer. Also it's a matter of how close you are to the flame. If you are close to it, you see a lot of fine structure which may be misleading. If you move back a distance, you finally say, "I cannot see the details and maybe I can treat it as a completely uniform volume." So in other words, you might have a chance of treating it as a stirred reactor region and in this case you could have a much smaller flame volume and higher combustion intensity. This might be too simple a model, but no one knows yet, so why put in complexity till you know it is necessary? This matter of a model is something that could be quite critical also in regard to flame stability, which is one of the concerns of the station superintendent, particularly if you are doing experiments on his boiler.

One reason why I raise this question of stirred reactors and combustion intensity is that a paper by Wasser, also given at the Atlantic City meeting in September 1971, showed the NO_x as a function of swirl, and displayed a rather intriguing pattern similar to that shown by Dr. Breen. The author said this pattern resembled a little what happended to smoke emissions from a heavy fuel oil burner reported by Drake and Hubbard (J. Inst. Fuel *39,* 98, 1966). In those experiments, that were conducted about 6 or 7 years ago, they were able to show that (a) if you treat the combustion chamber as being the classical pattern (Bragg Configuration) of a perfectly stirred reactor followed by a plug flow section, and (b) if you then measure the relative proportions of these sections, you will find that the size of the stirred section is first of all decreasing and then increasing again and the smoke emissions followed this variation. Wasser also measured the effective volumes of these two regions (stirred and plug flow) and found that the variation of the residence time of the gas in the stirred section was in fact the inverse of the variation of NO_x. The NO_x also increased exponentially with combustion intensity. Of course, the time factor is only one of the factors involved. But overall this would seem to suggest from the point of view of the qualitative concepts, and particularly if you like the detailed models that have been used for calculation, sufficient information is already available about NO_x formation for many purposes.

So where do we go from here? In many practical circumstances, we do not now require more knowledge about NO_x formation. Instead, we need to investigate means of reducing NO_x emissions by system design. In many cases, you do not have to calculate very accurately: an emission standard is a different animal from a precision design standard; it represents a inequality, not an equality, and you only have to calculate to a sufficient accuracy to be certain of an emission rate that is less than a maximum (legal) limit. It all comes back very largely to an interpretation of what is going on in a boiler flame. Do you really get these tremendous temperature excursions because of the little pockets? If such temperature excursions occur throughout the system, it means that more attention is needed to provide uniform mixing. So I suggest that what we need now is more attention to mixing (and perhaps at the same time to smoke, which I think is very closely associated) with emphasis in the mixing on getting it as uniform as possible. The NO_x formation problem has now had a good workout, and although there are many detailed problems remaining and some work must continue, it does seem that the time has come now to move on to the next problem.

B. P. Breen

I would like to thank Professor Essenhigh for his interpretative comments and for the opportunity, between himself and myself, to emphasize certain important points which he has brought up.

First of all, the operation of a utility boiler involves complex equipment interaction and safety consciousness. Added to the system complexities are the combustion and the nitric oxide formation complexities. Basically, all the classical combustion problems of fuel/air injection, mixing and ignition are involved in stable flame-holding, load-following and combustion efficiency considerations. However, in the case of nitric oxide formation I have tried, in a survey type paper, to point out the importance of the adiabatic temperature excursions of stoichiometric and near stoichiometric (slightly fuel-lean) eddies. These adiabatic excursions occur in the near injection region of the boiler before significant overall (bulk) heat loss has taken place and the details of the turbulence, stoichiometry and ignition of these eddies determine the worst nitric oxide formation. Nitric oxide can be formed to a lesser extent in other bulk gas regions but it is the percentage of fuel which combusts in the adiabatic regions which is of major significance.

This fact is shown in Figure 6 of my paper where dramatic nitric oxide reductions are presented versus fuel-richness *at the burner throat* for actual power plant operation. Thus, in this one curve is shown the effects of Professor Essenhigh's items: 2.) fuel-richness, 3.) staged combustion and 6.) low excess air. All of these techniques reduce to one consideration; i.e., what is the fuel-richness at the adiabatic throat condition? The difference in the techniques is only realized in practice when they are used along with item 7.), burner configuration, to control CO and smoke formation.

Secondly, I would like to agree with Professor Essenhigh on the point that the equivalence ratio location of peak nitric oxide can be explained by the difference between premixed and turbulently mixing diffusion flames. However, this difference points out the importance of looking at mixing details rather than stirred reactor models. At low loads, the decrease in nitric oxide formation is due to a decrease in mixing and turbulence in the adiabatic zones. The increase in nitric oxide at full load corresponds to the increased turbulence intensity through the constant diameter throat. As the title of my paper indicates, "Operational Control of Nitric Oxide," we are not so much concerned – where the *peaks* occur – as we are with the operational definition of where the *minimum* nitric oxide formation occurs, consistent with low CO and smoke formation.

CONTROL AND REDUCTION OF AIRCRAFT TURBINE ENGINE EXHAUST EMISSIONS

D. W. BAHR

General Electric Company, Lynn, Massachusetts/Cincinnati, Ohio

ABSTRACT

The aircraft turbine engine exhaust emissions in the category of air pollutants, that are generated to some degree, consist of carbon monoxide (CO), unburned or partially oxidized hydrocarbons (H/C's), carbon particulates as soot or smoke, oxides of nitrogen (NO_x) and sulfur oxides (SO_x). The primary concern associated with these emissions is their possible impact on the environments of major airport localities, where the exhaust emissions resulting from high volumes of localized aircraft operations may tend to be concentrated.

To minimize any adverse effects on the environments of airport localities, significant development efforts have been conducted and are underway within the industry and government to provide technology for the control and reduction of the levels of any objectionable emissions. To date, extensive engine evaluations have been conducted to determine the exhaust emissions characteristics of both production and development aircraft turbine engines. Significant progress has also been made in the development of technology for the design of engine combustors with reduced smoke emission levels. As a result of these latter efforts, combustors with virtually non-visible smoke emission levels have been developed and are being placed into service.

More recently, some promising results have been obtained in efforts to develop improved fuel atomization and primary zone stoichiometry control methods for reducing CO and H/C's emissions levels. Essentially, these investigations involve the development of methods for improving combustion efficiency performance at the low power operating conditions. It is anticipated that, with sufficient development, these efforts will result in the definition of combustor designs with improved low power

References p. 372

combustion efficiency performance and, therefore, that future engines will have significantly lower CO and H/C's emissions levels than those of current technology engines.

Because the NO_x emissions characteristics of any given engine are directly and strongly related to its combustor inlet air temperature levels, obtaining significant reductions in the levels of these emissions by combustor design techniques appears to be a more difficult task. Recent investigations have shown that some degree of suppression is attainable by design approaches which involve improved control of the primary combustion zone stoichiometry and gas residence time. Also, these investigations have shown that reductions in the levels of these emissions are attainable with the use of water injection into the combustor. However, extensive additional research and development efforts to provide NO_x emissions abatement technology for use in turbine engines appear to be needed.

INTRODUCTION

Within recent years, the number of turbine engine-powered aircraft in both commercial and military service has increased at an extremely rapid rate. This rapidly increasing usage of turbine engine-powered aircraft has logically resulted in increased interest in assessing the possible contributions of aircraft turbine engines to the air pollution problems confronting many metropolitan areas throughout the nation. Several studies, to evalute the extent of any air pollution resulting from the operation of aircraft turbine engines have been conducted. (1)

These studies have shown that the overall contributions of aircraft turbine engines to the air pollution problems of the nation and the world are quite small, as compared to those of other types of contributors. These studies have also shown that the exhausts of aircraft turbine engines generally contain very low concentrations of gaseous and particulate emissions considered to be in the category of air pollutants. These typically low exhaust concentrations of objectionable emissions are due to the continuous, well-controlled and highly-efficient nature of the combustion processes in turbine engines and to the use of fuels which contain very small quantities of inpurities.

Nonetheless, even though relatively low concentrations and total amounts are generated in most instances, the exhaust emissions in the category of air pollutants resulting from the operations of aircraft turbine engines are of possible concern. The specific emissions of possible concern from an air pollution standpoint, which are present to some degree in the exhaust gases of aircraft turbine engines, are carbon monoxide (CO), unburned or partially oxidized hydrocarbons (H/C's), carbon particulates as soot or smoke (C), oxides of nitrogen (NO_x), sulfur oxides (SO_x) and extremely minute traces of various metal compounds, chiefly metal oxides. The levels of these latter two categories of emissions are, however, normally very low at all engine operating conditions, because aircraft turbine engine fuels contain only very small quantities of sulfur compounds and other inorganic impurities.

The levels of the CO, H/C's, smoke particulates and NO_x emissions that are produced vary considerably with different engine operating conditions and engine types. Generally, the highest levels of these various emissions are generated at some specific engine operating conditions that typically occur in and around airports. At cruise operating conditions, the levels of these various emissions are normally very low. Also, at cruise operating conditions, any objectionable emissions that are generated are dispersed over wide geographic areas and generally at high altitudes. Thus, the total amounts of objectionable emissions resulting from aircraft cruise operations are relatively small and those that are generated are normally distributed widely in the atmosphere. For these reasons, the exhaust emissions resulting from aircraft cruise operations are, in general, not of concern. Thus, the only significant concern associated with aircraft turbine engine exhaust emissions involves their possible impact on the environments in and around busy metropolitan airport localities, where the exhaust emissions resulting from large volumes of aircraft flight and ground operations may tend to be concentrated.

To minimize any adverse effects on the environments of airport localities, significant development efforts have already been conducted within the aircraft turbine engine industry to develop technology for the control and reduction of the levels of these objectionable exhaust emissions. In particular, major efforts have been conducted to develop technology for the design of aircraft turbine engine combustors with low smoke particulates emission levels. As a result of these development efforts, combustors with virtually non-visible smoke emission levels have been designed and placed in service. Efforts are now underway to develop technology for the design of aircraft turbine engines with reduced CO, H/C's and NO_x emission levels. Efforts of this kind have been and are being carried out at General Electric.

This paper contains a summary of the results of these General Electric investigations and programs to determine the exhaust emissions characteristics of aircraft turbine engines, to ascertain the causes and sources of these exhaust emissions and to develop technology for the control and reduction of the levels of the objectionable exhaust emissions. Also included are discussions of the possible further reductions in the levels of these aircraft turbine engine exhaust emissions of concern – which may be attainable in the future with additional turbine engine combustor development efforts.

EXHAUST EMISSIONS CHARACTERISTICS OF AIRCRAFT TURBINE ENGINES

A typical illustration of the exhaust emissions characteristics of current operational aircraft turbine engines is presented in Fig. 1. As is illustrated in this figure, the peak

References p. 372

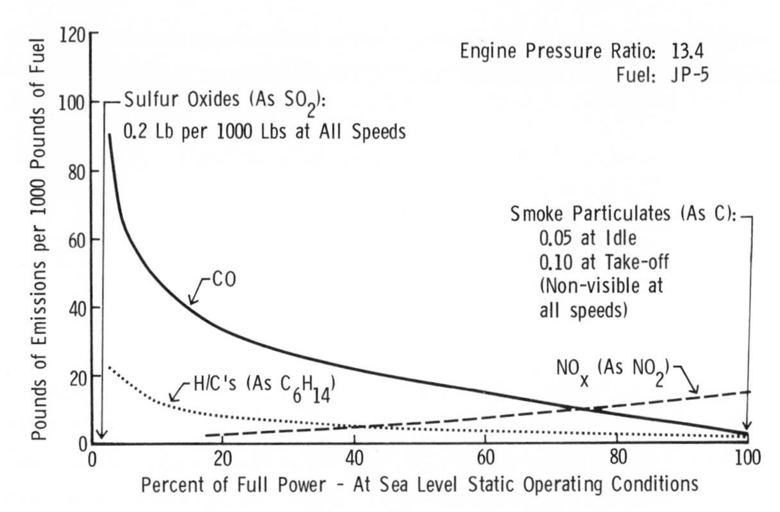

Fig. 1. Typical engine exhaust emissions characteristics.

CO and H/C's emissions levels occur at the idle and other low engine power operating conditions. At the higher power settings, the levels of these emissions are generally quite small.

The peak levels of the smoke particulates emissions, on the other hand, normally occur only at takeoff and other high power setting operating conditions. In some older technology engines, exhausts which contain visible concentrations of smoke particulates emissions are produced, particularly at the takeoff operating conditions. These visible smoke emissions can result even though the smoke particulates concentrations are very low. Although not generally regarded as a significant health hazard, these visible smoke particulates emissions are, of course, undesirable and can result in dirt and visibility problems in and around busy airports. Therefore, extensive efforts to reduce the levels of these emissions have already been carried out and most of the newer engines now in service or about to go into service have virtually non-visible smoke emission levels.

Like the smoke particulates emissions, the NO_x emissions characteristics of aircraft turbine engines are such that significant exhaust concentrations are normally generated only at full power operating conditions, particularly at takeoff. At the low engine power operating conditions, as well as at cruise conditions, the NO_x emissions levels of turbine engines are typically quite low.

Thus, the highest levels of these various emissions are normally generated either at idle and taxi operating conditions or at take-off and climb out operating conditions. At cruise operating conditions, the levels of all of these emissions are usually quite low. In addition, any emissions that are generated during cruise operations are usually dispersed over wide geographic areas. In view of these two considerations, aircraft turbine engine exhaust emissions generated during cruise operations are not generally considered to be a significant concern. Accordingly, the primary concerns associated with aircraft turbine engine exhaust emissions are their possible impact on the environments of major metropolitan airport localities, where the exhaust emissions resulting from large volumes of localized aircraft operations may tend to be concentrated.

In view of these considerations, the specific exhaust emissions characteristics of aircraft turbine engines which are of most concern are:

- Visible smoke particulates emissions.
- CO emissions – at idle operating conditions.
- H/C's emissions – at idle operating conditions.
- NO_x emissions – at full power operating conditions.

Sulfur oxides (SO_x) emissions are not included in this list of concerns, because the levels of these emissions are extremely low at all engine operating conditions. These SO_x emissions consist primarily of sulfur dioxide (SO_2). The typically low SO_2 emission levels are a direct consequence of the low sulfur contents of aircraft turbine engine fuels. Also, metal compounds emissions, such as metal oxides, do not appear to be of concern because they are present only in trace concentrations. The only significant sources of these latter emissions are any inorganic impurities in the fuel. Aircraft turbine engine fuels, however, are normally very clean and free of such contaminants. Thus, the magnitude of any metal compound emissions would be generally expected to be less than 10 parts per billion (by weight) of engine exhaust gases.

As an initial step in assessing the nature and degree of any reductions in the levels of the engine exhaust emissions of concern, it is helpful to assess the manner in which these emissions are generated in and around the airport environment. Since the primary concerns regarding these exhaust emissions are the total quantities of objectionable exhaust products that are discharged into a given metropolitan region (that is, in and around the airport complex of the metropolitan area), the extent of any emissions problem can be viewed in terms of the average levels of the emissions generated from aircraft operations in and around airports. Thus, an aircraft mission for the purpose of estimating the average emissions levels discharged into the airport region may be defined as all aircraft operations within the metropolitan area. In this form, the prescribed mission is somewhat equivalent to the average urban trip mission which is used to define exhaust emissions standards for automotive vehicles. The aircraft operations in this type of mission definition would, therefore, generally consist of the following operations: taxi, idle, takeoff, climb-out to 3500 feet, approach from 3500 feet, landing and taxi back.

To determine the average level of any specific exhaust constituent, the fuel consumed at a particular airport operating condition and the associated emission level at this engine operating condition are used to determine the total quantity of the

References p. 372

emission that is generated at each of the aircraft operating conditions of interest. The resulting quantities are then combined for all of the aircraft operations and divided by the total fuel flow consumed within the airport environment. This type of analysis is illustrated in Table 1, where the average H/C's emissions level of an advanced transport aircraft in the airport environment is computed.

TABLE 1

Average Airport Environment H/C's Exhaust Emissions Levels Of An Advanced Transport Aircraft

Aircraft Operation	Nominal Fuel Comsumption-Lbs.	Emissions (as C_6H_{14}) Lbs./1000 Lbs. Fuel	Lbs.
Idle & Taxi	1500	20.0	30.0
Takeoff	450	0.1	0.05
Climb-out (To 3500 Feet)	2200	0.1	0.2
Approach (From 3500 Feet)	1020	0.3	0.3
Landing	100	8.0	0.8
Totals	5270		31.4

Average Airport Environment Emission Level (Pounds Per 1000 Pounds Of Fuel): $= \frac{31.4 \times 1000}{5270} = 6.0$

Similarly, the estimated airport mission average CO and NO_x emissions levels for this aircraft are shown in Table 2.

TABLE 2

Estimated Average Airport Environment CO and NO_x Emissions Levels Of An Advanced Transport Aircraft

Emission	Emissions Level Pounds Per 1000 Pounds of Fuel: Idle	Takeoff	Airport Environment Average
CO	60.0	0.5	18
NO_x (as NO_2)	3.0	32.0	22

As shown by this type of analysis, the average airport environment emissions levels are considerably smaller than the peak levels. The use of this kind of analysis approach is, therefore, an appropriate method of defining an allowable peak emission level of any given exhaust constituent, once the desired average level for this emission in and around a given airport complex is established.

SMOKE PARTICULATES EMISSIONS

Many present day aircraft turbine engines do produce exhausts which contain visible quantities of smoke. These objectionable visible smoke conditions result even though the smoke concentrations of the engine exhausts are very low, typically less than 0.005 percent by weight and represent very small losses in combustion efficiency. However, because these smoke emissions are comprised of extremely small and finely divided particulate matter, these low smoke particulates concentrations can result in significant visual effects. Various studies conducted at General Electric and elsewhere have shown that a large proportion of the smoke particulates formed in turbine engines are low density agglomerates. The sizes of these agglomerates are generally less than one micron and, thus, are of the same order of magnitude as the wave length of visible light. For a given weight concentration of smoke, particulate matter in this size range produces a greater visual effect than would result from particulates with larger average size.

In commercial aircraft operations, engine exhausts containing visible quantities of smoke are of concern primarily because of air pollution and related atmospheric visibility reduction considerations. In military aircraft operations, the presence of visible smoke emissions can, in addition, result in an unsatisfactory condition because of tactical considerations. Accordingly, during recent years, significant developmental efforts have been conducted within the aircraft turbine engine industry to reduce the smoke emission levels of turbine engines.

For the past several years, investigations have been conducted at General Electric to develop technology for the design of low smoke emission combustors for use in a wide variety of turbine engine applications, including advanced engines with high pressure ratios and engines designed to operate with low grade distillate fuels, such as diesel fuel. A summary of some of these investigations is presented in Ref. 2. These investigations have established that the design of low smoke emission combustors involves providing both leaner average fuel-air mixtures within the primary combustion zone and more effective fuel and air mixing within the primary combustion zone, as compared to those of combustors with high smoke emission levels. These investigations have demonstrated that both of these provisions are needed to eliminate any fuel-rich mixtures within the primary combustion zone and, therefore, that both are of major importance.

Providing the required leaner fuel-air mixtures and improved mixing in the primary combustion zone has been found, however, to involve significant changes in the overall design approaches used in the combustors of present day engines. Also, combustor design features added to reduce smoke emission levels have been found, in some instances, to result in losses in other aspects of combustor performance, especially ignition performance. Thus, the design and definition of low smoke emission combustors generally entails careful, iterative development efforts to provide the required low smoke emission characteristics, as well as to meet all the usual

References p. 372

ignition, stability, efficiency, exit temperature distribution, life and other performance requirements.

One of the major results of the smoke reduction technology programs at General Electric has been the development of an axial swirler combustor dome design concept for use in fuel pressure-atomizing types of combustors. An illustration of this type of dome design and the smoke emission levels obtained with this design approach in the CF6-6 engine are presented in Fig. 2. This dome design concept has been successfully applied in the TF39, LM2500 and CF6 engines. Other illustrations of the

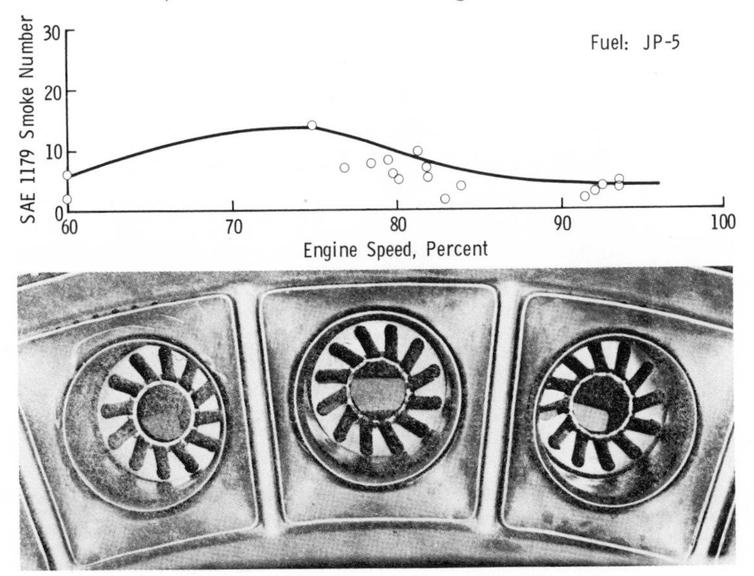

Fig. 2. Typical axial swirler combustor configuration and smoke levels.

Fig. 3. CF6 engine test at Peebles Test Site – engine operating at peak measured smoke emission level.

effectiveness of this design approach are in Figs. 3, 4 and 5. In the sea level static engine test photograph shown in Fig. 3, the engine is being tested at its peak measured smoke emission level operating condition. In the C-5 aircraft flight test photograph (Fig. 5), two of the engines are equipped with axial swirler combustors, while the other two engines are equipped with orignial production design-type TF39 combustors. These initial production design combustors are currently being replaced with axial swirler-type combustors in all existing TF39 engines, when the engines are overhauled. All new TF39 engines are being equipped with axial swirler-type combustors.

Fig. 4.CF6 engine – powered DC10 aircraft flight test.

Fig. 5. TF39 engine-powered flight test of C5 aircraft. Two engines are equipped with axial swirler combustors.

As is illustrated in Fig. 2 and in several subsequent figures, smoke particulates emission levels are expressed at General Electric in terms of the SAE ARP 1179 smoke number. On this scale, which runs from zero to 100, low smoke numbers indicate low smoke particulates emission levels. Visibility thresholds for various types of engines range from about 20 to 40 on this scale. The visibility threshold of any given engine is strongly dependent on its size. Thus, larger-size engines have lower smoke numbers as their visibility thresholds than smaller engines, because of the

References p. 372

larger sizes of their exhaust plumes. Larger size plumes result in greater path lengths and, accordingly, greater amounts of light scattering and absorption for the same smoke concentration in the exhaust gas. For engines in the larger size class, visibility threshold smoke numbers are normally in the range of 20 to 30. Engines in the smaller size class generally have visibility threshold smoke numbers in the range or 30 to 40. The core engines of high-bypass turbofan engines also generally have somewhat higher threshold smoke numbers than turbojet engines of a similar size, because of the dilution of the core engine exhaust provided by the bypass air. In any event, a SAE smoke number of about 20 is generally at or below the visibility thresholds of almost all aircraft turbine engines – regardless of size or the angle at which the exhaust plume is viewed.

More recently, other types of combustor design approaches have been developed at General Electric for use in several advanced engines. The combustors of these advanced engines are carbureting designs, in which the fuel is injected at low pressure into various types of premixing chambers where it is mixed with air prior to being delivered to the primary combustion zone. Because of this premixing of the fuel with part of the combustion air, carbureting combustor designs generally have low smoke emission levels. As an example, the smoke emission levels of the carbureting combustor of one of these advanced engines are shown in Fig. 6.

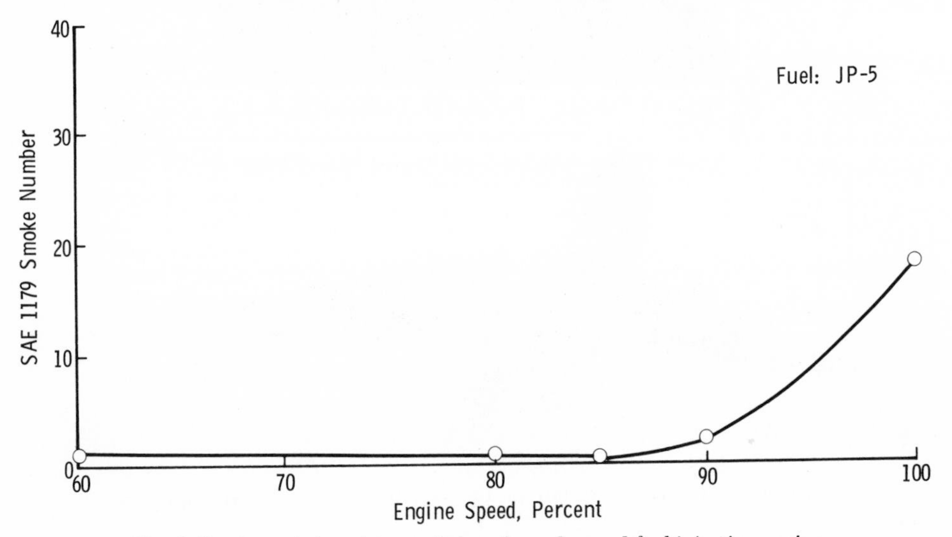

Fig. 6. Smoke emission characteristics of an advanced fuel injection engine.

Additional data on the smoke emissions characteristics of various General Electric engines are presented in Fig. 7. As is shown in this summary chart, the peak smoke emission levels of the advanced engines are generally below a SAE 1179 smoke number of 20 and, therefore, are below the threshold of visibility. These advanced engines encompass a wide range of sizes and cycle pressure ratios. Thus, technology for the design of low smoke emission combustors, which also fulfill engine ignition and other performance requirements, is reasonably well defined.

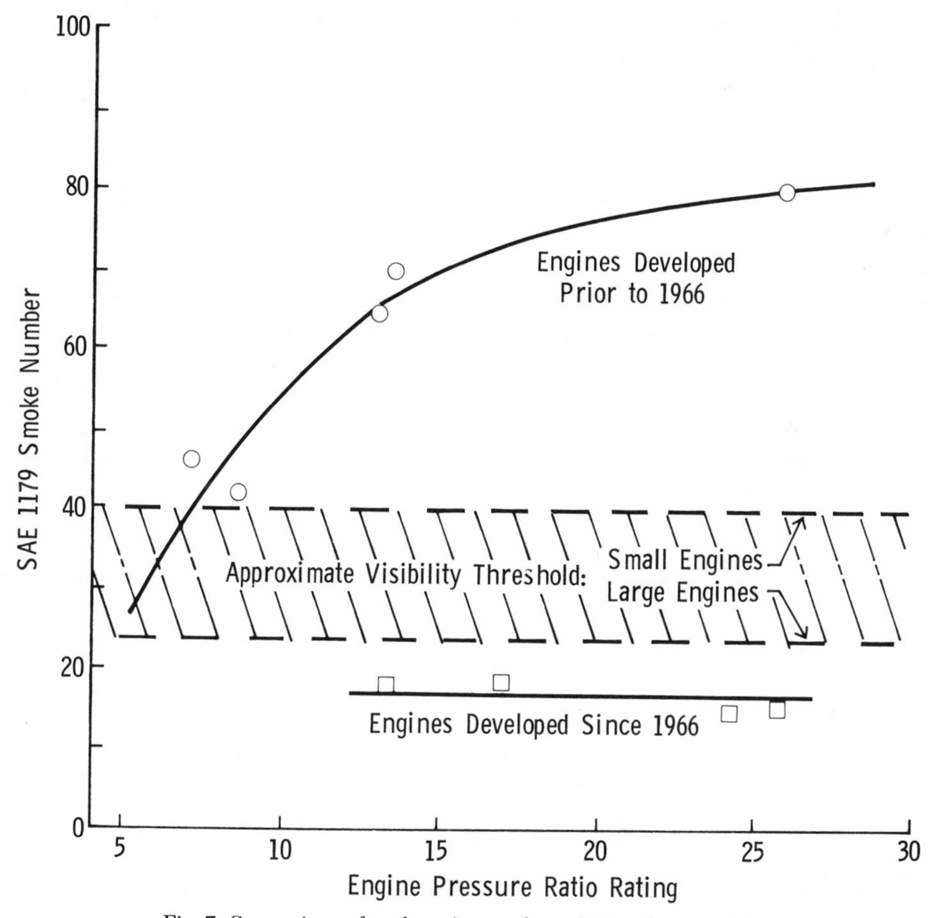

Fig. 7. Comparison of peak engine smoke emission characteristics.

With peak SAE 1179 smoke numbers of 20, the exhausts of almost all engines are not only non-visible regardless of viewing angle, but also the weight concentrations of smoke particulates are extremely small – less than 2 ppm of exhaust gas, by weight. Furthermore, even these relatively low smoke emission levels are usually reached only at sea level takeoff engine power settings. At lower power settings, the smoke levels are normally much lower. Thus, a peak smoke number of about 20 can be expected to result in very low average emission levels in and around airports and very low levels at cruise operating conditions, as well. Although it is probably possible to obtain even somewhat lower peak smoke emission levels, some sacrifice in engine ignition and altitude relight performance is likely to result. In view of these considerations, further development efforts to reduce the peak smoke emission levels of aircraft turbine

References p. 372

engines to SAE 1179 smoke numbers significantly less than 20 do not appear to be needed or appropriate.

CARBON MONOXIDE AND UNBURNED HYDROCARBONS EMISSIONS

The carbon monoxide (CO) and the unburned and partially oxidized hydrocarbons (H/C's) emissions are products of inefficient combustion. As is illustrated in Fig. 1, these emissions are produced only at idle and other low engine power operating conditions, because the combustion efficiency levels of most engines at these low power operating conditions are not optimum and are typically in the 90 to 95% range. At higher engine power settings, the combustion efficiency levels of most engines are generally well in excess of 99% and, therefore, the quantities of incomplete combustion products produced at these operating conditions are very small.

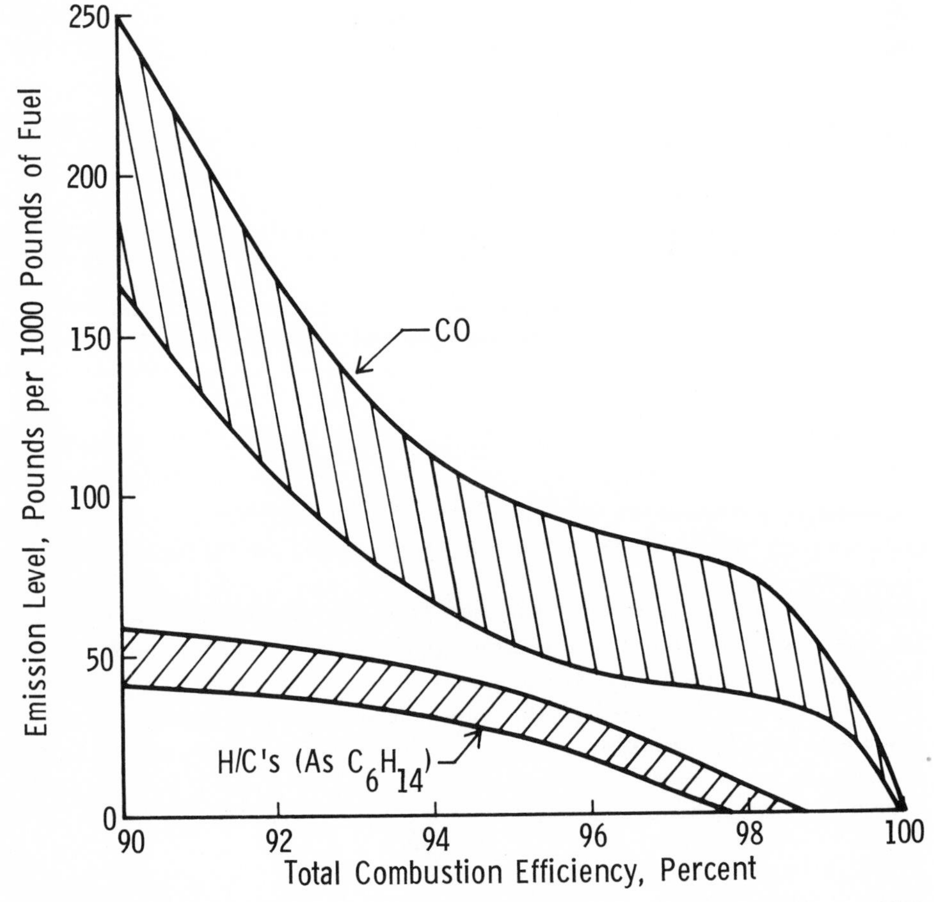

Fig. 8. Typical relationships between combustion efficiency and levels of CO and H/C's emissions.

The relationships between combustion efficiency performance and the levels of the CO and H/C's emissions are illustrated in Fig. 8. The ranges of emissions levels shown in this figure over the indicated range of combustion efficiency values encompass the CO and H/C's emissions levels that have been obtained in tests of various General Electric engines. Within the ranges shown, many combinations of CO and H/C's emissions levels are possible. However, at any given combustion efficiency value, the levels of these two emissions are interrelated since the individual efficiency losses attributable to the levels of each of these emissions must be equal to the total efficiency loss. Thus, at a given overall efficiency level, if the CO emission level is near the upper limit of the CO emission band shown in Fig. 8, the H/C's emissions level must be near the lower limit of the H/C's band – and vice versa.

At the idle power operating conditions, the weight ratios of CO and H/C's emissions levels have been found in tests of General Electric engines to range from about 2 to about 4. The specific ratios of CO to H/C's emissions levels of engines at idle power operating conditions appears to be dependent on engine type, engine size and the specific design features of the engine combustor. Generally, smaller size engines have been found to have lower weight ratios of CO to H/C's emissions levels than those of larger size engines – for the same idle power combustion efficiency value.

The H/C's emissions consist of a wide variety of compounds ranging from low molecular weight paraffin compounds to partially oxidized hydrocarbon compounds. Because the H/C's emissions are comprised of such a large number of individual compounds, with widely varying physical and chemical properties, measurements of the concentrations of each of the individual species are extremely difficult to obtain. Accordingly, this category of emissions is normally treated as a total quantity and the concentrations of specific hydrocarbon species are not determined. In some cases, however, the partially-oxidized hydrocarbons are measured and treated as a separate category of emissions.

The largest portion of these H/C's emissions is known from special tests, conducted at General Electric and elsewhere, to consist of paraffinic and olefinic hydrocarbon compounds similar to those contained in kerosene fuels. Smaller portions of these emissions are comprised of various types of lower molecular weight paraffins, olefins and, perhaps, aromatic compounds. Another portion is comprised of various aldehydes and ketones, which are partially oxidized hydrocarbons. Some tests conducted at General Electric to determine the emissions characteristics of turbine engines have included measurements of the levels of these partially oxidized hydrocarbons. In general, this latter category was found in these tests to amount to less than 20 percent, by weight, of the total H/C's emissions levels.

The H/C's emissions also include some quantities of condensible compounds. When the engine exhaust gases cool, these latter compounds are converted to liquid or, to a small extent, to solid particulates. Thus, these condensible compounds are another

References p. 372

source of particulates emissions. Relatively little data are available on the levels of this type of particulates emissions. In general, however, the levels of these condensible hydrocarbons emissions appear to be small. Some preliminary tests to determine the levels of these emissions have been conducted by the Aviation Exhaust Study Group of the Coordinating Research Council, Incorporated. Also, tests have been conducted by the General Electric Gas Turbine Operations in Schenectady, New York. In these test series, the total particulates emissions were typically found to consist of generally similar proportions of solvent soluble constituents (hydrocarbons particulates) and insoluble constituents (carbon smoke particulates). However, based on the experience obtained in these and in similar test series, the presently available analysis tools for these condensible hydrocarbons emissions appear to be inadequate for obtaining accurate and reproducible measurements of their levels. More research in this area of emissions measurement technology appears to be needed before detailed data on the levels of this category of emissions can be obtained.

The H/C's emissions are also the cause of the odors associated with aircraft turbine engine operations at low engine speeds. Although, the specific hydrocarbons causing such odors have not been identified, there are probably a number of compounds that contribute to the characteristic odors of engine exhaust products. The compounds that do produce such odors are probably present in extremely low concentrations – in the order of a few ppb of exhaust gas. The specific types of compounds primarily responsible for any such odors appear to be the aldehydes, ketones and possibly some of the aromatic and olefinic compounds which may be contained in trace amounts in engine exhaust gases. Considerably more research in this area of engine exhaust emissions is needed, however, to define the specific constituents which result in odors.

The somewhat reduced combustion efficiency performance of most existing aircraft turbine engines at idle and other low power operating conditions is due to the adverse combustor operating conditions that normally prevail at these operating engine conditions. At the low engine power operating conditions, the combustor inlet air temperature and pressure levels are relatively low, the overall fuel-air ratios are generally low and the quality of the fuel atomization and/or distribution is usually poor because of the low fuel and air flows. In any engine, all of these adverse combustor operating conditions are rapidly eliminated as the engine power setting is increased above idle power levels and, accordingly, its combustion efficiency performance is quickly increased to near-optimum levels.

The relationships between the CO and H/C's emissions levels and combustion efficiency performance have been determined in tests of several General Electric engines. Some of these engine test results are presented in Table 3. In this table, measured CO and H/C's emissions levels and combustion efficiency performance levels are shown for three engine power levels – idle, 50 percent power and full power (takeoff). As is shown in this tabulation, the CO and H/C's emissions levels are generally very low at the two higher power operating conditions.

TABLE 3

CO and H/C's Emissions Characteristics of General Electric Engines
Sea Level Static Operation with Kerosene Fuel (JP-5)

Engine Type	Carbon Monoxide (Pounds Per 1000 Pounds Of Fuel)	Total Unburned Hydrocarbons – As C_6H_{14} (Pounds Per 1000 Pounds Of Fuel)	Combustion Efficiency (Percent)
High Pressure Ratio Turbofan			
Idle	60.0	20.0	96.6
50% Power	3.0	0.3	99.9
Take-Off (Full Power)	0.5	0.1	99.98
Low Pressure Ratio Turbofan			
Idle	179.0	45.0	91.3
50% Power	75.0	2.0	96.3
Take-Off (Full Power)	39.0	0.7	98.4
Intermediate Pressure Ratio Turbojet			
Idle	130.0	40.0	93.0
50% Power	21.0	5.0	99.0
Take-Off (Full Power)	3.5	1.0	99.8
Low Pressure Ratio Turbojet			
Idle	150.0	80.0	88.5
75% Power	18.0	2.9	99.3
Take-Off (Full Power)	14.0	2.6	99.4

Some of the emissions levels presented in this tabulation are average values obtained in tests of more than one engine.

The data in this tabulation also show that the measured CO and H/C's emissions levels, at engine idle operating conditions, of the high-pressure ratio engines are considerably less than those of the lower-pressure ratio engines. These lower emissions levels are, of course, due to the higher idle power combustion efficiency performance of the high-pressure ratio engines. The idle power combustion efficiency performance levels shown in Table 3, along with other engine test data, are presented in Fig. 9 as a function of engine pressure ratio. This somewhat gross correlation generally indicates that the increased idle operation combustor inlet temperatures and pressures, associated with the higher pressure ratio cycles, result in improved combustion efficiency. The typical relationships of overall engine pressure ratio with the combustor inlet air temperature and pressure levels, at idle engine power levels, are shown in Fig. 10. Since the new generation of aircraft turbine engines currently going

References p. 372

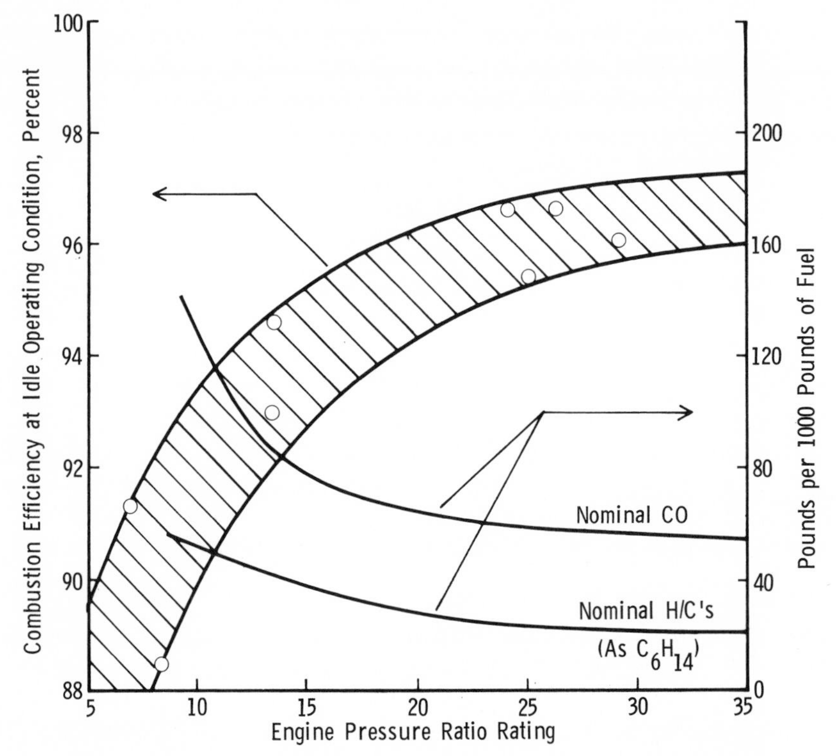

Fig. 9. Idle power combustion efficiency of performance and emissions trends.

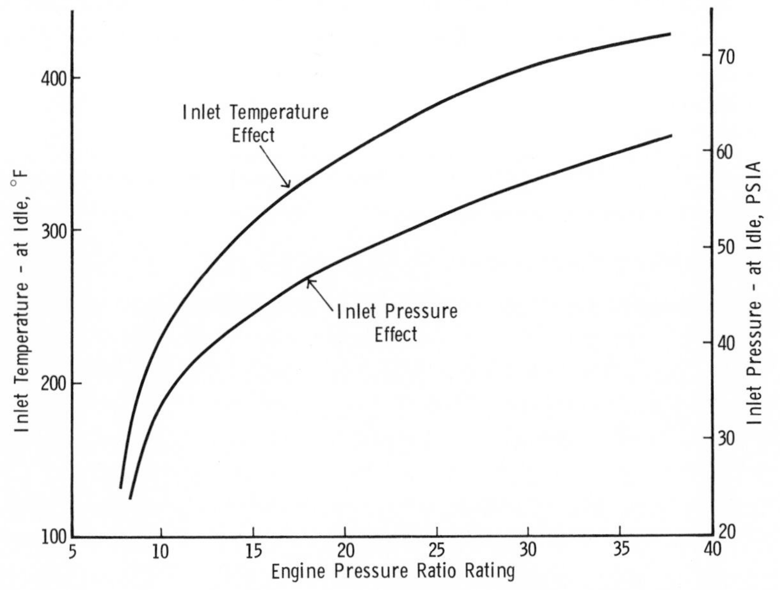

Fig. 10. Effect of engine pressure ratio on idle power combustor inlet air temperature and pressure.

into service to power the new wide-bodied commercial transport aircraft generally have much higher pressure ratios than their predecessors, these results show that this new generation of engines can generally be expected to have much lower CO and H/C's emissions levels than those currently in service.

In any specific engine, low speed combustion efficiency performance is also affected to a significant extent by the fuel-air ratio (weight ratio of fuel to air) in the primary combustion zone and the quality of the fuel atomization. At the low power operating conditions, the overall engine fuel-air ratios are generally low and, thus, the primary zone fuel-air ratios are also low – typically one-half of the stoichiometric value. Further, fuel atomization quality tends to be relatively poor at low engine power levels because of the low fuel flows and air flows. In pressure atomizing combustors, these low fuel flows result in low fuel nozzle pressure drops and, thus, poor atomization. In premixing combustors, in which the fuel is injected at low pressures and is atomized by the compressor discharge air, these low air flows result in low air velocities and, thus, poor atomization. In both types of combustors, poor atomization results in slow fuel vaporization and poor mixing of the fuel with the air.

In a given engine, therefore, improved low power combustion efficiency performance can be obtained by increasing the primary zone fuel-air ratio, by improving the quality of the fuel atomization or by a combination of both approaches. Thus, for use in both high and low pressure ratio engines, combustor design approaches are needed which result in both higher primary zone fuel-air ratios and improved fuel atomization – at the low engine power operating conditions. Therefore, programs to develop various combustor design approaches of this kind are underway at General Electric, for both atomizing and carbureting types of combustors. Some of the design approaches, which are being evaluated and developed for use at the idle and other low power operating conditions, are as follows:

- Improved fuel atomization.
- Locally richer mixtures in primary zone.
- Decreased primary zone air flow.
- Increased primary zone residence time and fuel-air mixing.

Some promising results have been obtained in these investigations. An advanced full annular combustor has been tested, at engine idle operating conditions, with modified fuel nozzles to determine whether improved fuel atomization would result in an increase in low power combustion efficiency performance. A significant improvement in combustion efficiency was obtained – from about 95.0% to about 97.8% at the nominal idle fuel-air ratio operating condition. These test results showed that the relatively poor atomization quality provided by the existing dual-orifice fuel nozzle design at the idle power operating conditions results in a loss of combustion efficiency. The results of this test series are shown in Fig. 11.

References p. 372

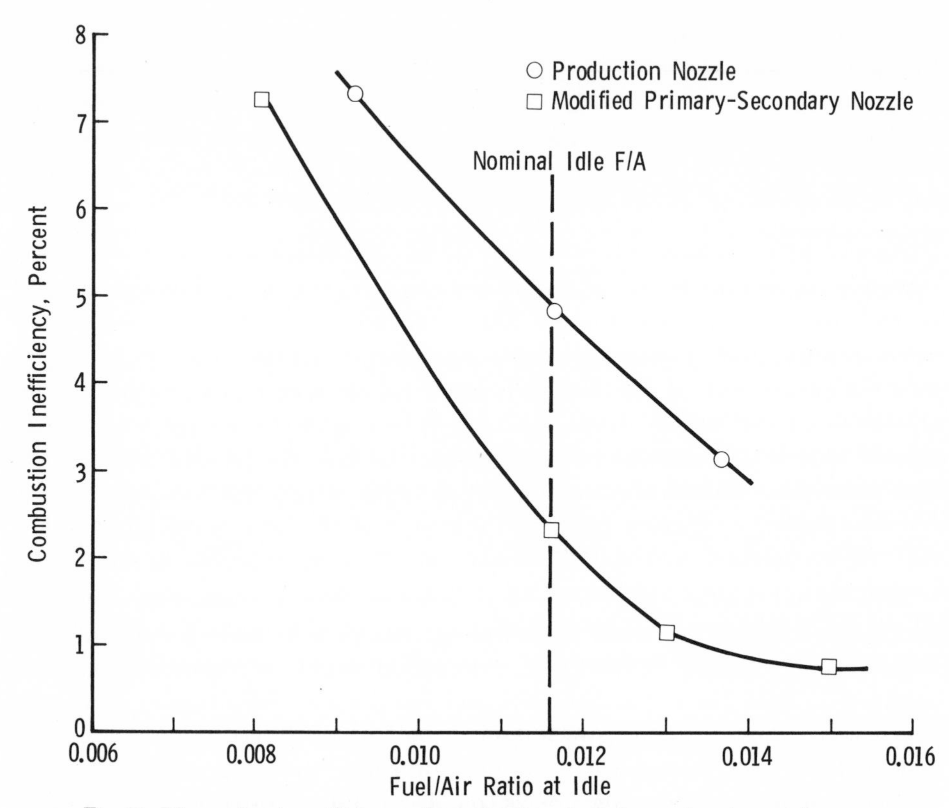

Fig. 11. Idle combustion efficiency characteristics of an advanced engine combustor.

Thus, some progress has already been made in the development of methods to reduce further the CO and H/C's emissions levels of aircraft turbine engines. Based on the studies conducted to date, idle power combustion efficiency performance levels as high as 97 to 98 percent appear to be attainable in future high performance and high pressure ratio aircraft turbine engines – with sufficient development effort. For aircraft turbine engines in the medium to large-size class, an idle power efficency value of 98 percent would be typically expected to result in maximum idle power CO and H/C's emissions levels in the order of 60 and 8 pounds per 1000 pounds of fuel, respectively. With idle power emissions levels of this order, the approximate average airport environment CO and H/C's emissions levels (as defined in a preceding section of this paper) would typically be expected to be in the order of 18 and 2 pounds per 1000 pounds of fuel, respectively. Thus, an idle combustion efficiency level of 98 percent generally results in quite low average CO and H/C's emissions levels in the airport environment.

The presently available combustor design technology for providing this degree of combustion performance improvement is, however, extremely limited. Extensive development efforts will, therefore, be required to provide combustor design features for approaching this degree of improvement, which are both practical and reliable. It is anticipated that these idle combustion efficiency performance improvement investigations will, in time, result in the definition of combustor designs with significantly improved idle power combustion efficiency performance, which will be suitable for use in various types of aircraft turbine engines, both low-pressure and high-pressure versions. Therefore, it is expected that engines to be developed and placed in service in future years will have significantly lower CO and H/C's emissions levels.

These reduced total H/C's emissions levels would, of course, be expected to result in very low levels of the partially-oxidized hydrocarbons emissions, since these emissions usually only represent a small portion of the total H/C's emissions. Further, since the greater part of the total particulates emissions of low smoke combustors appears to be due to the condensible constituents of the total H/C's emissions, reduced total H/C's emissions levels would also be expected to result in quite low total particulates emissions levels. In addition, reduced total H/C's emissions levels would probably be expected to lessen the degree of the odor problems associated with engine operations at low power levels, since these odors are due to some of the compounds which make up the total H/C's emissions. Thus, the achievement of lower CO and H/C's emissions levels in future engines is expected to result in several areas of improvement with regard to the impact of aircraft on airport environments.

OXIDES OF NITROGEN EMISSIONS

Small amounts of nitrogen oxides (NO_x) are generated, to some degree, in any combustion process. In the combustion systems of turbine engines, these emissions are formed within the primary combustion or flames zones and within the dilution zones immediately downstream of the primary zones. Oxides of nitrogen emissions result directly from the oxidation of atmospheric nitrogen rather than from the combustion of fuel constituents. These emissions consist primarily of nitric oxide (NO) together with small amounts of nitrogen dioxide (NO_2), and result from the formation of NO during the combustion process. The small amounts of NO_2 emissions result from the further oxidation of the NO that is formed in the combustion or flame zones. Once discharged into the atmosphere, however, the NO is gradually converted to NO_2.

The thermochemical equilibrium quantities of NO that can be generated are strongly dependent on the flame temperature levels of the hot combustion gases and on the availability of oxygen. Thus, these equilibrium quantities increase rapidly as the initial air temperature is increased and as the combustion zone fuel-air equivalence ratio (ratio of actual fuel-air ratio divided by stoichiometric fuel-air ratio) approaches values in the order of 0.9. An illustration of the thermochemical equilibrium quantities of NO_x which can be produced in combustion processes is presented in Fig. 12.

References p. 372

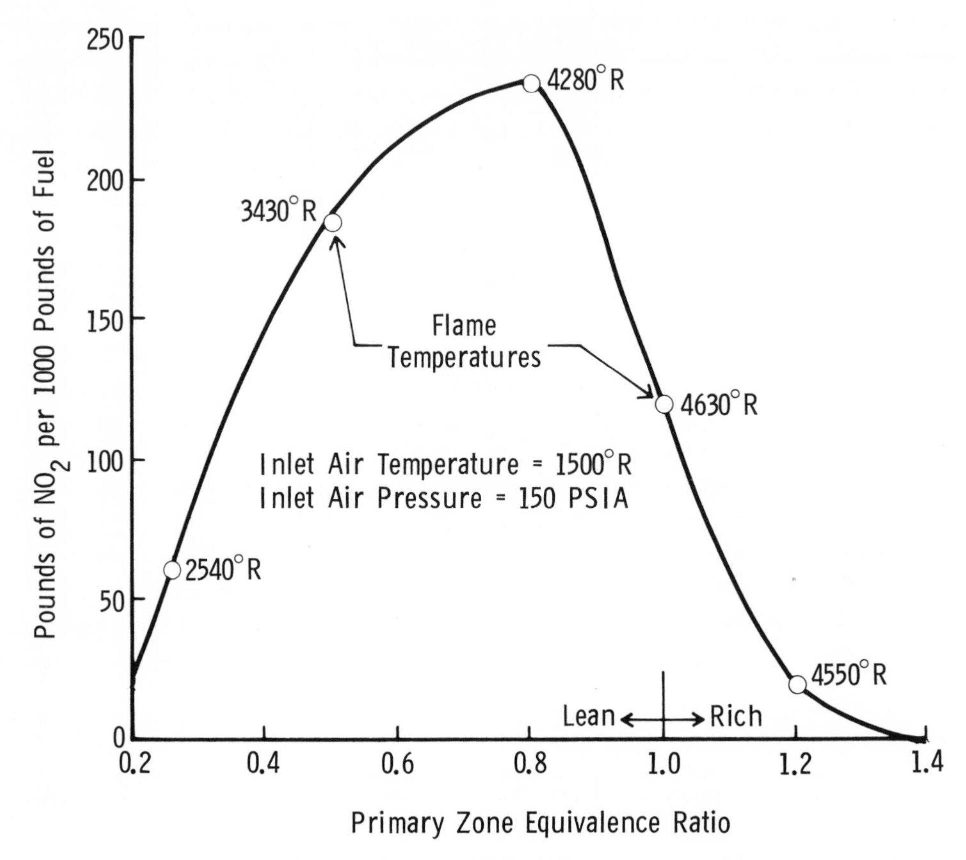

Fig. 12. Thermochemical equilibrium NO_x concentrations.

The formation rate of NO is also strongly dependent on the initial air temperature and on the fuel-air ratio value in the combustion zone. The rates increase rapidly as the initial air temperature level is increased and as the combustion zone fuel-air equivalence ratio approaches unity. Increases in the inlet air pressure level also increase the formation rate. Thus, the NO_x emissions levels of turbine engines generally reach levels of any significance only at the high power operating conditions, where the combustor inlet temperature and pressure levels are high. Even at the high inlet temperature and pressure operating conditions, however, the NO formation rates are relatively slow, as compared to the fuel combustion reactions. Therefore, the quantities of the NO_x emissions that are generated, even at the high engine power levels, are usually much less than the thermochemical equilibrium values, because they are limited by the short residence times of the hot combustion gases in the primary zones of the engine combustors. As a result, very high fuel combustion efficiencies can be obtained without the generation of thermochemical equilibrium NO concentrations. Some typical NO formation rate data are presented in Fig. 13.

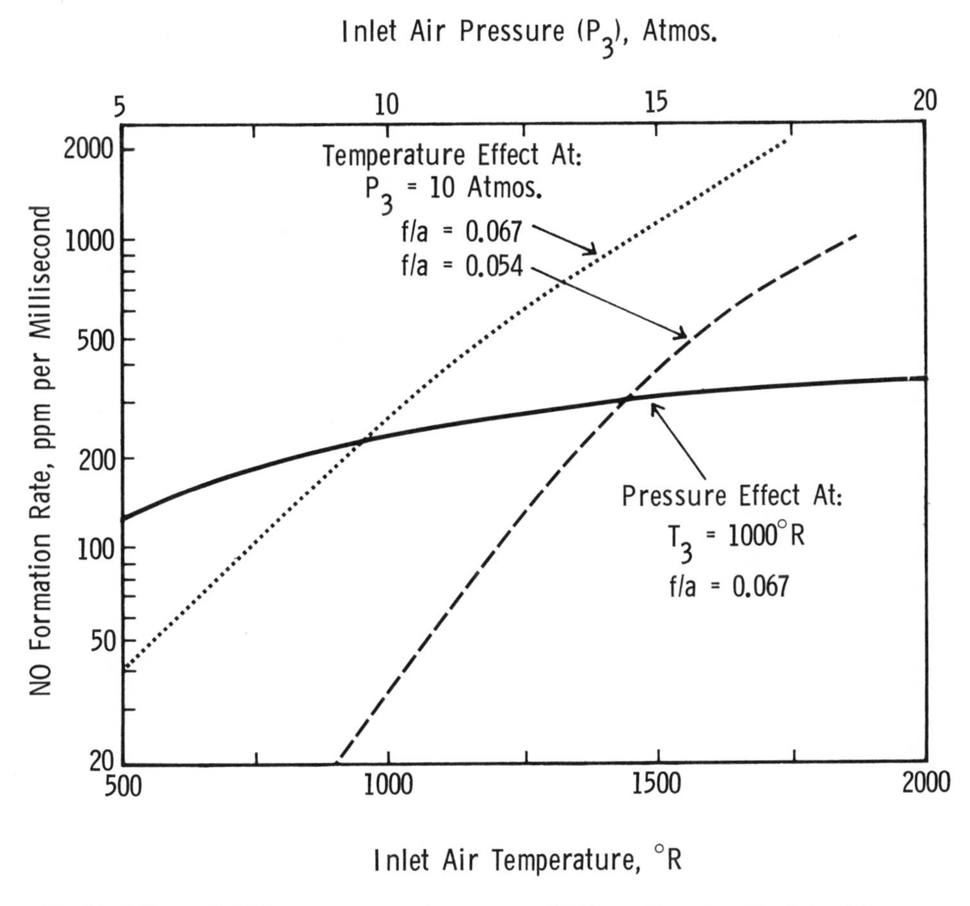

Fig. 13. Effects of initial temperature and pressure on NO formation rate of fuel-air mixtures.

Because of the very strong dependence of NO formation rate on the initial air temperature level and since turbine engine combustor inlet air temperature levels are a direct function of engine pressure ratio, the NO_x emissions characteristics of turbine engines are directly related to the overall engine cycle pressure ratio. The NO_x emissions characteristics of several General Electric engines have been determined in engine tests. Some of these engine test results are presented in Table 4. In this table, the NO_x emissions levels are shown for three engine power levels – idle, 50 percent power and full power (takeoff). As is shown, the emissions levels are generally low at the two lower power operating conditions. Also shown is that the full power NO_x emissions levels of the high-pressure ratio engines are greater than those of the low-pressure ratio engines. This strong relationship between NO_x emissions levels and pressure ratio is illustrated in Fig. 14, which is a plot of the data shown in Table 4 and other data.

References p. 372

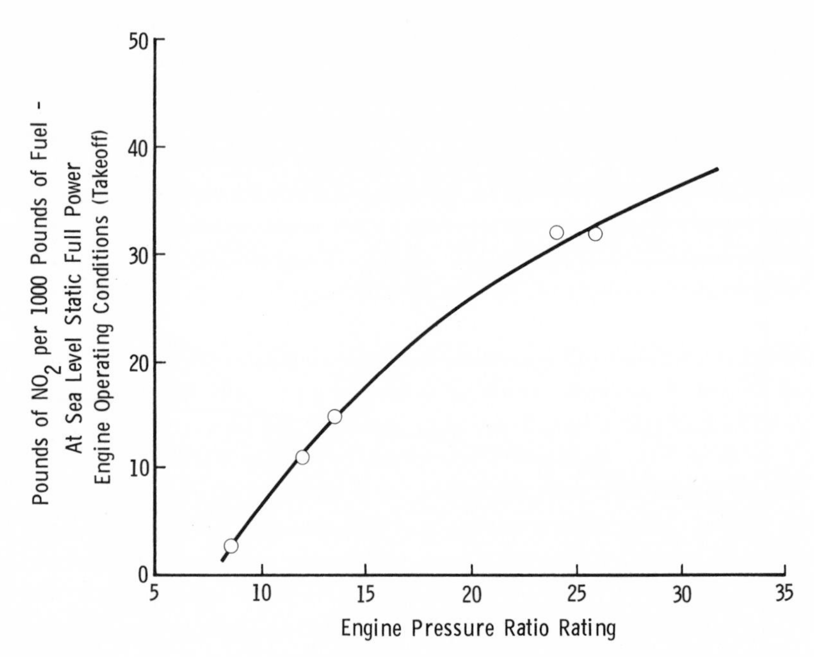

Fig. 14. NO_x emissions level trends typical of current technology engines.

TABLE 4

NO_x Emissions Characteristics of General Electric Engines. Sea Level Static Operation With Kerosene Fuel (JP-5)

Engine Type	Pounds of NO_2 Per 1000 Pounds of Fuel
High Pressure Ratio Turbofan	
Idle	2.7
50% Power	22.4
Take-Off (Full Power)	32.0
Low Pressure Ratio Turbofan	
Idle	1.5
50% Power	3.0
Take-Off (Full Power)	–
Intermediate Pressure Ratio Turbojet	
Idle	1.6
50% Power	6.4
Take-Off (Full Power)	14.4

Some of the emissions levels presented in this tabulation are average values obtained in tests of more than one engine.

At any specific inlet air temperature level, the NO_x emissions levels of a given combustor are also affected by the average and local fuel-air ratio values within the primary combustion zones and by the hot combustion gas residence times within this zone. To minimize the quantities of NO_x emissions formed in a given combustor, the residence times of the hot gases in its primary zone and the dilution zones immediately downstream should be minimized. Also, its primary combustion zone should preferably be designed to operate with average fuel-air ratios slightly greater than stoichiometric and its fuel-air ratio distribution should be as uniform as possible, since considerable amounts of NO can be formed in local pockets of lean mixtures. The qualitative importance of these factors is illustrated by comparing the NO_x emissions characteristics of high-smoke emission combustors with those of low-smoke emission designs. Low-smoke emission combustors generally have leaner average primary zone fuel-air ratios than high-smoke emission combustors and, therefore, some increase in their NO_x emissions levels might be expected. However, their primary zone fuel-air ratios are generally much more uniform than those of high-smoke emission designs. These effects tend to counterbalance each other and, therefore, the NO_x emissions characteristics of a given high-smoke emission combustor design are usually very similar to those of a low-smoke emission version of the combustor design. An example of this is shown in Fig. 15, where the NO_x emission characteristics of two combustors, the original design and a low-smoke emission design, for the same high-pressure ratio turbofan engine are compared.

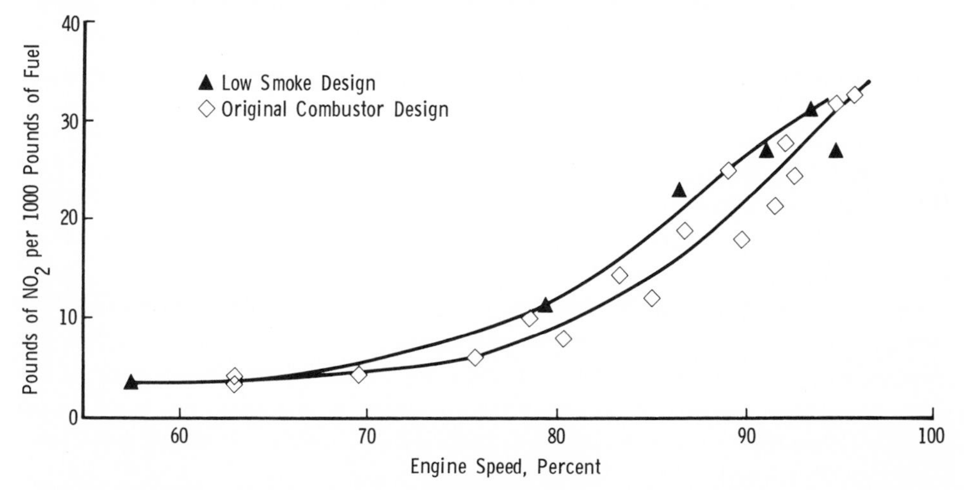

Fig. 15. Comparison of NO_x emissions characteristics of low smoke emission combustor with high smoke emission version of the same combustor.

Thus, once its overall cycle and operating conditions are established, control and reduction of the NO_x emissions levels of a given turbine engine design involve precise control of the average and local fuel-air ratios within the primary and dilution zones of the combustors and precise control of the residence times of the hot combustion

References p. 372

gases in these zones. The present quantitative technology and knowledge on how to obtain this degree of control of the NO formation process in engine combustors is, however, very limited. Thus, extensive development efforts to provide detailed design technology for the control of this NO formation process are required. Accordingly, analytical and experimental studies to provide a more quantitative description of the NO formation process in engine combustors and to define methods for reducing their NO_x emissions levels are underway at General Electric and elsewhere in the industry.

Some potentially promising results have been obtained in these studies. The use of an advanced fuel injection method, involving premixing of the fuel and air, has been found, for example, to result in somewhat lower NO_x emissions levels as compared to those of the more conventional pressure-atomizing combustor designs – at the same combustor operating conditions. In tests of advanced combustor designs of this kind, NO_x emissions levels reductions in the order of 20 to 30 percent have been obtained, as compared to those of the more conventional designs, at similar operating conditions. These lower NO_x emissions levels appear to be the result of improved primary combustion zone fuel-air ratio uniformity and shorter residence times of the combustion gases in the primary and dilution zones of the combustor.

Although significantly shorter primary zone residence times than those provided by these advanced combustor designs do not appear to be attainable, without losses in other areas of combustor performance, some further improvements in the primary zone fuel-air ratio distributions of these combustors do appear to be potentially obtainable. With such fuel-air distribution improvements, it is anticipated that the takeoff NO_x emissions levels of these advanced combustor designs might be reduced somewhat further, as compared to the takeoff NO_x emissions levels of the more conventional combustor designs for high pressure ratio engines. However, it is anticipated that to attain significant additional degrees of NO_x emissions level suppression, near-optimum primary zone fuel-air ratio distributions, along with minimized primary zone residence times, will be necessary. Obtaining these near-optimum operating characteristics may require the use of considerably more advanced and possibly complex combustor design approaches. These approaches, for example, may involve the use of variable geometry techniques to modulate the primary zone fuel-air ratios by control of the air flow into the primary zone or the use of modular designs in which the combustor is comprised of many small combustor modules, each equipped with fuel injection and fuel-air mixing provisions.

The use of water injection into the combustor has also been found to be another general method of reducing NO_x emissions levels. The major effect of the water is to reduce the flame temperatures and, thereby, to reduce the NO formation rates. Some results of analytical and experimental studies to define the effects of water injection are presented in Figs. 16 and 17. As is shown by these results, water injection in amounts of 1 to 2 percent – by weight of the combustor air flow – appears to provide considerable degrees of NO_x emissions level suppression. Based on results of

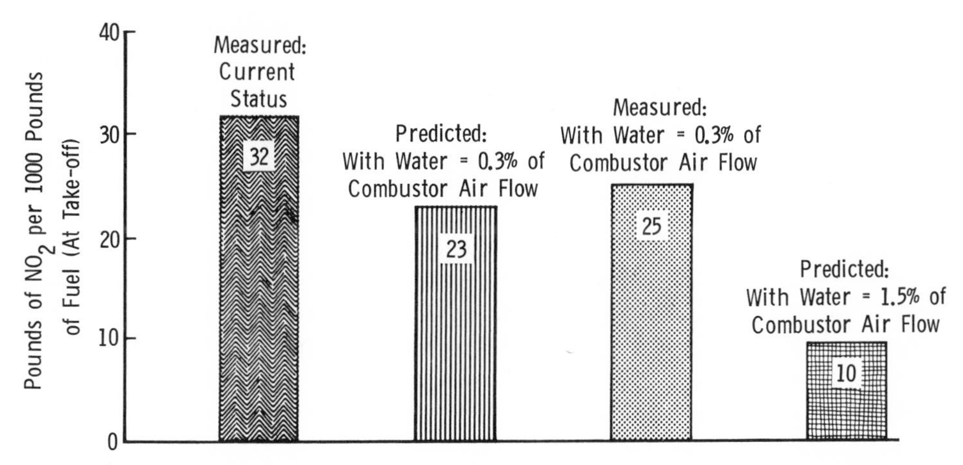

Fig. 16. NO_x emissions reduction by water injection into primary combustion zone of advanced turbofan engine combustor.

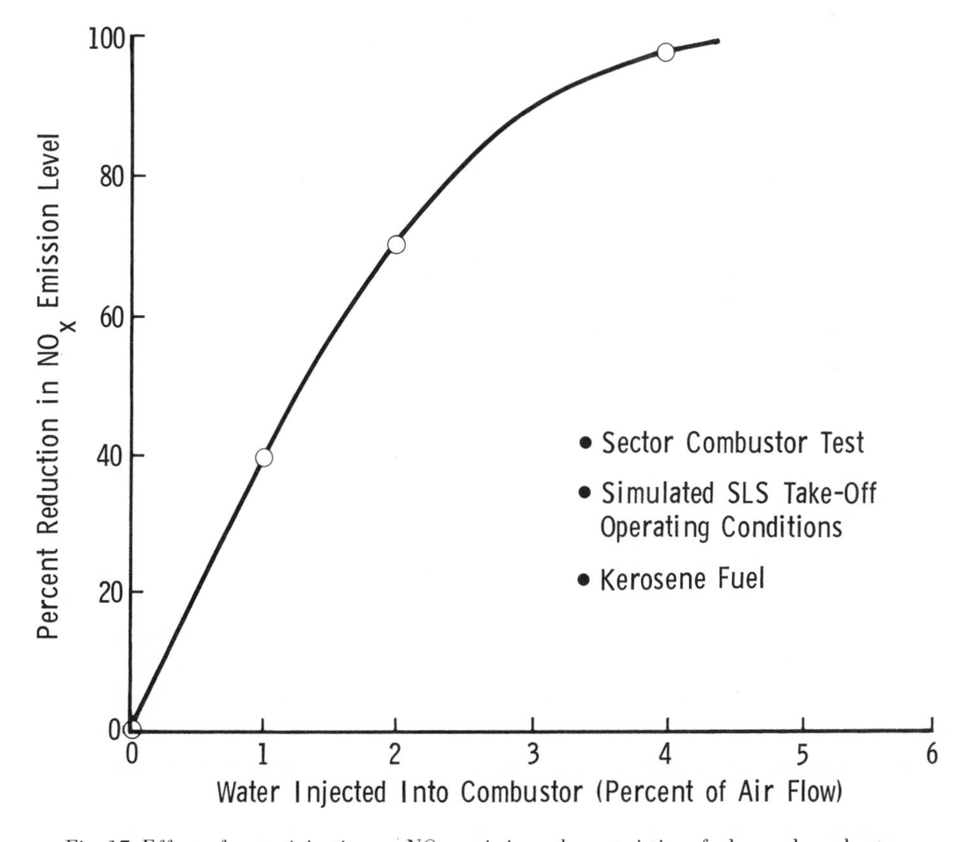

Fig. 17. Effect of water injection on NO_x emissions characteristics of advanced combustor.

References p. 372

this kind, water injection appears to be an effective method of obtaining significant reductions in NO_x emissions levels, particularly during takeoff operations where the NO_x emissions levels are normally at their maximum values.

Thus, some preliminary progress has also been made in the development of methods to suppress the NO_x emissions levels of turbine engines. The investigations conducted to date have also shown that the NO_x emissions of aircraft turbine engines will probably be the most difficult to control. The other exhaust emissions of primary concern – CO, H/C's and smoke particulates – are products of inefficient combustion. The levels of these emissions can be controlled and minimized to a large degree by various combustor design modifications. Technology for minimizing smoke emission levels to virtually non-visible concentrations is already well in hand. Combustor design approaches for improving low engine power combustion efficiency performance and, thereby, for significantly reducing CO and H/C's emissions levels of engines appear to be achievable. Control of the levels of the NO_x emissions of a given engine cycle and design, on the other hand, by combustor design techniques involves the more difficult problems of precisely controlling primary combustion zone operating characteristics such as residence time and fuel-air ratio distribution.

In spite of these apparent difficulties, it is anticipated that the NO_x emissions abatement technology investigations underway at General Electric and elsewhere will, in time, result in the development of NO_x emissions level reduction approaches suitable for use in a wide variety of engines. Thus, it is anticipated that the NO_x emissions levels of the advanced engines, especially the high-pressure ratio and high-performance engines, to be developed and placed into service in future years will be reduced to values significantly less than those being measured in current engine models with equivalent pressure ratio ratings. However, it appears at present that several years of combustion development efforts will be required to bring this NO_x emissions abatement technology to the point where it can be satisfactorily applied and significant NO_x emissions level reductions can be realized in future engines.

CONCLUSIONS

1.) The primary concern associated with aircraft turbine engine exhaust emissions is their possible impact on the environments in and around busy metropolitan airports. An appropriate method of assessing the extent of any possible aircraft engine emissions problems in specific airport localities is to determine the average levels of the emissions generated by aircraft operations in and around airports. The average level of any given exhaust emission may be defined as the average value resulting from all aircraft operations within the metropolitan area – taxi, idle, takeoff, climb-out to 3500 feet, approach from 3500 feet, land and taxi back. For any particular exhaust constituent, the average emission level determined in the manner is a more suitable value for assessments of the impact of aircraft operations on the airport environment than the peak emission level, since the peak levels of the engine exhaust emissions of possible concern are normally generated in significant amounts only at very specific engine operating conditions.

2.) Significant progress has been made at General Electric and elsewhere in the development of technology for the design of combustors with much reduced smoke emission levels. As a result of these efforts, combustors with virtually non-visible smoke emission levels at all engine operating conditions have been developed and placed into service. The peak SAE 1179 smoke numbers of these advanced combustors are typically in the order of 20 or less. SAE 1179 smoke numbers of this order correspond to very low concentrations of smoke particulates in engine exhausts – only about 2 ppm (by weight) of exhaust gas.

3.) Based on the results of various assessments and investigations conducted to date at General Electric, it appears that idle power combustion efficiencies in the order of 97 to 98 percent may be attainable in future high performance and high pressure ratio aircraft turbine engines, with sufficient development effort. It is anticipated, therefore, that the development efforts presently underway at General Electric and elsewhere will, in time, result in the definition of combustion design approaches, which provide improved idle combustion efficiency performance and which are suitable for use in various types of turbine engines. Several combinations of CO and H/C's emissions levels are possible with combustion efficiency values in the order of 98 percent. For turbine engines in the medium to large size class, however, an idle power efficiency value of 98 percent would be typically expected to result in maximum idle power CO and H/C's emissions levels in the order of 60 and 8 pounds per 1000 pounds of fuel, respectively. With idle power emissions levels of this order, the approximate average airport environment CO and H/C's emissions levels would typically be expected to be 18 and 2 pounds per 1000 pounds of fuel, respectively. Thus, an idle combustion efficiency level of 98 percent generally results in quite low average CO and H/C's emissions levels in the airport environment.

4.) Based on the results of tests to determine the emissions characteristics of engines and analytical studies of the NO_x formation process in turbine engine combustors, the most formidable emissions problem associated with aircraft turbine engines appears to be the NO_x emissions – particularly in the case of the advanced technology, high-pressure ratio engines. To date, some potentially promising combustor design approaches for providing some suppression of the NO_x emissions levels of high-pressure engines have been identified in exploratory development efforts underway at General Electric. These design approaches appear to be well suited for use in advanced engine combustors. Also, the use of water injection into combustors has been found in these studies to be a promising and effective method of reducing the NO_x emissions levels of engines. This latter approach appears to be particularly promising as a way of suppressing the peak NO_x emissions levels associated with takeoff operations. Based on the results obtained to date in these exploratory studies, therefore, it is anticipated that the investigations underway at

References p. 372

General Electric and elsewhere will, in time, result in the development of NO_x emissions level reduction approaches suitable for use in a wide variety of engine types, including both low and high pressure ratio engines. Thus, it is anticipated that the NO_x emissions levels of the advanced engines, especially the high pressure ratio and high performance engines, to be developed and put into service in future years will be reduced to values somewhat less than those being measured in current engine models with equivalent pressure ratio ratings. However, much additional research and development on the formation and control of NO_x emissions in turbine engine combustors appear to be needed.

5.) To realize significant further reductions in the exhaust emissions levels of future aircraft turbine engines, especially in the case of the NO_x emissions, extensive additional technology development efforts appear to be required. Except for the smoke particulates emissions, adequate emissions reduction design technology for the current generation of engines is not available, at present. To bring this needed emissions abatement technology to the point where it can be satisfactorily applied in engines will probably require several years of additional combustor development efforts. Thus, efforts to achieve further reductions in the objectionable exhaust emissions levels of aircraft turbine engines should preferably be targeted for engines to be certified and placed into service during the latter part of this decade.

REFERENCES

1. *Exhaust Emissions From Gas Turbine Aircraft Turbine Engines, Sub-Council Report, National Industrial Pollution Control Council, February, 1971.*
2. *D. W. Bahr, J. R. Smith, and M. J. Kenworthy, "Development Of Low Smoke Emission Combustors For Large Aircraft Turbine Engines," AIAA, Vol. 69, 1969, p. 493.*

DISCUSSION

W. R. Roudebush *(National Aeronautics and Space Administration)*

Mr. Bahr has provided a valuable survey of the aircraft turbine engine exhaust emissions problem. There are a few points on which I would like to comment.

With regard to smoke emissions the point is made that further efforts to obtain smoke numbers significantly less than 20 do not appear to be warranted. We should keep in mind, however, that from an air purity standpoint the goal of eliminating visible smoke is a somewhat arbitrary one. Furthermore, indications are that the approaches taken so far to reduce smoke might not yield much further improvement. Therefore, we should put some effort into study of novel concepts that could reduce carbon particles even further in case that should ever be required.

The author is optimistic about being able to obtain high idle efficiencies. Although I am also optimistic, there is little actual data yet available to show this can be done in a practical way. Preliminary data taken by NASA support the author's contention that idle efficiency can be significantly improved by better fuel atomization. However, idle efficiencies of 98% may not be gotten by this means alone. If altering primary zone stoichiometry should prove to be necessary to obtain such high efficiencies, the problem may be more severe than we presently expect it to be.

Very little systematic experimental work has been done to reduce nitric oxide. Therefore, I agree with the conclusion drawn in this paper that this is an area where extensive research is most likely to be required if major reductions are ever to be achieved through combustor design.

The discussion of specific hydrocarbon emissions was very interesting. I was not aware of data in this area and would like to have seen the results presented more quantitatively. It may become quite important to understand and control the formation of specific components of the hydrocarbon emission as more is learned of the photochemistry of pollution and of the generation of disturbing odors.

EMISSION CHARACTERISTICS OF CONTINUOUS COMBUSTION SYSTEMS OF VEHICULAR POWERPLANTS – GAS TURBINE, STEAM, STIRLING

W. R. WADE and W. CORNELIUS

General Motors Research Laboratories, Warren, Michigan

ABSTRACT

The low emission potential of continuous combustion systems has generated considerable interest in their application to vehicular powerplants. While the prime consideration for a vehicular combustion system is low exhaust emissions, emphasis must also be placed on the usual considerations of performance, reliability, durability and simplicity.

The General Motors Research Laboratories has had experience with three different types of continuous combustion powerplants; namely, gas turbines, steam engines and Stirling engines. Combustors for the steam and Stirling engines were patterned after the can-type burner of the regenerative gas turbine. The sizes and configurations of the burners varied markedly because of the differences in the heat release rates and the thermodynamic cycles of the powerplants which accounted for the widely-different operating ranges of inlet pressure, inlet temperature and fuel-air ratio of the three burners.

Unburned hydrocarbon (HC), carbon monoxide (CO) and nitric oxide (NO) emissions were measured from typical combustion systems of the three powerplants. Using these measurements, steady-state emission characteristics of the combustion systems were compared on the bases of exhaust gas concentration (ppm), emission index (gm/kg of fuel burned) and specific emission (gm/bhp-hr). While the emission index is exclusively an indicator of burner performance, the specific emission parameter reflects overall powerplant performance.

These test data indicated the major problem associated with each burner to be high emissions of oxides of nitrogen. The NO_X emission indices of the regenerative gas turbine and Stirling engine burners were comparable, reflecting the strong influence

References pp. 440-443

of high inlet temperature. The NO_X emission index of the steam engine burner was considerably lower due to operation at ambient inlet air temperature, but was still higher than required to meet the 1976 Federal standards. Emissions of unburned hydrocarbons and carbon monoxide from each burner were decidedly low, although operation of the steam engine burner at ambient inlet conditions necessitated considerable burner development to minimize these pollutant emissions. Exhaust smoke and aldehyde emissions of the burners have not constituted serious problems.

Emission control techniques were applied to each burner after reviewing the chemical kinetics of the pollutant species of concern. It was determined that two conflicting requirements must be satisfied for a low emission burner; namely, low temperature – short residence time for low NO emission and high temperature – long residence time for low HC and CO emissions.

The gas turbine type of burner was deemed to be more amenable to the control of NO emissions than the other two types of burners because of the availability of a large quantity of excess air that can be used for emission control and a reasonably-low outlet temperature that precludes continuing chemical reactions. Early quench and primary zone equivalence ratio modification have been effective means for reducing NO emissions from the conventional gas turbine burner. Theoretical studies reveal that other control techniques may be equally or more effective depending on the design and operating characteristics of the particular engine burner.

Vehicle mass emissions are influenced significantly not only by the design and operating conditions of the combustion system but also by the performance of the powerplant and the duty cycle of the vehicle. In seeking a vehicle that will meet the 1976 Federal mass emission requirements, it will be most advantageous to consider the combustion system and other major components of the powerplant as an integrated system. Since vehicle mass emissions (grams per mile) are dependent on the emission index of the burner as well as the fuel economy of the vehicle (miles per gallon of fuel burned), reductions in vehicle mass emissions can be obtained by either separate or preferably combined improvements in burner, engine and vehicle performances.

INTRODUCTION

The search for an alternate thermal powerplant to replace the spark ignition reciprocating engine in a vehicle has received considerable attention for many years. The General Motors Research Laboratories is in the unique position of having developed combustion systems for several alternate powerplants, namely the gas turbine, steam, and Stirling engines. These investigations have been prompted by various desirable characteristics of these engines. Recently, the low emission potentials of the continuous combustion systems employed in these three powerplants have received increased attention.

General Motors experience with the vehicular gas turbine dates back two decades when a simple cycle gas turbine was developed (1). Since then, extensive research has been devoted to improving the powerplant to make it competitive in today's engine market. The GT-309 gas turbine engine was the latest design in a series of heavy duty vehicular, dual shaft, regenerative gas turbine engines developed at the GM Research Laboratories (2). Specifically designed for heavy-duty vehicle applications, the GT-309 is rated at 280 hp on an 80 F day. The engine features a single-stage centrifugal compressor, a rotating regenerator, a single can-type direct-spray burner, and two axial turbines. To improve part load fuel economy and to provide engine braking, a novel system called Power Transfer is employed which connects the gasifier and power turbine shafts through a controlled torque coupling.

General Motors experience with the steam engine dates back to 1929 when two steam-powered buses were evaluated. More recently, the SE-101 steam powerplant was developed for installation in a modified 1969 Pontiac Grand Prix (3,4). This powerplant develops 160 gross horsepower with an expander inlet steam rate of 1800 lb/hr at a pressure of 800 psia and a temperature of 725 F. The expander of the SE-101 is a four-cylinder, single expansion, uniflow, single-acting type. The steam generator is a once through forced-circulation type with the expander inlet steam being controlled by a throttle valve. The combustion system consists of two gas turbine-type direct-spray burners supplied with air by a single combustion air blower.

Concurrently with the SE-101 project, Besler Developments under contract to General Motors, designed, built, and installed the SE-124 steam engine in a modified 1969 Chevelle. The expander of the SE-124 is a double-acting, compound, reciprocating, V2 configuration (4). The steam generator is a monotube forced-circulation design. The combustion system consists of a vortex type combustion chamber with an air-atomized fuel spray. The maximum road load speed of the car is only about 58 mph.

General Motors investigated Stirling engines in a cooperative program initiated with N. V. Philips Laboratories of the Netherlands (5). For more than a decade, General Motors has evaluated various sizes and types of Stirling engines (6). The GM Research Laboratories undertook the adaptation of the Stirling engine as part of a program of the U. S. Army Mobility Command's Engineer Research and Development Laboratories at Ft. Belvoir, Virginia, to seek a quiet powerplant. These ground power unit (GPU) packages included a single-cylinder engine using hydrogen as a working fluid at a design pressure of 1000 psi. The latest GPU-3 engine is rated at 10 hp at 3000 rpm. This closed-cycle external combustion engine uses No. 2 diesel fuel and operates with preheated combustion air.

Coincident with the development of a burner for the Stirling ground power unit, a combustion system was developed at the GM Research Laboratories to satisfy the requirement of a 100 hp output cylinder for a multi-cylinder Stirling engine. A scaled-up version of the 10 hp GPU burner was developed for this application.

References pp. 440-443

Several years ago, the 10 hp Ground Power Unit was used together with appropriate electrical components to build the Stir-Lec I (7), a Stirling hybrid electric system powering a modified 1968 Opel Kadett. Later, the electric drive system of the Stir-Lec I was modified and the vehicle was renamed the Stir-Lec II.

It was determined that each of these continuous combustion powerplants had some desirable emission characteristics (4,8,9). The nature of the continuous combustion system, as contrasted to the intermittently-firing spark-ignited reciprocating engine, offers potential for lower emissions. Continuous combustion of a very lean overall fuel-air mixture combined with minimal quench surfaces results in low HC and CO emissions.

This paper reviews the design and operating requirements of these three powerplant burners, the emission characteristics of each burner, powerplant, and vehicle, and the emission control techniques which are peculiarly effective for each burner. The similarities and differences of the three combustion systems are discussed.

BURNER OPERATING REQUIREMENTS, DESIGN AND PERFORMANCE PARAMETERS

Function of Burner in Engine Thermodynamic Cycle – The thermodynamic cycle of a continuous combustion engine includes either a heat addition process (for the internal combustion gas turbine engine) or a heat transfer process (for the external combustion steam and Stirling engines). In each case, the operating conditions of the burner are determined by the location of the heat addition or heat transfer process within the thermodynamic cycle. These operating conditions, in turn, determine the general design requirements of the burner. The burner operating conditions, together with the specific burner design, have a major influence on the emission characteristics of each burner.

The influence of the thermodynamic cycle on the burner operating requirements for each of the three powerplants are discussed below.

Gas Turbine – The Brayton thermodynamic cycle for a dual-shaft regenerative gas turbine is illustrated on a temperature versus entropy plot in Fig. 1. The total engine airflow is compressed to the peak cycle pressure during process 1-2. Process 2-3 represents heat addition during passage through the regenerator. The burner heat addition process is shown to occur during process 3-4 along a constant pressure line. The thermal energy of the burner exhaust gas is extracted by expansion through the turbine stages during process 4-5-6 followed by a constant pressure heat transfer process 6-7 in the regenerator. The design or full load burner operating conditions for the GT-309 gas turbine engine are presented in Table 1.

Steam – The Rankine thermodynamic vapor cycle is plotted in Fig. 2 on a temperature versus entropy plot for steam. The feedwater pump raises the pressure of

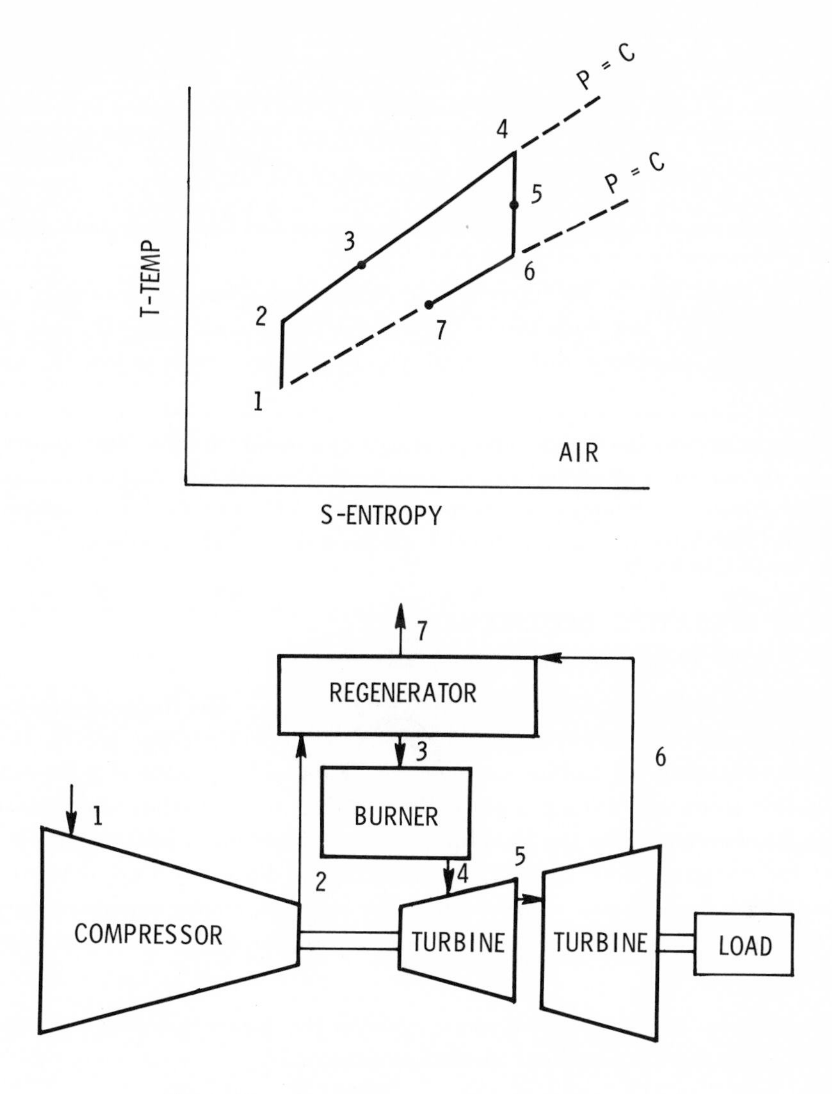

Fig. 1. Basic Brayton cycle engine (gas turbine).

the water during process 1-2. Heat is transferred from the high temperature combustion products to the working fluid during processes 2-3 (liquid state), 3-4 (saturated liquid-vapor state) and 4-5 (superheated vapor state). Useful power is generated by the expander during the expansion process (5-6). Steam is condensed during the process 6-1.

A temperature versus entropy plot for the external combustion system is also shown in Fig. 2. The combustion air blower raises the air pressure slightly during process 1A-2A. During the constant pressure combustion process 2A-3A, thermal

TABLE 1

Design Burner Operating Conditions

	Regenerative Gas Turbine (GT-309)	Steam Engine (SE-101)		Stirling Engine (GPU-3)
Inlet Temperature (F)	1091	100		1200
Outlet Temperature (F)	1700	2600		3360
Temperature Rise (F)	609	2500		2160
Inlet Pressure (Atm)	3.68	1.03		1.03
Differential Pressure, $\Delta p/p$ (Pct)	4.0	3.0		3.0
		Total	One of Two Burners	
Air Flow (lb/sec)	3.66	1.27	0.635	0.0345
Fuel Flow (lb/hr)	131.9	183.0	91.5	5.0
Heat Release (10^6 Btu/hr)	2.45	3.4	1.7	0.093
Fuel-Air Ratio	0.01		0.04	0.04

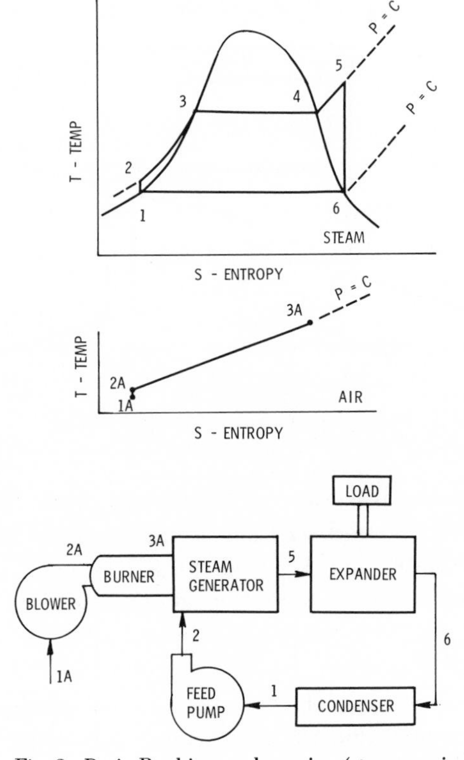

Fig. 2. Basic Rankine cycle engine (steam engine).

energy is generated by burning a fuel-air mixture within the burner. The thermal energy of the burner exhaust gas is extracted by a heat transfer process during passage of the gas through the steam generator. The design or full load burner operating conditions for the SE-101 steam engine are presented in Table 1.

Stirling – The Stirling thermodynamic cycle is shown in Fig. 3. An isothermal compression of the working fluid (hydrogen, helium, or other gases) occurs during process 1-2. Heat transfer to the working fluid occurs during the constant volume process 2-3 and during the isothermal expansion process 3-4 (power stroke). Constant volume cooling of the working fluid takes place during process 4-1. Since the final heat addition process is an isothermal process, the temperature of the combustion gases leaving the working fluid heat exchanger can be no lower than T_4. Therefore, an air preheater is required in a Stirling engine to lower the exhaust gas temperature to acceptable values by extracting energy from the exhaust gas, and thereby preheating the inlet air.

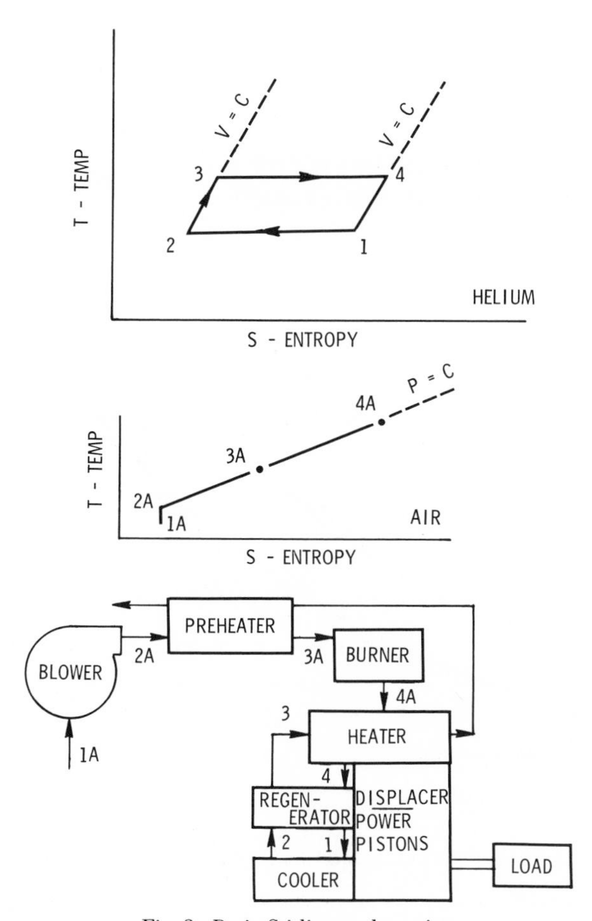

Fig. 3. Basic Stirling cycle engine.

References pp. 440-443

A temperature versus entropy plot for the external combustion system is also shown in Fig. 3. The combustion air blower raises the air pressure slightly during process 1A-2A. The burner exhaust gas preheats the combustion air during process 2A-3A. During the constant pressure combustion process 3A-4A, thermal energy is generated by burning a fuel-air mixture within the burner. This energy is subsequently transferred in the working fluid heat exchanger. The design burner operating conditions for the GPU-3 Stirling engine are presented in Table 1.

The burner heat addition processes for the combustion circuits of the three powerplants are compared on a temperature versus entropy plot shown in Fig. 4. The three processes are distinctly different with respect to pressure level, inlet temperature and temperature rise as determined by overall fuel-air ratio. The location of each process on the temperature entropy plot has an important influence on the exhaust emission characteristics of the particular powerplant as well as the design of its burner.

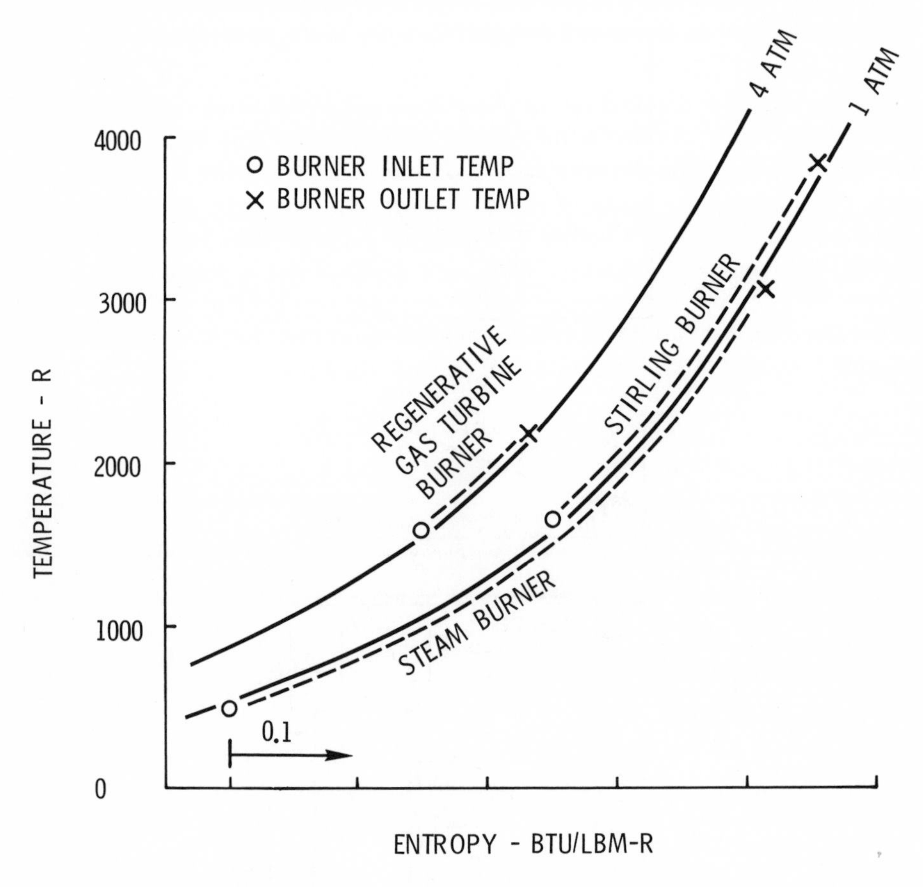

Fig. 4. Burner heat addition processes for the gas turbine, steam, and Stirling powerplants shown on a temperature versus entropy plot.

Burner Design Parameters – The overall burner size required for each of the three powerplants is determined primarily by the total heat release required and by the location of the heat addition process on the temperature versus entropy plot shown in Fig. 4. The configuration of the burner has a significant influence on its emission characteristics. The discussions in this paper will refer to can-type configurations of various designs.

To obtain a high combustion intensity and thereby minimum burner volume, vigorous mixing of the air with the fuel spray must be accomplished. To enhance mixing, the primary and dilution air jets must have sufficient penetration. To assure continuous combustion, a vigorous and stable recirculation flow pattern must be established within the primary zone. Adequate cooling of the burner wall must also be provided. The fulfillment of these combustion and cooling conditions is realized by incorporating a certain amount of burner pressure loss into the system (10-14). While such a pressure loss is required to obtain good combustion performance, it has a depreciating effect on the overall thermal efficiency of the powerplant.

The pressure loss required for air penetration and recirculation may be related to a pressure loss factor, defined as the ratio of static pressure drop across the burner to the dynamic pressure in the annulus (15). The three burners under consideration were each designed with a pressure loss factor of approximately 30 to assure good performance. The GT-309 burner was designed for a pressure loss of four percent. The SE-101 and GPU-3 burners were each designed for a pressure loss of three percent. The burner size expressed as cross-sectional area per unit of air flow for each of the three burners is listed in Table 2. The differences are due to variations in design pressure loss, inlet temperature and inlet pressure conditions.

TABLE 2

Burner Design Parameters

	GT-309	SE-101	GPU-3
Cross Sectional Area Per Unit Air Flow ($in^2/(lb/sec)$)	11.5	122.0	320.0
Hole Area per Unit Air Flow ($in^2/(lb/sec)$)	4.3	12.0	16.0
Combustion Intensity (10^6 Btu/Hr-Ft^3-Atm)	3.7	4.15	4.8
Burner Temperature Rise (R)	609	2500	2160

References pp. 440-443

The total air entry area and locations of the ports in the burner must be determined to provide the specified pressure loss and to control the distribution of the air in the burner. For each of the three burners, the burner hole area per unit of air flow is listed in Table 2. The differences are due to variations in the design pressure loss, inlet temperature, and inlet pressure conditions.

The heat release rate per unit of combustor volume and pressure, or combustion intensity, for diffusion flame burners of similar design is influenced by many design and operating variables. Combustion intensities of the three burners are listed in Table 2. The combustion intensities of the three systems are similar, even though the heat release processes occur at different locations on the temperature versus entropy plot (see Fig. 4).

The design temperature rises of the three burners operating at the design fuel-air ratios are listed in Table 2 (16). These clearly define the operating temperature ranges of the burners. It is apparent that the gas turbine burner has the lowest temperature rise and the leanest fuel-air ratio. On the other hand, the temperature rises and fuel-air ratios of the steam and Stirling burners are similar, with the slightly lower temperature rise of the Stirling burner being due to greater dissociation of the gases with operation at higher inlet air temperatures.

Design and Construction Factors – The burners of the GT-309, SE-101 and GPU-3 powerplants are illustrated in Fig. 5. In each case, effectual combustion performance

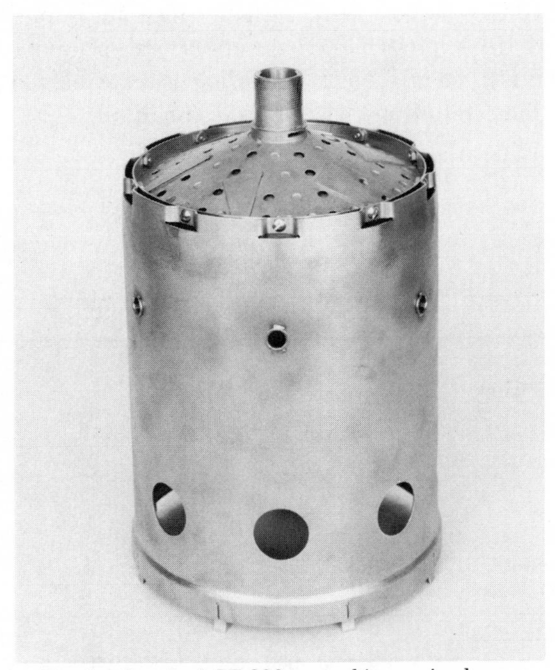

Fig. 5a. Standard GT-309 gas turbine engine burner.

was the prime design consideration. No serious attempt was made to minimize exhaust emissions, although a correlation does exist between effectual combustion performance and minimum combustible pollutant emissions. The SE-101 burner is similar to the GT-309 burner except for the absence of a row of dilution air holes. In both designs, air is admitted into the burner through openings in the conical dome and is swirled tangentially along the inner surface of the dome by means of internal baffles. The fuel nozzle and igniter are inserted into the burner through openings in the dome. Air to cool the burner body is admitted through a single annular opening located at the junction of the dome and the body. The primary air holes are bushed for improved penetration. The GPU-3 burner is much smaller than the other two burners. While it is basically similar in design to the SE-101 burner, some of its design details are different. Air is admitted through tangential slits in the simplified dome. The body is attached to the dome by means of the primary air bushings.

In the case of each burner, all of the air entering through the dome holes and the bushed primary holes in the burner body is considered to be available for reaction with the incoming fuel. While the air admitted through the annular opening is used primarily to cool the burner body, some of this air becomes involved in the combustion process. For the gas turbine burner, the air entering through the lower row of holes in the burner body mixes with the hot combustion products and thus reduces the temperature of the gas entering the gasifier turbine to a tolerable level.

An air-atomizing fuel nozzle was used in each burner. The effect of fuel spray on emission performance will not be considered in this paper. The igniters are modifications of an automotive-type spark plug. While a large variety of fuels have

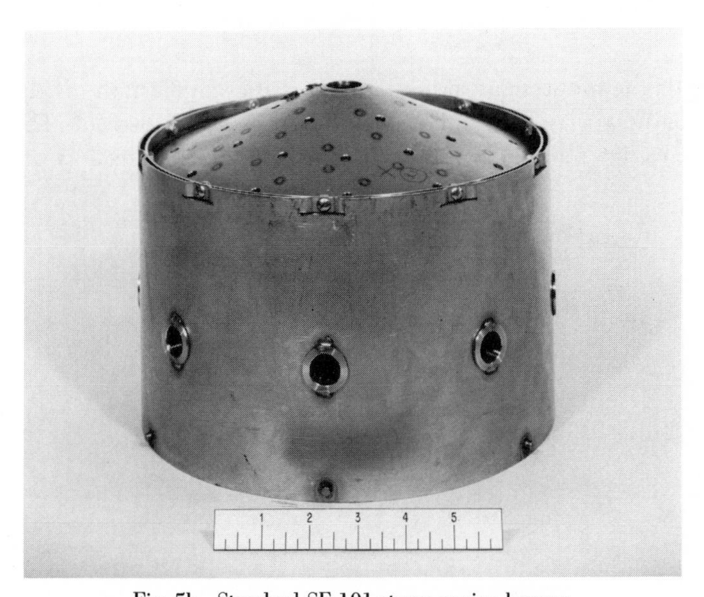

Fig. 5b. Standard SE-101 steam engine burner.

References pp. 440-443

been tested in these burners, the emission data reported in this paper were obtained while burning Diesel Fuel No. 2.

Burner Performance Parameters – In addition to the prime requirement of low exhaust emissions, a practical combustion system for a vehicular powerplant must prove satisfactory in the following additional performance areas:

1. Combustion Efficiency – One commonly used definition, the ratio of actual to ideal temperature rise, is strongly dependent upon the ability to measure accurately the elevated temperatures of the burner exhaust gases. Adiabatic combustion efficiency may be more accurately defined as the ratio of the actual heat release to the ideal heat release (17). Actual heat release may be determined by subtracting the enthalpies of combustion of the combustible components in the exhaust (exhaust emissions) from the enthalpy of combustion of the fuel. Thus, adiabatic combustion efficiency and exhaust emission levels are related as follows:

$$\eta_{comb} = \frac{\bar{h}_{fuel} - \Sigma (m_i \bar{h}_i)}{\bar{h}_{fuel}} \qquad (1)$$

where: i = combustible components in exhaust (HC, CO, H_2)

m_i = mole fraction of component "i"

$\bar{h}_i$ = enthalpy of combustible component "i" (BTU/mole)

$\bar{h}_{fuel}$ = enthalpy of combustion of the fuel (BTU/mole)

Using this more accurate procedure, adiabatic combustion efficiencies of the three burners were determined to be well above 99 percent. The commonly reported value for combustion efficiency may be determined by subtracting the

Fig. 5c. Standard GPU-3 Stirling engine burner.

burner heat loss per unit heat release from the adiabatic combustion efficiency. In this paper only adiabatic combustion efficiency values will be discussed.

2. Combustion Intensity – The design goal of each burner was the capability of releasing all of the fuel energy needed for maximum engine power development in the volume assigned to the burner. This goal of high heat release per unit volume was realized in each burner as evidenced by the high combustion efficiency which indicated essentially complete combustion at the burner exit.

3. Burner Pressure Loss – Because of its adverse effect on engine power, the design pressure loss in each burner was held to the minimum required for suitable combustion and cooling performance. For example, an additional pressure loss of one percent in the gas turbine burner can cause an increase in specific fuel consumption of one percent for the engine. Similar deteriorations can occur for the steam and Stirling engines.

4. Combustion Stability – Stable burner performance was experienced while operating each burner in the test facility and in the test engines at all steady-state operating conditions from idle to design point. No flame-outs or erratic combustion were encountered in any of the vehicle installed engines during rapid and frequent changes in operating conditions.

5. Carbon Deposition – Negligible amounts of carbon were deposited on the exposed surfaces of the burners, fuel nozzles and igniters during prolonged running of the several test engines.

6. Ignition – Reliable ignition of minimum time delay was realized over a wide range of ambient air temperature conditions in the test vehicles.

7. Durability – A major area of concern in these burner developments, particularly at elevated temperatures, is the ability to cool adequately the burner body with cooling air admitted through a single annular opening located at the junction of the body and dome. Analytical and experimental studies were conducted on the GT-309 burner to determine the parameters necessary to predict film cooling effectiveness as a function of design and operating parameters (18). Use of a single annular opening for the introduction of film cooling air proved to be adequate; good durability was achieved with each burner.

8. Outlet Gas Temperature Profiles – Uniform outlet gas temperature profiles are necessary to preclude hot spots which can cause damage to downstream components, either a turbine wheel or a heat exchanger. The T_{max}/T_{ave} ratios of gas temperature profiles were below 1.10 for all burners.

Emission Performance Parameters – The major gaseous air pollutant emissions of concern are unburned hydrocarbons (HC), carbon monoxide (CO) and oxides of nitrogen (NO_x). The emissions from combustors, engines and vehicles are measured in terms of the volumetric concentration (typically parts per million, ppm) of the specific pollutant in the exhaust gas. The instrumentation and gas sampling procedures employed at the GM Research Laboratories to obtain representative measurements of burner exhaust concentrations of the air pollutants are described in Appendix A. It is necessary to convert these concentration values into more meaningful mass emission performance parameters to make realistic comparisons of the emission performances of different combustors, engines, and vehicles.

Emission Index – For a valid comparison of the emission characteristics of burners, the pollutant emissions are expressed in terms of an emission index defined as grams of pollutant emitted per kilogram of fuel burned. Thus, the emission index of a burner describes how efficiently the burner consumes the fuel supplied to it in terms of the pollutant emitted. Using the measured emission concentration data and corresponding fuel-air ratios, the emission index is computed as follows:

$$EI_i = C_i \cdot 10^{-3} \cdot \frac{M_i}{M_{exh}} \cdot \frac{W_{exh}}{W_{fuel}} \tag{2}$$

where: i = pollutant species of interest

EI_i = emission index of pollutant species "i" (gm pollutant/kg fuel)

C_i = concentration of pollutant species "i" (ppm) as measured

M_i = molecular weight of pollutant species "i"

M_{exh} = molecular weight of exhaust (wet or dry depending upon the basis for C_i)

W_{exh} = exhaust mass flow rate (gm/hr)

W_{fuel} = fuel flow rate (gm/hr)

An approximate value of the emission index can be computed from the following simplified expression:

$$EI_i = C_i \cdot 10^{-3} \frac{M_i}{M_{exh}} \cdot \frac{1}{FA} \tag{3}$$

where: FA = overall fuel-air ratio

Eq. 3 indicates that the emission index is proportional to the pollutant concentration and inversely proportional to fuel-air ratio. Rearranging Eq. 3 indicates that pollutant concentration is proportional to the emission index and the fuel-air ratio.

A typical plot of the relationships between pollutant concentration, fuel-air ratio and emission index is shown in Fig. 6. The figure displays NO concentrations as a function of fuel-air ratio for a number of constant emission indices*. For a constant emission index, it will be seen that the permissible concentration increases for richer fuel-air ratios. Data points at the design conditions of the GT-309, SE-101 and GPU-3 burners are indicated on the plot. By comparing lines of constant emission index with these plotted burner data points, it can be concluded that measured concentrations cannot be used to compare the emission levels of different combustion systems unless related to the fuel-air ratio. On the other hand, an emission index does provide a valid comparison basis of combustion systems isolated from other powerplant considerations.

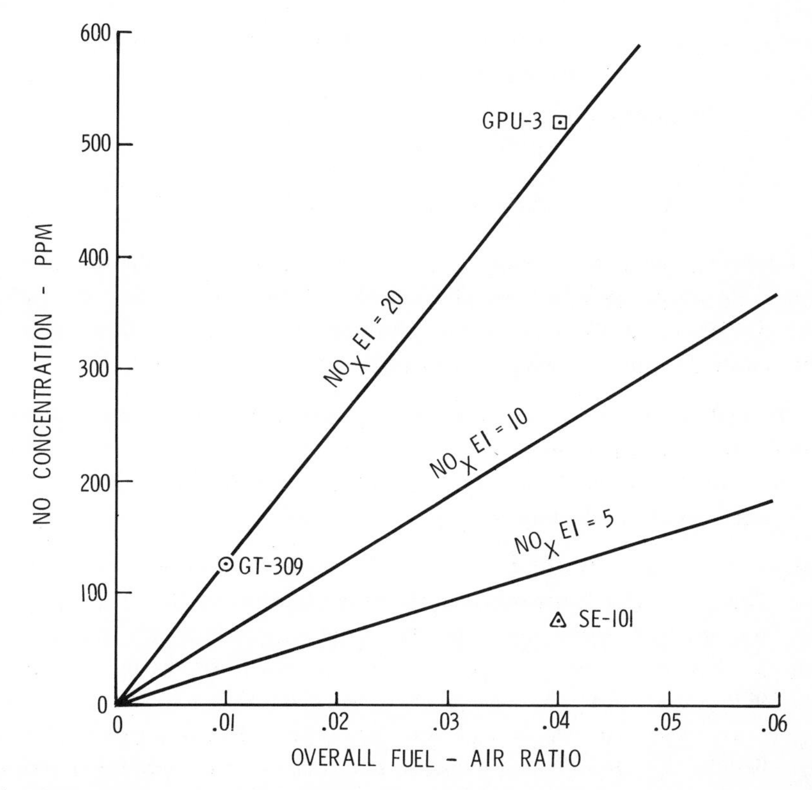

Fig. 6. Relationship between NO concentration and fuel-air ratio for constant emission indices.

Specific Emission – While the emission index provides an excellent means for rating the emissions performance of a burner, another standard of comparison is

* *In typical burner exhaust gas, the predominate species of oxides of nitrogen is nitric oxide (NO). Only nitric oxide (NO) concentrations have been measured and will be discussed in this paper. Mass emissions of oxides of nitrogen (NO_x) are always calculated in terms of the equivalent mass of nitrogen dioxide (NO_2).*

References pp. 440-443

needed when the burner is installed in an engine and the emission performance of the engine is considered. This engine standard must take into account the efficiencies of the combustion system as well as the efficiencies of all of the other powerplant components. Such an appropriate means for evaluating the exhaust emissions of powerplants is in terms of specific emissions defined as grams of pollutant per horsepower hour. Specific emissions are computed as follows:

$$SE_i = EI_i \cdot BSFC \cdot K \tag{4}$$

where: SE_i = specific emissions of pollutant species "i" (grams per horsepower hour)

EI_i = emission index of pollutant species "i" (grams per kg fuel burned)

BSFC = brake specific fuel consumption (lbs/bhp-hr) of the powerplant

K = unit conversion constant (kg/lb)

Eq. 4 indicates that specific emission values are proportional to the burner emission index and the brake specific fuel consumption (which is inversely proportion to thermal efficiency) of the powerplant. Thus, specific emission values describe the mass of pollutants emitted during the generation of a unit of work.

The thermal efficiencies of the three powerplants under consideration vary widely. At design point operation, measured values of BSFC for the regenerative gas turbine and Stirling engines are in the range of 0.5 lbs/bhp-hr while the BSFC for the steam engine is in excess of 1.0 lbs/bhp-hr.

Vehicle Emissions – A third parameter is needed to compare the emissions of vehicles. This parameter must embody the performances of the entire power train (engine, transmission, drive line) and the vehicle aerodynamic drag and rolling resistances when the vehicle is operated over a prescribed driving schedule. The Federal 1970 7-Mode, 7-Cycle test procedure was the one in force for 1971 and prior passenger cars (19). All vehicle emission data reported in this paper refer to this driving schedule. Another Federal emission test procedure is replacing it and applies to 1972 and later model year passenger cars (20). In both of these test procedures, engine exhaust concentrations of HC, CO and NO_x are measured while operating a car over a prescribed schedule of steady-state and transient conditions on a chassis dynamometer. The emission data are reduced to a single vehicle mass emission value for each pollutant expressed as grams per test mile.

California has legislated a test procedure for evaluating the emissions of heavy-duty trucks (21). The truck engine is operated at a series of different speed and load settings. The exhaust concentrations of HC, CO and NO_x, in addition to the usual engine test data, are measured at each operating condition. The modal data are weighted and reduced to specific emission values expressed as gms/bhp-hr.

In the case of passenger cars, the vehicle emission value expressed as grams of pollutant per mile of operation is related to the average emission index of the burner as follows:

$$VE_i = EI_i \cdot \frac{K}{(mpg)} \tag{5}$$

where: VE_i = vehicle emission value (grams per mile)

EI_i = average emission index (grams per kg fuel burned)

mpg = average fuel consumption of the vehicle during test (miles per gallon)

K = conversion factor determined by the specific gravity of the fuel burned (kg/gallon)

Fig. 7 shows graphically the relationships between the burner emission indices of the three air pollutants required to meet the 1976 Federal vehicle emission standards

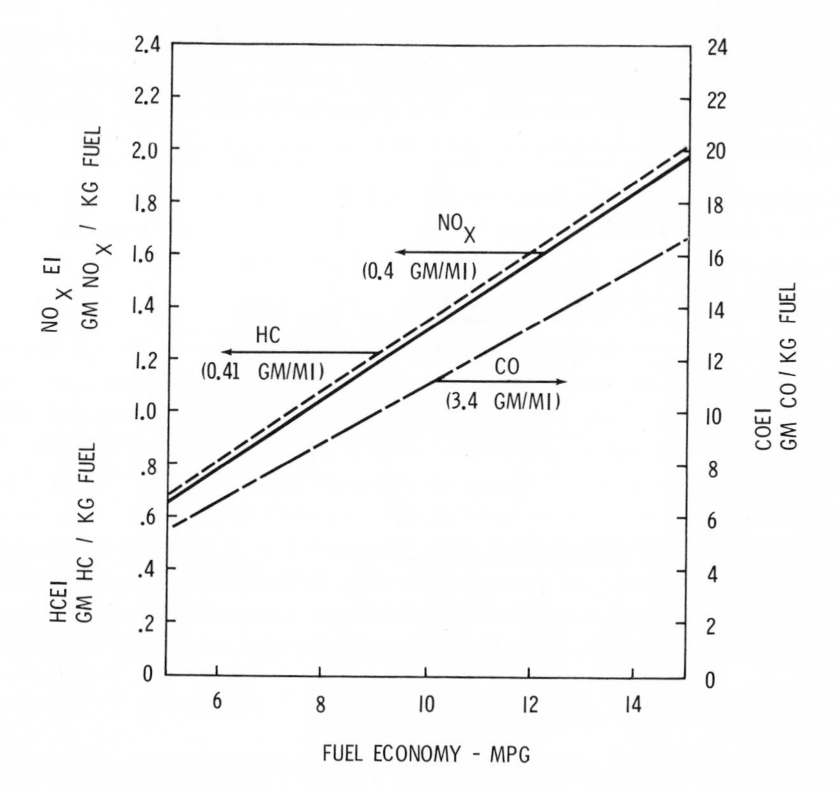

Fig. 7. Allowable emission indices required to satisfy 1976 Federal exhaust emission standards as a function of vehicle fuel economy.

References pp. 440-443

and the average fuel economy of the vehicle during operation over the test procedure. A fuel specific gravity of 0.8 was used for determining the relationships shown in Fig. 7.

Operation of a vehicle according to the prescribed Federal emission test procedure provides vehicle emission data expressed as grams per mile. Only a limited number of vehicles powered by continuous combustion powerplants have been available for testing at the General Motors Research Laboratories. These several test vehicles have not been comparable in size, weight and use. Testing of these vehicles has furnished valuable emission data. However, a comprehensive understanding of vehicle emission performance can be more readily obtained from an analysis and simulation of the elements that comprise the vehicle grams per mile values.

A computer simulation was developed for predicting driving schedule emissions from a passenger car powered by a dual shaft gas turbine engine using combustors for which test rig emission data were available. A diagram of the calculative procedure is shown in Fig. 8. The interrelationships of the driving schedule, transmission

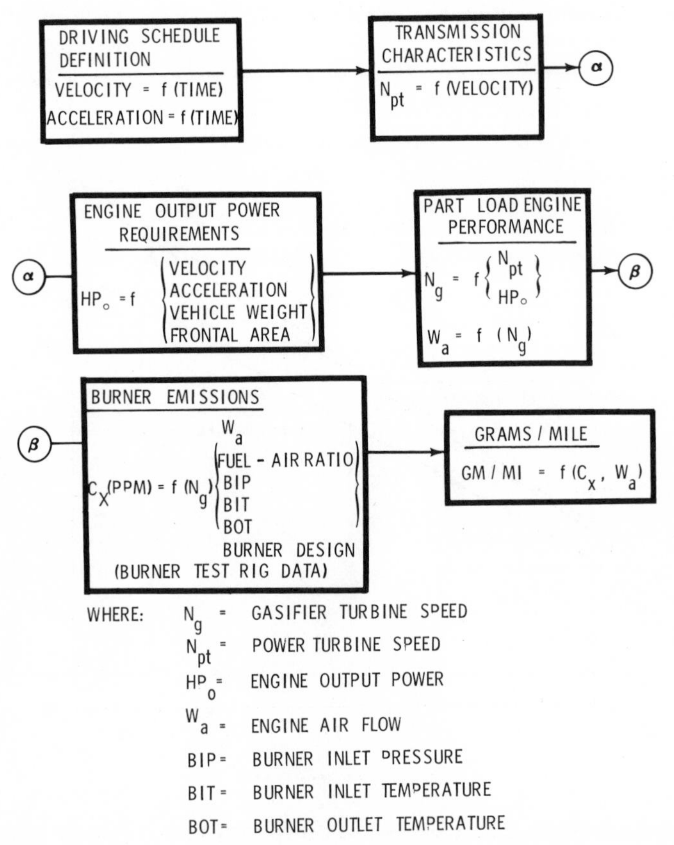

Fig. 8. Simulation of vehicle mass emissions from a two shaft gas turbine powered passenger car.

characteristics, vehicular characteristics, design and part load engine performances and burner emissions are illustrated. Details of the computation steps have been discussed elsewhere (8). This computer simulation provided emission values that correlated with experimental results obtained on a GT-309 engine-powered truck tractor. As a result, the vehicle simulation program has been used to calculate vehicle emission values of gas turbine installations in typical passenger cars using available steady-state burner test rig emission data.

At the present time, no test procedure has been promulgated for properly evaluating the mass emissions from a steam engine powered vehicle. A suitable procedure would have to account for the steam generator warmup time and the unpredictable occurrences of burner firing operation during the test driving schedule with an on-off control. The details of the special procedures, instrumentation and data reduction techniques that were used at the GM Research Laboratories for measuring the vehicle emission characteristics of steam cars are described in Ref. 4.

Using appropriate assumptions, a computer simulation was developed to predict driving schedule emissions from a passenger car powered by a steam engine. A diagram of this procedure is shown in Fig. 9. Employing steady-state burner test rig data in this simulation, the SE-101 vehicle emission performance was predicted with a good degree of accuracy except for a low prediction of hydrocarbon emissions. Hydrocarbon emissions are difficult to predict because of the on-off combustion system employed on the SE-101. The inter-relationships of the burner, vehicle, transmission, powerplant, and driving schedule for a Stirling engine powered vehicle are comparable to the steam powered vehicle relationships shown in Fig. 9.

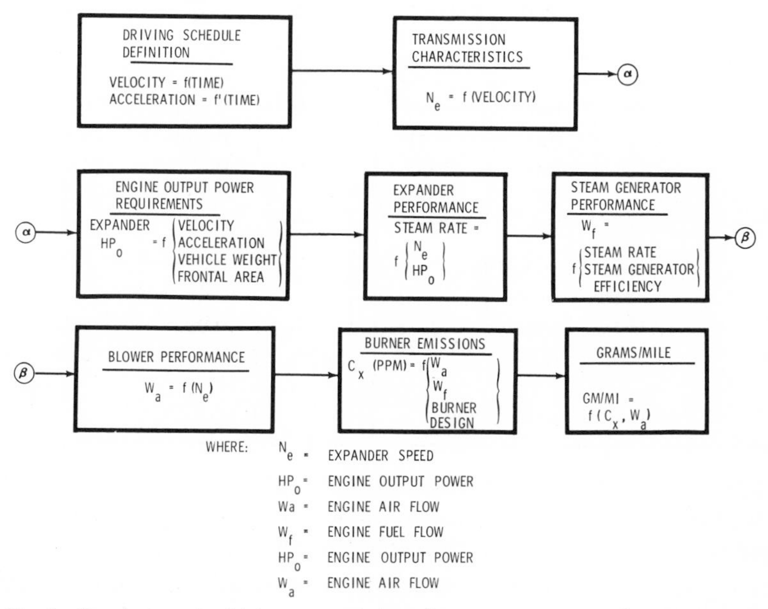

Fig. 9. Simulation of vehicle mass emissions from a steam powered passenger car.

References pp. 440-443

STEADY-STATE EMISSION CHARACTERISTICS OF THE STANDARD BURNERS

Exhaust Concentrations – Exhaust emission concentrations of the three standard engine burners were measured during simulated steady-state engine conditions in specially designed burner test facilities. The simulated engine operating conditions for each burner are listed in Table 3.

TABLE 3

Schedule of Burner Operating Conditions

Condition		Burner Inlet Temp.	Burner Outlet Temp.	Burner Inlet Pressure	Burner Airflow	Burner Fuel Flow	Fuel-Air Ratio
		F	F	in Hga	lb/sec	lb/hr	
GT-309		With Power Transfer					
	100(1)	1091	1700	110.2	3.66	131.9	0.0100
	90	1154	1700	92.1	3.06	99.1	0.0090
	80	1233	1700	75.6	2.48	69.1	0.0078
	70	1314	1700	62.3	1.99	46.2	0.0065
		Without Power Transfer					
	60	903	1132	50.8	1.78	22.6	0.0035
	50	964	1140	43.2	1.37	13.5	0.0027
	(1) Percent of design gasifier speed.						
SE-101	100(2)	100	2600	32.0	0.635	91.5	0.040
	80	100	2600	32.0	0.508	73.2	0.040
	60	100	2600	32.0	0.381	54.9	0.040
	40	100	2600	32.0	0.254	36.6	0.040
	20	100	2600	32.0	0.127	27.4	0.040
	(2) Percent of design fuel flow.						
GPU-3	100(3)	1350		32.0	0.038	5.42	0.040
	50	1210		32.0	0.021	2.86	0.040
	(3) Percent of design fuel flow.						

The HC, CO and NO exhaust concentrations (measured in parts per million, ppm) of the standard GT-309 burner are plotted in Fig. 10 as functions of percent of design burner load (i.e., percent of design fuel flow). Initial application of power transfer in the engine at approximately 35 percent of the design fuel flow condition results in

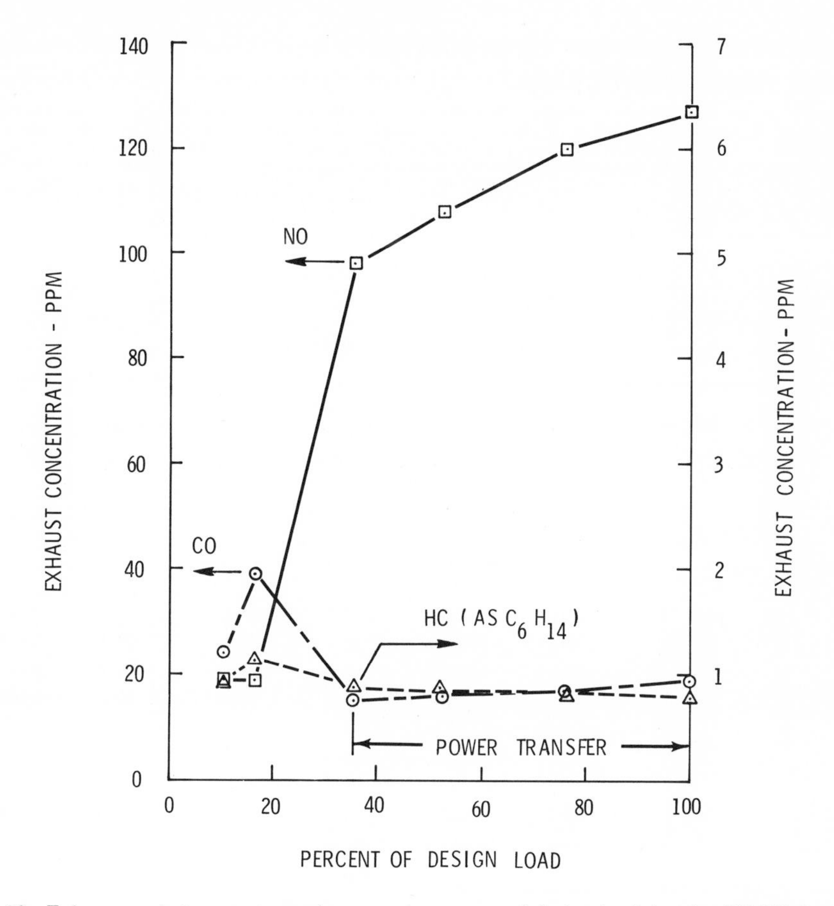

Fig. 10. Exhaust emission concentrations versus percent of design load for the GT-309 burner.

significant increases in burner inlet and outlet gas temperatures. Accompanying increases in burner air flow and inlet pressure occur normally as a result of increased compressor speed. Inlet temperature decreases slightly from a maximum value at the 35 percent condition with increased load due to the power transfer scheme of operation. As a result of elevated temperatures, NO concentrations remain at comparable high levels throughout the power transfer mode of operation but drop significantly to much lower levels at very light loads when power transfer is inoperative (corresponding to leaner overall fuel-air ratios and lower inlet temperatures). HC and CO concentrations are very low during power transfer operation attesting to the high combustion efficiency obtained with this burner at these operating conditions. Increased levels of HC and CO concentrations are experienced during non-power transfer operation.

The HC, CO and NO exhaust concentrations of the standard SE-101 burner are plotted in Fig. 11 as functions of percent of design burner load. At each part load condition investigated, the fuel and air flows were reduced proportionately to maintain a constant fuel-air ratio. Inlet pressure and temperature remain substantially constant at near ambient levels. As illustrated, exhaust concentrations of the three pollutants remain approximately constant throughout the load range, indicating the strong influence of fuel-air ratio and the weak influence of mass flow rate on the pollutant concentrations.

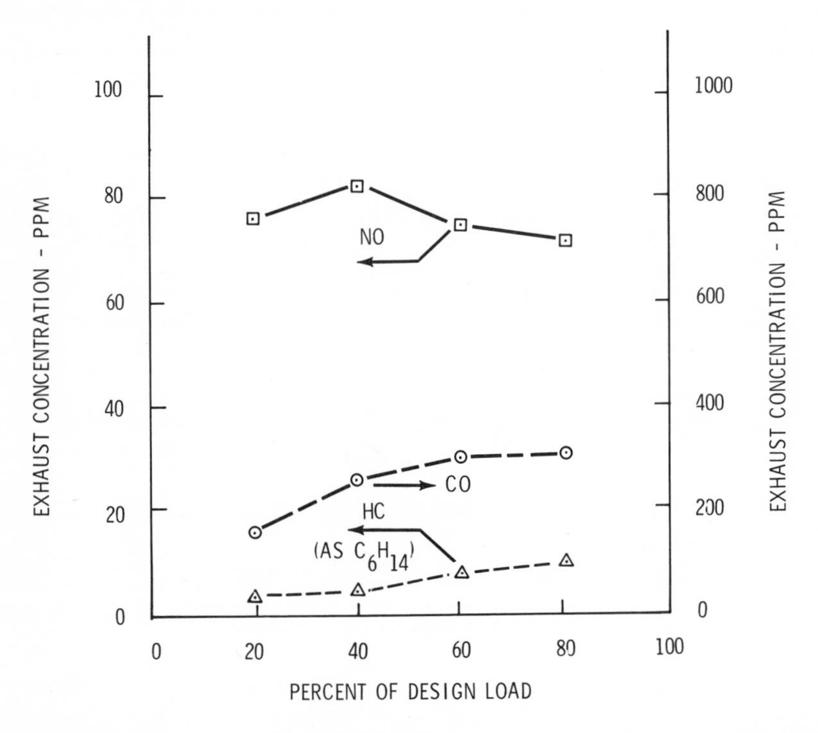

Fig. 11. Exhaust emission concentrations versus percent of design load for the SE-101 burner.

The HC, CO and NO exhaust concentrations of the standard GPU-3 Stirling burner are displayed in Fig. 12 as functions of percent of design burner load. The exhaust pollutant concentrations are not influenced significantly by a change in load because the fuel-air ratio is held constant throughout the load range. In comparison to the steam burner operating at a comparable fuel-air ratio, the NO concentrations of the Stirling burner are significantly higher while the HC and CO concentrations are lower, indicating the strong influence of high burner inlet air temperature.

The adiabatic combustion efficiencies of the three burners, determined from the exhaust pollutant concentrations, are plotted in Fig. 13 as functions of percent of the design fuel flows. As shown, the combustion efficiencies of the burners are in excess of 99.6 percent at all engine operating conditions investigated. The efficiencies of the SE-101 burner are generally lower than those of the GT-309 and the GPU-3 burner

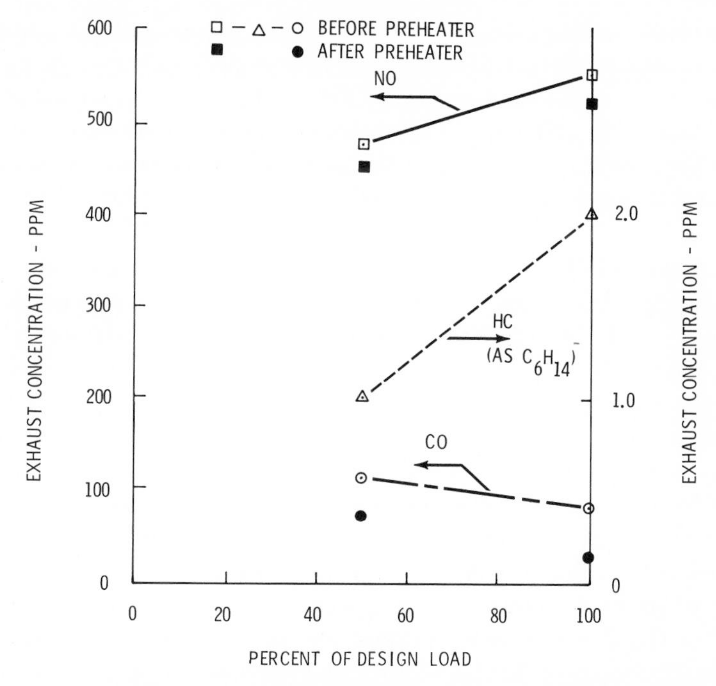

Fig. 12. Exhaust emission concentrations versus percent of design load for the GPU-3 burner.

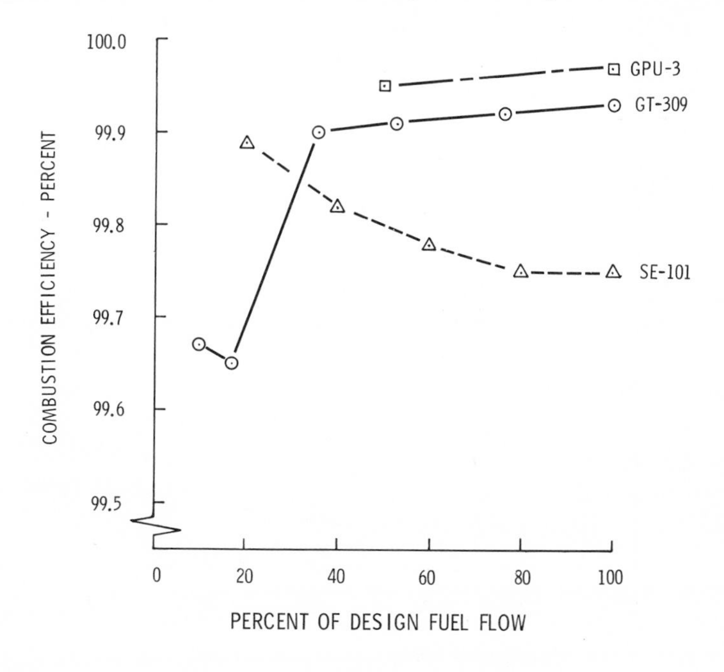

Fig. 13. Combustion efficiencies of the GT-309, SE-101, and GPU-3 burners versus percent of design fuel flow.

References pp. 440-443

throughout most of the operating range because of operation at approximately ambient inlet air temperature. However it is interesting to note that the efficiency of the SE-101 burner is higher than that of the GT-309 burner at the 20 percent fuel flow condition. This difference occurs because the SE-101 burner continues to operate at the design fuel-air ratio at the part load conditions whereas the GT-309 burner operates at significantly leaner fuel-air ratios at part load. The efficiencies shown in Fig. 13 do not take into account any heat losses from the burners.

Emission Indices – The HC, CO and NO_X emission indices of the three burners are compared in Figs. 14, 15 and 16, respectively as functions of the percent of design fuel flows. It will be seen in Fig. 14 that the HC emission indices of the Stirling burner are the lowest of the three burners, reflecting the high inlet temperature and the long residence times of the exhaust gas at high temperature that are typical of this combustion system. The HC emission indices of the gas turbine burner are not as low as the Stirling burner because even though the gas turbine burner operates at high inlet temperatures, the combustion products are quenched by dilution air sooner than the Stirling burner combustion products are cooled. The steam burner has the highest HC emission indices due to its operation at low inlet temperatures which are not as favorable for high combustion efficiencies as with preheated air. The relative variations of the HC emission indices with fuel flow rate for each burner correspond with the combustion efficiencies of that particular burner as shown in Fig. 13.

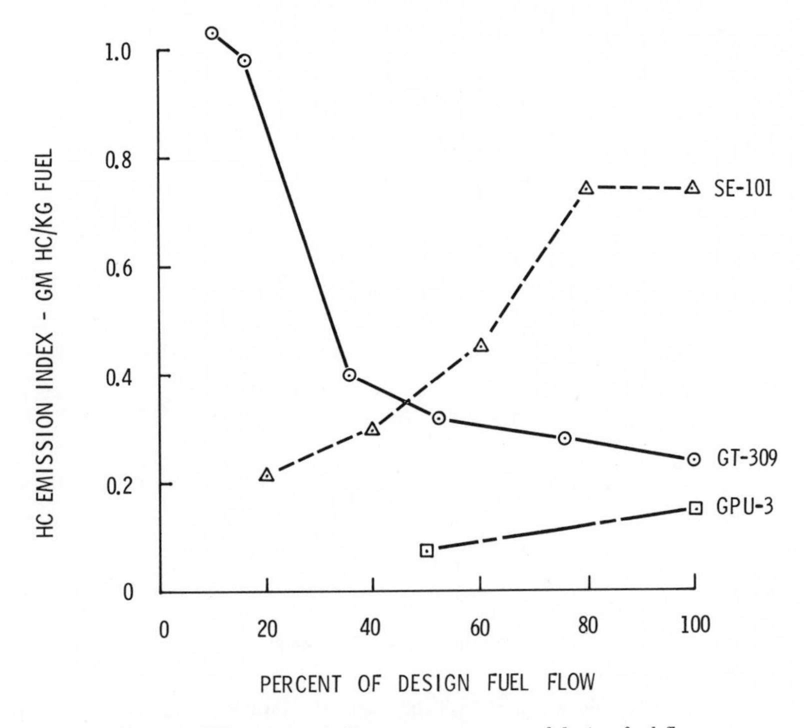

Fig. 14. HC emission indices versus percent of design fuel flow.

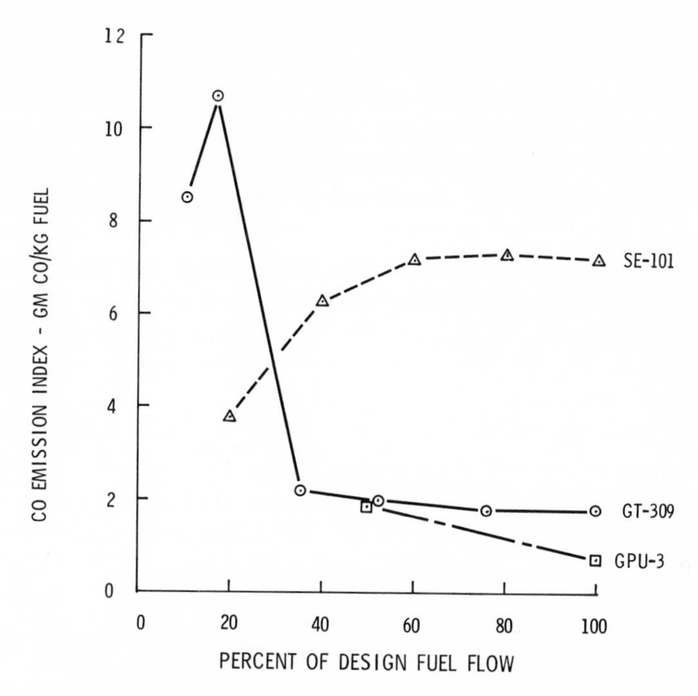

Fig. 15. CO emission indices versus percent of design fuel flow.

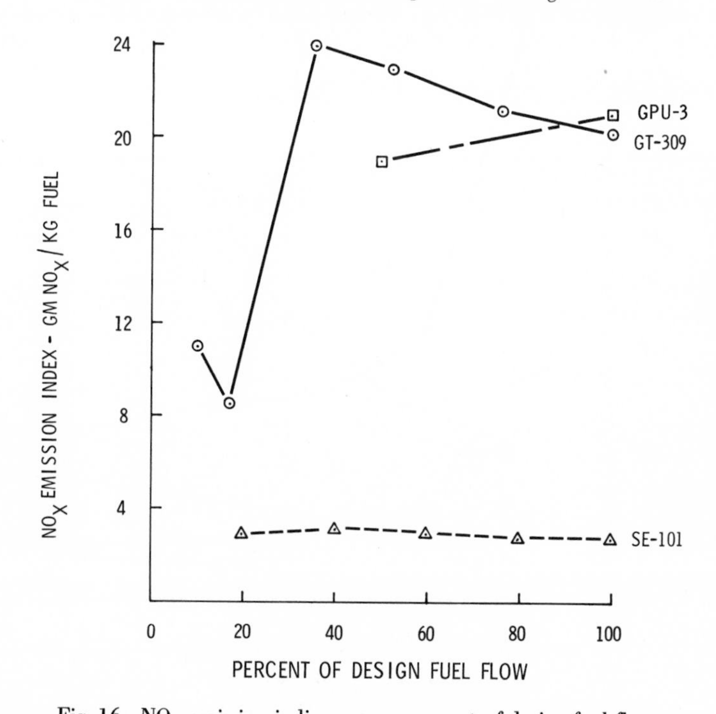

Fig. 16. NO_X emission indices versus percent of design fuel flow.

References pp. 440-443

The relative variations of the CO emission indices of the three burners as functions of percent of design fuel flow rates shown in Fig. 15 are comparable to the HC emission index data displayed in Fig. 14. The correspondence of these emission index data is not unexpected because the CO emissions are probably the result of similar inefficiencies in the combustion process. However, the actual levels of the CO emission indices are considerably higher than the related HC emission indices because of the relatively slower CO oxidation rate. As a result, a relatively larger quantity of carbon monoxide is present when the exhaust gas is quenched, thereby terminating the HC and CO oxidation processes.

Unlike the HC and CO emission index data, Fig. 16 shows the NO_X emission indices of the gas turbine and Stirling burners to be similar and of a much higher magnitude than the NO_X emission indices of the steam burner. Since the air supplied to the gas turbine and Stirling burners is preheated to similar temperatures, the flame temperatures (for similar equivalence ratios) in both burners are similar and significantly higher than in the steam burner which is supplied with air at an ambient temperature. The exponential effect of the reaction temperature on the NO formation rate is primarily responsible for the large differences in the NO_X emission indices between the steam burner and the gas turbine and Stirling burners. These differences are discussed in greater detail later (see Theoretical Considerations).

Specific Emissions – Brake specific HC, CO and NO_X emission values of the three powerplants are compared in Figs. 17, 18 and 19 as functions of the percent of design

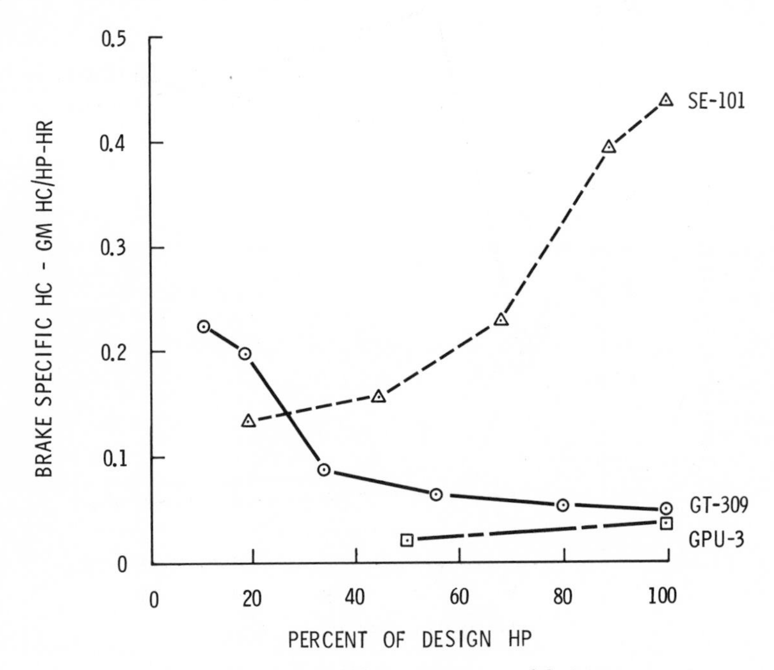

Fig. 17. Brake specific HC versus percent of design power.

engine power. The general trends of these specific emission values as functions of engine output are similar to their emission index counterparts. Higher thermal efficiencies (lower BSFC's) of the gas turbine and Stirling engines, as contrasted to that of the steam engine, cause their respective HC and CO specific emissions to be more widely separated from the corresponding values of the steam engine than was the case with the emission indices of the three burners (see Figs. 14, 15, and 16). In contrast, the relative differences between the NO_X emission levels of the steam powerplant and of the gas turbine and Stirling engines decrease when the emissions are expressed in terms of specific emissions rather than in terms of emission indices.

VEHICLE EMISSIONS

Gas Turbine – A GT-309 powered truck tractor was tested in such a manner as to duplicate the operation of a 4500 pound passenger car powered by the same engine. The test results are shown in Fig. 20. The vehicle HC, CO and NO_X emissions are indicated in terms of grams of pollutant per mile of operation over the 1970 Federal driving schedule. Because of non-representative drive line gear ratios, excessive drive line inertia and windage and friction losses, the emission values for this test vehicle are not representative of a typical passenger car.

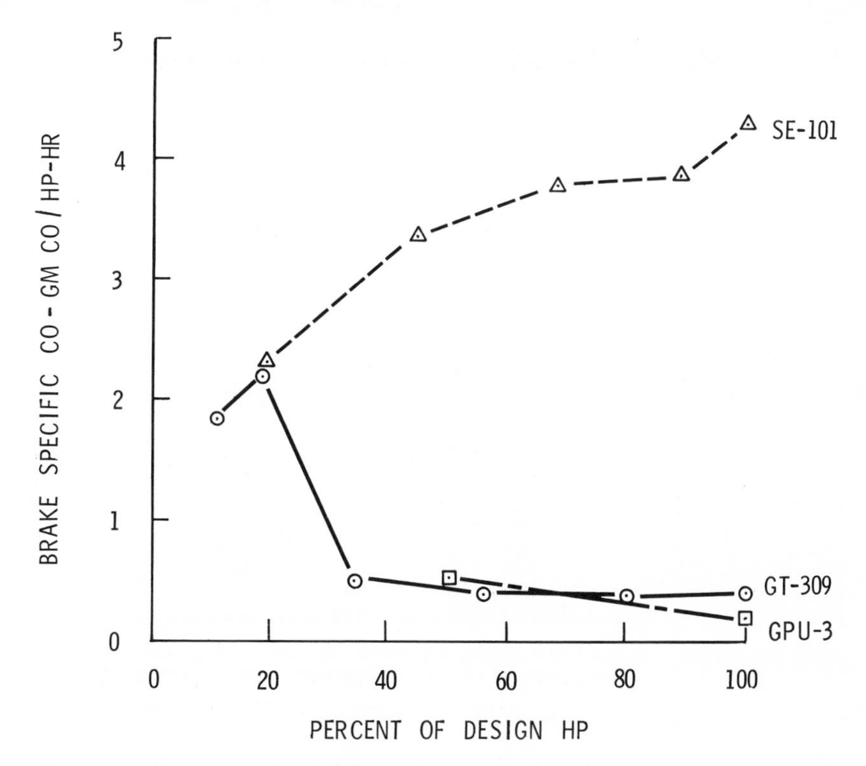

Fig. 18. Brake specific CO versus percent of design power.

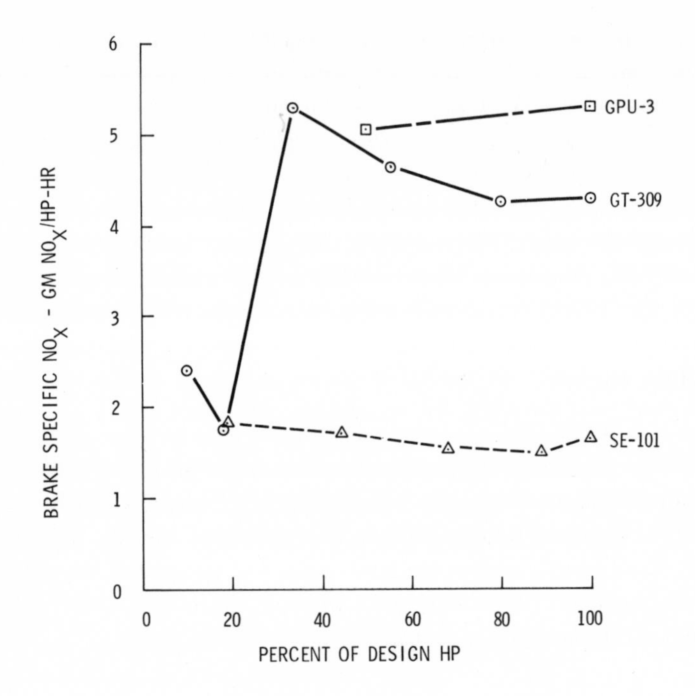

Fig. 19. Brake specific NO_x versus percent of design power.

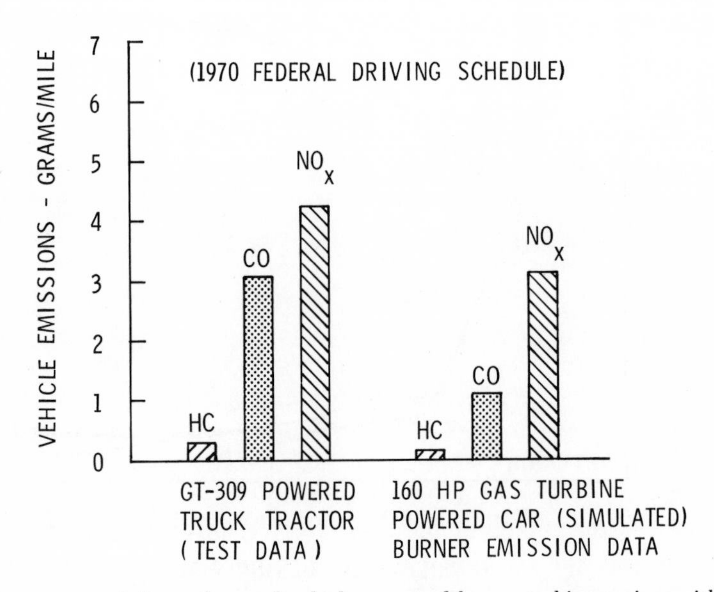

Fig. 20. True mass emissions of several vehicles powered by gas turbine engines with standard burners.

Shown also in Fig. 20 are similar emission data that were computed using the emission simulation (Fig. 8) to predict true mass emission values for a medium-size car weighing 4500 pounds and powered by an arbitrarily selected 160 hp scaled version of the GT-309. Cold start emissions cannot be estimated by the emission simulation program. The mass emissions of the three pollutants from this hypothetical car are shown in Fig. 20 to be much lower than the mass emissions from the truck tractor (50 percent lower HC emissions, 65 percent lower CO emissions and 25 percent lower NO_X emissions). While the HC and CO emission levels of both vehicles are less than the 1976 Federal emission standards, the NO_X emission levels of both vehicles are an order of magnitude higher than the standard. Although these true mass emission values are based on the 1970 driving schedule, similar conclusions would probably be reached if the 1972 driving schedule were used.

Steam – The measured true mass emissions from the fully warmed-up SE-101 powerplant installed in a 5500 pound modified 1969 Pontiac Grand Prix car operated over the 1970 Federal driving schedule are presented in Fig. 21. The measured

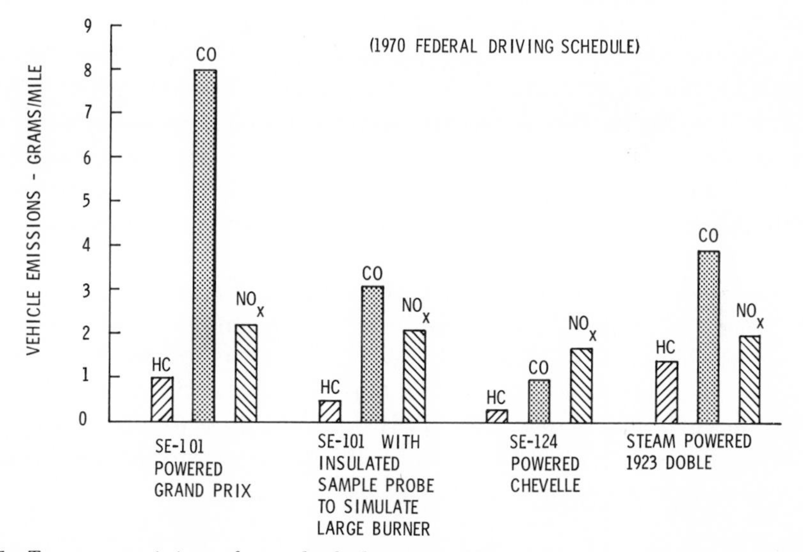

Fig. 21. True mass emissions of several vehicles powered by steam engines with standard burners.

emissions were higher than anticipated from burner test rig data. It was theorized that the combustion gases were quenched more rapidly in the powerplant by the steam generator than in the burner test rig. To investigate this possibility, a sampling probe was installed in the center of the steam generator. The probe was insulated in order to maintain the sampled combustion gases at a high temperature for a longer time prior to analyses to simulate the flow through the burner test rig. The results obtained while using this insulated sampling probe during vehicle operation over the Federal driving schedule are shown also in Fig. 21. The results indicate that the HC and CO

References pp. 440-443

oxidation processes were completed to a greater extent during passage of the gas through the insulated sampling probe than they were during passage of the exhaust gas through the actual steam generator. Thus, it was concluded that lower emissions could be realized from the powerplant if the standard combustion system were modified to provide greater residence time at high temperature for the HC and CO oxidation reactions to proceed.

True mass emissions of the fully warmed-up SE-124 steam engine installed in a 3500 pound modified 1969 Chevelle operated over the 1970 Federal driving schedule are shown also in Fig. 21. Immediately prior to the vehicle emission test, the powerplant was operated a sufficient time to develop the design steam temperature and pressure conditions in the steam generator. Also shown in the figure are the mass emission values for a similarly operated 1923 Doble Model E steam automobile weighing 6200 pounds. Note that the true mass NO_X emissions of the three steam powered vehicles are similar in magnitude and are well above the 1976 Federal emission standards. Similar conclusions would probably be reached if the 1972 driving schedule were used.

The differences in mass emissions of the three steam cars stem from variations in fuel economy and in pollutant exhaust concentrations. The fuel economy differences are discussed in Ref. 4. The pollutant concentrations depend on such combustion system variables as: burner design, combustion intensity, gas temperature level, and the number and quality of burner relights during the vehicle driving schedule. Two distinctly different types of burners were used in these steam powerplants. In the case of the SE-101, the combustion chamber was designed in accordance with gas turbine design principles for minimum size. It operated with a reasonably-high pressure loss. On the other hand, the SE-124 and the Doble powerplants each incorporated a very large volume, low pressure drop vortex-type combustion chamber. The combustion intensity of the SE-101 burner was about four times higher than that of the SE-124. Thus, the much smaller burner of the SE-101 provides the least time for oxidation of the hydrocarbons and carbon monoxide in the burner exhaust products before the exhaust gases enter the steam generator.

Stirling – The Stir-Lec II, a Stirling hybrid electric system powering a modified 1968 Opel Kadett (7) was the only vehicle powered by a Stirling engine that has been tested for emissions by the GM Research Laboratories. In this vehicle, the Stirling engine drives a generator which charges the batteries of the electric drive system which actually moves the car. This power transmission system is less efficient than the conventional mechanical drive systems employed in the gas turbine and steam powered vehicles which were evaluated. Because of the lower transmission efficiency of the Stir-Lec II, more fuel per unit of work output is consumed, resulting in higher vehicle mass emissions than if a mechanical drive system had been used. Nevertheless, the data are presented as observed.

The exhaust emissions of the fully-warmed up Stir-Lec II vehicle were measured during operation over seven 1970 Federal driving cycles. In addition, emissions were measured while the engine was operated at full load following the driving schedule to recharge the batteries to their original state of charge. True mass emission data indicated that approximately 58 percent of the total vehicle emissions were emitted during the recharging period. The results are shown in Fig. 22. The NO_X emissions of

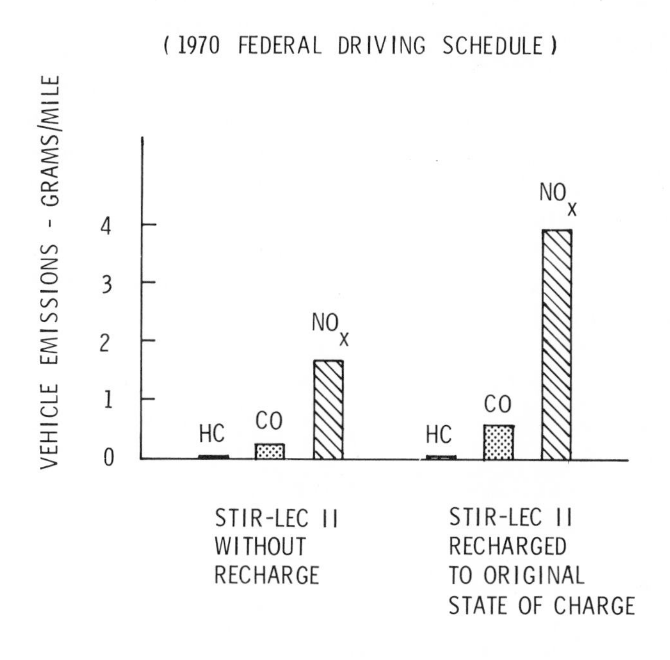

Fig. 22. True mass emissions of a vehicle powered by a Stirling engine with the standard burner.

this vehicle are similar to the GT-309 regenerative gas turbine emissions shown in Fig. 20. The HC and CO mass emissions are well below 1976 Federal emission standards for these pollutants. Similar conclusions would probably be reached if the 1972 driving schedule were used.

Comparison – Comparisons of the emissions of the gas turbine, steam and Stirling powered vehicles with the 1976 Federal emission standards reveal that none of these vehicles meet all three pollutant standards. NO_X emissions from all three powerplants are significantly higher (at least 5 times) than the 1976 Federal standards. The HC and CO mass emissions from all of the powerplants except one were below the future requirements. The exception is the SE-101 steam car with HC and CO emissions that are significantly higher than the 1976 Federal standards. In addition to emission measurements made after a powerplant has been warmed up, emissions that occur during an initial cold start of the engine must be considered. These additional evaluations are beyond the scope of this paper.

References pp. 440-443

EMISSION CONTROL TECHNIQUES

Theoretical Considerations – The control of exhaust pollutants from a continuous combustion system requires an understanding of the parameters governing the formation and elimination processes of the pollutants. Many investigators are using detailed analytical techniques for predicting emission levels in continuous combustion systems (22-26). A simplified analysis will be used in this paper to demonstrate the influence of changes in burner design parameters or operating conditions on burner emission levels. Only the design operating conditions for the gas turbine, steam, and Stirling burners will be considered.

Figs. 23, 24 and 25 present computed gas flow and temperature distributions for the standard GT-309 regenerative gas turbine, SE-101 steam and GPU-3 Stirling burners, respectively. Shown in each figure is a schematic of the burner configuration, a block diagram illustrating the relative locations of the major entries of air into the burner, and plots of gas temperature and velocity versus burner axial length. A nominal primary zone equivalence ratio of 0.8 was assumed as representative for each of these standard burner designs.

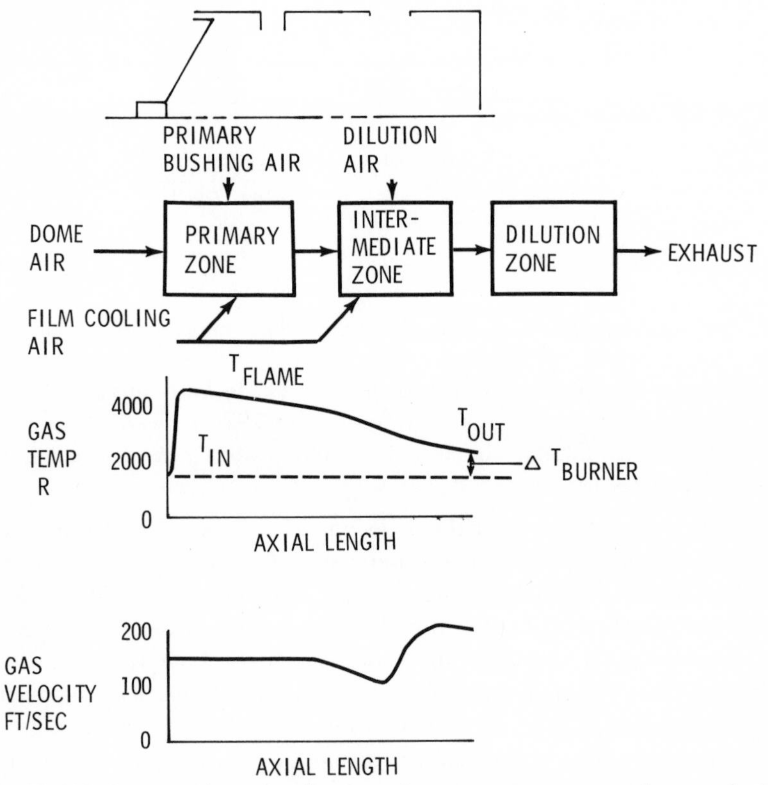

Fig. 23. GT-309 burner schematic showing air, gas temperature and gas velocity axial distributions.

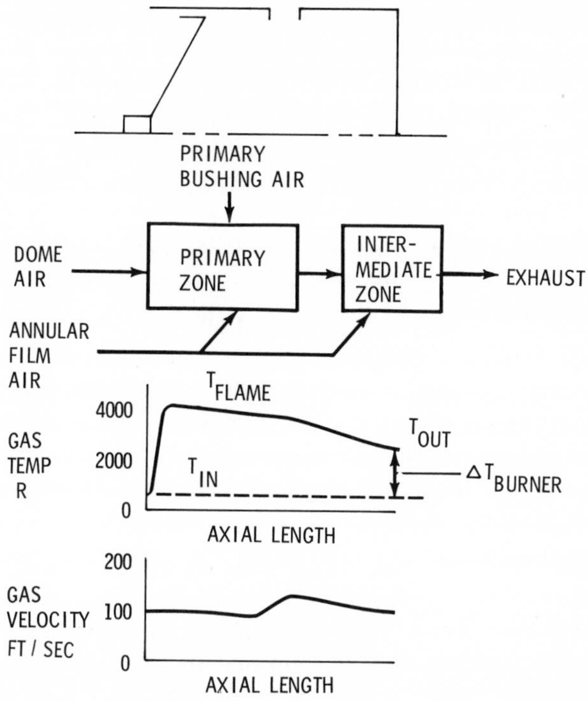

Fig. 24. SE-101 burner schematic showing air, gas temperature and gas velocity axial distributions.

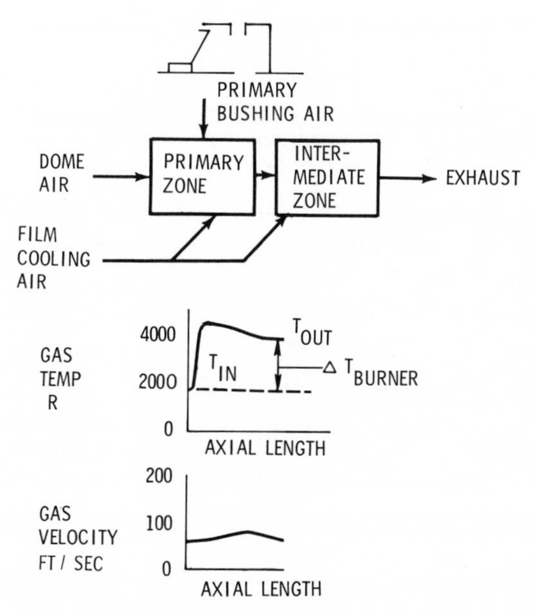

Fig. 25. GPU-3 burner schematic showing air, gas temperature and gas velocity axial distributions.

References pp. 440-443

As shown in Fig. 23, air enters the primary zone of the gas turbine burner through the dome openings, the bushed primary air holes and a portion of the film cooling gap. The film cooling air mixes with the primary zone combustion products in the intermediate zone. Additional air that is admitted through the dilution holes downstream of the intermediate zone mixes with the combustion products prior to the discharge of the gas from the burner. In the primary zone, a temperature rise due to combustion is experienced, followed by a progressive decline in temperature until the burner exit temperature is reached. The overall burner temperature rise, as indicated previously in Fig. 4, is the difference between the burner exhaust and inlet air temperatures. A gas velocity of approximately 150 ft/sec occurs in the primary zone followed by a higher velocity downstream in the dilution zone as additional air is added. This velocity profile is characteristic of a constant cross sectional area burner with progressive air admission. By combining the temperature-time history of a fluid particle in the burner with appropriate chemical kinetic considerations of the pollutant reactions, a procedure was developed to estimate the quantities of pollutants emitted from the burner.

Comparing Fig. 24 with Fig. 23 reveals that the general thermodynamic characteristics of the standard SE-101 steam burner are similar to the regenerative gas turbine burner with two notable exceptions. The temperature of the air admitted to the steam burner is approximately ambient and the intermediate zone combustion products are not diluted by the introduction of additional air downstream. Mixing of the primary zone combustion products and the film cooling air occurs prior to the discharge of the gas from the burner. Unlike the gas turbine burner, there is no need to reduce appreciably the temperature of the gas leaving the steam burner.

It will be observed in Fig. 25 that the general thermodynamic characteristics of the Stirling burner are similar to those of the steam burner. The sole exception is the significantly higher inlet air temperature that is comparable to that of the regenerative gas turbine burner.

Since it has been shown that nitric oxide is the most troublesome pollutant to control in a continuous combustion system, this pollutant species will be considered in more detail. Many investigators have postulated NO reaction schemes (27-32). The most commonly accepted Zeldovich scheme (33) will be used to provide sufficiently detailed information for burner comparisons. According to this scheme, NO production results from the following two reactions:

$$O + N_2 \underset{k_{1b}}{\overset{k_{1f}}{\rightleftarrows}} NO + N \tag{6}$$

$$N + O_2 \underset{k_{2b}}{\overset{k_{2f}}{\rightleftarrows}} NO + O \tag{7}$$

From these expressions, the NO rate of formation may be expressed as

$$\frac{d(NO)}{dt} = k_{1f}\,(O)\,(N_2) - k_{1b}\,(NO)\,(N) + k_{2f}\,(N)\,(O_2) - k_{2b}\,(NO)\,(O) \qquad (8)$$

Integration of Eq. 8 yields instantaneous NO concentrations as a function of time. Plots of NO concentrations versus time for various temperatures and pressures representative of the primary zone conditions of the three burners under consideration are shown in Fig. 26.

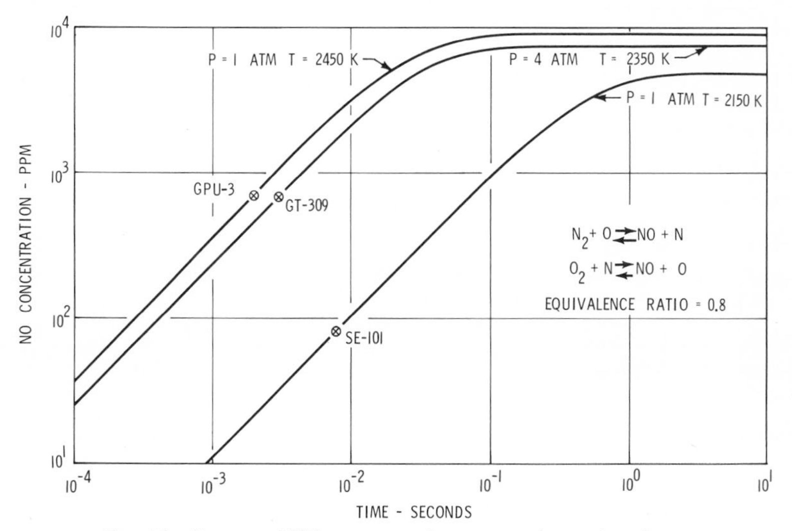

Fig. 26. Computed NO concentrations versus formation time.

A summary of the pertinent parameters affecting NO production in the three burners is presented in Table 4. Primary zone temperature and pressure are shown to affect the rate of NO formation, d(NO)/dt. The rates of NO formation in the gas turbine and Stirling burner are similar because of the similarity in reaction temperatures. The kinetic ratio is defined as the kinetically-limited primary zone NO concentration divided by the equilibrium NO concentration. The dilution ratio is defined as the ratio of the primary zone mass flow to the total mass flow rate.

A chamber pressure of 4 atmospheres and a flame temperature of 4250 R are representative of the GT-309 primary zone. A primary zone residence time of 3 msec may be estimated by integrating the velocity-distance plot shown in Fig. 23. At these conditions, it may be determined from Fig. 26 that a kinetically-limited NO concentration of 690 ppm would be expected at the end of the primary zone. This value represents only 9.2 percent of the equilibrium concentration which could occur at the primary zone conditions with sufficient time. The rapid drop in gas temperature following the primary zone quenches the NO formation process. Since

References pp. 440-443

TABLE 4

Nitric Oxide Formation Parameters

	GT-309	SE-101	GPU-3
Average Primary Zone Equivalence Ratio – ϕ	0.8	0.8	0.8
Average Flame Temperature – R	4250	3800	4360
Pressure – Atm	4.0	1.0	1.0
Equilibrium NO Concentration for Primary Zone – ppm	7541	4884	9124
NO Formation Rate d(NO)/dt – ppm/msec	210	10.4	302
Primary Zone Residence Time – msec	3.0	8.0	2.0
Kinetically Limited Primary Zone NO Concentration – ppm	690	82	700
Kinetic Ratio $\left[\frac{NO_{pz}}{NO_{eq}}\right]$	0.092	0.017	0.077
Dilution Ratio $\left[\frac{W_{pz}}{W_a}\right]$	0.185	0.74	0.74
Exhaust NO Concentration – ppm	128	61	520

the NO decomposition rate is very slow, no measurable reduction in NO concentration due to decomposition should occur. While the admission of air further downstream in the burner causes a decrease in the NO concentration value, the total mass of nitric oxide formed in the burner is not altered. For the GT-309 burner, about 18.5 percent of the total burner air enters the primary zone. Therefore, the NO concentration at the exit of the primary zone will be decreased by the ratio 0.185 to give a burner exhaust concentration of 128 ppm. Reference to Fig. 10 indicates that this simple analysis gives values which are comparable to measured concentrations.

Representative conditions for the primary zone of the steam burner are a chamber pressure of 1 atmosphere and a temperature of 3800 R. A primary zone residence time of 8 msec can be estimated from the information presented in Fig. 24. This residence time is longer than the comparable time for the GT-309 burner since a longer primary zone length results from a fixed L/D ratio and a larger diameter. At these conditions, it may be determined from Fig. 26 that a kinetically-limited NO concentration of 82 ppm would be realized at the exit of the primary zone. This value represented only 1.7 percent of the equilibrium concentration which could be reached at these primary zone conditions with sufficient time. The drop in temperature of the gas leaving the primary zone resulting from the mixing of film cooling air quenches the NO formation process. Applying the dilution ratio of 0.74 for the steam burner listed in Table 4 yields a burner exhaust value of 61 ppm. This exhaust concentration compares favorably with the experimental values shown in Fig. 11.

Representative conditions for the primary zone of the Stirling burner are a chamber pressure of 1 atmosphere and a temperature of 4360 R. A primary zone residence time of 2 msec was estimated from the information shown in Fig. 25. This time is about 1 msec less than the comparable time for the GT-309 burner. At these conditions, it may be determined from Fig. 26 that a kinetically-limited NO concentration of 700 ppm would result at the exit of the primary zone. This value represents 7.7 percent of the equilibrium value. The primary zone concentration is reduced by a dilution ratio of 0.74 to yield an exhaust concentration of 520 ppm. This calculated value compares favorably with the values shown in Fig. 12.

The second pollutant in order of concern is carbon monoxide. The oxidation of carbon monoxide as well as hydrocarbons should be completed in the intermediate or dilution zone of a burner. For this to occur, favorable oxidation conditions must prevail. Generally, HC oxidation rates are faster than CO oxidation rates. Thus, if the CO emissions can be controlled effectively, then adequate control of HC emissions should be expected.

Equilibrium values of CO concentrations were computed as a function of temperature for combustion products at an equivalence ratio of 0.8 and at two representative pressures of 1 and 4 atmospheres. These data are plotted in Fig. 27. If an equilibrium concentration were to be realized at typical engine exhaust temperatures, a CO concentration of less than 1 ppm would be expected. However, the oxidation of carbon monoxide is kinetically limited in the dilution zone of the gas turbine burner.

Most investigators agree that the rate-controlling reaction for the oxidation of carbon monoxide is the following (34, 35):

$$CO + OH \underset{k_b}{\overset{k_f}{\rightleftarrows}} CO_2 + H \qquad (9)$$

The CO oxidation rate is expressed as follows:

$$\frac{d[CO]}{dt} = k_b\ [CO_2]\ [H] - k_f\ [CO]\ [OH] \qquad (10)$$

References pp. 440-443

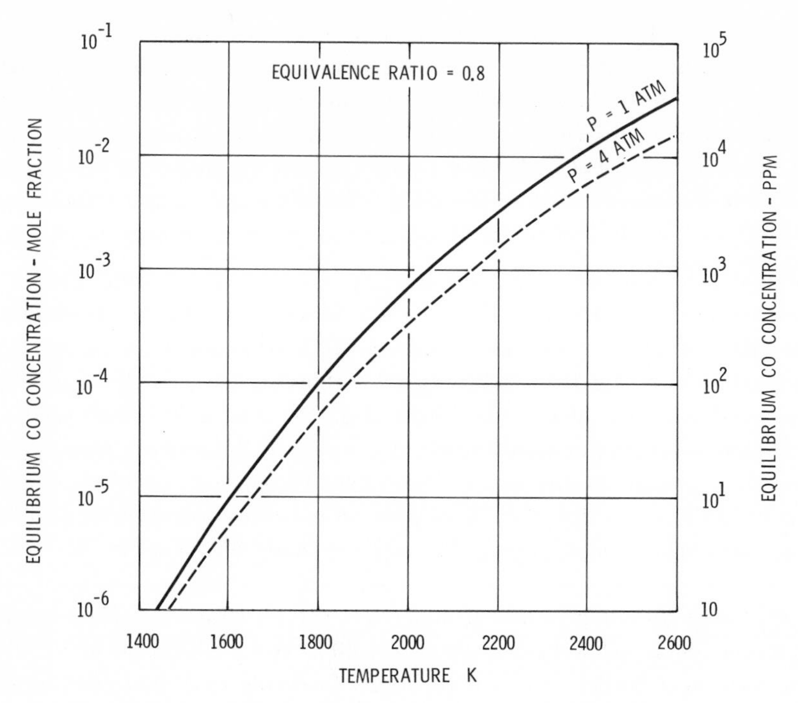

Fig. 27. Equilibrium CO concentrations versus temperature.

Integration of Eq. 10 yields CO concentrations as functions of time. Fig. 28 shows solutions of Eq. 10 for several different final temperatures. Each curve is based on an initial CO concentration of 1 percent which is somewhat typical of the exit primary zone concentration. Referring to the 1600 K data, it will be seen that an increase in pressure increases the rate of oxidation of carbon monoxide.

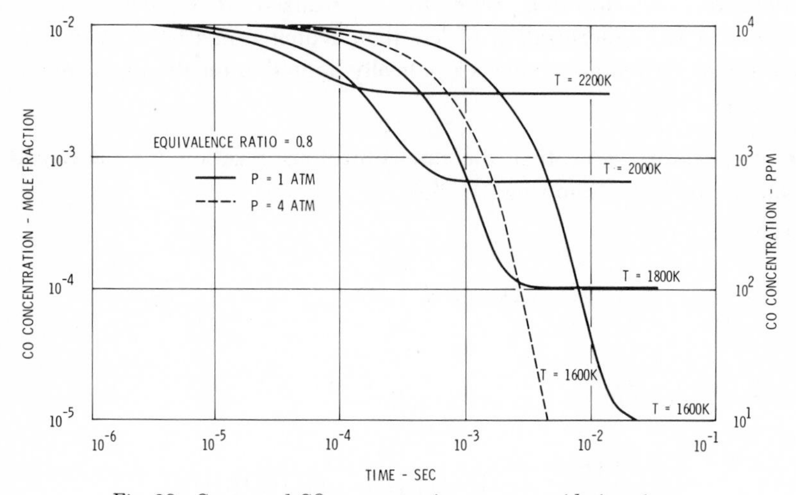

Fig. 28. Computed CO concentrations versus oxidation time.

Referring to Figs. 23, 24 and 25, it will be seen that when the combustion products enter the intermediate and dilution zones, the temperature of the combustion products is reduced by mixing with the film cooling air and dilution air. During the early phase of this cooling process, the CO oxidation rate is so fast that equilibrium values occur as the gas temperature decreases. However as the gas temperature continues to decrease to lower values, the CO oxidation rate is reduced sufficiently so that equilibrium conditions cannot be maintained. The reduction in the CO oxidation rate with decreasing temperature is illustrated in Fig. 28.

A summary of the pertinent parameters affecting CO oxidation in the three burners is presented in Table 5. A primary zone CO concentration of 1 percent was assumed for each case. The computed equilibrium CO concentrations at typical burner exhaust temperatures are indicated to be less than 1 ppm for all three burners. The hypothetical frozen CO concentrations of the primary zone products at the burner

TABLE 5

Carbon Monoxide Oxidation Parameters

	GT-309	SE-101	GPU-3
Average Primary Zone Equivalence Ratio – ϕ	0.8	0.8	0.8
Average Flame Temperature – R	4250	3800	4360
Pressure – Atm	4.0	1.0	1.0
Assumed Primary Zone Concentration – ppm	10,000	10,000	10,000
Exhaust Temperature (A/B) – R	2160/2160	3060/500	3830/1500
Equilibrium CO at Exhaust Temperature (B) – ppm	<1	<1	<1
Frozen CO Concentration in Primary Zone Flow – ppm	103	405	41
Kinetic Ratio $\left[\frac{CO_f - CO_{eq}}{CO_{pz} - CO_{eq}}\right]$	0.01	0.04	0.004
Dilution Ratio $\left[\frac{W_{pz}}{W_a}\right]$	0.185	0.74	0.74
CO Exhaust – ppm	19	300	30

(A) Before heat exchanger.
(B) After heat exchanger (steam generator or heater tubes).

References pp. 440-443

exit are determined from measured exhaust concentrations. This is a hypothetical value since, in reality, the primary zone products are mixed with dilution air before reaching the burner exit. An integration of Eq. 10, using the temperature-time history of the gas in the dilution zone, would be required to compute this value as is done in some theoretical combustor models. The kinetic ratio is defined as the ratio of the frozen CO concentration minus the equilibrium CO concentration divided by the difference of primary zone and exhaust equilibrium CO concentrations. In the limit, if the kinetic ratio is zero, the final CO concentration would be equal to the theoretical equilibrium concentration at the final conditions of pressure, temperature and equivalence ratio. The GPU-3 burner most closely approaches this condition with a kinetic ratio of 0.004. The dilution ratios for the primary zone products are identical to the values used for the NO analysis.

The analyses of NO and CO emissions have been discussed for two reasons. First, the analyses provide an understanding of why the measured values were obtained in the standard burner designs. Second, since the analyses identify the important parameters affecting these pollutant emissions, control techniques can now be directed toward modifications of these parameters.

Nitric Oxide Control in the Stirling Burner – High flame temperatures in the Stirling burner accelerate the formation of nitric oxide. As discussed previously, the NO_X emissions from the standard Stirling GPU-3 engine burner are an order of magnitude higher than the 1976 Federal emission standards. Three methods of controlling NO emissions by modifications of the burner operating conditions have shown promise. In each procedure, a decrease in the NO concentration is accomplished by lowering the peak combustion temperature, thereby penalizing to some extent the thermal efficiency of the powerplant. The degrees of improvement that these methods, applied either singly or in combination with one another, may provide the true mass emissions of a Stirling engine powered vehicle operated over the Federal emission test have not been evaluated. These methods of controlling NO emissions are described below:

1. Reduction in Burner Inlet Air Temperature – A direct means for lowering the flame temperature is to reduce the burner inlet air temperature. This results in decreases in both the theoretical equilibrium concentration and the rate of formation of nitric oxide. The influences of these two parameters on burner exhaust concentrations of nitric oxide are shown in Table 4.

   Fig. 29 shows the exhaust emission concentrations and engine thermal efficiency as functions of burner inlet air temperature for a constant air-fuel ratio of 25:1. NO concentrations exhibited a five-fold decrease while HC and CO concentrations exhibited significant increases as the burner inlet air temperature was reduced from 1200 F to 100 F. CO and NO measurements made upstream and downstream of the preheater are presented. Because of the sample temperature requirement for the heated FID hydrocarbon analysis

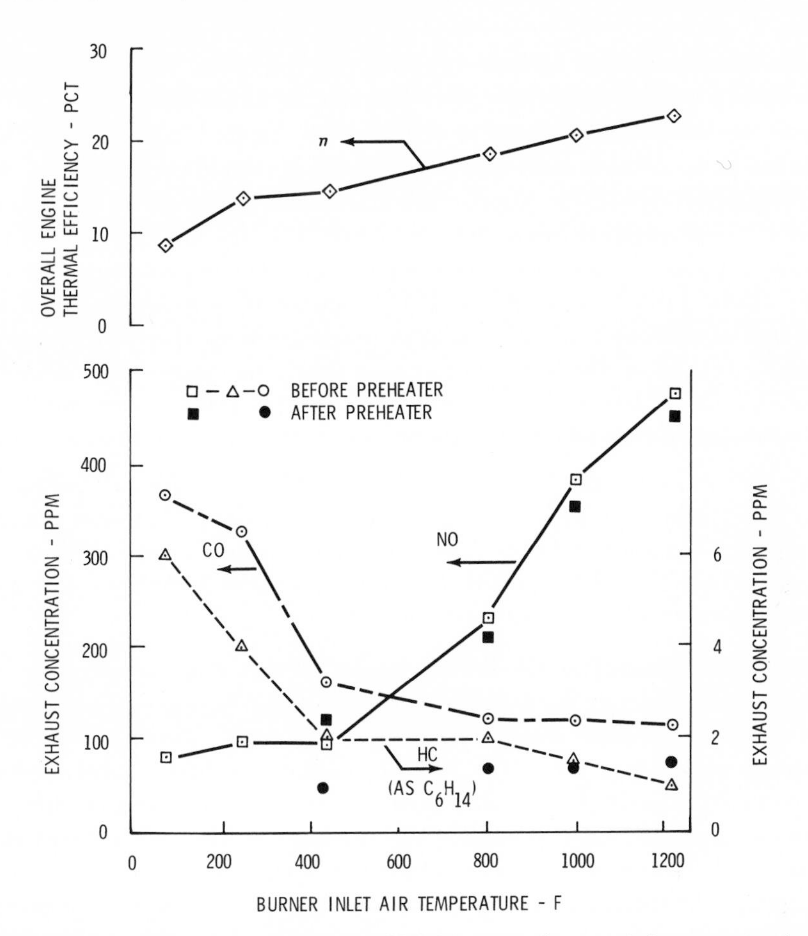

Fig. 29. Effect of inlet air temperature on GPU-3 burner exhaust emission concentrations and powerplant efficiency.

system used for these tests, the measurements of HC concentrations were restricted to the upstream or hot side of the preheater. Although no HC measurements were taken downstream of the preheater, it is expected that the HC concentrations would follow the same trend as the CO concentrations – namely, a decrease in concentration following the preheater due to added residence time for oxidation to occur. On the other hand, the NO concentration did not change significantly as the gas flowed through the preheater.

A decrease of more than 50 percent in engine efficiency accompanied the reduction in inlet air temperature from 1200 F to 100 F. Engine fuel flow would have to be approximately doubled to maintain the specified engine power output. As a result, the expected reduction in NO_X mass emissions from a vehicle would be appreciably less than the reduction in concentrations shown in Fig. 29.

2. Leaning the Mixture Ratio at Full Air Preheat – Peak combustion temperature decreases as the air-fuel ratio is leaned out from the stoichiometric ratio. The NO concentration in the burner exhaust gas decreases also because of the direct dependence of the rate of NO formation on the gas temperature. As the mixture ratio is leaned out, gas flow rate through the burner must be increased to maintain the engine power output constant. Since nitric oxide formed in a Stirling burner is kinetically limited, increasing the mass flow rate acts to shorten the residence time of the gas at which nitric oxide is formed. These two depreciating effects on NO formation, decreased gas temperature and residence time, are only slightly offset by the increase in oxygen concentration in lean combustion products which favors increases in NO concentration.

   As shown in Fig. 30, the exhaust concentrations of the three pollutants decreased with increasing air-fuel ratio at the investigated half-load engine operating condition. NO concentration showed the most significant change, decreasing over six-fold as the air-fuel ratio was increased from 20:1 to 40:1. Overall engine thermal efficiency decreased by 18 percent as the air-fuel ratio

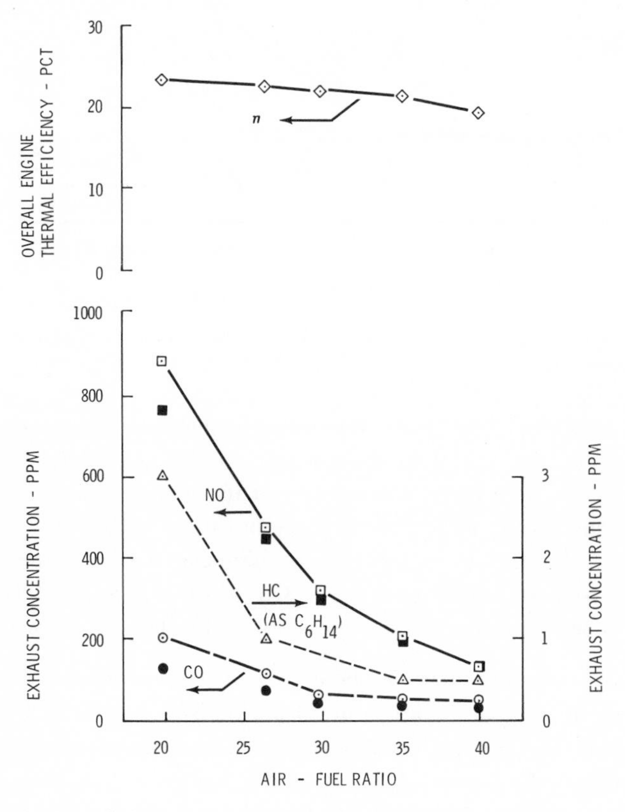

Fig. 30. Effect of air-fuel ratio on GPU-3 burner exhaust emission concentrations and powerplant efficiency.

was increased from 20:1 to 40:1; therefore the expected reduction in vehicle mass emissions would be appreciably less than the reductions in concentrations shown in Fig. 30.

3. Recirculation of Cooled Exhaust Gas – The external recirculation of cooled exhaust gas has been proposed as a means of controlling NO emissions from reciprocating and continuous combustion engines (36, 37). Exhaust gas recirculation is a method of introducing an inert (or almost inert) diluent into the combustion zone of a burner. An inert diluent acts as a non-reactive heat sink to reduce the flame temperature in the primary zone. The resulting lower flame temperature reduces the NO formation rate. Because there is an increase in total gas flow through the burner (constant fresh air flow rate), the residence time of the gas at high temperature is reduced, causing an additional decrease in the NO formation in the burner. Since heat is lost from the system during the external exhaust gas recirculation process, engine efficiency decreases with increasing recirculation rate.

The effect of exhaust gas recirculation was investigated on an 80 horsepower single-cylinder engine that was designed and built by the Philips Research Laboratories. Louvered and slit burners described in Ref. 9 were used during the study. European diesel fuel was burned. Ducting was installed on the engine to recirculate cooled engine exhaust gas. Exhaust gas from the outlet of the preheater was channeled back to the suction side of the combustion air blower. Because the recirculation duct was not insulated, the exhaust gas flowing through it was cooled several hundred degrees. This heat loss was responsible for the progressive decreases in engine thermal efficiencies with increasing recirculation rates that are illustrated in Figs. 31 and 32. Percent recirculation is defined as 100 times the ratio of the mass of recirculated exhaust gas to the mass of incoming fresh air. During these tests, all incompletely burned carbon compounds were reported in terms of "CO" (9).

As shown in Fig. 31, a five-fold decrease in NO concentration was realized when a 100 percent recirculation of the exhaust gas was instituted in the louvered burner. "CO" concentrations exhibited an unexplained dip at a 33 percent recirculation rate. Results of a similar study on the slit burner shown in Fig. 32 indicate that operation at recirculation rates beyond 33 percent had little effect on NO exhaust concentration. "CO" concentration increased steadily with increasing recirculation. Since the overall engine thermal efficiency decreases with increasing recirculation rates, the expected reduction in vehicle mass emissions would be appreciably less than the reductions in concentrations shown in Figs. 31 and 32.

Without any exhaust gas recirculation, the NO concentrations for the Philips louvered and slit burners are significantly lower than comparable concentrations for the standard Stirling GPU-3 engine burner. The low emission potential of these burners needs further evaluation.

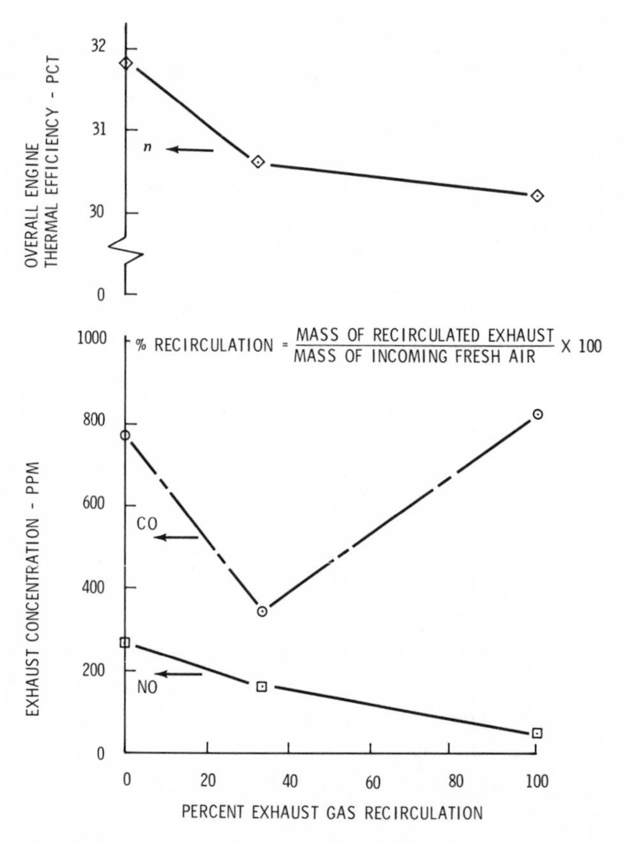

Fig. 31. Effect of exhaust gas recirculation on the Stirling louvered burner exhaust emission concentrations and powerplant efficiency.

Emissions Optimization in the Steam Burner — To expedite the development of a low emission combustion system for the SE-101 steam engine, the burner that had been previously developed for the multi-cylinder Stirling engine was selected as the starting point in developing an appropriate SE-101 burner. This Stirling burner had been designed to operate at an air-fuel ratio of 25:1 and at approximately ambient inlet air pressure conditions, both of which were also operational conditions of the SE-101 burner. In marked contrast, the Stirling burner was designed for an inlet air temperature of 1200 F as compared to the much lower temperature of 100 F established for the steam burner. The air distribution scheme as determined by the hole pattern of this Stirling burner is described in Table 6 under the designation of Design No. 1 burner.

The first major objective was necessarily to determine the effect of a decrease in burner inlet air temperature from 1200 F to 100 F on exhaust emissions from the Design No. 1 burner. The air flow rate and the air-fuel ratio were maintained at constant values as the inlet air temperature was reduced progressively. The test results

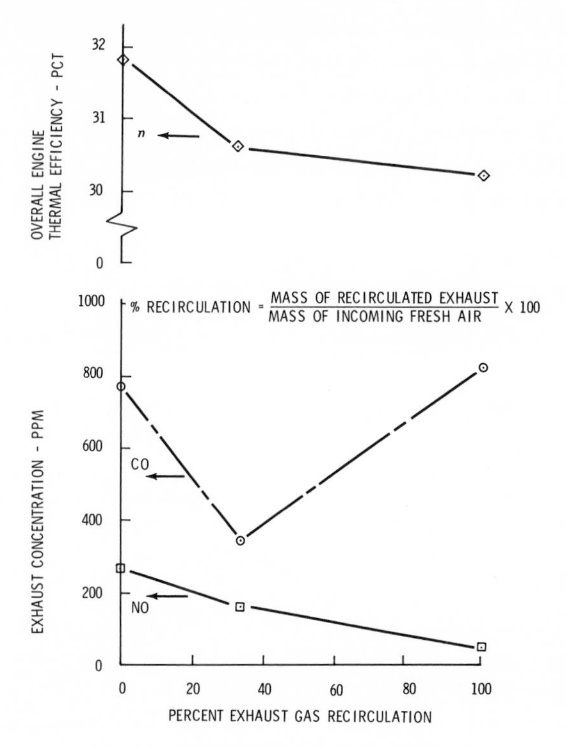

Fig. 32. Effect of exhaust gas recirculation on the Stirling slit burner exhaust emission concentrations and powerplant efficiency.

TABLE 6

Hole Area Proportions for
Experimental Steam Burner Designs

Burner Design No.	1	2	3
Dome			
No. Holes	30	15	30
Percent Total Area	33.1	18.0	18.0
Film Cooling Gap			
Percent Total Area	42.2	45.9	45.9
Primary Holes			
No. Holes	16	8	8
Percent Total Area	24.7	36.1	36.1
Total Area-in^2	3.89	3.58	3.58
Primary Zone Equivalence Ratio	0.91	0.96	0.96
Percent Primary Air	64.9	61.8	61.8

References pp. 440-443

are presented in Fig. 33. It will be seen that the HC and CO emissions were unaffected as the inlet air temperature was lowered from 1200 F to 600 F. Below 600 F, HC and CO concentrations increased significantly with decreasing air temperature. Conversely, the NO concentration decreased continually with decreasing burner inlet air temperature. The trends in concentration data are similar to the ones experienced when the inlet air temperature of the Stirling GPU-3 burner was reduced.

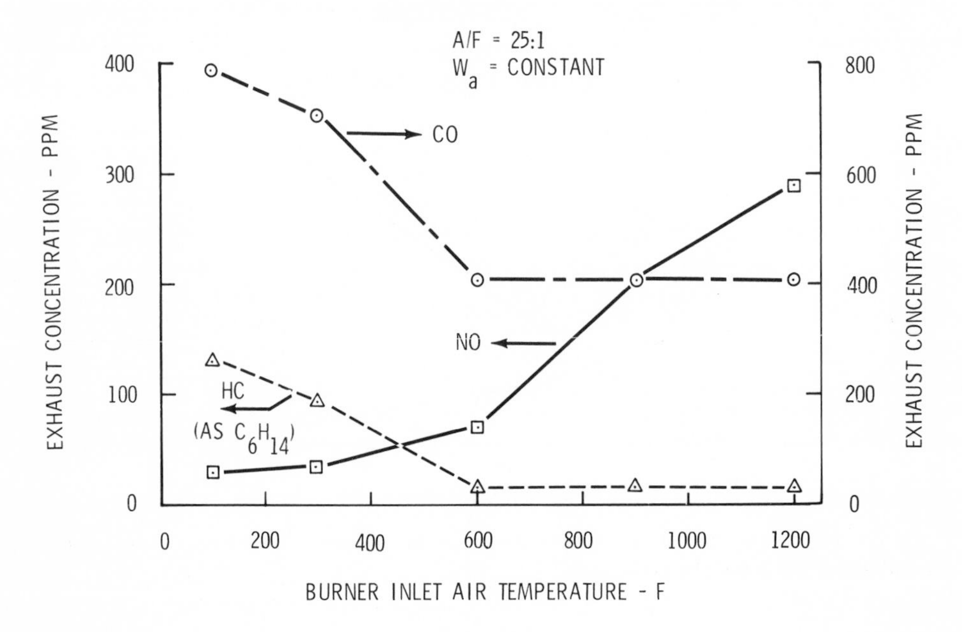

Fig. 33. Effect of burner inlet air temperature on exhaust emission concentrations of steam burner Design No. 1.

Fig. 34 presents pollutant concentration data for the Design No. 1 burner that were obtained at simulated operating conditions of the steam engine. Appreciable reductions in HC and CO emissions are shown to occur as engine load is decreased from full load. In contrast, the NO emissions remain relatively constant over the entire load range investigated.

Combustion efficiency at the full load operating conditions are plotted in Fig. 35 as a function of burner inlet air temperature. At the 1200 F design temperature of the Stirling burner, the combustion efficiency was calculated to be about 99.6 percent. However, at the 100 F design temperature for the SE-101 burner, the combustion efficiency decreases to about 98.4 percent. This decrease reflects the effect of the threefold increase in the HC and CO emissions. It was concluded from these test data that basic design modifications would have to be made to the Design No. 1 burner to improve combustion performance at the required 100 F burner inlet air temperature.

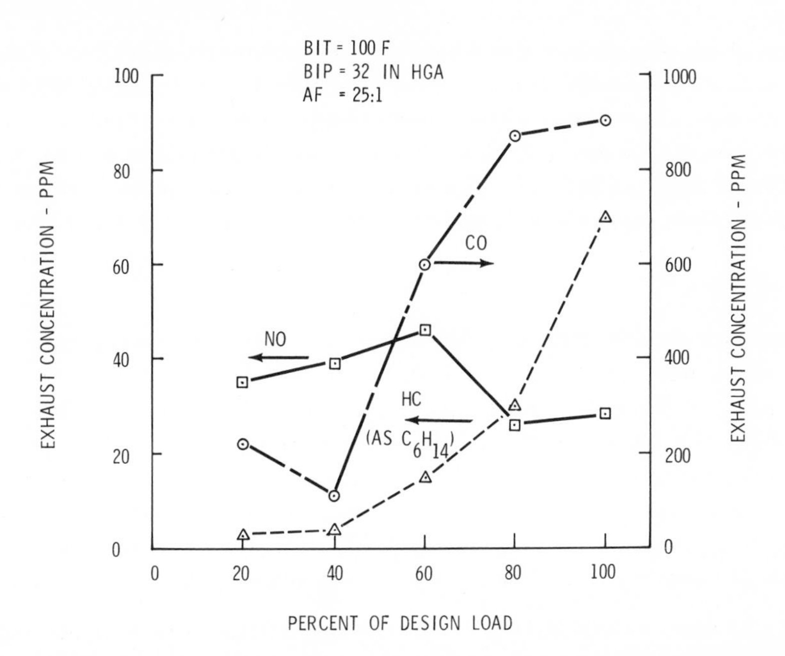

Fig. 34. Exhaust emission concentrations versus percent of design load for steam burner Design No. 1.

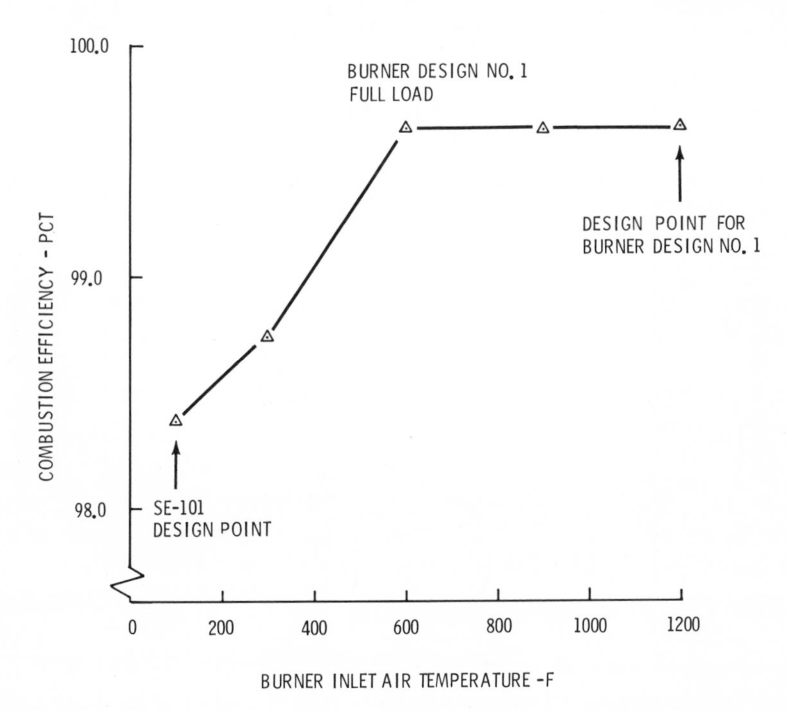

Fig. 35. Effect of inlet air temperature on combustion efficiency of steam burner Design No. 1.

References pp. 440-443

Visual flame observations and temperature measurements of the burner dome and body surfaces of Design No. 1 burner indicated the combustion process to be relatively slow. Acceleration of the combustion process would require an increase in turbulence in the forward end of the burner to accelerate fuel-air mixing. Modifying the primary zone air-fuel ratio would also improve combustion performance. To accomplish these goals, the hole pattern of the burner was altered to increase the air flow through the primary holes and to decrease proportionately the air flow through the dome holes.

Several burner hole patterns were investigated before improvements in emission performance were observed. The first burner which showed improved emission performance is designated Design No. 2 in Table 6. In this burner, the area of the dome holes was halved by plugging 15 of the existing 30 holes. The number of the primary air holes was decreased from 16 to 8. By decreasing the number of the primary holes, the penetration of air into the burner was increased and turbulent mixing of the fuel and air was enhanced. The total hole area was reduced about 8 percent. The burner pressure drop was not changed appreciably.

The part load emission data of the Design No. 2 burner that were obtained at the 100 F inlet air temperature are presented in Fig. 36. The HC and CO emissions are

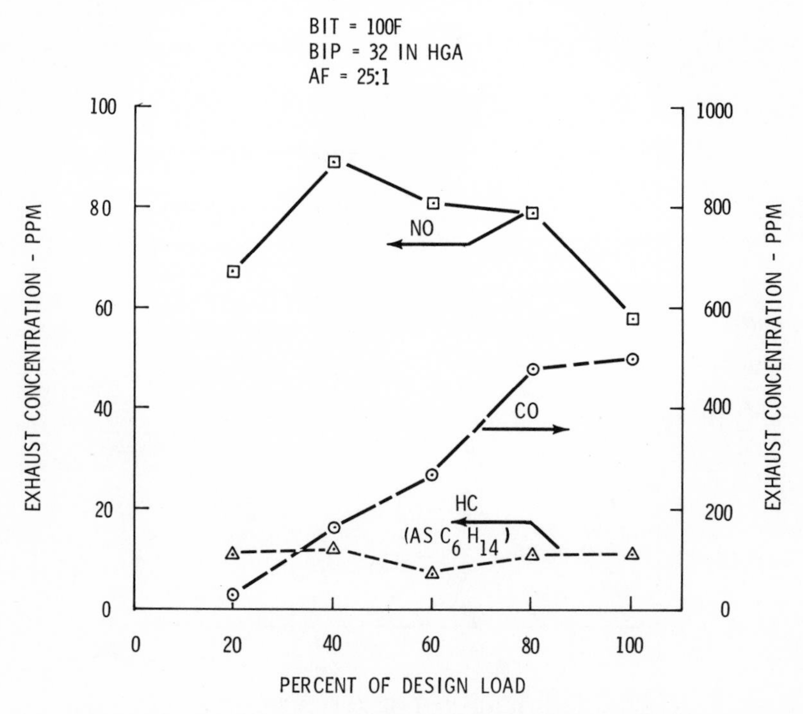

Fig. 36. Exhaust emission concentrations versus percent of design load for steam burner Design No. 2.

significantly decreased over the Design No. 1 burner values. CO emissions are lower by about 50 percent to a concentration level of less than 500 ppm at the full load condition. HC emissions are reduced to less than 12 ppm over the entire load range investigated. The NO concentrations are approximately doubled to a maximum value of about 80 ppm which is an indication of a higher combustion temperature. Some carbon soot was deposited on the surfaces of the dome baffles.

To prevent carbon depositing on the dome baffles and to further improve fuel-air homogeneity, the number of burner dome holes was increased from 15 to 30 while maintaining an equal total hole area. This burner is identified as the Design No. 3 burner in Table 6.

A photograph showing a view looking upstream into this burner during firing operation at full load is presented in Fig. 37. The essentially uniform flame fills the entire cross-sectional area of the burner. The fact that the primary holes are visible

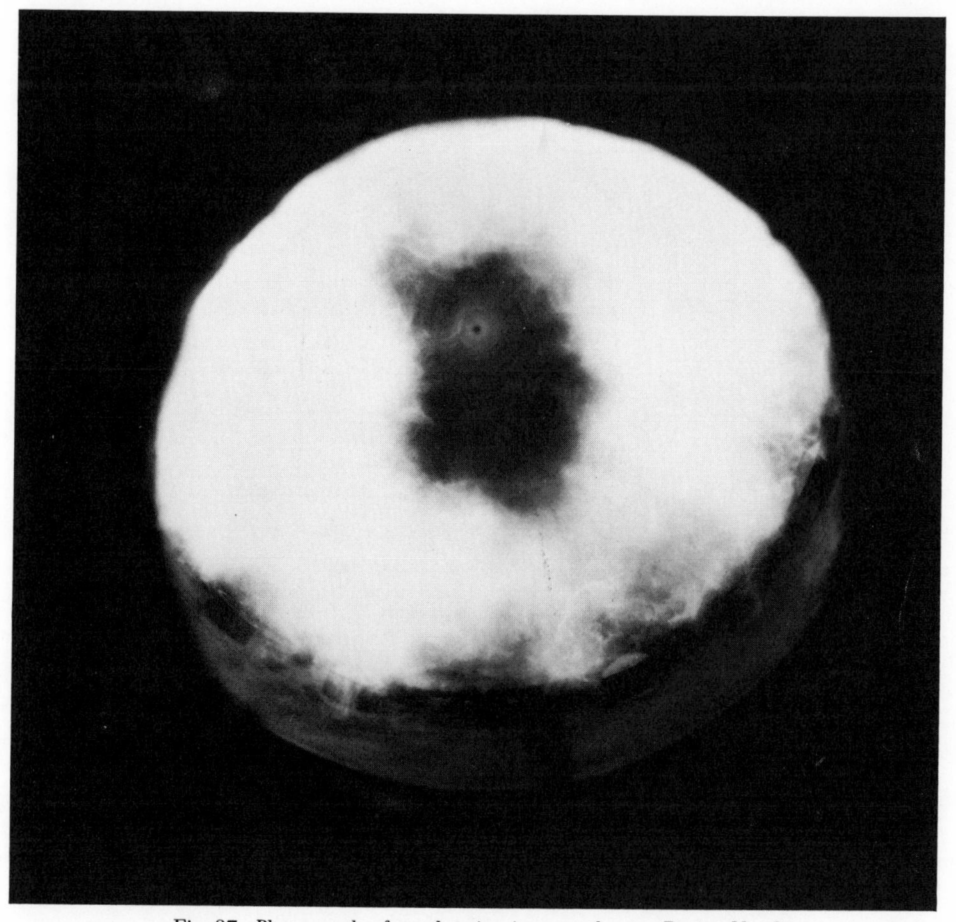

Fig. 37. Photograph of combustion in steam burner Design No. 3.

References pp. 440-443

indicates that the combustion process is probably completed at this downstream location. The slightly-off center position of the fuel nozzle is attributed to the angle at which the picture was taken.

Design No. 3 burner emissions that were measured over the entire operating range at the design inlet air temperature of 100 F are shown in Fig. 38. They indicate a significant improvement over the Design No. 2 burner data. CO emissions were reduced by about 50 percent to a level of 250 ppm at full load with similar percentage improvements at the part load conditions. HC emissions were reduced to under 10 ppm at all investigated load settings. NO concentrations were decreased slightly.

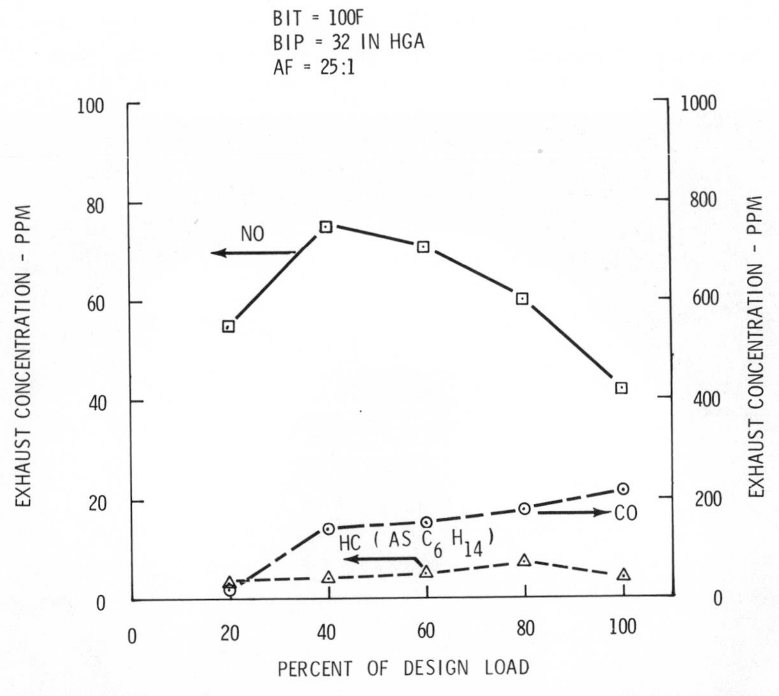

Fig. 38. Exhaust emission concentrations versus percent of design load for steam burner Design No. 3.

Metal temperatures of the Design No. 3 burner were significantly higher than those of the initial burner design. These elevated surface temperatures are desirable in an ambient inlet air temperature burner because a hot environment is needed to preclude wall quenching of the HC and CO oxidation reactions. Burner surface temperatures increase with increasing combustion intensity. No durability problem should exist with the burner operating at ambient inlet air temperatures. Based on the performance of the Design No. 3 burner, this burner was selected as the prototype design for the SE-101 steam engine.

One method of controlling the heat release rate of a steam engine burner is to modulate the fuel flow rate at a constant prescribed air flow rate. A range of air-fuel ratios from 20:1 to 60:1 was investigated with the air flow set at 40 percent of the design value. The emission results for Design No. 3 are presented in Fig. 39. It will be observed that the HC and CO concentrations pass through minimum values at an air-fuel ratio of approximately 30:1 whereas the NO concentration decreases progressively with increasing air-fuel ratio. Low CO emissions are realized only over a relatively narrow range of air-fuel ratios. A broader range is evidenced for the HC emissions.

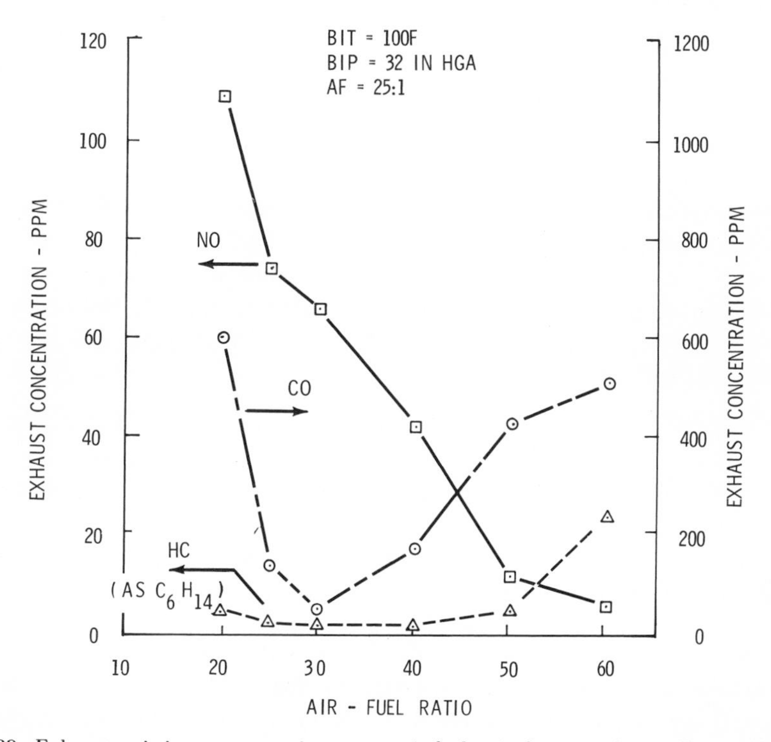

Fig. 39. Exhaust emission concentrations versus air-fuel ratio for steam burner Design No. 3.

A partial explanation for the observed minimum HC and CO concentrations with air-fuel ratio is obtained by examining the tabulation of air-fuel ratio data in Table 7. Both rich and lean primary zone air-fuel ratios are encountered while the overall design air-fuel ratio remains leaner than stoichiometric. The relatively high HC and CO emission levels experienced when operating at primary zone air-fuel ratios richer than stoichiometric are due to an insufficiency of oxygen. The relatively high HC and CO emission levels obtained on the lean side of the stoichiometric ratio are the result of poor combustion associated with the relatively low flame temperatures indicated by the measured low NO concentrations that are characteristic of burning lean fuel-air mixtures.

TABLE 7

Overall and Primary Zone Air-Fuel Ratios
of the Steam Burner Design No. 3

Overall A/F	Primary Zone A/F (nominal)
20:1	12.3:1
25:1 (design)	15.4:1
30:1	18.4:1
40:1	24.6:1
50:1	30.8:1
60:1	37.0:1

The overall design air-fuel ratio of 25:1 had been selected for the SE-101 engine to satisfy the maximum gas temperature capability of the steam generator. The minimum CO concentration was obtained at an overall air-fuel ratio of 30:1 which corresponds to a nominal primary zone air-fuel ratio of 18.5:1. This latter ratio is equal to the primary zone air-fuel ratio of the low emitting, highly efficient GT-309 regenerative gas turbine burner.

The Design No. 3 burner was operated at the design air-fuel ratio of 25:1 and grab samples of the exhaust gas were collected at several part-load settings. The samples were analyzed for formaldehyde content using the 3-methyl-2-benzothiazolone hydrozone (MBTH) procedure. The results indicated the formaldehyde level to be under 2 ppm over the entire engine operating range. This is an acceptable level based on experience with the GT-309 engine (18).

The influence of air-fuel ratio on the combustion efficiency of the prototype steam burner at the 40 percent air flow setting is illustrated in Fig. 40. Combustion efficiency is shown to be maximized at 99.95 percent at an air-fuel ratio of approximately 30:1. This constitutes a significant increase in combustion efficiency compared to the initial Design No. 1 burner (see Fig. 35). The improvement corresponds to appreciable reductions in HC and CO emissions but with some penalty in NO emissions.

The previous design series featured an annular film cooling gap which was a necessary part of the baseline burner when operating at inlet air temperatures of 1200 F. Film cooling is not necessarily required for ambient inlet air temperatures, although good performance was obtained with the use of annular film air. Burners operating at ambient inlet conditions without film air were also evaluated. Poor combustion performance was repeatedly obtained with burner designs having no annular film air. Therefore, it was surmised that the annular film air provides a gradual supplement to the combustion process, thereby assuring completion of the reaction with minimum wall quenching.

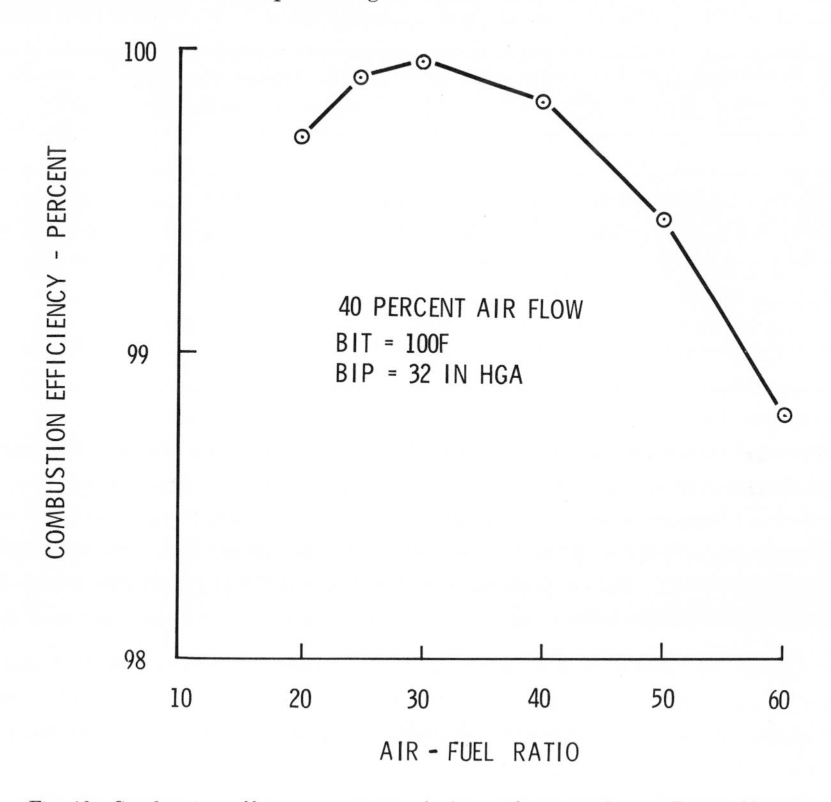

Fig. 40. Combustion efficiency versus air-fuel ratio for steam burner Design No. 3.

No further attempts were made to reduce simultaneously the HC, CO, and NO emissions from a steam burner due to altered priorities following the vehicle test evaluations of the SE-101 steam engine powered Grand Prix in which the two burners were scaled versions of the Design No. 3 burner. However, it is recognized that some of the projected techniques for reducing emissions from a gas turbine burner would be equally applicable to a steam burner. These techniques, such as fuel vaporization, premixing, and exhaust gas recirculation, are described later in this paper (see Advanced Burner Concepts).

Nitric Oxide Control in the Gas Turbine Burner — Several inherent characteristics of a gas turbine combustion system enhance its potential for achieving low NO emissions. It may be possible to utilize the large quantity of excess air passed through the burner to control NO emissions. The reasonably-low burner outlet gas temperature eliminates continuing chemical reactions beyond the confines of the burner. Since the prescribed burner outlet gas temperature is appreciably lower than the flame temperature, it should be possible to lower the flame temperature and reduce the NO formation rate without impairing overall engine performance. These attributes of the gas turbine burner do not exist in the Stirling and steam engine burners.

One obvious way to decrease NO formation in a burner is to lower the burner inlet air temperature and thereby decrease the combustion temperature in the primary zone. However a reduction in inlet air temperature to control NO formation is not acceptable because it has a direct adverse effect on engine efficiency (through the reduction of turbine inlet temperature). To maintain the power output of the engine constant, the burner outlet temperature (also termed the turbine inlet temperature) must be maintained constant. To accomplish this with a reduction in the burner inlet temperature requires that the fuel flow be increased.

The reductions in NO emissions from the standard GT-309 engine burner achieved by reductions in the burner inlet air temperature are shown in Fig. 41. Two different schemes are illustrated as applied to burner operation at the 80 percent gasifier turbine speed condition. By maintaining the air and fuel flows constant at the design conditions, an 80 percent reduction in NO concentration is realized when the burner inlet air temperature is decreased from the design value of 1233 F to 600 F. This, however, involves a substantial reduction in power which is not permissible if a specific duty cycle must be performed by the engine. If, on the other hand, as the burner inlet air temperature is reduced, the fuel flow is increased to maintain the

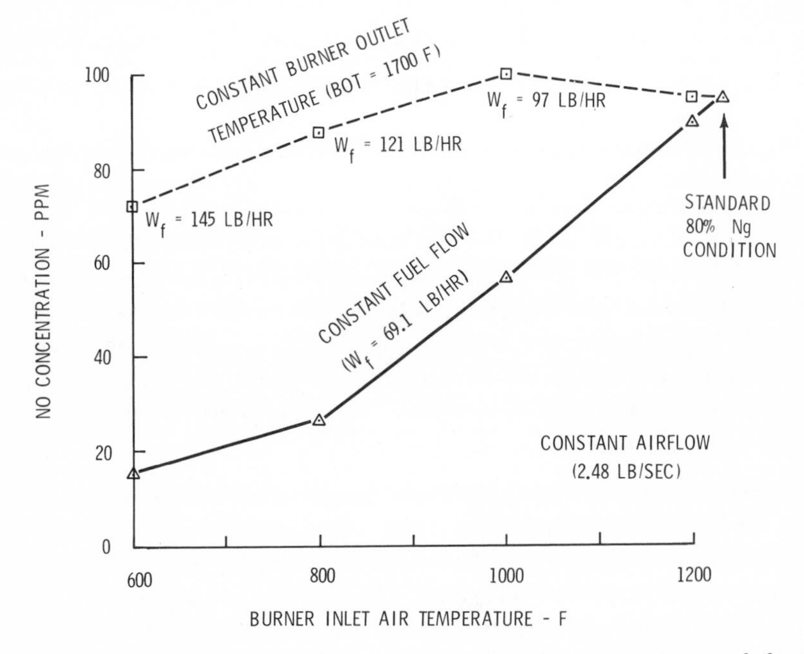

Fig. 41. Influence of inlet air temperature on exhaust NO concentrations of the GT-309 standard burner.

design burner outlet gas temperature of 1700 F and the design power, the NO concentration remains relatively constant down to a temperature of about 1000 F and then decreases as the temperature is lowered further. This latter scheme is the more realistic one but accounts for only a 27 percent decrease in NO concentration when the inlet air temperature is decreased from the design value of 1233 F to 600 F. This test was conducted using only a burner design that was optimized for an inlet temperature of 1233 F. For operation at an inlet temperature of 600 F, a different burner design would be needed if the primary zone equivalence ratio were to be held constant.

These two different trends in NO emissions with decreasing burner inlet air temperature have logical explanations. When the fuel flow rate is held constant, the flame temperature decreases linearly with decreasing burner inlet temperature. An exponential decrease in the NO concentration results. On the other hand, when the burner outlet temperature is held constant at the design value of 1700 F, the fuel flow rate must be increased as the burner inlet temperature is lowered. Because the air flow rate and distribution are fixed for both cases considered, a resultant enrichment of the primary zone fuel-air ratio occurs in the second case which causes an increase in flame temperature in spite of reduced inlet temperatures. These high temperatures account for the slight increase in NO concentrations as the inlet temperature is reduced. At a burner inlet temperature of approximately 800 F, the primary zone fuel-air ratio has been enriched to the stoichiometric value and a maximum flame temperature is realized.

Ref. 8 describes a study that was conducted at the GM Research Laboratories to determine the reductions in NO emissions that could be obtained with only minor modifications to the standard engine burner design and while operating the burner at the prescribed GT-309 engine conditions. The following three methods were investigated at that time:

1. Early quenching of the NO formation reaction in the primary zone of the burner.

2. Variation of the design primary zone equivalence ratio of the burner.

3. Combinations of Methods 1 and 2.

To achieve early quenching of the NO formation process, the primary and dilution air holes of the standard burner were moved progressively upstream in the burner. The total area of the burner air openings was not changed during the study. Hence the burner pressure drop and the primary zone equivalence ratio were maintained constant throughout the investigation. Fig. 42 shows the standard engine burner and a modified version of it which incorporates the most forward locations of primary and dilution air holes that were evaluated.

Fig. 42. GT-309 standard and early quench burners.

Fig. 43 shows the correlation obtained between the exhaust concentration and the time available for the NO formation process in the primary zone of the burner when operated at the engine design condition. Similar trends are displayed for the measured concentrations and the corresponding computed kinetically-limited concentrations. With Modification No. 1, the nominal residence time of the gas in the primary zone was reduced about 34 percent. The residence time of the gas in the primary zone of burner Modification No. 2 was decreased further to about 44 percent of that provided in the standard burner. An average decrease of about 40 percent in NO exhaust concentrations was obtained over the entire engine operating range with burner Modification No. 2.

To evaluate the influence of primary zone equivalence ratio, two modified versions of the standard burner were tested. These modified burners are shown in Fig. 44. One burner was designed for a lean primary zone equivalence ratio of 0.35. This was achieved by increasing the total area of the primary air holes at the expense of the dilution air holes. The other burner was designed for a rich primary zone equivalence ratio of 1.2. This was accomplished by decreasing the number and total area of the dome holes and increasing the total area of the dilution air holes an equal amount.

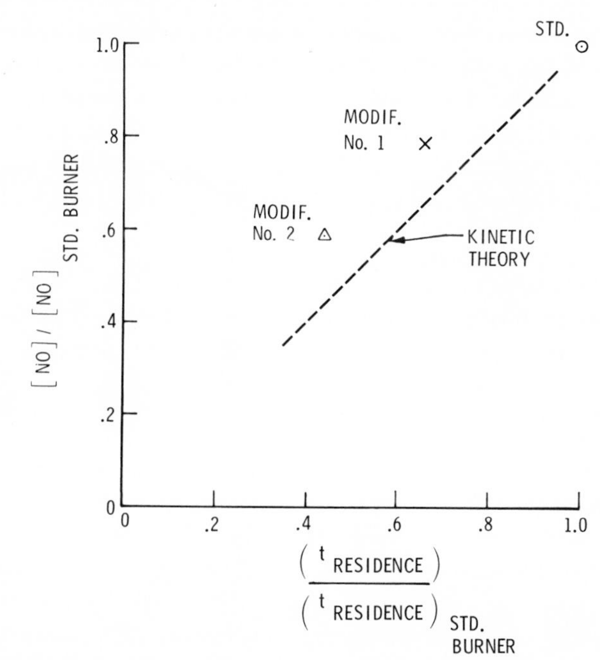

Fig. 43. Correlation of experimental and theoretical effects of primary zone residence time on exhaust NO concentration.

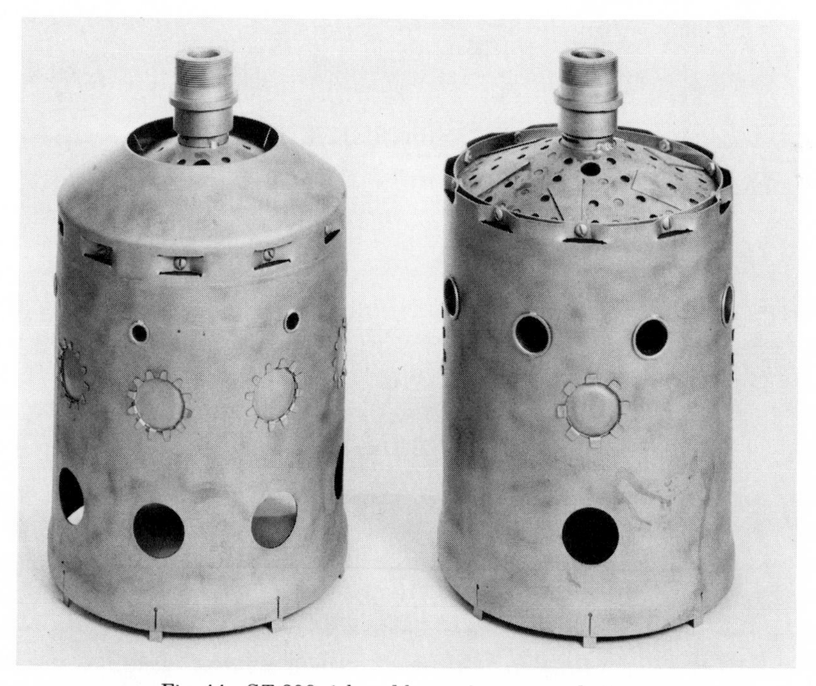

Fig. 44. GT-309 rich and lean primary zone burners.

References pp. 440-443

These primary zone equivalence ratios are nominal values based on the quantity of air assumed to flow through the zone of the fuel spray. They are approximately equal to the average combustion equivalence ratios only at the rated engine operating condition. Fuel spray configuration and air distribution throughout the burner change in many complex ways as the gas flow rate is decreased from the design value. As a result, it is difficult to define accurately a true part load primary zone equivalence ratio.

A difference exists between the primary residence times of the rich and lean burners. Since the air flow through the primary zone of the lean burner is greater than that for the rich burner, the residence time of the gas at high temperature in the lean burner is less than that in the case of the rich burner.

Fig. 45 illustrates the correlation that was obtained between NO exhaust concentration and the primary zone equivalence ratio when the burners were operated at the engine design condition. Reductions in NO emissions from both burner designs were measured. Similar trends are displayed for the actual measured concentrations and the corresponding theoretical computed concentrations that were determined by assuming equal kinetic ratios (see Table 4) for the burners being compared. A greater variation between the calculated and the experimental values was

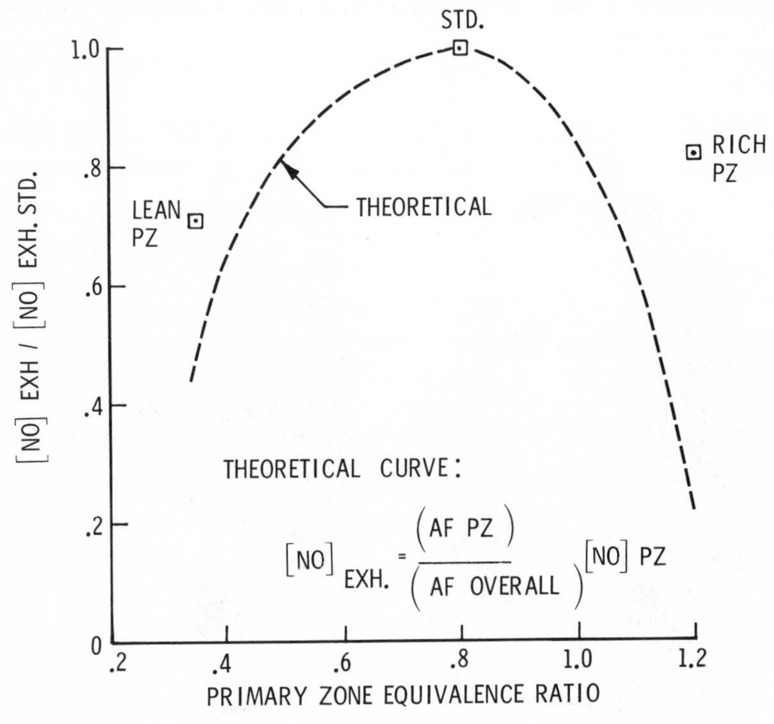

Fig. 45. Correlation of experimental and theoretical effects of primary zone equivalence ratio on exhaust NO concentration.

experienced in the case of the rich primary zone burner. A similar lack of correspondence between the calculated and experimental NO concentrations was observed by Newhall and Shahed (39) when burning a homogeneous fuel-air mixture richer than stoichiometric in a high-pressure combustion vessel of fixed volume.

Fig. 46 illustrates the reductions in NO emissions that were realized over the entire GT-309 engine operating range with the rich primary zone burner, the lean primary zone burner, and the Modification No. 2 early quench burner. Greater reductions in NO emissions were realized in the lean primary zone burner than in the rich primary zone burner. Reductions in NO emissions from the Modification No. 2 early quench burner were greater than either the rich or lean primary zone burners.

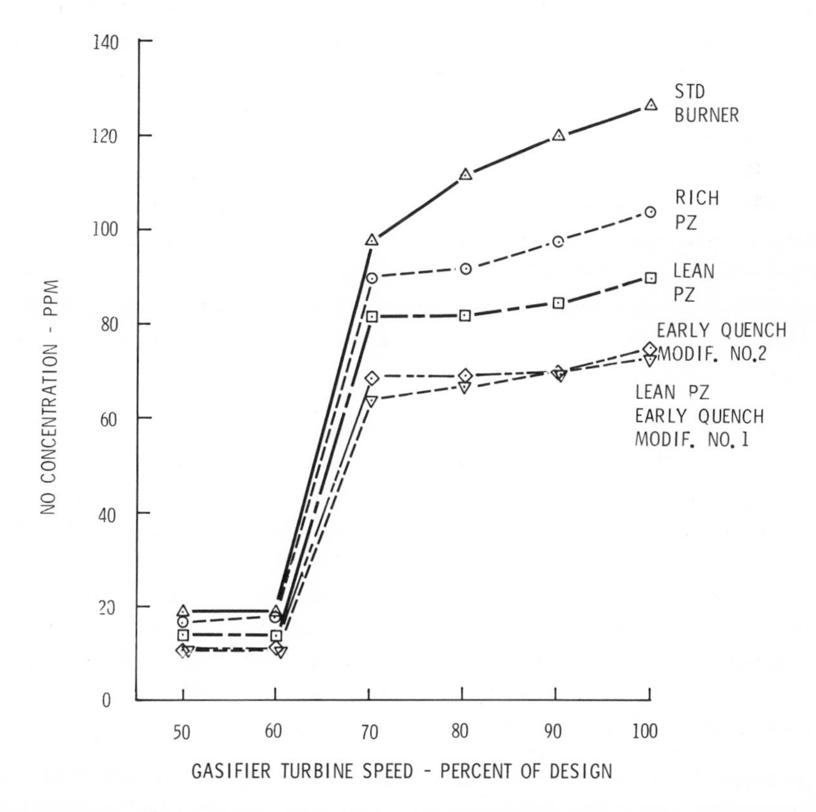

Fig. 46. Exhaust NO concentrations of modified GT-309 burners versus percent of design gasifier turbine speed.

Also shown in Fig. 46 are the reductions in NO emissions that were obtained by combining the lean primary zone configuration with the axial location of the burner holes as incorporated in the Modification No. 1 early quench burner. The reductions in NO emissions realized from this lean primary zone-early quench burner are similar to those of the Modification No. 2 early quench burner with a primary zone

References pp. 440-443

equivalence ratio of 0.8. Although not shown in the figure, a burner with a lean primary zone and employing the most forward locations of primary and dilution air holes (Modification No. 2) provided only marginal reductions in NO concentrations at the low gasifier turbine speed conditions.

The vehicle emission computer simulation program described earlier in the paper was used to compute true mass vehicle emission values for the experimental burners. Fig. 46 shows NO concentrations for these burners that were measured on the burner test rig. An arbitrarily-selected 160 hp scaled version of the GT-309 engine was assumed to power a 4500 lb passenger car over the 1970 Federal driving schedule. The computed true mass NO_X emission values of this simulated engine-vehicle combination are shown in Fig. 47 for several experimental burners. An NO_X mass emission value of 3.1 gm/mi is shown for the standard burner. A decrease in NO_X mass emissions of 33 percent is illustrated for the Modification No. 2 early quench burner. Intermediate reductions in NO_X mass emissions computed for the rich and lean primary zone burners with the standard residence time are illustrated. A minimum value of about 1.9 gm/mi is reached with the lean primary zone-early quench Modification No. 1 burner. Although this minimum value represents a 37 percent reduction in NO_X mass emissions compared to the standard burner, it is still about four times greater than the 1976 Federal emission standard for oxides of nitrogen. Although the computed true mass emission values are based on the 1970 Federal driving schedule, similar conclusions should be reached if the 1972 driving schedule were used.

Emissions from the Modification No. 2 early quench burner were also evaluated experimentally when the GT-309 engine powered truck tractor, incorporating this burner, was operated according to the 1970 Federal vehicle emission driving schedule. Reductions in NO_X emissions from the vehicle obtained by substituting this experimental burner for the standard engine burner were found to be similar in magnitude to the reductions measured on the burner test rig. Similar gains were obtained when the vehicle emission simulation program was used to make these burner comparisons. Having obtained good correlations of results with this simulation program and actual testing, other burner designs are being evaluated with increased confidence using this vehicle emission simulation program.

The computed vehicle emission values presented in Fig. 47 indicate that primary zone equivalence ratio and early quench techniques can be applied effectively to reduce the NO exhaust emissions of this particular burner configuration without impairing overall combustion and engine performances. Studies are in progress to extend these techniques to a broader range of equivalence ratio modifications.

ADVANCED BURNER CONCEPTS

Simple modifications to the conventional burner are apparently insufficient to reduce NO_X emissions to the 1976 Federal emission standards. To satisfy these future

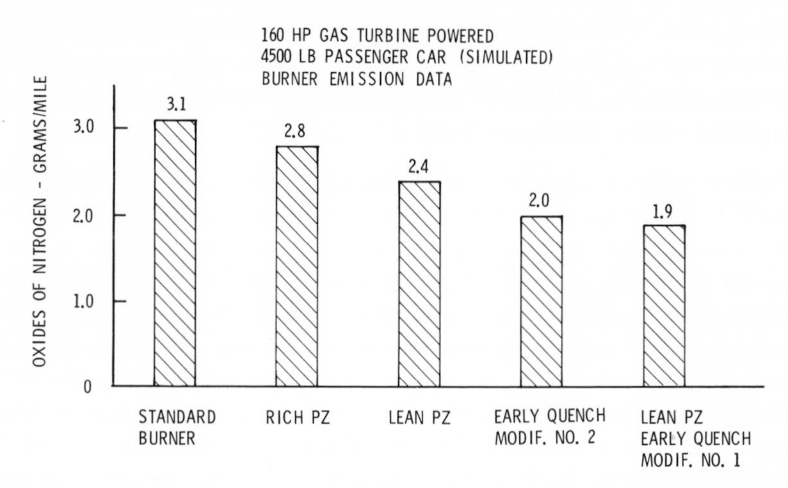

Fig. 47. True mass emissions of a simulated passenger car powered by a gas turbine engine with modified GT-309 burners.

standards with a regenerative gas turbine, or other continuous combustion engines, will require more complex burner changes. Several concepts which may show promise for additional reductions, applied either singly or in combinations with one another, are: (a) increase the homogeneity of the combustible fuel-air mixture; (b) recirculate some of the primary zone combustion products, (c) admit an inert diluent into the primary zone; (d) modify the engine thermodynamic cycle; (e) stage the combustion process; and (f) tailor the primary zone fuel-air ratio over the engine operating range.

Increased Homogeneity of Combustible Fuel-Air Mixture – Combustion of a lean homogeneous (premixed) fuel-air mixture should result in a significant reduction in NO emissions. Premixing would eliminate the undesirable characteristic of a diffusion flame which inherently has local near-stoichiometric regions (in spite of overall lean fuel-air ratios) in which the NO formation rate is high.

The NO emissions of a hypothetical diffusion flame and a premixed burner, both operating at an overall air-fuel ratio of 25:1 are compared in Fig. 48. All of the fuel admitted into the premixed burner is assumed to burn at an air-fuel ratio of 25:1. In the case of the diffusion flame burner, fuel droplets are injected into the combustion volume and combustion occurs at air-fuel ratios close to the stoichiometric value. For this example, it is assumed that the fuel burns at a air-fuel ratio of 18.5:1 (ϕ = 0.8) and the combustion products are diluted by the excess air to produce an overall air-fuel ratio of 25:1.

To estimate NO emissions for both burners, a portion of an equilibrium NO concentration versus air-fuel ratio plot is shown in Fig. 48. The effects of early quench will not be considered for this hypothetical example. An NO concentration of 1480 ppm is predicted for the premixed burner at an air-fuel ratio of 25:1. The NO concentration in the combustion products of the diffusion flame burner (3318 ppm)

References pp. 440-443

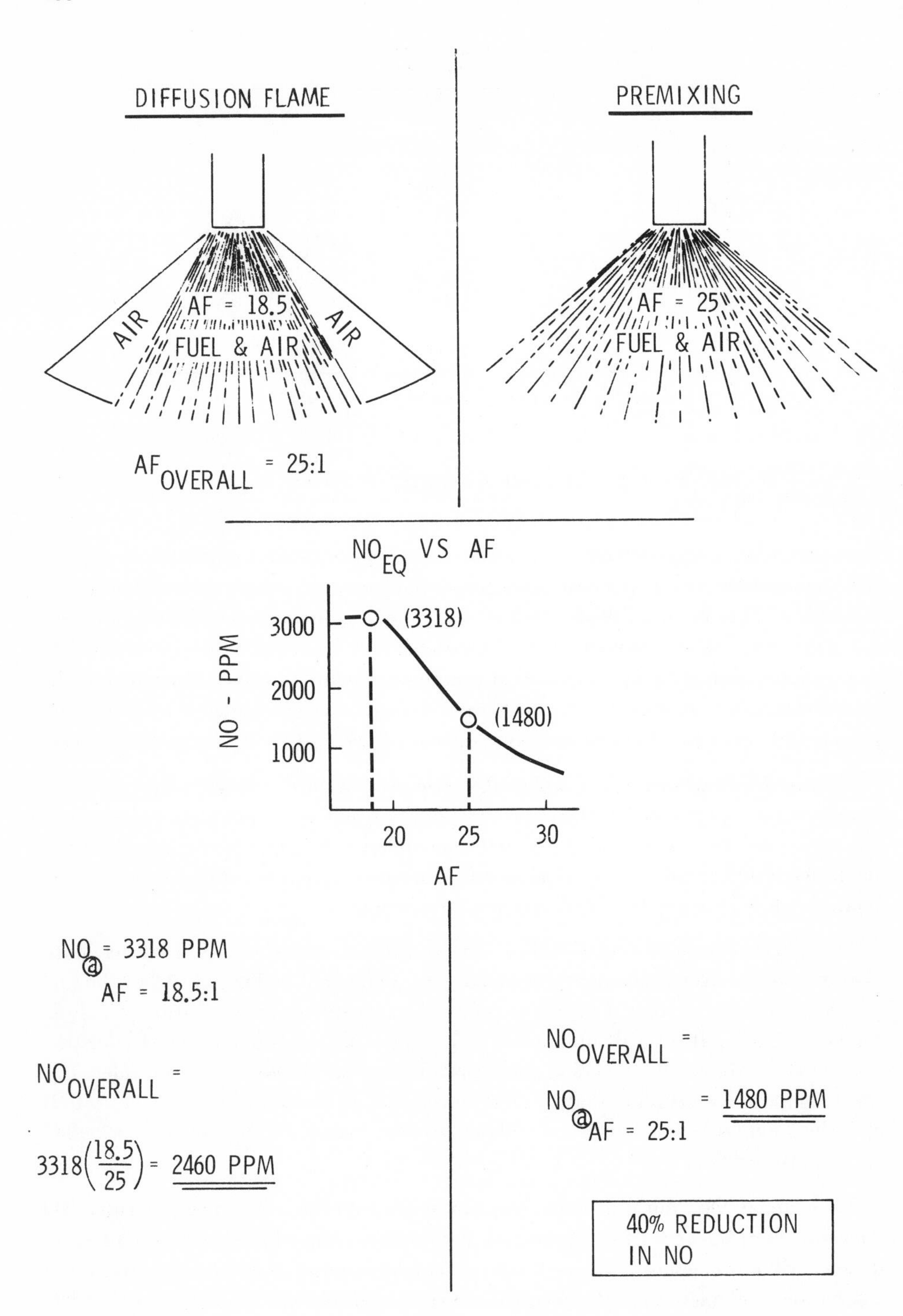

Fig. 48. Comparison of a simplified diffusion flame burner and a premixed burner.

is reduced by the dilution of the excess air by a factor of 0.74 to give an exhaust concentration of 2460 ppm. From this simple example, it may be concluded that, with both burners operating at an overall air-fuel ratio of 25:1, the NO exhaust concentration for the premixed burner should be 40 percent lower than for the diffusion flame burner.

A theoretical study was conducted concerning the effects of non-homogeneities on NO production in an environment similar to a gas turbine combustor. The study investigated the effects on NO production of both spatial and time variations of temperature, pressure and equivalence ratio. It was concluded that the non-linearity of kinetically limited NO concentrations with temperature causes the total NO produced in a combustor with timewise or spatial non-homogeneities in temperature to be greater than the NO produced at a uniform mean temperature. For example, it was computed that a temperature amplitude of 100 F would produce an NO concentration which is over 40 percent greater than for the case of a uniform mean temperature distribution. Pressure or equivalence ratio variations by themselves cause very small decreases in NO production compared to the uniform case. Non-homogeneities in temperature associated with equivalence ratio variations would be more likely to occur in a combustor than pressure variations. These theoretical conclusions have been qualitatively substantiated experimentally by Richter, Wiese, and Sage (40). These investigators concluded that the "quantity of oxides of nitrogen in a stable premixed flame is approximately ten percent of the quantity found to be typical in a turbulent diffusion flame under the same conditions."

Schemes which may be practical for achieving premixing are being studied. One scheme, using carbureting swirl-can combustor modules, is being investigated by the NASA Lewis Research Center (41). Many concepts for vaporizing burners may also be incorporated into premixed burner designs (42). A vaporizing burner design which has been used in the Stirling engine is discussed in Ref. 5. Although premixed burners seem to offer an advantage from a consideration of NO emissions, they also have disadvantages. Stability limits may be more restrictive than with a diffusion flame burner. Material durability may also be a consideration for vaporizing designs.

Recirculation of Primary Zone Combustion Products – The external recirculation of cooled combustion products has been shown to be effective in reducing NO emissions from a Stirling burner. Although this was accomplished with relative ease for the Stirling burner, difficulty would be encountered in recirculating the primary zone products in a gas turbine burner. Because of the extremely lean overall fuel-air ratios of the gas turbine burner, it appears that only the recirculation of cooled primary zone combustion products (rather than the overall exhaust) would be effective. Several schemes for achieving recirculation have been described (43-46). Several different oil burners have been devised which achieve internal recirculation. One incorporates a swirl chamber with reverse flow in the center core. In the other, an air jet is deflected by the Coanda effect and entrains the combustion products.

References pp. 440-443

Inert Diluent – If the flame temperature in the primary zone can be reduced below its adiabatic value, significant reductions in NO emissions can be obtained. This reduction stems not only from a reduction in the equilibrium value but also from a reduction in the rate of formation. A reduction in temperature of several hundred degrees can account for an order of magnitude reduction in the rate of NO formation (ppm/msec) as shown in Fig. 49. The introduction of a heat sink into the primary zone of a burner will result in a reduction of the flame temperature below its adiabatic value. Adding diluents to the intake charge of a spark ignition engine has produced significant reductions in NO emissions (47).

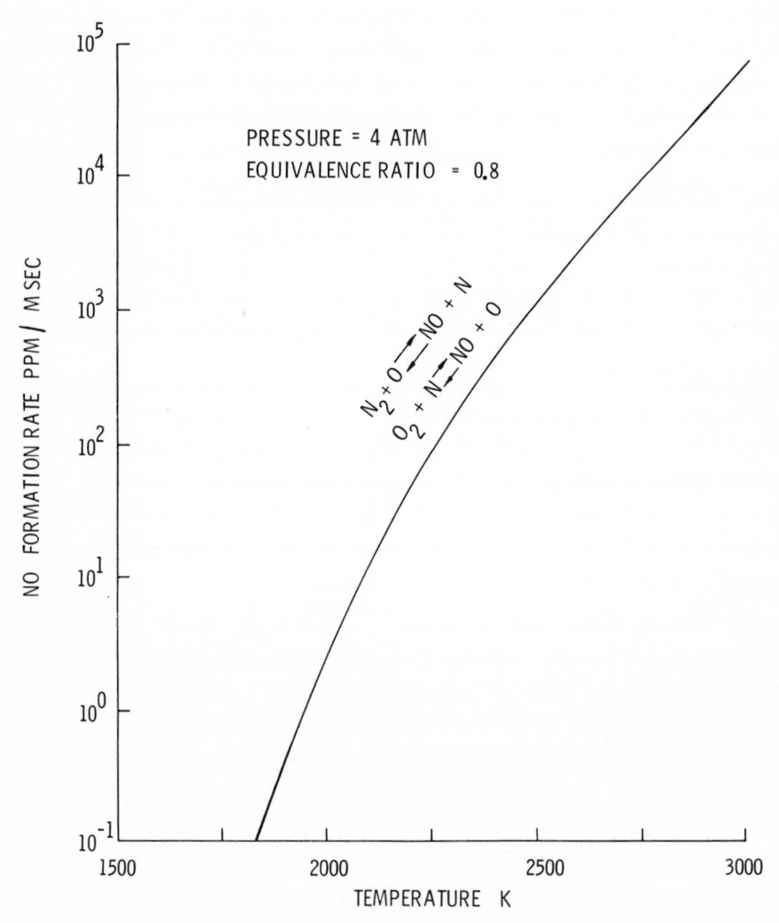

Fig. 49. Computed NO formation rate versus temperature.

One convenient fluid readily available for this use is cooled recirculated exhaust gas. Water may also be used as an inert diluent. One study (48) has indicated that significant reductions in NO emissions can be obtained with the injection of steam into large industrial gas turbine combustion chambers. The effect of water injection on NO emissions of the regenerative GT-309 gas turbine burner has been investigated experimentally at the GM Research Laboratories. A 50 percent reduction in vehicle NO_X mass emissions was obtained with a water-fuel ratio of 1.5. Several problems such as freezing, an additional tank, and a water pump make the application of this concept to vehicular engines appear undesirable.

Engine Thermodynamic Cycle – Until emissions became the important criteria of powerplant performance, thermal efficiency was considered the single most important criteria for evaluting an engine thermodynamic cycle. The low pressure, regenerative gas turbine engine has been widely accepted as the best compromise for a vehicular powerplant before emissions were a performance criteria. Since NO formation is so strongly dependent on burner inlet temperature, perhaps a cycle modification could lessen the difficulty experienced with high burner inlet temperatures. The influence of the thermodynamic cycle on the vehicle NO_X mass emissions has been investigated analytically in Ref. 49.

Staged Combustion – One concept for staged combustion consists of spraying the required amount of fuel into the burner at two separate locations (50). As a result, the local fuel-air ratio at each of these two locations should be leaner than if combustion of the total amount of fuel occurred at only one location. This offers a potential for reducing the temperature level in the burner which should be beneficial in reducing NO emissions.

An alternate concept for staged combustion consists of operating the first stage of the burner with the total fuel flow but with a deficiency of air (51). Combustion at rich fuel-air ratios should produce low NO concentrations. Additional air would then be added to the products from the first stage, which would contain very high HC and CO concentrations, and the oxidation of these combustible species would be completed.

Tailored Primary Zone Fuel-Air Ratio – It has been shown that NO formation is strongly influenced by the fuel-air ratio. The GT-309 engine operating schedule dictates a 3:1 turndown ratio of fuel-air ratio. Since a burner of fixed geometry can be designed for optimum operation at only one condition, a variable burner hole area geometry scheme would enable tailoring the primary zone fuel-air ratio to a desired value throughout the engine operating schedule. Many ideas for achieving variable geometry in a gas turbine burner have been patented. The patents listed in Refs. 52 and 53 describe some of these ideas.

Investigations of the potentials of all of the above concepts for reducing emissions in a vehicular gas turbine burner are being conducted at the GM Research Laboratories.

SUMMARY

Although originally designed for the usual considerations of performance, reliability, durability, and simplicity, the continuous combustion systems of the

References pp. 440-443

vehicular gas turbine, steam, and Stirling engines offer low emission potential. HC and CO emissions from each burner are decidedly low. Conversely, NO emissions from the regenerative gas turbine and Stirling burners are high, reflecting the strong influence of high inlet temperature.

Higher overall thermal efficiency of the powerplant, in addition to low burner emissions, is important in minimizing vehicle mass emissions. The HC and CO mass emissions of the SE-101 steam powered vehicle are significantly higher than for the gas turbine and Stirling engine powered vehicles. These higher vehicle mass emissions can be attributed to lower thermal efficiency of the powerplant (about 50 percent lower than the thermal efficiencies of the gas turbine and Stirling engines) and the inherently higher HC and CO burner emission indices resulting from operation at ambient inlet air temperatures. Lower NO_X burner emission indices are realized with the steam engine burner which is designed for operation at ambient inlet air temperatures. However, most of this advantage is reduced to a minimum when NO_X vehicle mass emissions are considered because of the higher fuel consumption of the steam powerplant. NO_X vehicle mass emissions from all three powerplants investigated using the standard burner are significantly higher than the 1976 Federal standards.

Emission control techniques have been applied to each burner after reviewing the chemical kinetics of the pollutant species of concern. The gas turbine burner was deemed to be more amenable to the control of NO emissions than the other two types of burners because of the availability of a large quantity of excess air that can be used for emission control and a reasonably low outlet temperature that precludes continuing chemical reactions. Minor changes to the conventional diffusion flame burner have not been sufficient to reduce the NO_X emissions to meet the 1976 Federal emission standard. More complex schemes, applied either singly or in combination, give renewed enthusiasm for controlling NO_X emissions from a gas turbine burner without penalizing fuel economy.

REFERENCES

1. *W. A. Turunen, "Pinwheels or Pistons?" SAE Trans., Vol. 63, pp. 72-83, 1955.*
2. *W. A. Turunen and J. S. Collman, "The General Motors Research GT-309 Gas Turbine Engine", SAE Trans., Vol. 74, Sec. 3, Paper No. 650714, pp. 357-377, 1966.*
3. *P. T. Vickers, C. A. Amann, H. R. Mitchell, and W. Cornelius, "The Design Features of the GM SE-101 – A Vapor-Cycle Powerplant", SAE Trans., Vol. 79, Sec. 1, Paper No. 700163, pp. 628-650, 1970.*
4. *P. T. Vickers, J. R. Mondt, W. H. Haverdink, and W. R. Wade, "General Motors' Steam Powered Passenger Cars – Emissions, Fuel Economy and Performance", SAE Trans., Vol. 79, Sec. 3, Paper No. 700670, pp. 2099-2126, 1970.*
5. *G. Flynn, W. H. Percival, and F. E. Heffner, "GMR Stirling Thermal Engine", SAE Trans., Vol. 68, pp. 665-684, 1960.*
6. *F. E. Heffner, "Highlights from 6500 Hours of Stirling Engine Operation", SAE Trans., Vol. 74, Sec. 2, Paper No. 650254, pp. 33-54, 1966.*

7. *P. D. Agarwal, R. J. Mooney, and R. R. Toepel, "Stir-Lec I, A Stirling Electric Hybrid Car", SAE Paper No. 690074, January 1969.*
8. *W. Cornelius and W. R. Wade, "The Formation and Control of Nitric Oxide in a Regenerative Gas Turbine Burner", SAE Trans., Vol. 79, Sec. 4, Paper No. 700708, pp. 2176-2202, 1970.*
9. *J. H. Lienesch and W. R. Wade, "Stirling Engine Progress Report: Smoke, Odor, Noise and Exhaust Emissions", SAE Trans., Vol. 77, Sec. 1, Paper No. 680081, pp. 292-307, 1968.*
10. *A. H. Lefebvre and E. R. Norster, "The Design of Tubular Gas Turbine Combustion Chambers for Optimum Mixing Performance", The Institution of Mechanical Engineers Proceedings 1968-1969, Volume 183, Part 3N.*
11. *C. C. Graves and J. S. Grobman, "Theoretical Analysis of Total Pressure Loss and Airflow Distribution for Tubular Turbojet Combustors with Constant Annulus and Liner Cross Sectional Areas", NACA Report 1373, September 1956.*
12. *W. R. Hawthorne and W. T. Olson, "Design and Performance of Gas Turbine Power Plants", Princeton University Press, 1960.*
13. *I. E. Smith, "Combustion in Advanced Gas Turbine Systems", Proceedings of an International Propulsion Symposium held at the College of Aeronautics, Cranfield, Pergammon Press, April 1967.*
14. *"Proceedings of the Joint Conference on Combustion", Institution of Mechanical Engineers, American Society of Mechanical Engineers, Boston, Massachusetts, June 1955, London, October 1955.*
15. *J. S. Clarke and S. R. Jackson, "General Considerations in the Design of Combustion Chambers for Aircraft and Industrial Gas Turbines", SAE Paper 444A, January 1962.*
16. *L. R. Turner and Donald Bogart, "Constant Pressure Combustion Charts Including Effects of Diluent Addition", NACA Report 937, March 1948.*
17. *D. L. Stivender, "Development of a Fuel-Based Mass Emission Measurement Procedure", SAE Paper No. 710604, June 1971.*
18. *W. Cornelius, D. L. Stivender, R. E. Sullivan, "A Combustion System for a Vehicular Regenerative Gas Turbine Featuring Low Air Pollutant Emissions", SAE Trans., Vol. 76, Sec. 4, Paper No. 670936, pp. 3140-3159, 1968.*
19. *Federal Department of Health, Education and Welfare, "Control of Air Pollution from New Motor Vehicles and New Motor Vehicle Engines", Federal Register, Vol. 33, No. 108 (June 1968), Part II.*
20. *Environmental Protection Agency, "Control of Air Pollution from New Motor Vehicles and New Motor Vehicle Engines", Federal Register, Vol. 36, No. 128 (July 2, 1971), Part II.*
21. *State of California Air Resources Board, "California Exhaust Emission Standards, Test and Approval Procedures for Diesel Engines in 1973 and Subsequent Model Year Vehicles Over 6001 Pounds Gross Vehicle Weight." Adopted Nov. 18, 1970, Amended Feb. 17, 1971.*
22. *D. C. Hammond, Jr., and A. M. Mellor, "Analytical Calculations for the Performance and Pollutant Emissions of Gas Turbine Combustors", AIAA Paper No. 71-711, June 1971.*
23. *J. B. Heywood, "Gas Turbine Combustor Modeling for Calculating Nitric Oxide Emissions", AIAA Paper No. 71-712, June 1971.*
24. *C. T. Crowe, D. T. Pratt, B. R. Bowman, T. W. Sonnichsen, "Prediction of Nitric Oxide Formation in Turbojet Engines for PSR Analysis", AIAA Paper No. 71-713, June 1971.*
25. *R. Edelman and C. Economos, "A Mathematical Model for Jet Engine Combustor Pollutant Emissions", AIAA Paper 71-714, June 1971.*
26. *R. Roberts, L. D. Aceto, and R. H. Kollrack, "An Analytical Model for NO Formation in a Gas Turbine Combustion Chamber", AIAA Paper 71-715, June 1971.*
27. *J. B. Heywood, J. A. Fay and L. H. Linden, "Jet Aircraft Air Pollutant Production and Dispersion", AIAA Paper No. 70-715, January 1970.*

28. *R. F. Sawyer, D. P. Teixeira, and E. S. Starkman, "Air Pollution Characteristics of Gas Turbine Engines", Journal of Engineering For Power, Trans. ASME, Series A, Vol. 91, No. 4, Oct. 1969, pp. 290-296.*
29. *E. S. Starkman, Y. Mizutani, R. F. Sawyer, and D. P. Teixeira, "The Role of Chemistry in Gas Turbine Emissions", ASME Paper No. 70-GT-81, May 1970.*
30. *P. Eyzat and J. C. Guibet, "A New Look at Nitrogen Oxides Formation in Internal Combustion Engines", SAE Trans., Vol. 77, Sec. 1, Paper No. 680124, pp. 481-500, 1968.*
31. *M. Camac and R. M. Feinberg, "Formation of NO in Shock Heated Air", Eleventh Symposium (International) on Combustion, The Combustion Institute, 1967, pp. 137-145.*
32. *C. P. Fenimore, M. B. Hilt, and R. H. Johnson, "Formation and Measurements of Nitrogen Oxides in Gas Turbines", ASME Paper No. 70-WA/GT-3, December 1970.*
33. *Zeldovich, Y. B., "The Oxidation of Nitrogen in Combustion Explosions", ACTA Physicochimica U.S.S.R., Vol. 21, pp. 577-628, 1946.*
34. *J. M. Singer, E. B. Cook, M. E. Harris, V. R. Rowe, J. Grumer, "Flame Characteristics Causing Air Pollution: Production of Oxides of Nitrogen and Carbon Monoxide", Bureau of Mines Report of Investigations RI 6958, 1967.*
35. *R. S. Brokaw and D. A. Bittker, "Carbon Monoxide Oxidation Rates Computed for Automobile Exhaust Manifold Reactor Conditions", NASA Technical Note D-7024, Lewis Research Center, Cleveland, 1970.*
36. *H. K. Newhall, "Control of Nitrogen Oxides by Exhaust Recirculation", SAE Trans., Vol. 76, Sec. 3, Paper No. 670495, pp. 1820-1836, 1968.*
37. *T. S. Bacon, "Production of Non-Corrosive Inert Gas by the Combustion of Fuel Gas", Petroleum Engineer, Vol. 11, No. 13, p. 51-56, September 1940.*
38. *A. W. Bell, N. Bayard de Volvo, and B. P. Breen, "Nitric Oxide Reduction by Controlled Combustion Processes", Paper presented at Western States Section, Combustion Institute, Spring Meeting, Berkeley, April 1970.*
39. *H. K. Newhall and S. M. Shahed, "Kinetics of Nitric Oxide Formation in High Pressure Flames", Paper presented at the Thirteenth Symposium (International) on Combustion, The Combustion Institute, University of Utah, August 1970.*
40. *G. N. Richter, H. C. Wiese, B. H. Sage, "Oxides of Nitrogen in Combustion, Premixed Flame," Combustion and Flame, Vol. 6, pp. 1-8, March 1962.*
41. *R. W. Niedywiecki and R. E. Jones, "Combustion Stability of Single Swirl-Can Combustor Modules Using ASTM-A1 Liquid Fuel", NASA TN D-5436, Lewis Research Center, Cleveland, October 1969.*
42. *A. W. Hussman and G. W. Maybach, "The Film Vaporization Combustor", SAE Trans., Vol. 69, pp. 563-574, 1961.*
43. *R. Kamo, P. W. Cooper, B. N. Glicksberg, and J. A. Fitzgerald, "Nonluminous Combustion of Heating Oil Through Recirculation", 1964 American Petroleum Institute Research Conference on Distillate Fuel Combustion, Conference, Chicago, Illinois. Paper CD 64-8, June 1964.*
44. *I. Reba, "Combustion of Residual Fuel with Massive Recirculation", IITRI Report M 6120, August 1967.*
45. *R. E. Schreter, L. G. Poe, and E. M. Kuska, "Industrial Burners Today and Tomorrow (State of the Art)", ASME Paper 69-WA/FU-7, November 1969.*
46. *D. W. Locklin, "Small Oil Burners – A Review of Some Recent Developments", ASME Paper 69-WA/FU-5, November 1969.*
47. *A. A. Quader, "Why Intake Charge Dilution Decreases Nitric Oxide Emission from Spark Ignition Engines," SAE paper No. 710009, January 1971.*
48. *N. R. Dibelius, M. B. Hilt, R. H. Johnson, "Reduction of Nitrogen Oxides from Gas Turbines by Steam Injection", ASME paper 71-GT-58, March 1971.*

49. *C. A. Amann, W. R. Wade, M. K. Yu, "Some Factors Affecting Gas Turbine Passenger Car Emissions", SAE paper No. 720237, January 1972.*
50. *A. H. Lefebvre, Personal Communication.*
51. *D. H. Barnhart and E. K. Diehl, "Control of Nitrogen Oxides in Boiler Flue Gases by Two Stage Combustion", Air Pollution Control Assoc. J., V. 10, 1960, p. 397-406.*
52. *E. D. Brown, "Gas Turbine Combustion Chamber with Variable Area Primary Air Inlet", U. S. Patent No. 2,655,787, Oct. 20, 1953.*
53. *C. Kind, "Automatic Air Regulating Device in Combustion Chambers of Gas Turbine Plants", U.S. Patent No. 2,837,894, June 10, 1958.*
54. *H. Niki, A. Warnick, and R. R. Lord, "An Ozone-NO Chemiluminescence Method for NO Analysis in Piston and Turbine Engines", SAE Paper 710072, January 1971.*

APPENDIX A

EXPERIMENTAL TECHNIQUES

Instrumentation – Rapid, on-the-spot, continuous measurements of the low concentrations of burner exhaust pollutants are desirable during steady-state test rig operating conditions and transient engine operating conditions. The development of continuous analytical techniques for determining these low concentrations led to the incorporation of appropriate analyzers in the compact portable console illustrated in Fig. A-1. Contained in the console are the following exhaust gas analyzers: (a) Beckman Model 108 Total Hydrocarbon Analyzer (Flame Ionization Detector); (b) Beckman Model 315AL Non-Dispersive Infrared Analyzer sensitized with nitric oxide; (c) Beckman Model 315AL Non-Dispersive Infrared Analyzer sensitized with carbon monoxide; and, (d) Beckman Model 315 Non-Dispersive Infrared Analyzer sensitized with carbon dioxide.

The capability of measuring NO concentration has recently been expanded with the development of the NO chemiluminescent analyzer (54). Prior to the commercial availability of this instrument, C. E. Quinn of the General Motors Research Laboratories constructed a chemiluminescent analyzer which has been used to measure the characteristically low NO concentrations in the exhaust of a regenerative gas turbine burner operated in the burner test facility. It was determined that this chemiluminescent analyzer is more sensitive than the long-path NO-NDIRA, and requires a significantly shorter time to reach full scale with a minimum of drift.

To improve the capability of measuring the concentrations of unburned hydrocarbons, a Beckman Model 402 heated flame ionization detector was acquired. This instrument maintains the sampling and analysis systems at an optimum temperature which provides the best compromise between loss of hydrocarbons from condensation and/or absorption at low temperatures and loss of hydrocarbons by oxidation at high temperature.

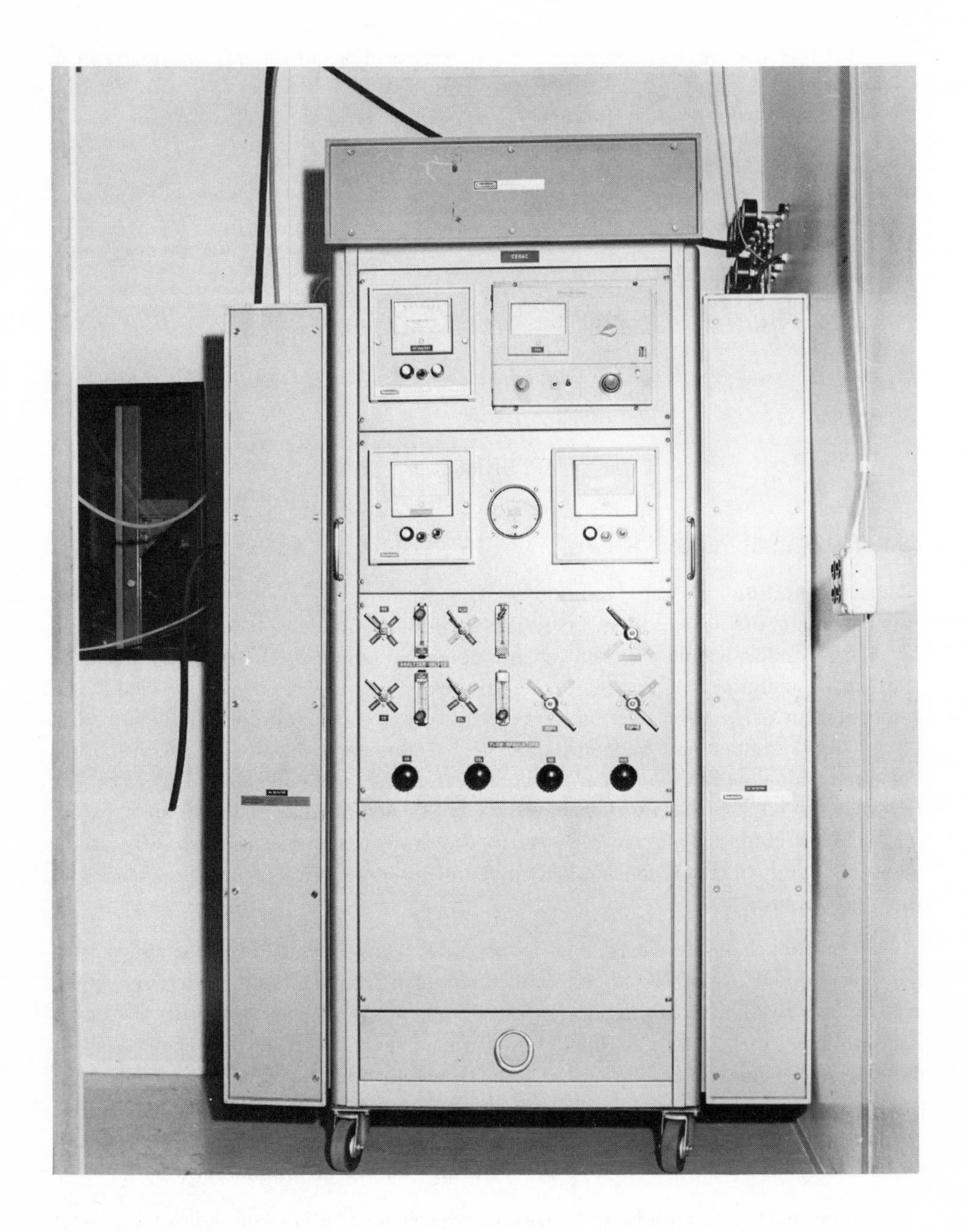

Fig. A-1. Exhaust gas analyzer console for continuous combustion system gas analysis.

Wet chemical colorimetric techniques, requiring the collection of grab samples of the burner exhaust gas, are employed to measure (a) oxides of nitrogen concentration by the modified phenol disulfonic acid (PDS) and modified Saltzman procedures and

(b) formaldehyde concentration by either the phenylhydrazine-ferricyanide photometric method or the 3-methyl-2-benzothiazolone hydrozone (MBTH) method. Exhaust smoke can be evaluated either visually or with a Von Brand smoke meter.

Burner Test Facilities – The three burners were evaluated in single burner test facilities prior to engine installations. These burner test rigs were designed to duplicate as nearly as possible the internal environment to which the burner is exposed in the actual engine installation. These rigs are currently being used for continuing research on low emission burner concepts. Specific details of most of these test facilities have been described elsewhere (8).

Sampling Procedures – For accurate measurement of exhaust pollutants, it is essential that the sampling point be close to the analyzers. This serves a two-fold purpose: (a) to prevent any change in composition or character of the gas prior to being analyzed and (b) to prevent contamination of the sample by the tubing being used.

To assure meaningful measurements, a representative exhaust gas sample from a burner test rig or an engine must be supplied to the instrument console. In the case of the GT-309 gas turbine burner facility, the gas sample is extracted from the burner exhaust section. The gas sample is removed through five aspirating type temperature probes that are equally spaced and mounted in this section of the rig. The gas extracted by each probe flows into a common mixing chamber and thence to the analyzer console. To obtain a representative "tailpipe" emission value from the SE-101 burner test facility, two manifolded multiple-hole probe tubes are mounted in the exhaust duct. Similar probe tubes placed at the outlet of each burner provide a representative gas sample for the evaluation of individual burner performance. The large Stirling burner test facility, where small-scale steam burner tests were conducted, also incorporates a multiple-hole probe tube for the extraction of a representative exhaust gas sample. Also, a combination gas sampling-thermocouple water-cooled probe is used for extracting local samples within a burner. The GPU-3 burner test facility incorporates sample extraction provisions that are similar to those of the larger Stirling burner test facility.

DISCUSSION

A. F. McLean *(Ford Motor Company)*

Once again, Wally Wade and Walt Cornelius are to be congratulated for a paper which will become a frame of reference in the field of continuous combustion systems. Credibility for this frame of reference stems from the very factual nature of the paper – continuous combustion systems were DESIGNED and BUILT for the gas turbine, steam, and Stirling engines; data and discussion are based on ACTUAL TESTS of these units.

Quantifying emissions is sometimes confusing. Emission concentrations are measured in parts per million; Federal standards are in gms/mile; the Advanced Automotive Power group of the Environmental Protection Agency have helpfully defined their goals in gms/kilogram of fuel. I have not yet reached the point where I can mentally manipulate "parts per million," "gms/kilogram of fuel," "fuel/air ratio," "miles per gallon," and "gms/mile." As such, I find Figs. 6 and 7 of the paper very useful, handy curves. Before publication, however, may I suggest showing a family of gms/mile curves on Fig. 7 rather than just the 1976 Federal Standard.

The authors correctly emphasize the importance of both combustor and powerplant characteristics. Table 1 of the paper shows the combustor operating conditions as well as overall powerplant characteristics, both of which influence vehicle mass emissions. While specific fuel consumptions for the gas turbine and Stirling engines appear reasonable at a value of about 0.5, the efficiency of the steam engine does not seem to reflect the best state-of-the-art components. Would the authors give their opinion on the "fairness of comparison" of the steam engines to gas turbine and Stirling engines?

Next I would like to discuss the problem of high combustor inlet temperature on emissions. Enthusiasts of the steam engine and the non-regenerative turbine engine hail the big advantage of ambient or relatively cool combustor inlet conditions; how real is this advantage?

First, let's consider a given combustor design. As shown in Fig. 1 for a given combustor, increasing combustor inlet temperature reduces oxides of nitrogen emissions.

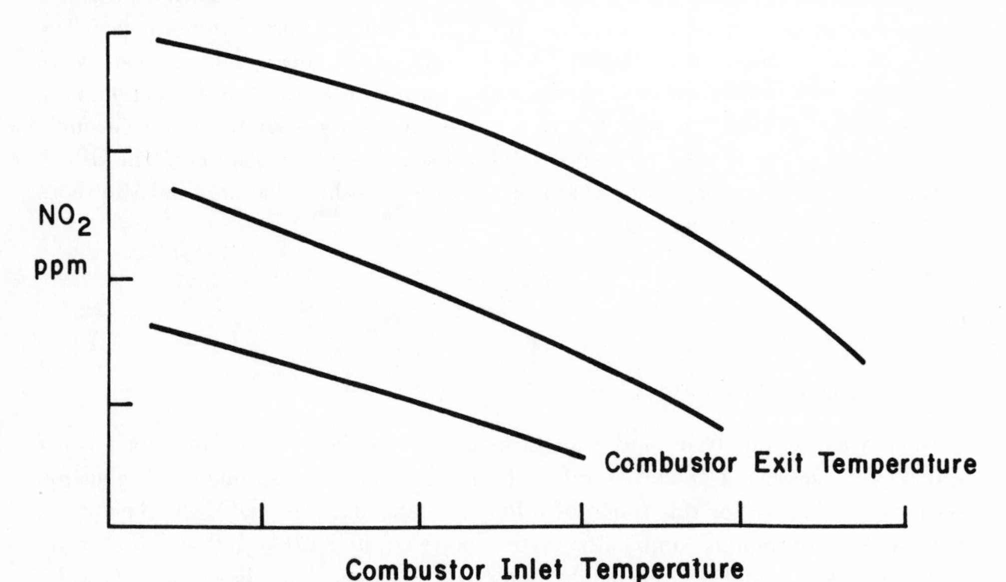

Fig. 1. Combustor temperature effects on NO_x emissions.

In light of what's been said at the Symposium, this may sound somewhat of an anomaly. The "trick", however, is understanding what is held constant while combustor inlet temperature is increased. Previous lecturers at the Symposium have always considered holding fuel/air ratio constant; in this case, as combustor inlet temperature is increased, combustor exit temperature also increases, as do oxides of nitrogen. In our case, however, we are generally interested in the influence of heat exchanger effectiveness in a gas turbine operating at a given turbine inlet temperature (since turbine inlet temperature defines the horsepower condition); in this instance as combustor inlet temperature increases, fuel/air ratio decreases and so do the concentration levels of nitrogen oxides.

Having considered the effect of increasing inlet temperature in a given combustor design, let's next consider combustor designs optimized for different inlet temperatures (i.e., different powerplants). The authors showed that computed values of nitric oxide agreed favorably with measurement thereby validating their assumption of a primary zone equivalence ratio of 0.8. However, the concept of the same primary zone equivalence ratio for each combustor design does not appeal to me. We know we want to operate as lean as possible to minimize nitric oxide formation and to minimize temperature spread. We also know that we must have a certain margin from the lean-blow-out condition to provide stability, especially for extreme transient operation. Ideally then, for a fair comparison, each combustor should be designed for a given "lean-blow-out-margin," not necessarily a given equivalence ratio. Perhaps the example of the varying combustor inlet temperature illustrated in Fig. 2 will demonstrate this point.

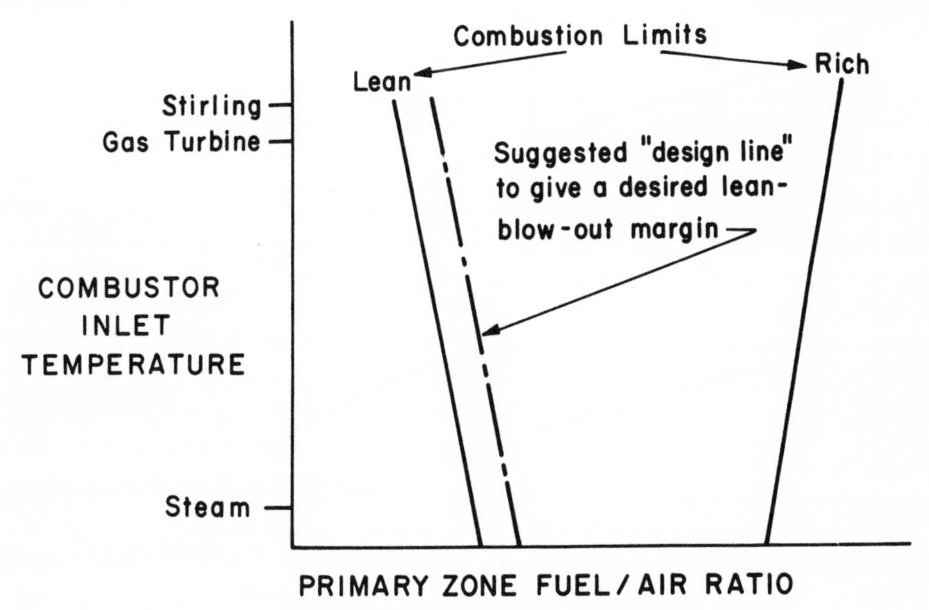

Fig. 2. Effect of combustor inlet temperature on lean-blow-out limit.

From this Figure, it can be seen that the higher the combustor inlet temperature, the lower the tolerable fuel/air ratio and resulting primary zone temperature rise. Some calculations, based on laboratory-determined flammability limits of methane, indicate that, for a given lean-blow-out-margin, peak flame temperature remains approximately constant as combustor inlet temperature increases. Would the authors comment on the ability to lean out the primary zone with increasing combustor inlet temperature in a practical combustion system?

Experimental work on low emissions continuous combustion systems is underway at Ford's Product Development Group. We are finding that tailoring any desired primary zone equivalence ratio has to consider, not only the bulk distribution of air in the combustor, but also local conditions surrounding fuel droplets. Therefore combustor geometry, fuel presentation and fuel/air mixing are all important in designing low emissions combustors with adequate lean-blow-out-margin. Currently at Ford, we are investigating the potential of many unconventional continuous combustion schemes. Work to date, as well as theoretical considerations give reason to believe that low nitric oxide emissions can be achieved; the real question is, how low and how soon?

W. R. Wade

I would like to express my appreciation to Mr. McLean for his comments on our paper. With regard to Mr. McLean's comment concerning Fig. 7 of our paper, this figure shows graphically the relationship between the burner emission indices of the three air pollutants required to meet the 1976 Federal vehicle emission standards and the average fuel economy of the vehicle during operation over the test procedure. The grams per mile values required by the 1976 Federal emissions standards were chosen for this figure as being most meaningful. The relationship shown in equation 5 can be used for calculating the required emission index values for any other grams per mile values of interest.

As noted by Mr. McLean, the concept of leaning out the primary zone with increasing burner inlet temperature is one of the promising concepts for a low emission regenerative gas turbine burner. When compared to the standard burner operating at an equivalence ratio of 0.8, significant reductions in NO emissions can be calculated theoretically for operation at fuel-air ratios near the lean blow out limit at elevated burner inlet temperatures. Some of the problems to be solved before this concept becomes a reality are: generation of a homogeneous lean fuel-air mixture with stable combustion and high combustion efficiencies, adequate turn-down ratio, good cold starting and relighting characteristics and durability.

Homogeneous lean operation will require complex burner design changes. As was shown in Fig. 45, reductions in NO emissions achieved by leaning out the nominal primary zone fuel-air ratio in a conventional type of burner are not of the magnitude expected theoretically because of the non-homogeneities of fuel-air ratio in the primary zone. It should be clarified that a nominal primary zone equivalence ratio of 0.8 was assumed for the computations of NO formation shown in Table 4 in order to simulate average conditions in the primary zones of the standard direct fuel spray burners. All of these standard burners had been designed before emissions were an important combustion consideration.

I would like to refer Mr. McLean's question concerning the value of the specific fuel consumption for the SE-101 steam engine to Mr. Paul Vickers.

P. T. Vickers *(Research Laboratories, GMC)*

The question is: "Could the efficiency of the SE-101 steam engine be improved?" Yes, to some extent. However, even with the incremental improvements which might be realized, I do not believe that the conclusion drawn from the measured emissions results for the SE-101 would be altered significantly.

H. P. Fredriksen *(Research Laboratories, GMC)*

The paper, "Emission Characteristics of Continuous Combustion Systems of Vehicular Powerplants – Gas Turbine, Steam, Stirling", by Mr. Wade and Mr. Cornelius, GMR, has shown that the thermodynamic cycle affects the exhaust emissions of different vehicular powerplants. One method of decreasing the exhaust emissions would be to optimize the thermodynamic cycle. When the thermodynamic cycle has been determined the combustion system can also be modified to decrease the exhaust emissions. SAE paper No. 700708, "The Formation and Control of Nitric Oxide in a Regenerative Gas Turbine Burner", by Mr. Cornelius and Mr. Wade, discussed some burner modifications which resulted in decreased nitric oxide exhaust emissions.

Fig. 1 shows a recent variation of the GMR diffusion flame gas turbine burner. It is basically a lean primary zone burner with a decreased burner diameter. The calculated primary zone equivalence ratio of this burner was 0.25 at full load. The flow area was decreased by 30% to that of the conventional burner. All the burner air was injected through the dome, the film cooling gap and the primary holes. There was no dilution air with this burner.

Fig. 2 shows the nitric oxide exhaust concentrations obtained with the conventional diffusion flame burner, the lean primary zone-early quench burner and the lean primary zone-decreased diameter burner. The lean primary zone-decreased diameter burner yields lower nitric oxide exhaust concentrations than other diffusion flame burners of the can type when evaluated over the GT-309 burner operating schedule. I think that the data indicate that significant decrease in the nitric oxide exhaust concentrations can be achieved by minor changes to burner design, and this without impairing burner performance or efficiency.

Fig. 1. Lean primary zone decreased diameter burner.

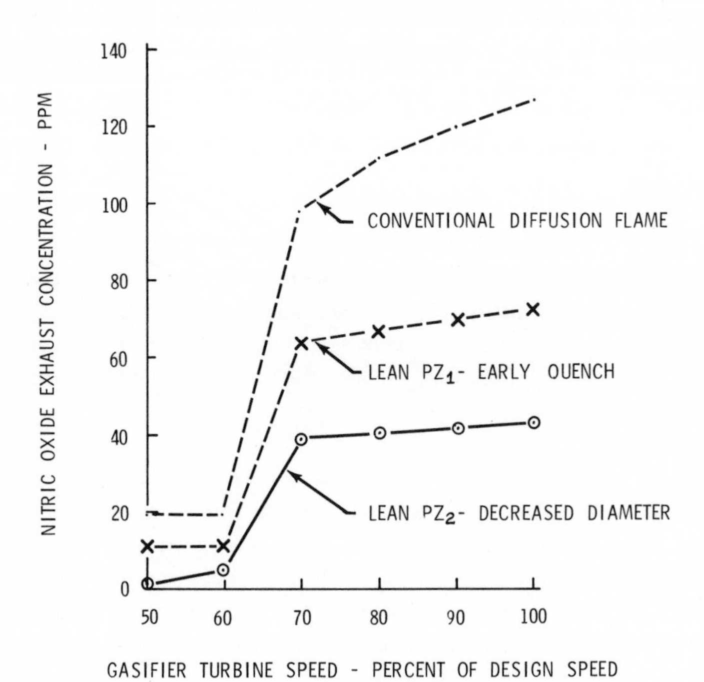

Fig. 2. Nitric oxide emissons from three diffusion flame burners.

GENERAL DISCUSSION

P. S. MYERS

This is the last session of this symposium. Thus, I think it is appropriate that the three papers we have heard this afternoon have to do with the application of knowledge to practical problems. The first two papers have been concerned with reduction of pollutants from existing operating systems – the third with reduction of pollutants from experimental laboratory systems. You have heard during the past two days about the relative effects of temperature, time and air-fuel ratio on the formation of oxides of nitrogen. You have heard this afternoon from these three authors how the different investigators have mixed together their witches brew from these three components to come up with a reduction in oxides of nitrogen in practical systems. At the risk of oversimplication I would like to show you two Figures that I think summarize the way in which these three papers have utilized the three tools available to them.

Fig. 1 is a plot of the rate of formation of NO versus time at different temperatures. It is applicable to the combustion systems that have been discussed here today. The curves were computed assuming that at time t = 0 the temperature jumped instantaneously to the indicated value and that all species except those involved in the oxides of nitrogen calculations were in equilibrium. The temperature was assumed to be held at a constant value and the pressure also constant at 300 psia. The graph shown is for $\phi = 1.0$ but similar graphs were prepared for other equivalence ratios.

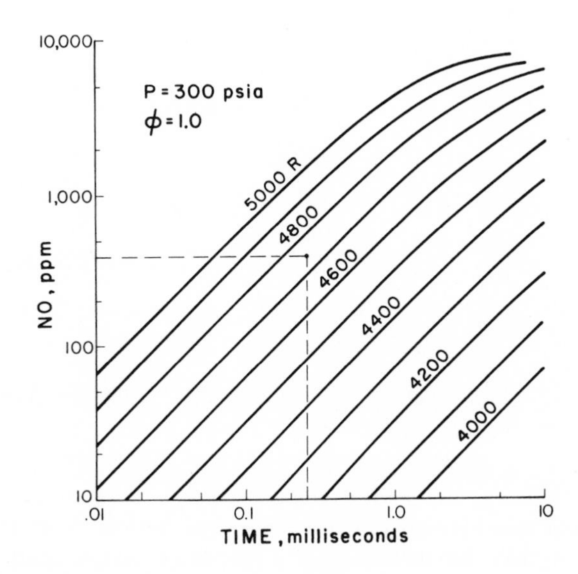

Fig. 1. Variation in NO formation rate with time for different temperatures.

In looking at this graph it is clear that if you wish a low level of oxides of nitrogen you must either:

(a) decrease the temperature in a very short time, or

(b) have a relatively low temperature after combustion.

In order to obtain the second graph let us assume the air-fuel mixture is at a given initial temperature and that we have adiabatic combustion in all cases. Under these circumstances you will end up with an adiabatic flame temperature that varies, of course, with equivalence ratio. Let us also assume that we desire a particular emission level – say, 350 ppm of oxides of nitrogen. With the oxides of nitrogen fixed at 350 ppm and the adiabatic flame temperature for $\phi = 1$ fixed at 4750°R we can read the time from the first graph for around 0.2 milliseconds. This says that we must cool the products of combustion after no longer than 0.2 milliseconds if we wish to end up with no more than 350 ppm NO in the exhaust. We can repeat the procedure for other equivalence ratios and from these results plot Fig. 2.

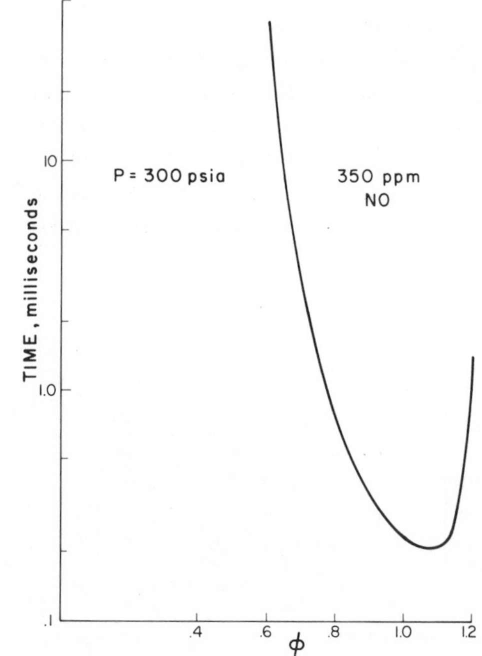

Fig. 2. Combinations of equivalence ratio and residence time to obtain a specified NO concentration.

From the standpoint of combustion stability we would like to operate at a mixture slightly richer than stoichiometric. However, as can be seen from the second graph, this gives minimum time before cooling if we are not to exceed 350 ppm NO. On the other hand, if we either go richer or leaner the permissible residence time increases significantly. Note that permissible residence time increases more rapidly on the rich side than on the lean side.

Fig. 2 illustrates the reasons why people utilize two-stage combustion. If you operate on the rich side you can have considerable residence time before you form large amounts of NO. By introducing additional air you can then cool the mixture and complete oxidation of hydrocarbons and CO formed during the rich combustion.

I think in attempting to reduce the oxides of nitrogen the papers presented this afternoon have all utilized the basic ideas shown in these two graphs.

W. Cornelius

We have been evaluating the exhaust emissions from aircraft and vehicular-type gas turbine combustors at the General Motors Research Laboratories since about 1955. When we commenced these studies, the effect of engine emissions on smog formation had not been confirmed. Since there was no reason at that time to publicize the low emission capability of the gas turbine, we reported the major advantages of a vehicular regenerative gas turbine to be as follows:

1. Small installed volume and low weight
2. Excellent torque and power characteristics
3. Good cold starting
4. Multi-fuel capability
5. Low oil consumption
6. Low maintenance
7. Low noise level
8. Low exhaust temperature

Then during the early 1960's, it became obvious that it would not be long before Federal and State exhaust emission standards would be established for vehicles. Now, truly appreciating the low emission potential of the gas turbine, we revised drastically our earlier list of advantages for the gas turbine to read as follows:

1. LOW
   EXHAUST
      EMISSIONS
2. Jet-smooth power
3. Low noise level
4. Good torque characteristics
5. Superior engine braking
6. Light weight – compact
7. Easy cold starting
8. Instant heat
9. Multi-fuel capability

S. G. Liddle *(Research Laboratories, GMC)*

I want to make a few comments regarding what happens when you apply burner ppm numbers to cars. These ppm numbers must be converted to grams per mile. I have been working on computer simulations for the EPA 1972 cycle for the past year and have found that the conversion is not an easy one. I found that variations in accessory load are important. The accessory load on a car may actually exceed the power requirements for driving a car down the road in many situations and in the EPA cycle this occurs about 35% of the time. Variations in the transmission input power characteristics will sometimes cause as much as a 3 to 1 variation in emissions because you change the operating routine of the engine. You can also have the same engine, the same vehicle and the same transmission, but a difference in the input power characteristics will result in a completely different emissions picture. I know of engine development programs where the target is 10 miles per gallon for the 1972 EPA car test cycle and so you compute pollutant ppm values requirements to meet the 1976 standards. You will not be sure that your car is going to get 10 miles per gallon until you actually drive it over the emission test cycle. In some cases you can actually increase the car NO_x emissions by decreasing the ppm of nitrogen oxides. What can happen here is that the technique you are using to reduce the ppm reduces the thermal efficiency so much that you have to increase the air flow through the engine to get the same horsepower level. Since the total emission is a function of both the concentration and the air flow, you can actually increase the grams per mile while you are decreasing the ppm. These are a few points that I think people should keep in mind when they are trying to meet the Federal grams per mile car standards.

N. A. Henein *(Wayne State University)*

During this symposium we heard a few comments that whatever you do to decrease the CO and hydrocarbon emissions in continuous combustion systems results in an increase in the nitric oxide emissions. In a recent study made at Wayne State University, we found that exhaust gas recirculation in a Stirling engine resulted in a decrease in both the incomplete combustion products and the NO emissions. I would like to present to you some of the results reached. The details of this work will be presented at the SAE October meeting at St. Louis, Missouri.

Fig. 1 shows a cross section in the burner and air preheater of the Stirling engine used. After combustor, the exhaust gases pass across the hydrogen heating tubes, and into the air preheater before they leave the engine. The air flows into the preheater and around the burner-can, before it enters the burner primary zone. We measured the temperature and took samples at the three locations (1), (2) and (3) shown in this figure. The tests were made at constant engine horsepower and speed and variable fuel-air ratio.

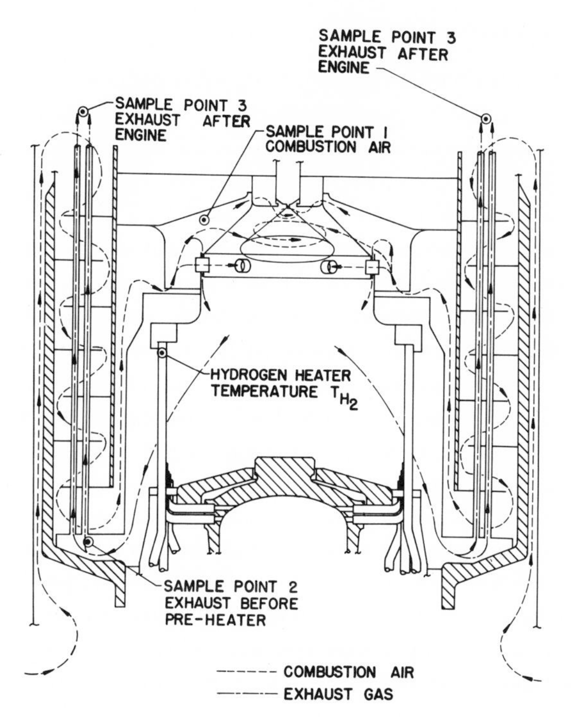

Fig. 1. Schematic of actual Stirling burner and air preheater.

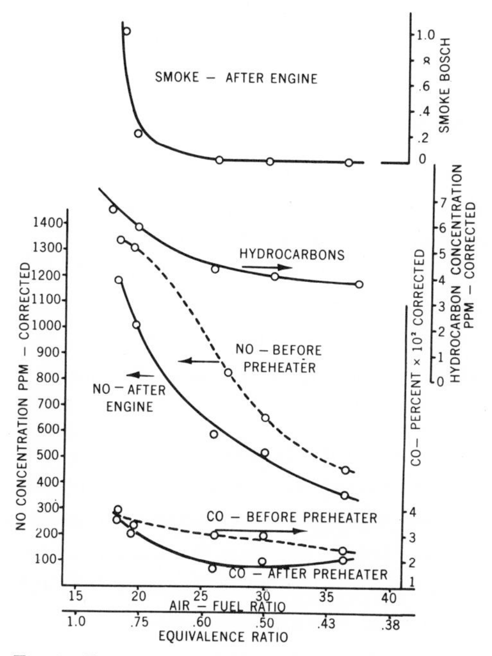

Fig. 2. Emissions at 5 horsepower without EGR.

Fig. 2 shows the base line data without exhaust gas recirculation at 5 horsepower and 3000 rpm. The hydrogen temperature in the heater header was kept constant at about 1320°F for all the runs. The results for CO show that the flow of the gases into the preheater reduced the CO at different degrees over the whole range of the air-fuel ratio. At 17.66 $\frac{A}{F}$, the reduction was 8%. It increased to 43% at an $\frac{A}{F}$ of 26:1, and decreased to 20% at an $\frac{A}{F}$ of 36.4:1. The factors that contributed to this behavior are the oxygen concentration, the residence time, and the average exhaust temperature in the air preheater.

The decrease in NO concentration after the preheater is believed to be due to the elimination reactions which took place in the preheater. The continuous decrease in the hydrocarbon and CO emissions with the increase in $\frac{A}{F}$ ratio is due to the increase in the exhaust gas temperature at the higher air fuel ratios. The continuous decrease in NO emission is believed to be mainly due to the shorter residence time in the burner at the higher $\frac{A}{F}$ ratios.

Fig. 3 shows the effect of the increase in percentage of exhaust gas recirculation on nitric oxide and carbon monoxide emissions. A 20% EGR reduced NO emissions from about 40% to 50%. This decrease continued with increasing EGR up to 60%, which is the highest percentage EGR we investigated. It is interesting to note that the increase in EGR resulted also in a decrease in CO emissions.

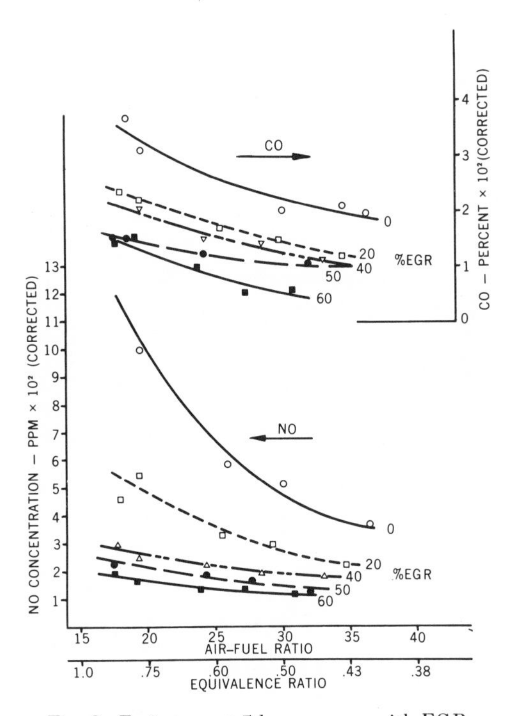

Fig. 3. Emissions at 5 horsepower with EGR.

The reason for this is shown in Fig. 4 which shows the continuous drop in the engine brake thermal efficiency with the increase in EGR. In order to keep the horsepower constant, more fuel has to be injected to produce the same amount of work, or the heat rejection increases. This results in an increase in the exhaust gas temperature as shown at the top of Fig. 4. Those who are working with exhaust gas reactors for gasoline engines can realize the effect of an increase of 100°F in the gas temperature on the reactor efficiency.

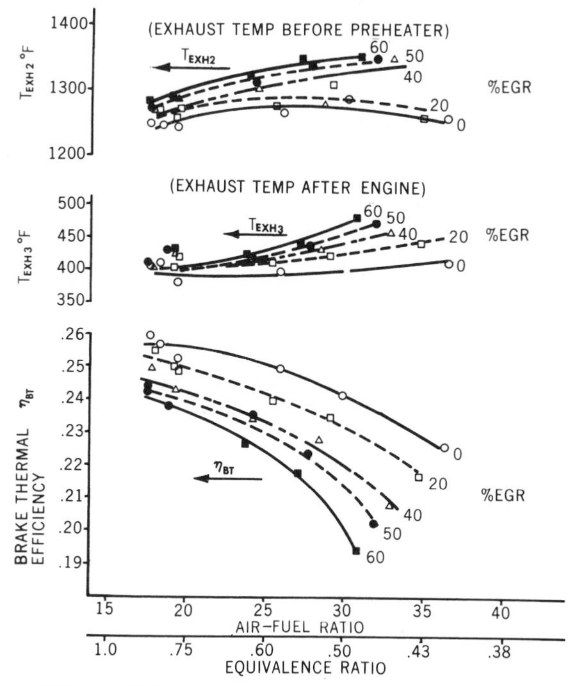

Fig. 4. Effect of EGR on engine efficiency and exhaust temperature.

From these results it is clear that EGR reduces both NO and CO emissions in the Stirling engine. However this is achieved at the expense of the fuel economy.

SESSION SUMMARY

P. S. MYERS
University of Wisconsin, Madison, Wisconsin

This brings to a close the technical part of the 1971 General Motors Symposium on Emissions from Continuous Combustion Systems. Consequently, I think this is an appropriate time for a representative of the group to recognize the part that General Motors has played in this Symposium. They have identified an important and interesting topic. They have surveyed the workers in the field and have brought together those who are active and making contributions. They have taken care of the myriad of details necessary to maximize communication and transfer knowledge within this group. I think it is appropriate to express our appreciation to Bill Agnew and his co-chairman, Walter Cornelius, and all the others who have made this symposium a success.

At a meeting like this, one is impressed with the detailed critical objective examination that is given by engineers and scientists to studies of technical problems conducted by their colleagues. I think we must increasingly examine some of the technological decisions made by society in this detailed critical objective way if our ecological needs are going to be met.

Let me illustrate what I mean. A small percentage of the population puts out more CO when they breathe than is permitted to be present in the atmosphere by the new Federal Air Quality regulations. I deduce from this that the mouth-to-mouth resuscitation you had hoped to practice on that pretty young girl may well be illegal in the future – your thoughts are probably already illegal! Futhermore, the NO_x air quality standard for NO_x apparently was set by examination of data from a preliminary study conducted to determine what type of research studies should be conducted! The standard was admittedly set at a level that was judged to cause no one – young or old, healthy or sick – any harm. I recognize the necessity to err on the safe side in dealing with human life and health. However, I also recognize the absurdity and unreality of programs offering zero risk to human life. I wonder if our group technological decisions concerning air quality have had the same detailed critical objective study we have given to the topics of the last two days?

You might deduce from these last remarks that I am anti-pollution. On the contrary, I made the remark because I am seriously concerned about pollution. Solutions to the pollution problem are not cheap. For example, we currently spend some 25-30 billion dollars each year on gasoline. If a reduction in oxides of nitrogen from automobiles to meet the '76 standards causes an increase of 20% in fuel consumption (a conservative estimate) this is an increased fuel cost of some 5-6 billion dollars per year irrespective of capital cost. If we have determined first that such a reduction is necessary and secondly that the most economic place to make the reduction is via the automobile then I am all for it. The thing that bothers me is that I see no evidence that the needed economic studies have or are being made – in fact, the opposite seems true. For example, by law, reductions of oxides of nitrogen in everything except automobiles are to proceed at a rate commensurate with "best demonstrated technological capabilities". In contrast, permissible oxides of nitrogen emission levels from automobiles have been set by legislation irrespective of technological capabilities or costs. I see no evidence in this procedure of the detailed critical objective study of the problem that I think is needed.

Let me make it clear why I make these comments. If we fail to find economically sound solutions to our pollution problems I think the resulting undue economic burden will result in a pollution "back lash" against needed pollution reductions. For example, a recent article in Newsweek pointed out that there is increasing question at all levels as to whether or not we can afford pollution reductions. I think we can't afford not to have needed reductions – both now and in the future. I fear for future reductions, however, if we technical people fail our responsibility of insisting that group technological decisions be made on a sound technological and economic basis and be subjected to the same critical objective examination we give to our individual studies.

SYMPOSIUM SUMMARY

W. G. AGNEW

General Motors Research Laboratories, Warren, Michigan

I would like to spend a minute or two reviewing what we have actually covered in the last two days. Here are things that were highlights to me in this session.

We started out, if you remember on Monday morning, with the subject of modeling, or "muddling" as it was called at one point. After listening to three fine speakers on this subject, I think we came to an agreement that the thing we have in our models so far is a framework – a framework with which we can detect and predict some trends, but one which is still nothing but a skeleton, which will not do with great refinement many of the things that we would like it to do. There are two parts to a model, as we talked about them – the flow part and the reaction kinetic part. We had some requests that these two aspects be kept in balance. We heard that the computational methods are fairly well developed. We did not hear a whole lot concerning correlations between the modeling work and actual experiment. This appears to me to be an area that needs a lot more work.

One of the things I noticed in the modeling sessions, in general, was that practically all the modeling work that has been done so far has been on relatively conventional systems. This is not all bad. If we get a mathematical model of what the engineer has already proven by experiment, then we can put it in a textbook. Everybody feels a whole lot better when they see it in a textbook. We will have a solid understanding then of what is conventional, and we have a base of operations from which to branch out and explore new systems. But the fact is, that these conventional systems, as has been shown in the remainder of this conference, are not adequate to the needs. Particularly the oxides of nitrogen from conventional systems are far above where we have to be. Having established some basic principles with our current modeling efforts, we are going to have to broaden our outlook a little to somewhat more unconventional systems as well.

Our second session had to do with chemical formation and destruction of pollutants. I think it was comforting in that section to hear that some of our experiments show pretty much the kind of formation of pollutants in flames which we expect from our kinetics. We seem to have a pretty good handle on a good part of this, and we can predict the appearance and the quenching of many of these pollutants in flames. Perhaps prompt NO was an exception. It appears to me to be extremely difficult to set up an experiment which will be conclusive in this prompt NO area. I gather that if we refine our kinetic treatment sufficiently we can in fact predict this kind of happening. But is the issue settled? As near as I can tell, it is not at this Symposium. Bowman says yes and Kaufman says no. I do not know where the rest of you stand, but it looks like there is more argument on that subject to come in the future. I asked myself and some of my engineer friends, "Is it important whether there is prompt NO or not? Is this just a thing that the kineticists tease each other with, or is it really going to have a practical effect on NO control?" I have gotten two kinds of answers. One is that if we ever get to the very low concentration levels of NO that we are being asked to reach in our practical systems, it may be that there is nothing left but the prompt NO and that will be the all-important issue. The other answer is that we are certainly learning a lot about the systems in our pursuit of prompt NO and that is certainly an advantage.

Carbon formation – soot formation, I gather, is not well defined kinetically, but at least in the group here there does not seem to be a great deal of concern over the kinetic description of soot formation. There seem to be engineering solutions and these are where the effort seems to be put now. I also gather that there are still some kinetic problems with NO on the rich side of stoichiometric. Probably, a lot more work needs to be done there.

One subject is missing from this Symposium. I did not hear anything about it, and I knew I was not going to hear anything about it because when we were organizing the Symposium we could not find anybody in the country who was willing to talk about the matter. This is the subject of oxygenate formation, the aldehydes. I feel that oxygenates are a serious consideration in all combustion systems – in particular in many of the continuous combustion systems that we are working with. In conventional gas turbine burners, we have pretty well eliminated oxygenates, so the odor problem does not particularly concern us; but we have also had some very real experiences indicating that if things go off design point very far you can get into some very bad odor problems which could be serious pollution problems in the future. Also, as we begin to get away from conventional burners to correct the NO problem, and if we are not careful, we can very well fall into a very large hole with oxygenate formation. So I would certainly like to see some attention given to that subject area in the future.

We talked about engineering parameters in the third session. Soot, hydrocarbons and carbon monoxide on the one side, and oxides of nitrogen control on the other side. At least through a large deal of the work that we talked about, these were opposing control measures and they are going to present apparently endless problems in that respect. We did cover a lot of the engineering parameters – pressures, temperatures, and air-fuel ratios – and I think it became very obvious that we have to talk of local fuel-air ratios for a proper understanding of pollutant emissions. We do have a fairly good understanding of the effect that these variables have on the various pollutants. I think it comes out that temperature is probably the most important factor for oxides of nitrogen control and also for hydrocarbon and CO burnout. Temperature, of course, depends on a number of other parameters, most particularly on local fuel-air ratio.

I begin to get the following picture in the engineering control of pollutant emissions. We have to carry out our combustion process in a temperature window. We have to keep the temperature low enough so that we do not form oxides of nitrogen, and yet we have to keep it high enough so that we burn up the hydrocarbons and the CO. As I see it, the temperature window in which we have to operate is between 1500 K and 1800 K, or 2200 F and 2800 F. Now maybe these numbers are kind of tight. I think the range is going to be fairly narrow, and it's going to be difficult to stay in there in all the operating modes and in all the locations in these various combustion systems. But it looks like that is the game, to provide a combustion system which will operate in that temperature window. Now there are many ways that we have heard to do that. They involve fuel atomization, evaporation, mixing, air introduction, heat abstraction by cooling, pre-mixing, variable geometry, air atomization – of one kind or another, and the introduction of swirls of various kinds. We have to do these things, while at the same time we maintain all the other necessary attributes – flame stability, large turndown ratios, no hot spots, high combustion intensity, low pressure loss. All of these things are required to get an adequate combustion system – a combustion system that retains the power level and the efficiency that we need but comes out with low emissions. I suspect there will turn out to be a variety of ways that we can do this, I don't think we're going to end up with one optimum combustion process for all applications.

In the last session, we discussed the practical applications. This is very interesting, because this is where the combustion scientists' ideas run into the realities of producing power and doing it efficiently. Dr. Breen had an apparently good handle on the stationary powerplant. His game is oxides of nitrogen in parts per million which is a thing he can measure fairly easily. So at least, he knows when he has won the game without too much trouble and when he is losing the game. The thing that surprises me about his presentation is that awe-inspiring inferno that he has to work with, the practical problems of safety, the massive control systems, and the design times. These appear to be his problems, applying known principles to an extremely large system. It does appear that modeling can be very useful in this system and is getting attention.

In the aircraft field, smoke seems to be under control. Hydrocarbon and CO are problems at idle, and it looks like there are ideas for correcting these problems but they have not been put into real commercial use yet. In the case of oxides of nitrogen, it makes you sad to hear that we are going to have to use something like water injection. That just looks like a very unsophisticated way out. It must hurt the combustion scientists tremendously to have to throw water into the combustion process to lower the NO. But there don't seem to be many other ideas. There was a hint that premixing might be used, and perhaps the aircraft people will pick up a little information about premixing from some of the other applications of gas turbine burners, as well as other approaches which might help to get the NO down without dumping water in the system.

As far as the automobile is concerned, I think that the interesting thing in the paper by Wade and Cornelius is the translation from combustion characteristics to grams per mile on a car driving cycle. Again, it has been pretty much the application of known principles to emission control. Modeling has been very useful in this automotive work, although Mr. Wade did not get into that too deeply in his presentation. However the written paper contains considerably more information. I think a very significant thing comes out of the automobile application of continuous combustion processes that perhaps many of you don't fully realize, and that is how low 0.4 of a gram per mile of oxides of nitrogen is on an automotive testing cycle. Much of the work that we have done with modeling, and everything we have investigated on conventional systems, has given emission values that are far above that level. Everything that Wade and Cornelius have investigated in their attempt to get the automotive gas turbine down there has been inadequate – and they have investigated almost every idea that has been mentioned at this Symposium. There is a long way to go.

We have in this Symposium revealed most of the problems which we have to deal with in emissions from continuous combustion systems. We have established many of the principles which govern emissions from this kind of system. We have also exposed most of the ideas that anybody has had about how to fix things up. Now all of us are on about the same level at this point. I think everyone knows pretty well what everyone else is doing. The channels are now open for the theoreticians, the modelers, and the kineticists to make their work more useful – to provide more understanding, to make the models more realistic, and to provide more prediction capability. Of course, the challenge for the engineers is to apply these ideas that have been talked about at this Symposium, many of which are only embryo ideas. There is a lot of engineering work to be done to put those things into actual use and find out whether they are going to pan out. So, I think the Symposium has been very useful. I am looking for big things to come out of it in the future. I think we ought to be able to make the big leap at this point and everyone should make a lot more progress.

All of the work that has been presented here, all of the papers, all of the discussions will be available in publication just as soon as we can get them printed. They will be available to you and to the technical community and the public at large. I am sure that this Symposium will be remembered as a jumping-off place, and this is what we wanted it to be from the start. We are very grateful to the organizers who helped us put this Symposium together. They were your session chairmen, Jack Longwell, Glenn Williams, and Arthur Lefebvre along with Ernie Starkman. We thank Phil Myers for filling in as chairman for the last session, and we thank all of you for coming, giving your time, preparing the papers, and entering into the very active discussions. We thank all of you very much, and we hope you have a good trip home. We hope to see a lot more of you in the future.

PARTICIPANTS

P. D. Agarwal
Research Laboratories, GMC

W. G. Agnew
Research Laboratories, GMC

W. R. Aiman
Research Laboratories, GMC

C. A. Amann
Research Laboratories, GMC

W. S. Anderson
U. S. Army Tank-Automotive Center
Warren, Michigan

C. E. Angell
Rochester Products Division, GMC

D. W. Bahr
General Electric Company
Cincinnati, Ohio

W. Bartok
Esso Research and Engineering
Linden, New Jersey

E. K. Bastress
Northern Research and Engineering Corp.
Cambridge, Massachusetts

S. E. Beacom
Research Laboratories, GMC

J. M. Beér
University of Sheffield
Sheffield, England

A. H. Bell
Engineering Staff, GMC

J. E. Bennethum
Research Laboratories, GMC

J. W. Bjerklie
Mechanical Technology Incorporated
Latham, New York

C. T. Bowman
United Aircraft Research Laboratories
East Hartford, Connecticut

F. V. Bracco
Princeton University
Princeton, New Jersey

B. P. Breen
KVB Engineering, Inc.
Tustin, California

R. S. Brokaw
NASA-Lewis Research Center
Cleveland, Ohio

W. G. Burwell
United Aircraft Research Laboratories
East Hartford, Connecticut

A. V. Butterworth
Research Laboratories, GMC

J. D. Caplan
Research Laboratories, GMC

L. S. Caretto
Imperial College of Science and Technology
London, England

P. F. Chenea
Research Laboratories, GMC

D. E. Cole
University of Michigan
Ann Arbor, Michigan

E. N. Cole
General Motors Corporation

J. S. Collman
Research Laboratories, GMC

J. M. Colucci
Research Laboratories, GMC

W. Cornelius
Research Laboratories, GMC

C. T. Crowe
Washington State University
Pullman, Washington

R. Davies
Research Laboratories, GMC

D. W. Dawson
Environmental Protection Agency
Ann Arbor, Michigan

D. L. Dimick
Engineering Staff, GMC

G. L. Dugger
Johns Hopkins University
Silver Springs, Maryland

T. Durrant
Rolls Royce, Ltd.
Derby, England

H. C. Eatock
United Aircraft of Canada Limited
Lonqueiul, Quebec, Canada

M. K. Eberle
Research Laboratories, GMC

R. B. Edelman
General Applied Science Laboratories, Inc.
Westbury, New York

R. H. Essenhigh
Pennsylvania State University
University Park, Pennsylvania

C. P. Fenimore
G.E. Research & Development Center
Schenectady, New York

G. Flynn, Jr.
Research Laboratories, GMC

O. I. Ford
Aerojet Nuclear Systems Company
Sacramento, California

J. P. Franceschina
Chrysler Corporation
Detroit, Michigan

H. P. Fredriksen
Research Laboratories, GMC

R. M. Fristrom
Johns Hopkins University
Silver Springs, Maryland

N. J. Friswell
Shell Research Limited
Chester, England

D. Goalwin
Rocketdyne
Canoga Park, California

J. S. Grobman
NASA-Lewis Research Center
Cleveland, Ohio

D. C. Hammond, Jr.
Purdue University
Lafayette, Indiana

J. L. Hartman
Research Laboratories, GMC

H. R. Hazard
Battelle Memorial Institute
Columbus, Ohio

N. A. Henein
Wayne State University
Detroit, Michigan

R. Herman
Research Laboratories, GMC

D. J. Henry
Research Laboratories, GMC

J. B. Heywood
Massachusetts Institute of Technology
Cambridge, Massachusetts

R. F. Hill
Research Laboratories, GMC

J. B. Howard
Massachusetts Institute of Technology
Cambridge, Massachusetts

J. K. Hulbert
Williams Research Corporation
Walled Lake, Michigan

S. C. Hunter
Air Research Manufacturing Co.
Phoenix, Arizona

R. W. Hurn
Bureau of Mines
Bartlesville, Oklahoma

F. P. Hutchins
Environmental Protection Agency
Ann Arbor, Michigan

F. E. Jamerson
Research Laboratories, GMC

H. L. Julien
Research Laboratories, GMC

S. Katz
Research Laboratories, GMC

F. Kaufman
University of Pittsburgh
Pittsburgh, Pennsylvania

J. C. Kent
Research Laboratories, GMC

J. R. Kliegel
KVB Engineering, Inc.
Tustin, California

R. B. Krieger
Research Laboratories, GMC

C. LaPointe
Ford Motor Company
Dearborn, Michigan

A. H. Lefebvre
The Cranfield Institute of Technology
Cranfield, England

L. L. Lewis
Research Laboratories, GMC

S. G. Liddle
Research Laboratories, GMC

J. H. Lienesch
Research Laboratories, GMC

W. H. Lipkea
Research Laboratories, GMC

J. P. Longwell
Esso Research and Engineering Company
Linden, New Jersey

C. Marks
Engineering Staff, GMC

R. A. Matula
Drexel University
Philadelphia, Pennsylvania

W. L. McGaw
United Aircraft Corp.
Farmington, Connecticut

J. G. McGowan
University of Massachusetts
Amherst, Massachusetts

A. F. McLean
Ford Motor Company
Dearborn, Michigan

A. M. Mellor
Purdue University
Lafayette, Indiana

H. J. Mertz, Jr.
Research Laboratories, GMC

G. A. Miles
Detroit Diesel Allison Division, GMC

W. Mirsky
University of Michigan
Ann Arbor, Michigan

J. Moore
General Electric Company
Schenectady, New York

N. L. Muench
Research Laboratories, GMC

L. J. Muzio
Columbia University
New York, New York

P. S. Myers
University of Wisconsin
Madison, Wisconsin

B. E. Nagel
Research Laboratories, GMC

H. K. Newhall
University of Wisconsin
Madison, Wisconsin

E. R. Norster
The Cranfield Institute of Technology
Cranfield, England

P. E. Oberdorfer
Sun Oil Company
Marcus Hook, Pennsylvania

J. Odgers
Laval University
Quebec, Canada

G. Opdyke, Jr.
AVCO Corporation
Stratford, Connecticut

H. B. Palmer
Pennsylvania State University
University Park, Pennsylvania

D. J. Patterson
University of Michigan
Ann Arbor, Michigan

W. H. Percival
Research Laboratories, GMC

C. E. Polymeropoulos
Rutgers University
New Brunswick, New Jersey

D. T. Pratt
Washington State University
Pullman, Washington

A. A. Quader
Research Laboratories, GMC

W. R. Roudebush
NASA Headquarters
Washington, D. C.

C. J. Sagi
Research Laboratories, GMC

A. F. Sarofim
Massachusetts Institute of Technology
Cambridge, Massachusetts

R. F. Sawyer
University of California
Berkeley, California

R. M. Schirmer
Phillips Research Center
Bartlesville, Oklahoma

W. L. Schultz
Ford Motor Company
Dearborn, Michigan

R. Schulz
Environmental Protection Agency
Ann Arbor, Michigan

R. L. Scott
Research Laboratories, GMC

T. Sebestyen
Environmental Protection Agency
Ann Arbor, Michigan

D. J. Seery
United Aircraft Research Laboratories
East Hartford, Connecticut

J. R. Shekleton
International Harvester Company
San Diego, California

R. Shinnar
New York City College
New York, New York

C. W. Shipman
Worchester Polytechnic Institute
Worchester, Massachusetts

R. M. Siewert
Research Laboratories, GMC

T. Singh
Wayne State University
Detroit, Michigan

W. A. Sirignano
Princeton University
Princeton, New Jersey

H. R. Smith
Diesel Equipment Division, GMC

A. F. Soby
Shell Oil Company
Wood River, Illinois

G. Sovran
Research Laboratories, GMC

D. B. Spalding
Imperial College of Science and Technology
London, England

R. C. Stahman
Environmental Protection Agency
Ann Arbor, Michigan

E. S. Starkman
Environmental Activities Staff, GMC

R. F. Stebar
Research Laboratories, GMC

D. L. Stivender
Research Laboratories, GMC

R. E. Sullivan
Detroit Diesel Allison Division, GMC

D. P. Teixeira
Southern California Edison Company
Rosemead, California

R. F. Thomson
Research Laboratories, GMC

J. G. Tomlinson
Detroit Diesel Allison Division, GMC

C. S. Tuesday
Research Laboratories, GMC

A. D. Tuteja
University of Wisconsin
Madison, Wisconsin

F. J. Verkamp
Detroit Diesel Allison Division, GMC

P. T. Vickers
Research Laboratories, GMC

C. W. Vigor
Research Laboratories, GMC

W. R. Wade
Research Laboratories, GMC

T. O. Wagner
American Oil Company
Whitney, Indiana

E. F. Weller
Research Laboratories, GMC

G. C. Williams
Massachusetts Institute of Technology
Cambridge, Massachusetts

M. K. Yu
Research Laboratories, GMC

E. B. Zwick
The Zwick Company
Huntington Beach, California

AUTHORS AND DISCUSSORS INDEX

Bold Face Type refers to papers and Symposium and Session Summaries

SUBJECT INDEX